Invariant Theory in All Characteristics

Volume 35

CRM PROCEEDINGS & LECTURE NOTES

Centre de Recherches Mathématiques
Université de Montréal

Invariant Theory in All Characteristics

H. E. A. Eddy Campbell
David L. Wehlau
Editors

The Centre de Recherches Mathématiques (CRM) of the Université de Montréal was created in 1968 to promote research in pure and applied mathematics and related disciplines. Among its activities are special theme years, summer schools, workshops, postdoctoral programs, and publishing. The CRM is supported by the Université de Montréal, the Province of Québec (FCAR), and the Natural Sciences and Engineering Research Council of Canada. It is affiliated with the Institut des Sciences Mathématiques (ISM) of Montréal, whose constituent members are Concordia University, McGill University, the Université de Montréal, the Université du Québec à Montréal, and the Ecole Polytechnique. The CRM may be reached on the Web at www.crm.umontreal.ca.

American Mathematical Society
Providence, Rhode Island USA

The production of this volume was supported in part by the Fonds pour la Formation de Chercheurs et l'Aide à la Recherche (Fonds FCAR) and the Natural Sciences and Engineering Research Council of Canada (NSERC).

2000 *Mathematics Subject Classification.*
Primary 13A50, 14L24, 14L30, 14L35, 13P99, 20C20, 54H15.

Library of Congress Cataloging-in-Publication Data

Invariant theory in all characteristics / H. E. A. Eddy Campbell, David L. Wehlau, editors.
p. cm. — (CRM proceedings & lecture notes, ISSN 1065-8580; v. 35)
Includes bibliographical references.
ISBN 0-8218-3244-1 (alk. paper)
1. Invariants–Congresses. I. Campbell, H. E. A. Eddy (Harold Edward Alexander Eddy), 1954–. II. Wehlau, David L., 1960–. III. Title. IV. Series. V. CRM proceedings & lecture notes (American Mathematical Society); v. 35.

QA+ 2004 2004046175

∞ The paper used in this book is acid-free and falls within the guidelines
established to ensure permanence and durability.
This volume was submitted to the American Mathematical Society
in camera ready form by the Centre de Recherches Mathématiques.
Visit the AMS home page at `http://www.ams.org/`

10 9 8 7 6 5 4 3 2 1 09 08 07 06 05 04

Contents

List of Participants

Letitia Banu
Romania

Karin Baur
Switzerland

Chris Brav
USA

Bram Broer
Canada

H. E. A. (Eddy) Campbell, Editor
Canada

James Carrell
Canada

Shizhen Cheng
China

Calin Chindris
USA

Jaydeep Chipalkatti
Canada

Jianjun Chuai
Canada

Brian Coolen
Canada

Charles Cuell
Canada

Harm Derksen
USA

Anna Duff
Canada

Ruth Enoch
USA

Peter Fleischmann
England

Gene Freudenburg
USA

M.Angeles Gmez Molleda
Spain

Julia Hartmann
Germany

Loek Helminck
USA

Evelyne Hubert
France

Ian Hughes
Canada

Markus Hunziker
USA

Wilberd van der Kallen
Netherlands

Dikran Karagueuzian
USA

Kiumars Kaveh
Canada

Nondas Kechagias
Greece

Gregor Kemper
Germany

Friedrich Knop
USA

Vera Koerkel
Germany

Tomaz Kosir
Slovenia

Hanspeter Kraft
Switzerland

Alexander Kuehn
Germany

Jochen Kuttler
Switzerland

Jorge Lauret
USA

Nicole Lemire
Canada

Jamie Mingo
Canada

Lucy Moser-Jauslin
France

David Murphy
USA

Mara Neusel
USA

Christian Ohn
France

Alessandra Pantano
USA

Jawahar (Jay) Pathak
USA

David Pollack
Canada

Vladimir Popov
Russia

Zinovy Reichstein
Canada

Marc Renault
USA

Leslie Roberts
Canada

Leonid Rybnikov
Russia

Yasmine Sanderson
USA

Gerald Schwarz
USA

Joel Segal
England

Mufit Sezer
USA

R. James Shank
England

Alexander Smirnov
Russia

Bart Van Steirteghem
USA

Peter Symonds
England

Nicolas Thiery
France

Adam Van Tuyl
Canada

David Wehlau, Editor
Canada

Michael Weingart
USA

Oxana Yakimova
Russia

List of Speakers

• First week •

Harm Derksen
USA

Peter Fleischmann
England

Hanspeter Kraft
Switzerland

Gerald Schwarz
USA

• Second week •

Jaydeep Chipalkatti
Canada

Jianjun Chuai
Canada

Julia Hartmann
Germany

Loek Helminck
USA

Markus Hunziker
USA

Wilberd van der Kallen
Netherlands

Dikran Karagueuzian
USA

Kiumars Kaveh
Canada

Nondas Kechagias
Greece

Gregor Kemper
German

Friedrich Knop
USA

Alexander Kuehn
Germany

Lucy Moser-Jauslin
France

Mara Neusel
USA

Vladimir Popov
Russia

Leonid Rybnikov
Russia

Yasmine Sanderson
USA

Joel Segal
England

Mufit Sezer
USA

R. James Shank
England

Alexander Smirnov
Russia

Nicolas Thiery
France

Oxana Yakimova
Russia

Preface

This is the proceedings volume of a workshop on Invariant Theory, held at Queen's University, April 8–19, 2002. All of the articles in this volume have been refereed.

Our purpose in organizing the workshop was to bring together two mostly separate communities of researchers in invariant theory, those working in characteristic 0 and those working in positive characteristic. The workshop was part of the theme year on "Groups and Geometry" being held under the auspices of the Centre de recherches mathématiques (CRM) in Montreal.

The first week of the workshop was devoted to introductory lectures aimed at graduate students, younger mathematicians and for those who wished to acquire a breadth of knowledge across invariant theory of both kinds. During the first week, four distinguished mathematicians gave a series of lectures. They were Professors P Fleischmann (Kent), Hanspeter Kraft (Basel), G W Schwarz (Brandeis), and H Derksen (Michigan). Three of them have contributed expository articles to this volume based on their lectures.

The second week was devoted to a more specialized lectures on Invariant Theory, for which many of the younger mathematicians stayed to listen. The speakers listed elsewhere gave talks of 50 minutes duration, bearing in mind the audience. In this second week, the focus was on current problems in keeping with the workshop nature. Most of the articles in this volume expand upon these lectures. In addition, Reg Wood contributed an article although he was unable to attend at the last minute.

During both weeks, organized problem sessions took place, and three refereed articles in this volume summarize those sessions. These describe a number of open problems of great interest to invariant theorists in both communities.

The CRM was our chief source of funding and we wish to express our gratitude to the centre and its then director, Jacques Hurtibise. We also wish to thank our other sponsors, the Royal Military College of Canada, Queen's University, the Natural Sciences and Engineering Research Council of Canada, and the National Science Foundation of the USA. Thanks to these sponsors, we raised close to $100,000 in support of the workshop.

In our view, the workshop was a great success, and we thank all of those who participated, contributed articles, as well as all of the referees. We thank Louise Letendre and Andre Montpetit of the CRM for their help in preparing this volume for publication. We are grateful to Ian Hughes, our friend, colleauge and collaborator for his support. We also express our gratitude to those who helped us run the workshop, Marge Lambert, Charlene Janeway, and especially Jennifer Read.

E. Campbell
D. Wehlau

Kingston, Ontario, August, 2003.

Centre de Recherches Mathématiques
CRM Proceedings and Lecture Notes
Volume **35**, 2004

A Local Study of Embeddings of Complexity One

G. Bousquet and L. Moser-Jauslin

ABSTRACT. The object of this paper is to show how a combinatorial description by Timashev of complexity one embeddings can be used to study local geometric properties of certain smooth embeddings.

1. Introduction

Given a complex reductive group G and a closed subgroup H, an *embedding* of the homogeneous space G/H is a normal algebraic G-variety having an open (dense) orbit isomorphic to G/H. Luna and Vust in [**10**] developed a combinatorial method to classify embeddings of homogeneous spaces. The method is especially effective for homogeneous spaces of complexity 0, also known as spherical varieties. The complexity of the homogeneous space is the minimal codimension of an orbit of a Borel subgroup of G in the space G/H. Spherical varieties have been studied by Brion, Knop, Luna, Vust and others (see, for example, [**4**] or [**6**]). The most well-known case is when G is an algebraic torus, in which case the theory of embeddings is precisely the theory of toric varieties.

The classification of Luna and Vust can also be used for homogeneous spaces of complexity one. However, the combinatorics of the classification can be difficult to interpret. In 1997, Timashev introduced a new way to look at the classification using "hypercones." The goal of this paper is to show how the description of Timashev can be used in a special case to reduce the study of local geometric properties to those of toric varieties.

We then apply the result to the study of embeddings of SL(2) modulo a finite subgroup. For the case that H is trivial, the classification was given explicitly in [**10**], and the general case can be easily deduced from this case. However, it was in general not clear from this description how to determine the geometric properties of the varieties obtained from the classification. In [**12**] the smooth embeddings for $H = \{\mathrm{Id}\}$ and $H = \{\pm\mathrm{Id}\}$ were determined. The methods developed there could be used to determine the smooth embedding for some cases where H is cyclic. However, for the other finite subgroups, a criterion for smoothness was not given. Using ideas developed here, the complete description of smooth embeddings of SL(2) modulo a finite subgroup is given in [**3**] and is the subject of a forthcoming article.

2000 *Mathematics Subject Classification.* Primary: 14M17; Secondary: 14L30.
This is the final version of the paper.

2. The local structure of embeddings

One of the basic tools to study local results of embeddings is a result due to Brion, Luna and Vust. Let G be a reductive group G and X a G-variety with a complete orbit Y. That is, the isotropy group G_y of a point of $y \in Y$ is a parabolic subgroup. Denote by P the opposing parabolic subgroup of G with respect to a maximal torus. Denote by P_u the unipotent radical of P.

THEOREM 2.1 ([**5**]). *There is an affine P-stable open subvariety X_0 of X which intersects Y, a Levi subgroup L of P, and an L-stable closed subvariety S of X_0 such that the map $P_u \times S \to X_0$ given by multiplication is a P-equivariant isomorphism. P_u acts on P_u by left multiplication, and L acts on S by the given action and on P_u by conjugation.*

The set $Y \cap X_0$ is a P-orbit. Also, the orbit Y intersects S in an L-fixed point $y_0 \in S$, and the study of local properties of X near Y is reduced to the study of the fixed point $y_0 \in S$. The other L-orbits of S correspond to P-orbits of X_0.

In the case where X is an embedding of G/H, the field of rational functions of X is $\mathbb{C}(G/H)$, which can be identified with $\mathbb{C}(G)^H$ (see, for example, [**2**, II.6.5]). The variety S from the theorem has the field of rational functions $\mathbb{C}(G)^{P_u \times H}$, where P_u acts on the left, and H acts on the right. Also, the ring of regular functions of S is given by $\mathbb{C}[X_0]^{P_u}$.

We review the set-up and notation from [**18**]. If X is an embedding of G/H of complexity one, then we have $\mathbb{C}(X) = \mathbb{C}(G)^H$, and for a Borel subgroup B of G, the B-invariants form a field $\mathbb{C}(X)^B = \mathbb{C}(t)$. Let U be the unipotent subgroup of B and T be a maximal torus. The character group of T is a free abelian group, and we can consider the subgroup $\Xi(X)$ of characters for which there is a semi-invariant in $\mathbb{C}(X)$. It is a free abelian group of rank r generated by $\chi_1, \ldots, \chi_r$. Denote by $e_i \in \mathbb{C}(X)^U$ a semi-invariant of T of weight χ_i, for $i = 1, \ldots, r$. Now we will show that $\mathbb{C}(X)^U = \mathbb{C}(t, e_1, \ldots, e_r)$. First of all, $\mathbb{C}(t, e_1, \ldots, e_r)$ is the subfield of $\mathbb{C}(X)$ generated by the B-semi-invariants, since the quotient of two semi-invariants of the same weight is invariant, and therefore is in the field generated by t. Now, it is not hard to see that this subfield is in fact $\mathbb{C}(X)^U$. Indeed, X has an affine B-stable open subvariety X_0 (see for example [**6**, Theorem 2.2 (a)]). Now $\mathbb{C}[X_0]^U$ is generated by the B-semi-invariants in $\mathbb{C}[X_0]$, and we will show that $\mathbb{C}(X)^U = \mathbb{C}(X_0)^U$ is the quotient field of $\mathbb{C}[X_0]^U$. If $f \in \mathbb{C}(X_0)^U$, there exists $g \in \mathbb{C}[X_0]$ such that $fg \in \mathbb{C}[X_0]$. Then Ug generates a finite-dimensional U-module in $\mathbb{C}[X_0]$ (see, for example [**7**, II.2.4]). Since U is unipotent, this U-module contains an element $g_0 \in \mathbb{C}[X_0]^U$. Also, $g_0 f \in \mathbb{C}[X_0]^U$.

3. Reduction to toric varieties

We will consider the case where X is a complexity one G/H-embedding and Y is an orbit of X isomorphic to G/B. By the result of Brion, Luna and Vust cited in the previous section, there is a B-stable open affine subvariety X_0 of X which intersects Y and a closed T-stable subvariety S of X_0 such that $U \times S \cong X_0$. We will call X_0 an affine B-chart. Denote by y_0 the T-fixed point given by $Y \cap S$. The field of rational functions on S is given by $\mathbb{C}(S) = \mathbb{C}(t, e_1, \ldots, e_r)$, and the ring of regular functions is an algebra of finite type contained in $\mathbb{C}(S)$ generated by semi-invariants whose quotient field is $\mathbb{C}(S)$. We will show that for many cases, including all smooth cases, the variety S is locally toric for a torus which includes

T. Thus, in particular, the question of whether Y consists of smooth points of X reduces to determining the smoothness of a fixed point in an affine toric variety.

To show this, we consider two possibilities. Following the notation of [**18**], we call an orbit Y of type I if an infinite number of B-orbits of X do not contain Y in its closure. Otherwise, we call Y of type II. If Y is of type I, then the ring of regular functions on S contains a non-constant T-invariant function, which, as we will see in Proposition 3.1, one can suppose is t. In this case, $\mathbb{C}[S]^T$ is a localisation of the ring $\mathbb{C}[t]$. If Y is of type II, then $\mathbb{C}[S]$ contains no non-constant invariants, and thus $\mathbb{C}[S]^T = \mathbb{C}$.

PROPOSITION 3.1. *If Y is an orbit of type* I *isomorphic to G/B, then there exists a normal affine $T \times \mathbb{C}^*$-variety S' which has an open orbit, and which has an open T-stable subvariety isomorphic to an open T-stable subvariety S_0 of S which contains y_0.*

This proposition shows that locally any orbit of type I intersects an open affine B-chart which is isomorphic to $U \times S$ where S is a T-stable open subvariety of a toric variety.

PROOF. Recall that $\mathbb{C}(S)^T = \mathbb{C}(t)$. Since the orbit Y is of type I, there is a non-constant rational function $\alpha(t) \in \mathbb{C}[S]^T$. In fact, one can suppose that $\alpha(t) = t$. To see this, note that since S is normal, $\mathbb{C}[S]$ is an intersection of valuation rings. Then we show that there exists $a \in \mathbb{C}$ such that either $t-a$ or $(t-a)^{-1}$ is in $\mathbb{C}[S]$. If not, then for all $a \in \mathbb{C}$ there are valuations v_1 and v_2 which are positive or zero on all of $\mathbb{C}[S]$ and such that $v_1(t-a) > 0$ and $v_2(t-a) < 0$. But then the degree of $\alpha(t)$ must be strictly negative, but the multiplicity of $t-a$ must be strictly positive. This cannot hold for all a. Also, by replacing t by a translation, we can suppose that $t(y_0) = 0$.

Since t separates generic orbits of S by Rosenlicht's theorem (see for example [**15**, IV, 2.2]), the generic fiber of the quotient map $t : S \to \mathbb{C}$ is irreducible. Define S_0 to be the open affine subvariety obtained by eliminating certain fibers. First we eliminate all the reducible fibers of t except for $t = 0$, and then we also eliminate any fiber such that the divisor of one of the rational functions e_i $(i = 1, \dots, r)$ has support in the fiber with the exception of $t = 0$. Denote by $\{a_1, \dots, a_k\}$ the complement of $t(S_0) \subset \mathbb{C}$. Let A_0 be the ring of regular functions on S_0. Let A' be the subalgebra of A_0 generated by the set $\{t^{n_0} e_1^{n_1} \cdots e_r^{n_r} | n_0, \dots, n_r \in \mathbb{Z}\} \cap A_0$. Now we show that $A_0 = A'[1/(t-a_1), \dots, 1/(t-a_k)]$. One direction of the inclusion is clear. For the converse, since A_0 is generated by T-semi-invariants, it suffices to show that any semi-invariant in A_0 is in $A'[1/(t-a_1), \dots, 1/(t-a_k)]$. A semi-invariant in A_0 is of the form $f = g(t) t^{n_0} e_1^{n_1} \cdots e_r^{n_r}$, where $g(t) = \prod_{a \in \mathbb{C}^*} (t-a)^{m_a}$ where all but finitely many $m_a = 0$. We can suppose that $m_a = 0$ for $a = a_1, \dots, a_k$ since the factors $(t - a_i)$ are units. Now consider the divisor of f in S_0. By construction, the divisor has the irreducible component $t^{-1}(a)$ with multiplicity m_a for all $a \in \mathbb{C}^* \setminus \{a_1, \dots, a_k\}$. Thus $m_a \geq 0$, and furthermore, $f/g(t)$ is also in A_0, and therefore in A'. This proves the claim. Also, we see that A' is finitely generated. This is because A_0 is generated by a finite number of T-semi-invariants, and by what we have seen, we can choose them to be $1/(t-a_1), \dots, 1/(t-a_k)$ and functions of the form $t^{n_0} e_1^{n_1} \cdots e_r^{n_r}$. If we eliminate the first k functions, we have a finite set of generators of A'. Thus we choose $S' = \operatorname{Spec} A'$. Then S' has an action of $T \times \mathbb{C}^*$ where $\mathbb{C}^*$ fixes $e_1 \ldots, e_r$ and acts on t by multiplication. Thus, S' is a toric variety. □

PROPOSITION 3.2. *If Y is an orbit of type* II *isomorphic to G/B, and if $y_0 \in S$ is smooth, then $S \cong \mathbb{C}^{r+1}$, and T acts linearly on S.*

This proposition can be proven using the Luna Slice Theorem [**8**]. One can also prove it directly.

PROOF. Denote by $\Xi(S)$ the characters of T for which there is a non-constant semi-invariant in $\mathbb{C}[S]$. It is a semi-group of the character group of T, and it does not contain 0. Thus there exists a one-dimensional subgroup T_1 such that $\mathbb{C}[S]^{T_1} = \mathbb{C}$. The character group $\Xi(T_1)$ of T_1 is isomorphic to $\mathbb{Z}$, and all characters of T_1 which have non-constant semi-invariants in $\mathbb{C}[S]$ are of the same sign. We can assume they are positive. We have that $\mathbb{C}[S] = \oplus_{i=0}^{\infty} M_i$ where M_i denotes the set of semi-invariants of weight i. Denote by $\mathcal{M}$ the maximal ideal of y_0 in $\mathbb{C}[S]$. We have that $\mathcal{M} \subset \oplus_{i=1}^{\infty} M_i$, and therefore, for any $k \in \mathbb{N}$, we have $\mathcal{M}^k \subset \oplus_{i=k}^{\infty} M_i$.

If y_0 is smooth in S, then its local ring is generated by $r+1$ elements $f_0, \ldots, f_r$. We can suppose that they are semi-invariants for T and that they are in $\mathbb{C}[S]$. Now suppose $f \in \mathbb{C}[S]$. Then for any n, f can be expressed of the form $f = P_n(f_0, \ldots, f_r) + h_{n+1}$, where P is a polynomial of degree n and $h_{n+1} \in \mathcal{M}^{n+1}$. Suppose f is a semi-invariant for T_1 of weight n. Then, by projection on M_n, we see that we can suppose all the monomials in P_n are of weight n, and h_{n+1} is zero. Since $\mathbb{C}[S]$ is generated by semi-invariants, $\mathbb{C}[S] = \mathbb{C}[f_0, \ldots, f_r]$. □

REMARK 1. If Y is of type II and S is not smooth, then it cannot in general be viewed as a toric variety. In fact, the singularities that arise are not in general toric.

4. Embeddings of $\mathbf{SL(2, \mathbb{C})}$ modulo a finite subgroup

Consider the case where $G = \mathrm{SL}(2, \mathbb{C})$ and H is a finite subgroup. The classification of all G/H-embeddings was given in [**10**] for $H = \{\mathrm{Id}\}$, and the classification for $H =$ a finite group was deduced in [**11**]. It is also given in [**18**]. The smooth embeddings for $H = \{\mathrm{Id}\}$ and $H = \{\pm \mathrm{Id}\}$ were given in [**12**]. For the general case, one can determinine which embeddings are smooth using the results of the previous section. Let X be a G/H-embedding, and let Y be an orbit. If Y is either of dimension 2 or 3, then it is in the set of smooth points of X, because X is normal. If Y is a fixed point, then it is a singular point. (This can be seen using the Luna slice theorem [**8**]. See for example [**13**], Lemma 8, page 818.) The only possibility left is that $Y \cong G/B$, and the two propositions apply. Thus we have

COROLLARY 4.1. *If X is a smooth complete embedding of* $\mathrm{SL}(2)/H$, *where H is a finite subgroup, then X is covered by open affine charts isomorphic to $\mathbb{C} \times (\mathbb{C}^2 \setminus \{\cup_{i=1}^{k} \ell_i\})$ where $\mathbb{C}^2$ has a linear $\mathbb{C}^*$-action, and $\ell_1, \ldots, \ell_k$ are one-dimensional closed orbits.*

PROOF. Since X is complete, any closed orbit Y is isomorphic to $\mathrm{SL}(2)/B$, and Propositions 3.1 and 3.2 apply. If the orbit Y is of type II, then there is an affine chart intersecting Y isomorphic to $\mathbb{C}^3$, and Y is covered by two charts isomorphic to $\mathbb{C}^3$: the B-stable one and a translation of it. If it is of type I, there is an affine chart isomorphic to $U \times S$, where U is isomorphic to $\mathbb{C}$ and, by Proposition 3.1, S is isomorphic to $\mathbb{C}^2 \setminus \{$a finite set of closed one-dimensional T-orbits$\}$. □

REMARK 2. If X is smooth but not complete, it can be completed in a smooth way. Since $G = \mathrm{SL}(2, \mathbb{C})$, we can see this by checking the list of smooth embeddings of [**3**]. (This can, however, be proven in a more general setting. By a result of Sumihiro, [**17**], for any connected reductive group G and closed subgroup H, a normal G/H-embedding can be completed to a normal complete embedding, and if the original embedding was smooth, then the completion can be resolved to a smooth completion by [**1**, Theorem 13.2 (2)ii]. See also [**14**, Remark 3.3]). Thus the corollary also holds in this case, if one allows ℓ_1 to be any one-dimensional closed $\mathbb{C}^*$-stable subset of $\mathbb{C}^2$.

For orbits of type I, it is not hard to determine for which torus the variety S is toric, and thus one can find the list of smooth embeddings. However, for type II orbits, it is not evident from the classification how to check if the variety S is isomorphic to $\mathbb{C}^2$. In order to do this, we determinine first which are toric, and then among them, we check which are smooth. This calculation is carried out completely in the first author's thesis [**3**], and is the topic of a forthcoming article. To do this, we use the description of Timashev of B-charts given by hypercones. Here, we describe the method and show how it works for two examples.

5. Description of hypercones

In [**18**] a combinatorial description of affine B-charts of complexity one embeddings is described. We give here a brief sketch of this description. For more precise information, see the article of Timashev or [**3**].

The main idea of this description comes from the article [**10**]. It is shown there that the ring of an affine B-chart is completely determined by the B-stable divisors in it. These are divided into those which are G-stable and those which are not. Each irreducible divisor determines a valuation ring, and these rings are codified by combinatorial data. This imitates the classification of spherical varieties. As we will see from the examples, the main point of the present article is that, in order to use the description of toric varieties, one should not distinguish among the B-stable divisors those which are G-stable.

Choose $t \in \mathbb{C}(G)^{B\times H}$ such that $\mathbb{C}(t) = \mathbb{C}(G)^{B\times H}$, and choose $e_1, \dots, e_r$ as before.

We represent all geometric B-stable valuations of the field $\mathbb{C}(G)^H$. First, we define a "hyperspace" $\mathcal{H}$ as follows. Let $\mathbb{Q}$ be the field of rational numbers, and let $\mathbb{Q}_+$ be the subset of non-negative rational numbers. Set $\mathcal{H} = (\mathbb{Q}^r \times \mathbb{Q}_+)\times\mathbb{P}^1/\sim$, where the equivalence relation is given by $(x, 0, a) \sim (x, 0, b)$ for all $x \in \mathbb{Q}^r$ and $a, b \in \mathbb{P}^1$. That is, there is a half-space for each element $a \in \mathbb{P}^1$, and they are glued together along the base. All B-stable valuations of $\mathbb{C}(G)^H$ will be represented in $\mathcal{H}$.

First we consider the G-stable geometric valuations. Such a valuation v is determined by the restriction of its values on $\mathbb{C}(t)$ and by $v(e_i)$, $i = 1, \dots, r$. We represent v by the ray in $\mathcal{H}$ generated by $(v(e_1), \dots, v(e_r), \delta, a)$ where δ and a are given as follows. If the restriction of v on $\mathbb{C}(t)$ is trivial, then $\delta = 0$ and $a \in \mathbb{P}^1$ is arbitrary. That is, the ray is in the intersection of all the half-spaces of $\mathcal{H}$. If there exists $c \in \mathbb{C}$ with $v(t - c) > 0$, then $\delta = v(t - c)$ and $a = c$. That is, the ray is in the half-space of $\mathcal{H}$ determined by c. Finally, if $v(1/t) > 0$, then $\delta = v(1/t)$ and $a = \infty \in \mathbb{P}^1$. The ray completely determines the G-valuation (up to normalisation).

Secondly, we represent the B-stable valuations of G/H in the hyperspace. For any B-stable irreducible divisor D in G/H, consider the B-stable valuation v_D. Let $\mathcal{D}^B$ be the set of irreducible B-stable divisors of G/H. Consider the (not necessarily injective) map $\rho : \mathcal{D}^B \to \mathcal{H}$ given similarly to the G-valuations. That is, $\rho(D) = \big(v_D(e_1), \ldots, v_D(e_r), \delta, a\big)$ where $\delta = v_D(t-a)$ if $v_D(t-a) > 0$ and $a \in \mathbb{C}$, and $\delta = v_D(1/t)$ and $a = \infty$ if $v_D(t) < 0$. One represents D by a small circle at $\rho(D)$.

Finally, for an affine B-chart X_0 of an embedding, let $\mathcal{W}$ be the set of G-stable valuations corresponding to G-stable divisors of X which intersect X_0, and let $\mathcal{R}$ be the set of irreducible B-stable but not G-stable divisors in X_0. The affine B-chart X_0 is represented by a colored hypercone, which is defined as follows. Let $C(\mathcal{W}, \mathcal{R})$ be the bidual to $\mathcal{W}$ and $\rho(\mathcal{R})$. That is, $C(\mathcal{W}, \mathcal{R})$ is the set of elements of the hyperspace on which all functionals which are non-negative on $\mathcal{W}$ and $\rho(\mathcal{R})$ are non-negative. This hypercone is "colored" by filling in the circles corresponding to the elements of $\rho(\mathcal{R})$. This colored hypercone determines the B-chart.

There are many rules that such a cone must satisfy. This can be seen in detail in [**10**] or [**18**]. Here we will not discuss this.

It should be noted that different choices can be made for t and $e_1, \ldots, e_r$, and this can change the shape of the hypercone. Timashev deals with this by fixing once and for all a choice. However, we leave this choice open. Changing the choice of t simply permutes the cones corresponding to the different points of $\mathbb{P}^1$. More importantly, changing the choices of $e_1, \ldots, e_r$ changes the shape. In fact, one can show that X_0 is isomorphic to $U \times S$ with S toric if and only if one can choose $e_1, \ldots, e_r$ in such a way that all but two of the cones are identical (not taking into account the colors). In this case, the torus action on S comes from the action of the maximal torus T on S and a $\mathbb{C}^*$-action which fixes $e_1, \ldots, e_r$ but acts on t. We will see two examples in the next section.

6. Two examples

We fix some notation. $\mathrm{SL}(2) = \{\left(\begin{smallmatrix} x & y \\ z & w \end{smallmatrix}\right) | xw - yz = 1\}$. Let B be the subgroup defined by $y = 0$. Then the field of B-invariants in $\mathbb{C}(\mathrm{SL}(2))$ is given by $\mathbb{C}(x, y)$. We have $U = \{\left(\begin{smallmatrix} 1 & 0 \\ z & 1 \end{smallmatrix}\right) | z \in \mathbb{C}\}$, and we choose $T = \{\left(\begin{smallmatrix} x & 0 \\ 0 & x^{-1} \end{smallmatrix}\right) | x \in \mathbb{C}^*\}$.

First consider the case where $H = \{\pm \mathrm{Id}\}$. We can choose $t = y/x$, and $e_1 = x^2$. There is an affine B-chart whose colored cone is given by Figure 1a. In all of the figures, the shaded region of the hyperspace represents the cone $C(\mathcal{W}, \mathcal{R})$, the black circles represent the elements of $\rho(\mathcal{R})$ and the empty circles represent the elements of $\rho(\mathcal{D}^B)$ which are not in $\rho(\mathcal{R})$. Here the hypercone has three cones which are not like the generic ones. However, if we change the choice of e_1 to $x(x-y)$, the hypercone is given in Figure 1b, where one sees that all the cones are alike except for one. Thus, in particular, the orbit has an affine chart which is toric.

In fact, one can show that this example corresponds to a B-chart of the embedding $\mathbb{P}^1 \times \mathbb{P}^1 \times \mathbb{P}^1$, where $\mathrm{SL}(2)$ acts on each copy of $\mathbb{P}^1$ in the standard way. There is a unique closed orbit given by the diagonal, and the B-chart of Figure 1 is the unique affine B-chart which intersects this orbit. It is isomorphic to $U \times S$, where S is isomorphic to $\mathbb{C}^2$, and the T-orbits of S are shown in Figure 1c. Three of them are the intersections of $\mathrm{SL}(2)$-orbits with S, and they all intersect in the unique fixed point y_0. The surface S is toric, that is, it is an embedding of $T \times \mathbb{C}^*$, however, not all the G-orbits are stabilized by $T \times \mathbb{C}^*$.

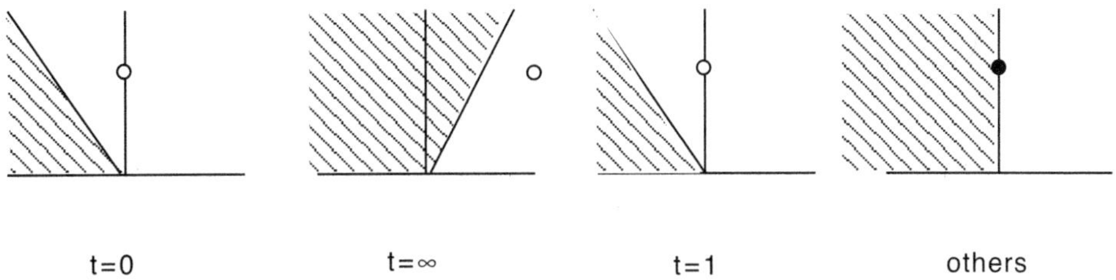

(a) $e_1 = x^2$

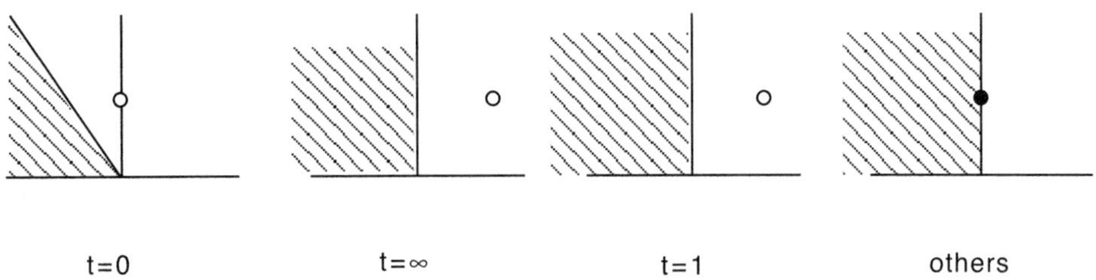

(b) $e_1 = x(x - y)$

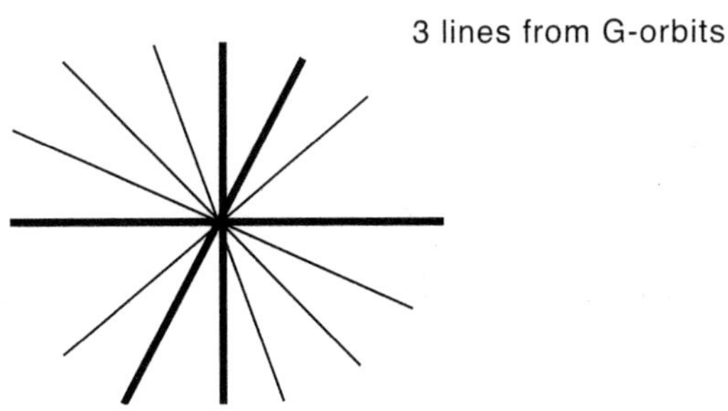

(c) T-orbits of S

FIGURE 1.

For a second example, consider the case where H is the cyclic group of four elements generated by $\left(\begin{smallmatrix} i & 0 \\ 0 & -i \end{smallmatrix}\right)$. We can choose $t = (y/x)^2$, and a choice given in [**18**], Section 5 is $e_1 = (x + y)^2(x - y)^2/(xy)$. There is an affine B-chart whose hypercone is given in Figure 2a. If we change the choice of e_1 to xy, the hypercone becomes Figure 2b, and again we see that there is a toric affine chart.

For this case, the B-chart comes from the embedding $\mathbb{P}^2 \times \mathbb{P}^1$, where the action of SL(2) on $\mathbb{P}^2$ is given by the adjoint action, and the action on $\mathbb{P}^1$ is the standard multiplication. There is one closed orbit, and the example is the unique B-chart intersecting this orbit. Again, the B-chart is isomorphic to $U \times S$, where S is isomorphic to $\mathbb{C}^2$, however, the action of T on S is different. The T-orbits are shown in Figure 2c. There are two T-orbits which are the intersections of S and

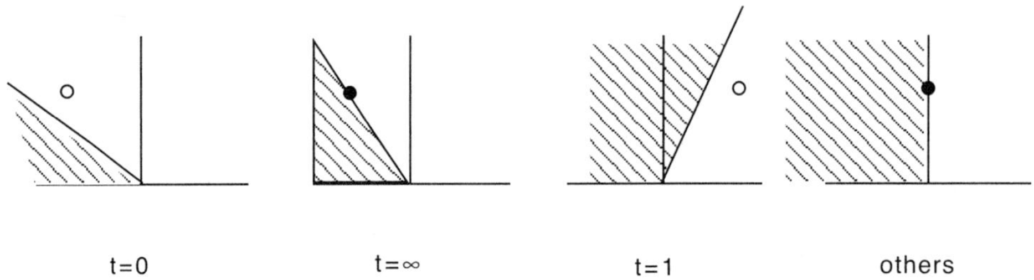

(a) $e_1 = (x+y)^2(x-y)^2/xy$

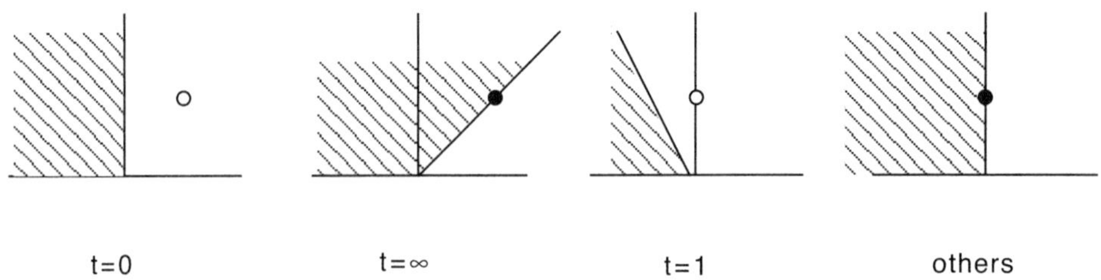

(b) $e_1 = xy$

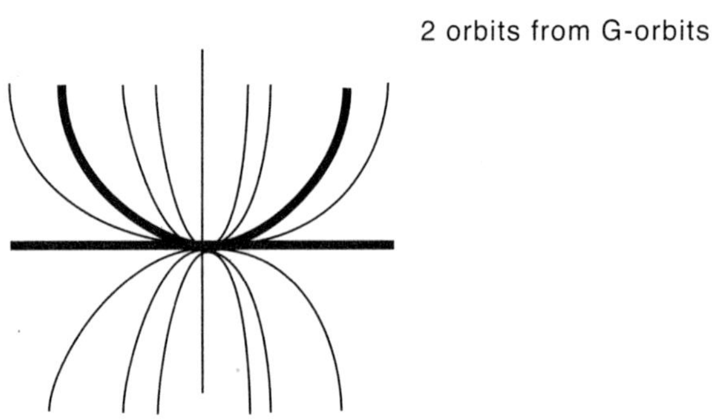

(c) T-orbits of S

FIGURE 2.

the two irreducible G-stable divisors, and they are tangent. The variety S is toric for the torus $T \times \mathbb{C}^*$, but this action does not fix the G-orbits. For this reason we do not distinguish between the G-stable valuations and those which are only B-stable to find the $T \times \mathbb{C}^*$-action.

7. A result of Luna on wonderful varieties

An embedding X of G/H is called wonderful of rank r if X is smooth and complete, the complement of the open orbit has r irreducible components D_i, $i = 1, \ldots, r$, all smooth of codimension 1 intersecting transversely, and X has 2^r orbits

X_I where I is a subset of $\{1, \dots, r\}$ and $\overline{X_I} = \cap_{i \in I} D_i$ ($X_\emptyset$ = the open orbit). In particular, X has one closed orbit.

In [**9**], Luna showed that all wonderful varieties are spherical. In order to prove this result, he reduced the problem to showing that if there were a wonderful variety of complexity strictly bigger than 0, then one could produce a wonderful variety X of dimension three with $G = \mathrm{SL}(2)$. That is, X would be a smooth complete embedding of SL(2) modulo a finite subgroup. Luna used a topological argument to show that such a wonderul variety does not exist. Here we will give another argument that this cannot happen, using Proposition 3.2 and some facts about SL(2)-embeddings which can be found in [**10**] and [**11**] or in [**9**]. We use the following facts, which are also used by Luna.

1. If X is an embedding of $\mathrm{SL}(2)/H$ where H is a finite subgroup, then there is and embedding $\widetilde{X}$ of SL(2) such that X is the quotient of $\widetilde{X}$ by H.
2. If X is a complete embedding with one closed orbit, then so is $\widetilde{X}$.
3. A complete embedding of SL(2) with exactly one closed orbit has at least three orbits of dimension 2.
4. If X has r codimension one orbits, then all but r B-stable divisors of $\mathrm{SL}(2)/H$ contain the closed orbit of X in its closure.

Now suppose X were a wonderful $\mathrm{SL}(2)/H$-embedding. It must be of rank two, because the closed orbit must be complete and consisting of smooth points of X, and therefore, it must be isomorphic to $\mathbb{P}^1$. Thus X would have four orbits: the open orbit, two orbits of dimension 2, and the closed orbit, which we call Y. By the result of Brion, Luna and Vust, there is an affine B-chart X_0 and a T-stable closed subvariety S of X_0 such that $U \times S$ is isomorphic to X_0. The T-orbits of S are the intersections of S with the B-orbits of X_0. We have that S has a unique fixed point, and we are in the situation of Proposition 3.2. Since X is smooth, S is isomorphic to $\mathbb{C}^2$ with a linear action of T.

Consider now the intersections of the two two-dimensional G-orbits of X with S. They must be two T-orbits which intersect transversely at the fixed point of S. The other T-orbits of S are all isomorphic to $\mathbb{C}^*$, and the action is given by a common character of T. Thus for all divisors which are B-stable but not G-stable and whose closures contain the unique closed orbit in X, the B action is isomorphic. If we identify the open orbit with G/H, the B-stable divisors which are not G-stable are exactly the closures of B-stable divisors of G/H, which are H-orbits of $G/B \cong \mathbb{P}^1$.

We recall all the possible finite subgroups of SL(2) (see, for example [**13**] or [**16**]). They are either cyclic, binary dihedral, binary tetrahedral, binary octohedral or binary icosahedral. For all but the cyclic case, there are three B-stable divisors for which the action of B is not like the generic case. They correspond to the vertices, edges and faces of the tetrahedron, octahedron or icosahedron, or, in the dihedral case, the three non-generic orbits. However, by condition (4), all but two B-divisors of $\mathrm{SL}(2)/H$ must be in X_0. The only possibility, considering the remarks above is that H is cyclic, and the two B-divisors which are not in X_0 correspond to the fixed points of the cyclic action on $\mathrm{SL}(2)/B$. Thus, in $\widetilde{X}$, all but two B-stable divisors of SL(2) contain the closed orbit. Using (4) again, we see that $\widetilde{X}$ must have two SL(2)-orbits of codimension 1. But by (3), this is impossible, and therefore, no $\mathrm{SL}(2)/H$-embedding is a wonderful variety.

REMARK 3. Note that in the second example of Section 6, there is a complete embedding with the correct number of orbits, but the two G-divisors do not intersect transversely. There is one B-divisor for which the character is different than for the others, and this divisor, with one of the G-divisors play the role of the axes for the linear action of T on S.

References

1. E. Bierstone and P. D. Milman, *Canonical desingularization in characteristic zero by blowing up the maximum strata of a local invariant*, Invent. Math. **128** (1997), no. 2, 207–302.
2. A. Borel, *Linear algebraic groups*, 2nd ed., Springer-Verlag GTM, vol. 126, New York, 1991.
3. G. Bousquet, *Plongements homogènes de* SL(2) *modulo un sous-groupe fini*, Thèse de doctorat, Université de Bourgogne, 2000.
4. M. Brion, *Variétés sphériques*, Notes de la session de la S.M.F. Opérations Hamiltoniennes et Opérations de Groupes Algébriques, Institut Fournier, Grenoble, 1997.
5. M. Brion, D. Luna, and Th. Vust, *Espaces homogènes sphériques*, Invent. Math. **84** (1986), 617–632.
6. F. Knop, *The Luna–Vust theory of spherical embeddings*, Proc. of the Hyderabad Conference on Algebraic Groups (1989), Manoj Prakashan, Madras, 1991, pp. 225–249.
7. H. P. Kraft, *Geomtrische Maethoden in der Invariantentheorie*, Vieweg, Braunshweig, 1985.
8. D. Luna, *Slices Étales*, Bull. Soc. Math. France, Mémoire **33** (1973), 81–105.
9. ______, *Toute variété magnifique est sphérique*, Transform. Groups **1** (1996), no. 3, 249–258.
10. D. Luna and Th. Vust, *Plongements d'espaces homogènes*, Comment. Math. Helv. **58** (1983), 186–245.
11. L. Moser-Jauslin, *Normal embeddings of* $SL(2)/\Gamma$, Thèse de doctorat, Université de Genève, 1987.
12. ______, *Smooth embeddings of* SL(2) *and* PGL(2), J. Algebra **132** (1990), no. 2, 384–405.
13. V. Popov, *Quasi-homogeneous affine algebraic varieties of the group* SL(2), Izv. Akad. Nauk SSSR Ser. Mat **37** (1973), no. 4, 793–831.
14. Z. Reichstein and B. Youssin, *Essential dimensions of algebraic groups and a resolution theorem for G-varieties*, Canad. J. Math. **52** (2000), no. 5, 1018–1056.
15. T. A Springer, *Aktionen reduktiver Gruppen auf Varietäten*, Algebraic Transformation Groups and Invariant Theory (Kraft, Slodowy and Springer, eds.), DMV Seminar, vol. 13, 1989, Birkhäuser, pp. 1–39.
16. ______, *Invariant theory*, Lecture notes in Math., vol. 585, Springer-Verlag, 1977.
17. H. Sumihiro, *Equivariant completion*, J. Math. Kyoto Univ. **14** (1974), 1–28.
18. D. A. Timashev, *Classification of G-varieties of complexity* 1, Izv. Math. **61** (1997), no. 2, 363–397.

UNIVERSITÉ DE BOURGOGNE, IMB CNRS—UMR 5584, 9 AVENUE ALAIN SAVARY - B.P. 47870 21078 DIJON CEDEX, FRANCE.

E-mail address, G. Bousquet: `gilles.bousquet@u-bourgogne.fr`

E-mail address, L. Moser-Jauslin: `moser@u-bourgogne.fr`

Centre de Recherches Mathématiques
CRM Proceedings and Lecture Notes
Volume **35**, 2004

Constructive Invariant Theory

H. Derksen

In these lecture notes, we will discuss the constructive aspects of modern invariant theory. A detailed discussion on the computational aspect of invariant theory can be found in the books [**9, 39**]. The invariant theory of the nineteenth century was very constructive in nature. Many invariant rings for classical groups (and SL_2 in particular) were explicitly computed. However, we will not discuss this classical approach here, but we will refer to [**25, 33**] instead. Our starting point will be a general theory of invariant rings. The foundations of this theory were built by Hilbert. For more on invariant theory, see for example [**24, 26, 36**].

1. Hilbert's first approach

Among the most important papers in invariant theory are Hilbert's papers of 1890 and 1893 (see [**15, 16**]). Both papers had an enormous influence, not only on invariant theory but also on commutative algebra and algebraic geometry. Much of the lectures will be centered around these two papers, and their place within modern invariant theory.

1.1. Hilbert's Basissatz.

THEOREM 1.1 (cf. [**15**]). *If K is a field, then the polynomial ring $K[x_1, x_2, \dots, x_n]$ is* Noetherian, *i.e., every ideal of $K[x_1, \dots, x_n]$ is finitely generated.*

The statement in this theorem is not constructive, because it does not tell us how to find the generators of an ideal. Today, exploiting new and important methods using Gröbner Bases, we see that many problems concerning ideals have constructive solutions. There are many books on Gröbner bases methods (for example [**1, 3, 6, 28, 40**]), so we will keep the discussion here to a minimum.

DEFINITION 1.2. An *admissible monomial ordering* is an ordering "$<$" on the monomials with the properties:

1. $<$ is a total ordering,
2. $1 < m$ for every monomial m not equal to 1,
3. If $m_1 < m_2$ then $mm_1 < mm_2$ for all monomials m_1, m_2, m.

2000 *Mathematics Subject Classification.* Primary: 13A50; Secondary: 13P10.
The author was partially supported by NSF, grant DMS 0102193.
This is the final version of the paper.

PROPOSITION 1.3. *If "$<$" is an admissible ordering, then it is a* well-ordering: *Every nonempty set of monomials has a smallest element.*

EXERCISE 1.4. Use Hilbert Basissatz (Theorem 1.1) to prove Proposition 1.3.

An important ordering is the *lexicographic ordering* defined by:

$$x_1^{a_1}x_2^{a_2}\cdots x_n^{a_n} < x_1^{b_1}x_2^{b_2}\cdots x_n^{b_n} \iff \text{there is a } k \text{ such that } a_i = b_i \text{ for } i < k \text{ and } a_k = b_k.$$

Other monomial orderings such as the reverse lexicographic total degree ordering may be more efficient in practice, but we will not discuss them here.

Fix an admissible ordering $<$. If f is a polynomial in $K[x_1, \ldots, x_n]$, then the *leading monomial* $\mathrm{lm}(f)$ is the largest monomial appearing in f. If c is the coefficient of $\mathrm{lm}(f)$ in f, then $\mathrm{lc}(f) := c$ is called the *leading coefficient* and $\mathrm{lt}(f) := c\,\mathrm{lm}(f)$ is called the *leading term* of f. If I is an ideal, then let $\mathrm{lm}(I)$ be the (monomial) ideal generated by all $\mathrm{lm}(f)$, $f \in I$.

DEFINITION 1.5. A subset $\mathcal{G} = \{f_1, f_2, \ldots, f_r\}$ of I is called a Gröbner basis (with respect to the ordering $<$) if $\mathrm{lm}(I)$ is generated by $\{\mathrm{lm}(f_1), \ldots, \mathrm{lm}(f_r)\}$.

It is clear that every ideal has a finite Gröbner basis, since $\mathrm{lm}(I)$ is a finitely generated ideal by Theorem 1.1.

PROPOSITION 1.6. *If $\mathcal{G}$ is a Gröbner basis of I, then $\mathcal{G}$* generates *I as an ideal.*

PROOF. Supppose that $h \in I$ is nonzero. Since $\mathrm{lm}(h) \in I$ and $\mathcal{G}$ is a Gröbner basis, we get that $\mathrm{lm}(h)$ is divisible by $\mathrm{lm}(f_{i_0})$ for some i_0. We define $h_0 := h$ and

$$h_1 := h - a_0 f_{i_0}$$

where $a_0 = \mathrm{lt}(h_0)/\mathrm{lt}(f_{i_0})$. Note that $\mathrm{lm}(h_1) < \mathrm{lm}(h)$. By repeating this process we have

$$h_{j+1} = h_j - a_i f_{i_j}$$

with $\mathrm{lm}(h_{j+1}) < \mathrm{lm}(h_j)$ for all j. We must have $h_k = 0$ for some k. Otherwise

$$\{\mathrm{lm}(h_0), \mathrm{lm}(h_1), \ldots\}$$

is strictly decreasing and this set would not have a smallest element, contradicting Proposition 1.3. From this it follows that $h \in (f_1, f_2, \ldots, f_r)$. This proof gives an algorithm for writing h as a combination of $f_1, \ldots, f_r$. □

If an ideal is given by a finite set of generators, then the so-called *Buchberger algorithm* can find a Gröbner basis for it. Although the general complexity of the Buchberger algorithm is very bad, one is still able to compute a Gröbner basis in many interesting examples.

The following Exercise shows that the lexicographic ordering is useful for the *elimination of variables* $x_1, \ldots, x_k$ in an ideal I of $K[x_1, \ldots, x_n]$.

EXERCISE 1.7. If $\mathcal{G}$ is a Gröbner basis with respect to the lexicographic ordering, then $\mathcal{G} \cap K[x_{k+1}, x_{k+2}, \ldots, x_n]$ is a Gröbner basis of $I \cap K[x_{k+1}, x_{k+2}, \ldots, x_n]$.

1.2. Algebraic groups. For convenience, we will work over an algebraically closed field. Radical ideals of the polynomial ring correspond to Zariski closed subsets of the affine space. Note that such a correspondance is based on Hilbert's Nullstellensatz which was proven in the 1893 paper (see [**16**]).

DEFINITION 1.8. A linear algebraic group is an affine variety G with a fixed element $e \in G$, and morphisms $m : G \times G \to G$ (multiplication) and $i : G \to G$ (inverse) such that G satisfies the usual group axioms.

A homomorphism of algebraic groups $\phi : H \to G$ is a morphism of affine algebraic varieties which is also a homomorphism of groups. Such a homomorphism is called an isomorphism if ϕ is an isomorphism of affine algebraic varieties.

Often a linear algebraic group is defined as a Zariski closed subgroup of GL_n for some n. One can show that our definition is equivalent to this. Exercise 1.9 below shows that a Zariski closed subgroup of GL_n is a linear algebraic group according to our definition. Remark 1.23 shows that any group that is a linear algebraic group according to our definition is indeed isomorphic to a Zariski closed subgroup of GL_n for some n.

Finite groups, the orthogonal group O_n, the special linear group SL_n are examples of algebraic groups.

EXERCISE 1.9. We can identify GL_n with the Zariski closed subset of $\mathrm{M}_n(K) \times K$ defined by

$$\{(A, b) \mid \det(A)b = 1\}$$

(here $\mathrm{M}_n(K)$ is the set of $n \times n$ matrices with entries in K). Using this identification, show that GL_n is a linear algebraic group. (What are e, m and i?) Also prove that every Zariski closed subgroup of GL_n is a linear algebraic group.

In particular the multiplicative group $K^\star = \mathrm{GL}_1$ is an algebraic group. The r-dimensional torus is defined as $(K^\star)^r$ and this group is of course an algebraic group as well.

DEFINITION 1.10. A *rational action* of G on an affine variety X is a morphism

$$\mu : G \times X \to X$$

which satisfies the usual axioms of an action of a group on a set. Instead of $\mu(g, x)$ we will write $g \cdot x$. If X is a vector space and G acts by linear transformations, then X is called a *representation space*. An affine variety with a rational action of G is called an *affine G-variety*.

If G acts rationally on an affine variety X, then G also acts on the coordinate ring $K[X]$ as follows. If $f \in K[X]$ and $g \in G$, then $g \cdot f$ is defined by

$$(g \cdot f)(x) := f(g^{-1} \cdot x), \quad x \in X.$$

(Note that we have to put in the inverse to make it an action.) Now the *invariant ring* is defined by

$$K[X]^G = \{f \in K[X] \mid g \cdot f = f \text{ for every } g \in G\}.$$

EXAMPLE 1.11. Suppose that $G = S_n$, the symmetric group, and V is a K-vector space with a basis $v_1, v_2, \ldots, v_n$. We can let G act on V by:

$$\sigma \cdot v_i = v_{\sigma(i)}.$$

Let $x_1, x_2, \ldots, x_n$ be the coordinate functions which are dual to $v_1, \ldots, v_n$. The coordinate ring of V is $K[x_1, x_2, \ldots, x_n]$. It is well-known that the invariant ring is equal to

$$K[V]^G = K[e_1, e_2, \ldots, e_n]$$

where e_k is the *k-th elementary symmetric function* defined by

$$e_k = \sum_{1 \leq i_1 < i_2 < \cdots < i_k \leq n} x_{i_1} x_{i_2} \cdots x_{i_k}.$$

EXAMPLE 1.12. Let us assume that $\operatorname{char} K = 0$. Suppose $G = \mathrm{SL}_2$. Define V_d, the *binary forms of degree d* by

$$V_d = \{a_0 X^d + a_1 X^{d-1} Y + \cdots + a_d Y^d \mid a_0, a_1, \ldots, a_d \in K\}.$$

Now V_d is a representation of SL_2 where SL_2 acts by

$$\begin{pmatrix} a & b \\ c & d \end{pmatrix} \cdot f(X, Y) = f(aX + cY, bX + dY).$$

Let t_i be the linear function on V_d that maps a polynomial $a_0 X^d + a_1 X^{d-1} Y + \cdots + a_d Y^d$ to the coefficient a_i. We can identify $K[V_d]$ with $K[t_0, t_1, \ldots, t_d]$. Gordan proved (before Hilbert's general finiteness theorem) that $K[V_d]^{\mathrm{SL}_2}$ is always finitely generated (see [**14**]). For $d = 2$ we have

$$K[V_2]^{\mathrm{SL}_2} = K[t_1^2 - 4t_0 t_2].$$

The term $t_1^2 - 4t_0t_2$ may be recognized as the discriminant of a quadratic equation.

EXAMPLE 1.13. Let $K^\star$ be the multiplicative group. We let $K^\star$ act on $V = K^3$ by

$$\lambda \cdot (x_1, x_2, x_3) = (\lambda^{-2} x_1, \lambda x_2, \lambda^3 x_3).$$

One can check that the invariant ring is equal to

$$K[x_1, x_2, x_3]^{K^\star} = K[x_1 x_2^2, x_1^2 x_2 x_3, x_1^3 x_3^2].$$

1.3. Hilbert's finiteness theorem. Nagata showed in [**31**] that invariant rings are not always finitely generated (answering Hilbert's fourteenth problem, [**17, 18**]). Nevertheless, invariant rings are finitely generated for a large class of groups, so-called *linearly reductive groups.*

Besides the notion of linear reductivity, there are also other notions like *group-theoretical reductivity* and *geometric reductivity* (these two notions actually coincide). Many results such as finite generation of invariant rings and geometric properties of invariant rings also hold for these more general notions of reductivity, but usually the proofs are more complicated. In characteristic 0, all notions of reductivity coincide. We will stick to linear reductivity and not further discuss the other notions of reductivity.

DEFINITION 1.14. A Reynolds operator for a linear algebraic group G is a K-linear map $\mathcal{R} : K[G] \to K$ such that $\mathcal{R}(1) = 1$ and $\mathcal{R}(g \cdot f) = \mathcal{R}(f)$ for all $g \in G$ and $f \in K[X]$ (G acts on itself rationally by left multiplication, so G also acts on $K[G]$). An algebraic group with a Reynolds operator is called *linearly reductive.*

Often an algebraic group is called linearly reductive if every representation is a direct sum of irreducible representations (i.e., representations without non-trivial proper subspaces which are stable under the action). It can be shown that this definition is equivalent to our definition.

EXAMPLE 1.15. A finite group G is linearly reductive as long as the characteristic of the base field K does not divide the group order G. Let us define $\mathcal{R}$ as the averaging over the group

$$\mathcal{R}(f) = \frac{1}{|G|} \sum_{g \in G} g \cdot f.$$

It is straightforward to check that $\mathcal{R}$ satisfies the conditions for a Reynolds operator (see Definition 1.14).

EXAMPLE 1.16. The group $K^\star$ is linearly reductive. Indeed, the coordinate ring of $K^\star$ can be identified with the ring of Laurent polynomials

$$K[x, x^{-1}].$$

For any $f \in K[x, x^{-1}]$, we define $\mathcal{R}(f)$ as the constant coefficient.

EXERCISE 1.17. Check that $\mathcal{R}$ in the previous example is indeed a Reynolds operator.

EXAMPLE 1.18. For classical groups over $\mathbb{C}$, the Reynolds operator is averaging over a maximal compact subgroup. For example, $\mathrm{GL}_n(\mathbb{C})$ has the subgroup $\mathrm{U}_n(\mathbb{C})$. It is known that the unitary group is Zariski-dense in $\mathrm{GL}_n(\mathbb{C})$. Then we can define

$$\mathcal{R}(f) = \int_{U_n(\mathbb{C})} f \, d\mu, \quad f \in \mathbb{C}[\mathrm{GL}_n(\mathbb{C})],$$

where $d\mu$ is a so-called Haar-measure for $U_n(\mathbb{C})$ (a measure that is invariant under the group action which we will normalize such that $\int d\mu = 1$). It is clear that $\mathcal{R}(1) = 1$ and that $\mathcal{R}(g \cdot f) = \mathcal{R}(f)$ for $g \in U_n(\mathbb{C})$. Using the fact that $U_n(\mathbb{C})$ lies Zariski dense in $\mathrm{GL}_n(\mathbb{C})$ one gets that $\mathcal{R}(g \cdot f) = \mathcal{R}(f)$ for all $g \in \mathrm{GL}_n(\mathbb{C})$.

A more algebraic definition of the Reynolds operator for GL_n and SL_n is in terms of the so-called omega process. This method was already known in the 19th century, but we will not discuss it here (see [**39**, 4.3]).

If G is a linearly reductive group acting rationally on an affine variety X, then the Reynolds operator also "acts" on the coordinate ring of X. Indeed, the action $\mu : G \times X \to X$ corresponds to a ring homomorphism

$$\mu^\star : K[X] \to K[G] \otimes K[X].$$

If we compose this with

$$\mathcal{R} \otimes \mathrm{id} : K[G] \otimes K[X] \to K \otimes K[X] \cong K[X]$$

where id is the identity, we get a linear map $(\mathcal{R} \otimes \mathrm{id}) \circ \mu^\star$ which we will denote by $\mathcal{R}_X$ (or simply again by $\mathcal{R}$ if there is no confusion).

EXERCISE 1.19. Suppose that G is a linearly reductive group acting rationally on an affine variety X. Prove that:

(a) $\mathcal{R}_X$ is a $K[X]^G$-module homomorphism, i.e.,

$$\mathcal{R}_X\Big(\sum_i a_i f_i\Big) = \sum_i \mathcal{R}_X(a_i) f_i$$

if $a_i \in K[X]$ and $f_i \in K[X]^G$ for all i,

(b) $\mathcal{R}_X(f) = f$ for all $f \in K[X]^G$,

(c) if $V \subseteq K[X]$ is a G-stable subspace, then $\mathcal{R}_X(V) = V^G$.

EXERCISE 1.20. Suppose that X and Y are affine G-varieties and suppose that

$$\psi : X \to Y$$

is a G-equivariant morphism. This means that $\psi(g \cdot x) = g \cdot \psi(x)$ for all $x \in X$ and all $g \in G$. The morphism ψ corresponds to a ring homomorphism $\psi^\star : K[Y] \to K[X]$.

(a) Let $\mathcal{R}_X : K[X] \to K[X]^G$ and $\mathcal{R}_Y : K[Y] \to K[Y]^G$ be the Reynolds operators. Show that the diagram

$$\begin{array}{ccc} K[Y] & \xrightarrow{\psi^\star} & K[X] \\ {\scriptstyle \mathcal{R}_Y}\downarrow & & \downarrow{\scriptstyle \mathcal{R}_X} \\ K[Y]^G & \xrightarrow[\psi^\star]{} & K[X]^G \end{array}$$

commutes, i.e., $\psi^\star \circ \mathcal{R}_Y = \mathcal{R}_X \circ \psi^\star$.

(b) Suppose that $\psi : X \to Y$ is a closed immersion (i.e., $\psi^\star$ is surjective). Use (a) to prove that the restriction of $\psi^\star$ to $K[Y]^G$ gives a surjective ring homomorphism $K[Y]^G \to K[X]^G$. In particular, if $K[Y]^G$ is finitely generated, then so is $K[X]^G$.

We will now prove Hilbert's finiteness theorem (stated in a slightly more general way).

THEOREM 1.21. *Suppose that V is a representation of a linearly reductive group G. Then $K[V]^G$ is finitely generated.*

PROOF. Define I as the ideal in $K[V]$ generated by all homogeneous invariants of positive degree. By Hilbert's Basissatz, I has finitely many generators. Each of these generators lies in an ideal generated by *finitely many* homogeneous invariants of positive degree. It follows that I can be generated by finitely many homogeneous invariants, say $f_1, \ldots, f_r$. We claim that $K[V]^G = K[f_1, \ldots, f_r]$. Suppose that $h \in K[V]^G$ is homogeneous of degree d. We will prove by induction on d that $h \in K[f_1, \ldots, f_r]$. The case $d = 0$ is trivial. If $d > 0$ then we can write

$$h = \sum_{i=1}^{r} a_i f_i \tag{1.1}$$

with $a_1, a_2, \ldots, a_r \in K[V]$. We may assume that a_i is homogeneous of degree $d - \deg(f_i)$ for all i. We apply the Reynolds operator to (1.1) to obtain

$$h = \mathcal{R}(h) = \sum_{i=1}^{r} \mathcal{R}(a_i) f_i.$$

Now $\mathcal{R}(a_i)$ is a homogeneous invariant of degree $d - \deg(f_i) < d$ for all i (since $\mathcal{R}$ preserves degree by Exercise 1.19). By induction we have $\mathcal{R}(a_1), \ldots, \mathcal{R}(a_r) \in K[V]^G$, hence $h \in K[V]^G$. ☐

LEMMA 1.22. *If X is an affine G variety, then there exists a representation Y of G and a G-equivariant closed embedding $\psi : X \to Y$.*

PROOF. Let $W \subset K[X]$ be a finite dimensional subspace such that W generates $K[X]$. Let Z be the span of all $g \cdot w$ with $g \in G$ and $w \in W$. It is clear that Z is G-stable and that Z generates $K[X]$. We claim that Z is still finite dimensional. Let $\mu^\star : K[X] \to K[G] \otimes K[X]$ be the homomorphism corresponding to the regular action $\mu : G \times X \to X$. We can find finite dimensional subspaces $A \subseteq K[G]$ and $B \subseteq K[X]$ such that $\psi^\star(W) \subseteq A \otimes B$. In particular, we have for fixed $g \in G$ and $w \in W$ that $(g \cdot f)(x) = \mu^\star(f)(g, x)$ (seen as a function on x only) lies in B. Since B is finite dimensional we get that Z is finite dimensional. The action of G on Z is again rational and linear, so Z is a representation.

We can view Z as the linear functions in the coordinate ring $K[Z^\star]$ of $Z^\star$. The inclusion $Z \to K[X]$ extends to a ring homomorphism $K[Z^\star] \to K[X]$ which respects the action of G. This homomorphism is surjective since Z generates $K[X]$. The homomorphism corresponds to a G-equivariant closed embedding of affine G-varieties $X \to Z^\star$. We can take $Y = Z^\star$. □

REMARK 1.23. We can view G as a G-variety by letting G act on itself by left multiplication. By Lemma 1.22 there exists a G-equivariant embedding $\psi : G \to Y$ where Y is a representation of G. Now ψ induces a morphism $\phi : G \to \operatorname{End}(Y)$. It is not hard to show that $\phi(G)$ is Zariski closed and that ϕ induces an isomorphism between G and $\psi(G)$. In particular G can be identified with a closed subgroup of GL_n where $n = \dim Y$.

COROLLARY 1.24. *If X is an affine G-variety, and G is a linearly reductive group, then $K[X]^G$ is finitely generated.*

PROOF. Let $\psi : X \to Y$ be a closed G-equivariant embedding where Y is a representation as in Lemma 1.22. We know that $K[Y]^G$ is finitely generated by Theorem 1.21. Hence $K[X]^G$ is finitely generated by Exercise 1.20. □

1.4. An algorithm for computing generating invariants. The **1890** proof of Hilbert's finiteness theorem is not constructive, and it was criticized for this. Nevertheless it was shown in [**7**] that it is possible to make an algorithm for computing generating invariants based on this proof.

PROPOSITION 1.25. *Suppose that V is a representation of a linearly reductive group. Let again I be the ideal of $K[V]$ which is generated by all homogeneous invariants of positive degree. If $I = (f_1, f_2, \dots, f_r)$, then $K[V]^G = K[\mathcal{R}(f_1), \dots, \mathcal{R}(f_r)]$.*

PROOF. Let $\mathfrak{m}$ be the maximal homogeneous ideal of $K[V]$. Since I and $\mathfrak{m}I$ are ideals which are stable under the action of G, we have $\mathcal{R}(I) \subseteq I$ and $\mathcal{R}(\mathfrak{m}I) \subseteq \mathfrak{m}I$. Now $\mathcal{R}$ induces a map

$$I/\mathfrak{m}I \to I/\mathfrak{m}I$$

which is just the identity, because I is generated by invariants. In particular this means that $I/\mathfrak{m}I$ is also generated by the images of $\mathcal{R}(f_1), \dots, \mathcal{R}(f_r)$. By the homogeneous Nakayama lemma (see Exercise 1.26 below) it follows that

$$I = \big(\mathcal{R}(f_1), \dots, \mathcal{R}(f_r)\big)$$

and by the proof of Hilbert's finiteness theorem it follows that

$$K[V]^G = K[\mathcal{R}(f_1), \dots, \mathcal{R}(f_r)]. \quad \square$$

EXERCISE 1.26. Prove the homogeneous Nakayama lemma: Suppose that $R = \bigoplus_{d=0}^{\infty} R_d$ is a finitely generated graded algebra over $R_0 = K$ and suppose that M is a finitely generated graded R-module. (For example, $R = K[V]$ and $M = I \subset R$ is a homogeneous ideal.) Let $\mathfrak{m} = \bigoplus_{d=1}^{\infty} R_d$ be the maximal homogeneous ideal. Suppose that the images of some homogeneous elements $f_1, \dots, f_r \in M$ span the vector space $M/\mathfrak{m}M$. Then M is generated by $f_1, f_2, \dots, f_r$.

Hint. Let N be the submodule generated by $f_1, \dots, f_r$ and suppose that $M \neq N$. Let g be an element of $M \setminus N$ of minimal degree and deduce a contradiction.

Let $B \subseteq V \times V$ be the Zariski closure of the subset $\{(v, g \cdot v) \mid v \in V, g \in G\}$ and let $\mathfrak{b}$ be the vanishing ideal of B. The set B is sometimes called the *graph of the action.* The graph of an action is useful for studying invariant rings (see [**36**, 2.3], [**26**, pp. 20–24] and [**30**]). Once we have generators for the ideal, one can easily compute the Hilbert ideal I from Proposition 1.25 as we will explain below. We can identify the coordinate ring $K[V \times V]$ with the polynomial ring in $2n$ variables, $K[x_1, \dots, x_n, y_1, \dots, y_n]$. We will sometimes use the abbreviations $x = (x_1, x_2, \dots, x_n)$ and $y = (y_1, y_2, \dots, y_n)$.

THEOREM 1.27. *We have*

$$\big(\mathfrak{b} + (y_1, \dots, y_n)\big) \cap K[x_1, \dots, x_n] = I.$$

In particular, if $\mathfrak{b} = \big(f_1(x,y), f_2(x,y), \dots, f_r(x,y)\big)$, *then*

$$I = \big(f_1(x,0), f_2(x,0), \dots, f_r(x,0)\big).$$

PROOF. If f is a homogeneous invariant of positive degree, then $f(x) - f(y) \in \mathfrak{b}$ since it vanishes on B. This proves "$\supseteq$".

We will now prove the other inclusion, "$\subseteq$". Suppose that $h(x,y) \in \mathfrak{b} + (y_1, y_2, \dots, y_n)$. We can write

$$h(x) = b + \sum_i a_i(x) f_i(y) \tag{1.2}$$

with $b \in \mathfrak{b}$ and f_i homogeneous of positive degree for all i. We view $V \times V$ as a representation of G where G acts trivially on the first factor and nontrivially on the second. To this action we associate a Reynolds operator

$$\mathcal{R} : K[V \times V] \to K[V \times V]^G = K[y_1, \dots, y_n]^G[x_1, \dots, x_n].$$

We apply $\mathcal{R}$ to (1.2) to obtain

$$h(x) = \mathcal{R}(b) + \sum_i a_i(x) \mathcal{R}\big(f_i(y)\big).$$

Now we will substitute $y = x$. Note that $\mathcal{R}(b) \in \mathfrak{b}$ since $\mathfrak{b}$ is G-stable (see Exercise 1.19). Now $\mathcal{R}(b)$ will vanish on the diagonal of $V \times V$, which means that $\mathcal{R}(b)$ becomes 0 after the substitution $y = x$. We get

$$h(x) = \sum_i a_i(x) \mathcal{R}\big(f_i(x)\big).$$

Now $\mathcal{R}\big(f_i(x)\big)$ is a homogeneous invariant of positive degree for all i, so we have shown that $h(x) \in I$. □

The ideal $\mathfrak{b}$ can be obtained in a constructive way. First, since G is a linear algebraic group, we may view G as a Zariski closed subset of K^s for some s, so the coordinate ring of G is $K[z_1, z_2, \dots, z_s]/I_G$ where $I_G = \big(h_1(z), \dots, h_t(z)\big)$ is the vanishing ideal of G. The representation of G on V is a homomorphism $G \to \mathrm{GL}(V)$

given by the $n \times n$ matrix

$$\begin{pmatrix} a_{1,1}(z) & a_{1,2}(z) & \cdots & a_{1,n}(z) \\ a_{2,1}(z) & a_{2,2}(z) & & a_{2,n}(z) \\ \vdots & & \ddots & \vdots \\ a_{n,1}(z) & a_{n,2}(z) & \cdots & a_{n,n}(z) \end{pmatrix}$$

where $a_{i,j}(z) \in K[z_1, \dots, z_s]$.

Let us define

$$\Gamma \subseteq G \times V \times V \subseteq K^s \times V \times V$$

by

$$\Gamma = \{(g, v, g \cdot v) \mid g \in G, v \in V\}.$$

The vanishing ideal

$$I_\Gamma \subseteq K[z_1, \dots, z_s, x_1, \dots, x_n, y_1, \dots, y_n]$$

of Γ is given by

$$I_\Gamma = \left(h_1(z), \dots, h_t(z), \left\{ y_i - \sum_{j=1}^{n} a_{i,j}(z) x_i \,\middle|\, i = 1, 2, \dots, n \right\} \right).$$

Note that

$$K[z_1, \dots, z_s, x_1, \dots, x_n, y_1, \dots, y_n]/I_\Gamma$$
$$\cong K[z_1, \dots, z_s, x_1, \dots, x_n]/\big(h_1(z), \dots, h_t(z)\big) \cong K[G] \otimes K[V]$$

is a reduced ring and that I_Γ is therefore a radical ideal. Now B is the Zariski closure of the projection of Γ onto $V \times V$. It follows from this that

$$\mathfrak{b} = I_\Gamma \cap K[x_1, \dots, x_n, y_1, \dots, y_n].$$

Using Buchberger's algorithm, one can find a Gröbner basis of the ideal I_Γ with respect to the lexicographic ordering. In this way one also obtains a Gröbner basis of $\mathfrak{b}$ (see Exercise 1.7), so in particular one has a set of generators of $\mathfrak{b}$ (see Proposition 1.6).

EXAMPLE 1.28. For illustration, we give an easy example. Let G be the cyclic group of order 2 generated by σ. As an algebraic variety G is just a set of two points, so we can identify G with $\{\pm 1\} \subset K$. Now the coordinate ring of G is equal to $K[z]/(z^2 - 1)$.

Consider now the representation of G in K^2, defined by

$$\sigma \cdot (x_1, x_2) = (x_2, x_1).$$

This representation is given by the matrix

$$\begin{pmatrix} \frac{z+1}{2} & \frac{1-z}{2} \\ \frac{1-z}{2} & \frac{z+1}{2} \end{pmatrix}.$$

For $z = 1$ we get the identity matrix and for $z = -1$ we get

$$\begin{pmatrix} 0 & 1 \\ 1 & 0 \end{pmatrix}$$

which represents the action of σ. We form the ideal

$$I_\Gamma = \left(z^2 - 1, y_1 - \frac{z+1}{2} x_1 - \frac{1-z}{2} x_2, y_2 - \frac{1-z}{2} x_1 - \frac{z+1}{2} x_2 \right).$$

We use the Buchberger algorithm with the pure lexicographic ordering such that

$$z > x_1 > x_2 > y_1 > y_2$$

in the ring $K[z, x_1, x_2, y_1, y_2]$ and we obtain a Gröbner basis

$$\{z^2 - 1, x_2 z - y_2 z + x_2 - y_2, y_1 z - y_2 z + 2x_2 - y_1 - y_2, x_1 + x_2 - y_1 - y_2,\\ x_2^2 - x_2 y_2 - x_2 y_1 + y_1 y_2\}.$$

We only take the elements in this Gröbner basis which lie in $K[x_1, x_2, y_1, y_2]$. This gives us a Gröbner basis of $\mathfrak{b}$:

$$\{x_1 + x_2 - y_1 - y_2, x_2^2 - x_2 y_2 - x_2 y_1 + y_1 y_2\}.$$

Now we substitute $y_1 = y_2 = 0$ and we obtain generators of Hilbert's ideal

$$I = (x_1 + x_2, x_2^2).$$

To find generators of $K[x_1, x_2]^G$ we have to apply the Reynolds operator to these to elements. For finite groups, the Reynolds operator is just averaging over the group (see Example 1.15). We obtain $\mathcal{R}(x_1 + x_2) = x_1 + x_2$ because $x_1 + x_2$ was already invariant. Also we have that $\mathcal{R}(x_2^2) = (x_2^2 + x_1^2)/2$. We obtain

$$K[x_1, x_2]^G = K\left[x_1 + x_2, \frac{x_1^2 + x_2^2}{2}\right].$$

Clearly we run into trouble if K has characteristic 2, but then G is no longer linearly reductive.

2. Hilbert's second approach

Hilbert's first proof of 1890 was being criticized for not being constructive (although as we have just seen, it can be made constructive). In 1893 Hilbert gave a second, more constructive, proof (see [**16**]). This new proof is more geometric in nature. The correspondence between (radical) ideals and algebraic sets is of fundamental importance to algebraic geometry. This correspondance is based on Hilbert's *Nullstellensatz* which was proven in his invariant theory paper of 1893. (We actually already used algebraic geometry and therefore implicitly the Nullstellensatz throughout the previous section.) We again will assume that K is algebraically closed. The reader is assumed to be familiar with Hilbert's Nullstellensatz, the correspondance between affine algebraic varieties and finitely generated rings over the field K, and with the correspondance between radical ideals and Zariski closed subsets of affine varieties. We will stick to the following notation. If $J \subseteq K[X]$ is an ideal, then we call

$$\mathcal{Z}(J) = \{\alpha \in X \mid f(\alpha) = 0 \text{ for all } f \in J\}$$

the *zero set of* J. If $S \subseteq X$ is a subset, then the vanishing ideal of S is

$$\mathcal{I}(S) = \{f \in K[X] \mid f(s) = 0 \text{ for all } s \in S\}.$$

This section will be in the spirit of Hilbert's paper of 1893.

2.1. Geometry of invariant rings. Suppose that G is a linearly reductive group, and that X is an affine G-variety. The invariant ring $K[X]^G$ is finitely generated, and corresponds to an affine variety which will be denoted by $X/\!\!/G$. The inclusion $K[X]^G \subseteq K[X]$ induces a dominant morphism $\pi : X \to X/\!\!/G$. This map π will be called the *categorical quotient* (since it has certain universal properties in the category of affine varieties). Sometimes the affine variety $X/\!\!/G$ itself is called the categorical quotient. We will study some of the nice geometric properties of this quotient map.

If $J = \mathcal{I}(Y) \subseteq K[X]^G$ is the vanishing ideal of a Zariski closed subset $Y \subseteq X/\!\!/G$, then $\sqrt{K[X]J}$ is the vanishing ideal of $\pi^{-1}(Y) \subseteq X$. If $J = \mathcal{I}(Y) \subseteq K[X]$ is the vanishing ideal of $Y \subseteq X$, then $J^G = J \cap K[X]^G$ is the vanishing ideal of $\overline{\pi(Y)}$, the Zariski closure of $\pi(Y) \subseteq X/\!\!/G$.

EXERCISE 2.1. Prove that for a linearly reductive group and an affine G-variety X the categorical quotient $\pi : X \to X/\!\!/G$ has the following universal property. If $\psi : X \to Y$ is a morphism of affine varieties which is constant on orbits, then there exists a unique morphism of affine varieties $\varphi : X/\!\!/G \to Y$ such that $\varphi \circ \pi = \psi$.

PROPOSITION 2.2. *The categorical quotient $\pi : X \to X/\!\!/G$ has the following properties:*

(a) *$\pi : X \to X/\!\!/G$ is surjective.*
(b) *If $Y_1, Y_2 \subseteq X$ are G-stable and Zariski closed, then*

$$\overline{\pi(Y_1)} \cap \overline{\pi(Y_2)} = \overline{\pi(Y_1 \cap Y_2)}.$$

(c) *If $Y \subseteq X$ is G-stable and Zariski closed, then $\pi(Y)$ is Zariski closed.*
(d) *The topology of $X/\!\!/G$ is the quotient topology.*
(e) *For every $x \in X/\!\!/G$, $\pi^{-1}(x)$ contains exactly one closed orbit. Every orbit closure in $\pi^{-1}(x)$ contains the closed orbit.*

PROOF. (a) Let $x \in X/\!\!/G$ and let $\mathfrak{m}_x = \mathcal{I}(\{x\}) \subset K[X]^G$ be the corresponding maximal ideal. The vanishing ideal of $\pi^{-1}(x)$ is $\sqrt{K[X]\mathfrak{m}_x}$. If $\pi^{-1}(x) = \emptyset$, then $1 \in K[X]\mathfrak{m}_x$, say

$$1 = \sum_{i=1}^{r} a_i f_i \tag{2.1}$$

with $a_1, \dots, a_r \in K[X]$ and $f_1, \dots, f_r \in \mathfrak{m}_x \subset K[X]^G$ by Hilbert's Nullstellensatz. We apply the Reynolds operator $\mathcal{R}_X : K[X] \to K[X]^G$ to (2.1):

$$1 = \mathcal{R}_X(1) = \sum_{i=1}^{r} \mathcal{R}_X(a_i) f_i.$$

The right-hand side clearly lies in $\mathfrak{m}_x$ since $\mathcal{R}_X(a_i) \in K[X]^G$ for all i. We see that $1 \in \mathfrak{m}_x$ which gives a contradiction to our assumption that $\mathfrak{m}_x$ is a maximal ideal.

(b) Suppose that $I_1 = \mathcal{I}(Y_1), I_2 = \mathcal{I}(Y_2) \subset K[X]$ are the vanishing ideals of Y_1 and Y_2 respectively. We have

$$(I_1 + I_2)^G = \mathcal{R}_X(I_1 + I_2) = \mathcal{R}_X(I_1) + \mathcal{R}_X(I_2) = I_1^G + I_2^G.$$

Now the zero set of $(I_1 + I_2)^G$ is $\overline{\pi(Y_1 \cap Y_2)}$ and the zero set of $I_1^G + I_2^G$ is $\overline{\pi(Y_1)} \cap \overline{\pi(Y_2)}$.

(c) Suppose that $x \in \overline{\pi(Y)} \setminus \pi(Y)$. Now $\pi^{-1}(x)$ and Y are G-stable and Zariski closed. By (b) we have

$$x \in \overline{\pi(\pi^{-1}(x))} \cap \overline{\pi(Y)} = \overline{\pi(\pi^{-1}(x) \cap Y)} = \emptyset$$

and we get a contradiction.

(d) This is clear, since $Y \subset X/\!\!/G$ is Zariski closed if and only if $\pi^{-1}(Y)$ is Zariski closed by (c).

(e) If $\pi^{-1}(x)$ contains at least 2 distinct closed orbits say $G \cdot y_1$ and $G \cdot y_2$, then

$$x \in \pi(G \cdot y_1) \cap \pi(G \cdot y_2) = \pi((G \cdot y_1) \cap (G \cdot y_2)) = \emptyset$$

by (b) and (c) which is a contradiction. Therefore $\pi^{-1}(x)$ has at most 1 closed orbit.

Now we will show that $\pi^{-1}(x)$ has at least one closed orbit. First we note that orbits are always locally closed, i.e., if $x \in X$, then the orbit $G \cdot x$ is a Zariski open subset of its closure $\overline{G \cdot x}$. To see this, we remark that as the image of a morphism, $G \cdot x$ is constructible. Therefore, $G \cdot x$ contains an nonempty open subset U of $\overline{G \cdot x}$. It is clear that $G \cdot x = G \cdot U$. But $G \cdot U \subseteq \overline{G \cdot x}$ is open since it is a union of open subsets.

Choose $y_1 \in \pi^{-1}(x)$. We define $y_2, y_3, \ldots, \in \pi^{-1}(x)$ by induction as follows. For every i, choose $y_{i+1} \in \overline{G \cdot y_i} \setminus G \cdot y_i$. If for some i, $G \cdot y_i$ is closed, then we have proven that $\pi^{-1}(x)$ has a closed orbit (and $\overline{G \cdot y_1}$ contains that closed orbit $G \cdot y_i$). Otherwise, we get an infinite sequence

$$\overline{G \cdot y_1} \supseteq \overline{G \cdot y_2} \supseteq \cdots . \tag{2.2}$$

Also observe that $\overline{G \cdot y_{i+1}}$ is contained in $\overline{G \cdot y_i} \setminus G \cdot y_i$ because $G \cdot y_i$ is open in $\overline{G \cdot y_i}$. In particular, (2.2) is a strictly descreasing sequence of Zariski closed sets, which contradicts the Noetherian property. □

EXAMPLE 2.3. Let $G = \mathbb{C}^\star$ be the multiplicative group acting on $V := \mathbb{C}^3$ as follows:

$$\lambda, (x_1, x_2, x_3) = (\lambda^{-2}x_1, \lambda x_2, \lambda^3 x_3).$$

The invariant ring $K[V]^{\mathbb{C}^\star}$ is generated by $x_1x_2^2$ $x_1^2x_2x_3$ and $x_1^3x_3^2$. We have

$$K[V]^{\mathbb{C}^\star} = K[x_1x_2^2, x_1^2x_2x_3, x_1^3x_3^2] \cong K[y_1, y_2, y_3]/(y_2^2 - y_1y_3).$$

We consider the quotient map

$$\pi : \mathbb{C}^3 \to \mathbb{C}^3/\!\!/\mathbb{C}^\star \cong \{(y_1, y_2, y_3) \in \mathbb{C}^3 \mid y_2^2 - y_1y_3 = 0\}.$$

We get

$$\begin{aligned}\pi^{-1}(0) &= \{(x_1, x_2, x_3) \in \mathbb{C}^3 \mid x_1x_2^2 = x_1^2x_2x_3 = x_1^3x_3^2 = 0\} \\ &= \{(x_1, x_2, x_3) \in \mathbb{C}^3 \mid x_1 = 0 \text{ or } x_2 = x_3 = 0\}.\end{aligned}$$

If $x_1 \neq 0$ then the orbit of $(x_1, 0, 0)$ is not closed (take the limit $\lambda \to \infty$ and one sees that $(0, 0, 0)$ lies in the orbit closure). If $x_2 \neq 0$ or $x_3 \neq 0$, then the orbit of $(0, x_2, x_3)$ is not closed (take the limit $\lambda \to 0$ and one sees again that $(0, 0, 0)$ lies in the orbit closure). The only closed orbit in $\pi^{-1}(0)$ is $\{0\}$. For any nonzero $y \in \mathbb{C}^3/\!\!/C^\star$, one can check that the fiber $\pi^{-1}(y)$ is connected and one dimensional. Since $\pi^{-1}(y)$ does not contain any zero dimensional orbits, $\pi^{-1}(y)$ consists of a single orbit.

EXERCISE 2.4. A quotient $\pi : X \to X/\!\!/G$ is called geometric, if the fibers of π are exactly the orbits in X. Suppose that G is a linearly reductive group and X is an affine G-variety. Show that the categorical quotient $\pi : X \to X/\!\!/G$ is geometric if and only if all orbits in X are Zariski closed.

EXERCISE 2.5. Consider the group $\mathrm{GL}_n(\mathbb{C})$ acting on the $n \times n$ matrices $\mathrm{M}_n(\mathbb{C})$ by conjugation. The categorial quotient $\pi : \mathrm{M}_n(\mathbb{C}) \to \mathbb{C}^n$ is equal to

$$A \mapsto \big(s_1(A), s_2(A), \ldots, s_n(A)\big)$$

where $s_i(A)$ is defined by

$$\det(\lambda I + A) = \lambda^n + s_1(A)\lambda^{n-1} + \cdots + s_n(A).$$

Using the Jordan normal form, describe the orbits in the fibers of this quotient map. Identify the closed orbit in each fiber.

2.2. Hilbert's nullcone. Suppose now that V is a representation of a linearly reductive group G. Let $I \subseteq K[V] \cong K[x_1, \ldots, x_n]$ be again the ideal generated by all homogeneous invariants of positive degree. We define

$$\mathcal{N} = \mathcal{Z}(I).$$

Geometrically, $\mathcal{N}$ is equal to $\pi^{-1}\big(\pi(0)\big)$ where $\pi : V \to V/\!\!/G$ is the categorical quotient. The only closed orbit in $\pi^{-1}\big(\pi(0)\big)$ must be the orbit $\{0\}$. By Proposition 2.2 it follows that

$$\mathcal{N} = \pi^{-1}\big(\pi(0)\big) = \{v \in V \mid 0 \in \overline{G \cdot v}\}.$$

This set $\mathcal{N}$ is often called *Hilbert's nullcone.*

DEFINITION 2.6. A 1-*Parameter SubGroup* (1-PSG) $\lambda : K^\star \to G$ is a morphism which is also a homomorphism of groups.

One parameter subgroups give an easy way to describe Hilbert's Nullcone because of the following result.

THEOREM 2.7 (Hilbert–Mumford criterion). *Suppose that $v \in V$, then*

$$v \in \mathcal{N} \iff \textit{there exists a } \text{1-PSG} \textit{ such that } \lim_{t \to 0} \lambda(t) \cdot v = 0.$$

The limit $\lim_{t\to 0} \lambda(t) \cdot v$ may not make any immediate sense because we are working over an arbitrary algebraically closed field K, not necessarily $\mathbb{C}$. However, note that $\lambda(t) \cdot v$ can be seen as a vector of Laurent polynomials. For a Laurent polynomial $p(t) \in K[t, t^{-1}]$ we say that $\lim_{t\to 0} p(t)$ exists if $p(t)$ is actually a polynomial. In that case we define $\lim_{t\to 0} p(t) = p(0)$.

EXAMPLE 2.8. Suppose that $G = \mathrm{SL}_2$ and

$$V_d = \{a_0X^d + a_1X^{d-1}Y + \cdots + a_dY^d \mid a_0, \ldots, a_d \in K\}$$

is again the binary forms of degree d. We can define a 1-PSG $\lambda : K^\star \to G$ by

$$\lambda(t) = \begin{pmatrix} t & 0 \\ 0 & t^{-1} \end{pmatrix}.$$

If

$$f(X, Y) = a_0X^d + a_1X^{d-1}Y + \cdots + a_dY^d$$

then

$$\lambda(t) \cdot f(X, Y) = a_0X^dt^d + a_1X^{d-1}Yt^{d-2} + \cdots + a_dY^dt^{-d}.$$

Therefore,

$$\lim_{t\to\infty} \lambda(t)\cdot f(X,Y) = 0$$

if and only if $a_i = 0$ for $i \geq d/2$. This condition is equivalent to $Y^{\lceil (d+1)/2\rceil}$ divides $f(X,Y)$. Every 1-PSG for SL_2 is actually conjugate to

$$t \mapsto \begin{pmatrix} t^k & 0 \\ 0 & t^{-k} \end{pmatrix}$$

for some k. It follows that

$$f(X,Y) \in \mathcal{N} \iff f(X,Y) \text{ has a zero of multiplicity } > \tfrac{d}{2}.$$

PROPOSITION 2.9. *Suppose that $f_1, f_2, \ldots, f_r \in K[V]^G$ are homogeneous invariants of positive degree such that $\mathcal{Z}(f_1, f_2, \ldots, f_r) = \mathcal{N}$. Then $K[V]^G$ is a finite $K[f_1, \ldots, f_r]$-module, i.e., there exist finitely many $u_1, \ldots, u_t$ such that $K[V]^G = Fu_1 + \cdots + Fu_t$ where $F = K[f_1, \ldots, f_r]$.*

PROOF. Let $h_1, \ldots, h_s$ such that $K[V]^G = K[h_1, \ldots, h_s]$. We can choose an l such that for all i we have

$$h_i^l \in K[V]f_1 + \cdots + K[V]f_r$$

and therefore

$$h_i^l = \mathcal{R}_V(h_i^l) \in K[V]^G f_1 + \cdots + K[V]^G f_r.$$

The $K[f_1, \ldots, f_r]$-module

$$K[V]^G/(f_1, \ldots, f_r)K[V]^G$$

is spanned by the images of all $h_1^{a_1} h_2^{a_2} \cdots h_s^{a_s}$ with $a_1, \ldots, a_s \in \{0, 1, 2, \ldots, l-1\}$. From the homogeneous Nakayama Lemma (Exercise 1.26) it follows that these elements generate $K[V]^G$ as a $K[f_1, \ldots, f_r]$-module. □

2.3. Homogeneous systems of parameters.

DEFINITION 2.10. A set of homogeneous polynomials $p_1, \ldots, p_s \in K[V]^G$ is called a *homogeneous system of parameters* (HSOP) if

1. $K[V]^G$ is a finite module over $P := K[p_1, \ldots, p_s]$,
2. $p_1, \ldots, p_s$ are algebraically independent.

LEMMA 2.11. *A set of homogeneous polynomials $p_1, \ldots, p_s \in K[V]^G$ is a* HSOP *if and only of $\mathcal{Z}(p_1, \ldots, p_s) = \mathcal{N}$ and $s = \dim K[V]^G$.*

EXERCISE 2.12. Prove the previous lemma.

The following result is a corollary of the Noether normalization lemma (although it was already used by Hilbert):

THEOREM 2.13. *The invariant ring $K[V]^G$ has a homogeneous system of parameters.*

PROOF. We will sketch the proof. Choose $f_1, \ldots, f_r \in K[V]^G$ homogeneous such that $\mathcal{Z}(f_1, \ldots, f_r) = \mathcal{N}$ (for example take homogeneous nonconstant generators of the invariant ring). Define $d_i := \deg(f_i)$ for all i and $d = \operatorname{lcm}(d_1, \ldots, d_r)$. Also define $\hat{f}_i = f_i^{d/d_i}$ for all i. Note that $\hat{f}_1, \ldots, \hat{f}_r \in K[V]^G$ are all of degree d, and $\mathcal{Z}(\hat{f}_1, \ldots, \hat{f}_r) = \mathcal{N}$. Instead of looking at Zariski closed subsets in V, we will look at Zariski closed subsets in $V/\!\!/G$. For polynomials $g_1, \dots, g_s \in K[V]^G$ we will denote their common zero set in $V/\!\!/G$ by $\mathcal{Z}_{V/\!\!/G}(g_1, \dots, g_s)$. Since

$$\mathcal{Z}(\hat{f}_1, \dots, \hat{f}_r) = \mathcal{N} = \pi^{-1}(0)$$

and π is surjective, we have that

$$\mathcal{Z}_{V/\!\!/G}(\hat{f}_1, \dots, \hat{f}_r) = \{0\}.$$

Here $0 \in V/\!\!/G$ is defined as $\pi(0)$, the image of $0 \in V$ under π. We will define p_i $(0 \leq 1 \leq s = \dim K[V]^G)$ by

$$p_i = \sum_{j=1}^{r} \alpha_{i,j} \hat{f}_j$$

for some suitable $\alpha_{i,j} \in K$. Now for suitable choices of $\alpha_{1,1}, \dots, \alpha_{1,r}$ we have that $\mathcal{Z}_{V/\!\!/G}(p_1)$ has dimension $s-1$. Then for suitable choices of $\alpha_{2,1}, \dots, \alpha_{2,r}$ we get $\dim \mathcal{Z}_{V/\!\!/G}(p_1, p_2) = s-2$. So eventually we will get $\dim \mathcal{Z}_{V/\!\!/G}(p_1, \dots, p_s) = 0$. Scalar multiplication on V also induces an action of $K^\star$ on $V/\!\!/G$. If $x \in \mathcal{Z}_{V/\!\!/G}(p_1, \dots, p_s)$ then also $\lambda \cdot x \in \mathcal{Z}_{V/\!\!/G}(p_1, \dots, p_s)$ for all $\lambda \in K^\star$, because $p_1, \dots, p_s$ are homogeneous. Since the dimension of $\mathcal{Z}_{V/\!\!/G}(p_1, \dots, p_s)$ is 0, x must be a fixed point of $K^\star$. The only fixed point in $V/\!\!/G$ is 0, so it follows that

$$\mathcal{Z}_{V/\!\!/G}(p_1, \dots, p_s) = \{0\}$$

and

$$\mathcal{Z}(p_1, \dots, p_s) = \pi^{-1}(0) = \mathcal{N}.$$

From Lemma 2.11 it follows that $p_1, p_2, \dots, p_s$ is a HSOP. □

3. The Cohen–Macaulay property

Suppose that $R = \bigoplus_{d=0}^{\infty} R_d$ is a graded ring of dimension s with $R_0 = K$. The maximal homogeneous ideal of R is $\mathfrak{m} = \bigoplus_{d=1}^{\infty} R_d$. Let M be a finitely generated graded R-module. The dimension $\dim(M)$ of the module M is defined as the dimension of $R/\operatorname{Ann}(M)$ where

$$\operatorname{Ann}(M) = \{f \in R \mid f \cdot a = 0 \text{ for all } a \in M\}$$

is the annihilator of M.

A regular sequence of length r for the module M is a sequence of nonzero homogeneous $f_1, f_2, \dots, f_r \in \mathfrak{m}$ such that multiplication by f_i is injective on the module $M/(f_1, f_2, \dots, f_{i-1})M$ for $i = 1, 2, \dots, r$. We define $\operatorname{depth}(M)$ as the largest integer r for which there exists a regular sequence of length r for the module M. It can be shown that $\operatorname{depth}(M) \leq \dim(M)$ (see [**11**, Proposition 18.2]). The module M is called *Cohen–Macaulay* if $\operatorname{depth}(M) = \dim(M)$. For more details on Cohen–Macaulay modules, the reader is referred to [**5**]. The ring R is called Cohen–Macaulay if it is Cohen–Macaulay as a module over itself. The following result will not be proven here.

LEMMA 3.1. *If $R = \bigoplus_{d=0}^{\infty} R_d$ is a finitely generated graded Cohen–Macaulay ring over $R_0 = K$ of dimension s and $p_1, \dots, p_s \in R$ is a* HSOP, *then $p_1, p_2, \dots, p_s$ is a regular sequence.*

PROOF. See for example [**2**, Theorem 4.3.5], [**23**] or [**37**]. □

The following theorem is the main result in this section.

THEOREM 3.2 (Hochster–Roberts, [**20**]). *If G is linearly reductive, then $K[V]^G$ is Cohen–Macaulay.*

We will sketch a proof in this section. More detailed proofs can be found in [**5**,**20**] and [**9**, 2.5.2]. Note that if $p_1, \dots, p_s \in K[V]^G$ is a HSOP, then by Lemma 3.1 we have that $p_1, \dots, p_s$ is a regular sequence. The following result we will prove later:

LEMMA 3.3. *If $p_1, \dots, p_s \in K[V]^G$ is a* HSOP, *then $K[V]^G$ is a* free *P-module, with $P = K[p_1, \dots, p_s]$, i.e.,*

$$K[V]^G = Ph_1 \oplus \cdots \oplus Ph_l$$

for certain homogeneous $h_1, \dots, h_l \in K[V]^G$.

PROOF OF THEOREM 3.2. We will now start the sketch of the proof of the Hochster–Roberts theorem. Recall that we have the Reynolds operator $\mathcal{R} : K[V] \to K[V]^G$. Therefore it will be enough to prove the following theorem:

THEOREM 3.4. *Let $R := K[x_1, \dots, x_n]$ and let $S \subseteq R$ be a graded subring. Also suppose that $\varphi : R \to S$ is an S-module homomorphism with $\varphi(s) = s$ for all $s \in S$. Then S is Cohen–Macaulay.*

PROOF. We will prove the theorem only in positive characteristic! The theorem is true also in characteristic 0, but in order to prove it in characteristic 0 one has to pass to positive characteristic and this step we will not present here.

Assume that K is an algebraically closed field of characteristic $p > 0$.

DEFINITION 3.5. If q is a power of p, and $I = (f_1, \dots, f_r) \subseteq R$ is an ideal, then $I^{[q]}$ is defined as

$$I^{[q]} = (f_1^q, \dots, f_r^q).$$

EXERCISE 3.6. Prove that $I^{[q]}$ is well-defined, i.e., the definition does not depend on the choice of the generators $f_1, \dots, f_r$.

If $I \subseteq R$ is an ideal and $f \in R$ then the *colon ideal* $(I : f)$ is defined by

$$(I : f) = \{a \in R \mid af \in I\}.$$

PROPOSITION 3.7. *We have*

$$(I^{[q]} : f^q) = (I : f)^{[q]}.$$

PROOF. The inclusion $\supseteq$ is easy: If $af \in I$, then $(af)^q \in I^{[q]}$, so $a^q \in (I^{[q]} : f^q)$.

We will prove the inclusion $\subseteq$. Suppose that $I = (f_1, \dots, f_r)$. Let us assume that $a \in (I^{[q]} : f^q)$. We have

$$af^q = \sum_{j=1}^{r} a_j f_j^q.$$

We can write

$$a = \sum_i b_i m_i$$

where $b_i \in K[x_1^q, \dots, x_n^q]$ for all i, and the m_i run over all monomials $x_1^{e_1} \cdots x_n^{e_n}$ with $0 \le e_1, \dots, e_n \le q-1$. Since $K[x_1, \dots, x_n]$ is a free module over $K[x_1^q, \dots, x_n^q]$, generated by the m_i's, it follows that the b_i's are unique. In a similar fashion we can write

$$a_j = \sum_i b_{i,j} m_i$$

for all j. We obtain

$$\sum_i (b_i f^q) m_i = \sum_i \left(\sum_{j=1}^r b_{i,j} f_j^q \right) m_i.$$

Comparing coefficients gives us

$$b_i f^q = \sum_{j=1}^r b_{i,j} f_j^q$$

for all i. We have assumed that K is algebraically closed. In particular we can choose $c_i, c_{i,j} \in K$ such that $c_i^q = b_i$ and $c_{i,j}^q = b_{i,j}$ for all i and j. So we have

$$(c_i f)^q = c_i^q f^q = \sum_{j=1}^r c_{i,j}^q f_j^q = \left(\sum_{j=1}^r c_{i,j} f_j \right)^q.$$

Hence we get

$$c_i f = \sum_{j=1}^r c_{i,j} f_j.$$

It follows that $c_i \in (I : f)$ and $b_i \in (I : f)^{[q]}$ for all i. We conclude that $a \in (I : f)^{[q]}$. □

DEFINITION 3.8. Suppose that $I \subseteq R$ is an ideal. The *tight closure* $I^\star$ of I is the ideal of all $f \in R$ for which there exist a nonzero $c \in R$ and a constant N such that $cf^q \in I^{[q]}$ for every p-power q with $q \geq N$.

THEOREM 3.9. *For any ideal I of R we have $I^\star = I$.*

PROOF. Suppose that $f \in I^\star$. There exists $c \in R \setminus \{0\}$ such that $cf^q \in I^{[q]}$ for $q \gg 0$ and q a p-power. Then we have

$$c \in \bigcap_{q \gg 0} (I^{[q]} : f^q) = \bigcap_{q \gg 0} (I : f)^{[q]} \subseteq \bigcap_{q \gg 0} (I : f)^q.$$

If $(I : f) \neq R$ then $\bigcap_{q \gg 0} (I : f)^q = (0)$ and we have a contradiction because c is nonzero. Therefore $(I : f) = R$, and $f \in I$. □

We continue with the proof of Theorem 3.4. Suppose that $f_1, \ldots, f_s$ is a homogeneous system of parameters of S. We write $(f_1, \ldots, f_s)_S$ for the ideal in S generated by $f_1, \ldots, f_s$ and $(f_1, \ldots, f_s)_R$ for the ideal in R generated by $f_1, \ldots, f_s$. We will show that $f_1, f_2, \ldots, f_r$ is a regular sequence in S. Suppose that $a_r f_r = \sum_{i=1}^{r-1} a_i f_i$ with $a_1, \ldots, a_r \in S$. We will prove that $a_r \in (f_1, \ldots, f_{r-1})_S$. It suffices to prove that $a_r \in (f_1, \ldots, f_{r-1})_R$ because by applying ϕ it would follow that $a_r \in (f_1, \ldots, f_{r-1})_S$. Define $A := K[f_1, \ldots, f_s]$. Choose $g_1, \ldots, g_l \in S$ linearly independent over the field $K(f_1, \ldots, f_s)$. Define

$$F = Ag_1 \oplus \cdots \oplus Ag_l.$$

There exists an element $c \in A$ such that $cR \subseteq F$. Now

$$c(a_r f_r)^q = c\sum_{i=1}^{r-1} a_i^q f_i^q \in (f_1^q, \dots, f_{r-1}^q)F.$$

Because F is a free module and $f_1^q, f_2^q, \dots, f_r^q$ is a regular sequence for this module, we obtain

$$ca_r^q \in (f_1^q, \dots, f_{r-1}^q)F$$

for all p-powers q. It follows that

$$ca_r^q \in (f_1^q, \dots, f_{r-1}^q)_R = (f_1, \dots, f_{r-1})^{[q]}.$$

Therefore,

$$a_r \in (f_1, \dots, f_{r-1})^\star = (f_1, \dots, f_{r-1}). \quad \square$$

4. Hilbert series

4.1. Basic properties of Hilbert series. Suppose that $V = \bigoplus_{d=a}^{\infty} V_d$ is a graded vector space such that $\dim V_d < \infty$ for all d. Then we can define the formal power series $H(V,t)$ by

$$H(V,t) = \sum_{d=a}^{\infty} (\dim V_d) t^d.$$

The series $H(V,t)$ is called the *Hilbert series* of V (also called *Poincaré series* or *Molien series*). We will be particularly interested in the Hilbert series of invariant rings.

EXAMPLE 4.1. Let $R = K[x_1, \dots, x_n]$. We can define a grading on R by $\deg(x_i) = d_i$. Now R is spanned by the monomials $x_1^{a_1} \cdots x_n^{a_n}$ with $a_1, \dots, a_n \geq 0$. Therefore, the Hilbert series of R are equal to

$$H(R,t) = \sum_{a_1,\dots,a_n \geq 0} (t^{d_1})^{a_1} \cdots (t^{d_n})^{a_n} = \frac{1}{(1-t^{d_1})\cdots(1-t^{d_n})}.$$

If V, V', V'' are graded vector spaces

$$0 \to V' \to V \to V'' \to 0$$

is an exact sequence of graded vector spaces (with the appropiate finiteness conditions) then $H(V,t) = H(V',t) + H(V'',t)$. This is useful to for computing Hilbert series.

If $J \subseteq R := K[x_1, \dots, x_n]$ is a graded ideal, and $<$ is a monomial ordering compatible with the grading on R, then

$$H(J,t) = H\big(\mathrm{lm}(J), t\big)$$

where $\mathrm{lm}(J)$ is the ideal of leading monomials as before. Indeed, the reader may check that $\dim J_d = \dim \mathrm{lm}(J)_d$ for all d.

In his 1890 paper, Hilbert showed that every ideal $J \subseteq R$ has a finite free resolution. In particular this implies that $H(J,t)$ is a rational function for any graded ideal J of R (also $H(M,t)$ is a rational function for any finitely generated graded R-module M).

EXERCISE 4.2. Prove by induction on the number of generators, then $H(I,t)$ is a rational function for any monomial ideal I of R, thus proving that $H(J,t)$ is rational for any graded ideal J of R.

Suppose that $f_1, \dots, f_r \in K[x_1, \dots, x_n]$ are homogeneous. Put $d_i = \deg(f_i)$. Consider the polynomial ring $K[y_1, \dots, y_r]$ with $\deg(y_i) = d_i$ for all i. We can define a surjective ring homomorphism

$$K[y_1, \dots, y_r] \to K[f_1, \dots, f_r]$$

by $y_i \mapsto f_i$ for all i. Let J be the kernel of this homomorphism. We have an exact sequence

$$0 \to J \to K[y_1, \dots, y_r] \to K[f_1, \dots, f_r] \to 0.$$

This implies that

$$H(K[f_1, \dots, f_r], t) = H(K[y_1, \dots, y_r], t) - H(J, t).$$

We have seen that $H(K[y_1, \dots, y_r], t)$ and $H(J, t)$ are rational, therefore $H(K[f_1, \dots, f_r], t)$ is rational.

To compute $H(K[f_1, \dots, f_r], t)$ in practice, we have to be able to find the ideal J. This can be done again using Gröbner basis techniques. Consider the ideal

$$\mathfrak{a} \subseteq K[x_1, \dots, x_n, y_1, \dots, y_r]$$

defined by

$$\mathfrak{a} = (f_1 - y_1, \dots, f_r - y_r).$$

Then J is the intersection of $\mathfrak{a}$ with $K[y_1, \dots, y_r]$. To compute this intersection, we can use Gröbner bases as in Exercise 1.7.

In particular if V is a representation and G is a linearly reductive group, then $K[V]^G$ is finitely generated by homogeneous invariants, so $H(K[V]^G, t)$ is a rational function. If we have found generators of $K[V]^G$, then $H(K[V]^G, t)$ can be computed as above. This is however not the way we go about it in practice. As we will see later, $H(K[V]^G, t)$ can be computed *a priori*, i.e., without knowing the generators. This is quite useful. Suppose that $f_1, \dots, f_r \in K[V]^G$ are homogeneous, but we are not sure if they are generators. Then $f_1, \dots, f_r$ generate $K[V]^G$ if and only if

$$H(K[f_1, \dots, f_r], t) = H(K[V]^G, t).$$

So comparison of Hilbert series gives us a criterion whether we have found a set of generating invariants of the invariant ring.

We give another application of Hilbert series. We will prove Lemma 3.3.

PROOF OF LEMMA 3.3. Suppose that V is a representation of a linearly reductive algebraic group G. Let $p_1, \dots, p_s$ be a homogeneous system of parameters of $K[V]^G$ and put $P = K[p_1, \dots, p_s]$. The ring $K[V]^G$ is Cohen–Macaulay and $p_1, p_2, \dots, p_s$ is a regular sequence. The ring $K[V]^G/(p_1, \dots, p_s)$ is finite dimensional. Choose $h_1, \dots, h_l \in K[V]^G$ homogeneous such that their images span $K[V]^G/(p_1, \dots, p_s)$. By the Homogeneous Nakayama Lemma (see Exercise 1.26) we have

$$K[V]^G = Ph_1 + Ph_2 + \cdots + Ph_l.$$

We claim that the sum on the right-hand side is direct. It suffices to show that

$$H(K[V]^G, t) = H(Ph_1 \oplus Ph_2 \oplus \cdots \oplus Ph_l, t).$$

First we evaluate the right-hand side. Let $d_i = \deg(p_i)$ for all i and $e_j = \deg h_j$ for all j. We have

$$H(P, t) = \frac{1}{(1 - t^{d_1})(1 - t^{d_2}) \cdots (1 - t^{d_s})}$$

and

$$H(Ph_1 \oplus \cdots \oplus Ph_l, t) = \frac{t^{e_1} + t^{e_2} + \cdots + t^{e_l}}{(1 - t^{d_1})(1 - t^{d_2}) \cdots (1 - t^{d_s})}. \tag{4.1}$$

Now we will compute $H(K[V]^G, t)$. Consider the exact sequence

$$0 \to p_r K[V]^G/(p_1, \ldots, p_{r-1}) \to K[V]^G/(p_1, \ldots, p_{r-1}) \to K[V]^G/(p_1, \ldots, p_r) \to 0.$$

Note that since $p_1, \ldots, p_r$ is regular, we get

$$H\big(p_r K[V]^G/(p_1, \ldots, p_{r-1}), t\big) = t^{d_r} H\big(K[V]^G/(p_1, \ldots, p_{r-1}), t\big).$$

We have

$$\begin{aligned} H\big(K[V]^G/(p_1, \ldots, p_r), t\big) &= H\big(K[V]^G/(p_1, \ldots, p_{r-1}), t\big) \\ &- H\big(p_r K[V]^G/(p_1, \ldots, p_{r-1}), t\big) = (1 - t^{d_r}) H\big(K[V]^G/(p_1, \ldots, p_{r-1}), t\big). \end{aligned}$$

By induction we obtain

$$H\big(K[V]^G/(p_1, \ldots, p_s), t\big) = (1 - t^{d_1})(1 - t^{d_2}) \cdots (1 - t^{d_s}) H(K[V]^G, t). \tag{4.2}$$

Since the images of $h_1, \ldots, h_l$ in $K[V]^G/(p_1, \ldots, p_s)$ form a basis, we have

$$H\big(K[V]^G/(p_1, \ldots, p_s), t\big) = t^{e_1} + t^{e_2} + \cdots + t^{e_l}. \tag{4.3}$$

Combining (4.2), (4.3) and (4.1) gives us

$$H(K[V]^G, t) = \frac{t^{e_1} + t^{e_2} + \cdots + t^{e_l}}{(1 - t^{d_1})(1 - t^{d_2}) \cdots (1 - t^{d_s})} = H(Ph_1 \oplus \cdots \oplus Ph_l, t). \quad \square$$

4.2. Computing Hilbert series a priori. For a linearly reductive group G and a representation V it is possible to compute the Hilbert series $H(K[V]^G, t)$ *a priori*, i.e., without knowing explicit generators of the invariant ring $K[V]^G$. For finite groups, there is *Molien's formula* for the invariant ring. For connected reductive groups there is a formula due to Weyl. There is a particularly nice formula for the SL_2 by Springer (see [**38**]). Littelmann and Procesi used this formula to compute the Hilbert series of binary forms up to a large degree ([**4**]).

To understand the formula's for the Hilbert series of invariant rings for connected reductive groups in general, one needs to understand the representation theory of G and Weyl's character formula in particular. This is somewhat beyond the scope of this paper. We will discuss here Molien's formula and we also will show how to compute the Hilbert series of the invariant ring for the multiplicative group $K^\star$.

Suppose that G is a finite group and $\rho_V : G \to GL(V)$ is a representation over a field K of characteristic 0. Now $I - t\rho_V(g)$ is an endomorphism of V with a parameter t in it. The expression $\det\big(I - t \cdot \rho_V(g)\big)$ is a polynomial of degree $n := \dim V$ in t.

THEOREM 4.3 (Molien's formula).

$$H(K[V]^G, t) = \frac{1}{|G|} \sum_{g \in G} \frac{1}{\det(I - t\rho_V(g))}.$$

PROOF. Suppose that $\rho_W : G \to GL(W)$ is a representation of G. Then

$$\frac{1}{|G|} \sum_{g \in G} \operatorname{trace}\big(\rho_W(g)\big) = \operatorname{trace}\bigg(\frac{1}{|G|} \sum_{g \in G} \rho_W(g)\bigg) = \dim W^G$$

because

$$\frac{1}{|G|}\sum_{g\in G}\rho_W(g)$$

is just the projection onto W^G (the Reynolds operator).

For any $g \in G$, $\rho_V(g)$ can be diagonalized (since it has finite order). We may assume that

$$g \cdot x_i = \lambda_i x_i$$

for $i = 1, \ldots, n$. Since $K[V]_d$ is spanned by all monomials of degree d, we get the following formula of the trace of the action of g on $K[V]_d$:

$$\operatorname{trace}\bigl(\rho_{K[V]_d}(g)\bigr) = \sum_{\substack{a_1,\ldots,a_n\geq 0\\ a_1+\cdots+a_n=d}} \lambda_1^{a_1}\cdots\lambda_n^{a_n}.$$

We multiply this with t^d and sum it over all d to get

$$\begin{aligned}\sum_{d\geq 0}\operatorname{trace}(\rho_{K[V]_d(g)})t^d &= \sum_{a_1,a_2,\ldots,a_n\geq 0} \lambda_1^{a_1}\cdots\lambda_n^{a_n}t^{a_1+\cdots+a_n}\\ &= \frac{1}{(1-\lambda_1 t)\cdots(1-\lambda_n t)} = \frac{1}{\det(I-t\rho_V(g))}.\end{aligned}$$

Now finally

$$\begin{aligned}H(K[V]^G,t) = \sum_{d\geq 0}(\dim K[V]_d^G)t^d &= \sum_{d\geq 0}\frac{1}{|G|}\sum_{g\in G}\operatorname{trace}\bigl(\rho_{K[V]_d}(g)\bigr)t^d\\ &= \frac{1}{|G|}\sum_{g\in G}\frac{1}{\det(I-t\rho_V(g))}. \qquad \square\end{aligned}$$

EXAMPLE 4.4. Consider the action of the alternating group A_4 on $V := K^4$ by permuting the coordinates. The expression $\det\bigl(I - t\rho_V(g)\bigr)$ only depends on the conjugacy class of g. For $e \in A_4$ we get

$$\det\bigl(I-t\rho_V(e)\bigr) = \det(I-tI) = \det\bigl((1-t)I\bigr) = (1-t)^4.$$

If g is equal to $(1\ 2)(3\ 4)$ or conjugate to it, then

$$\det\bigl(I-t\rho_V(g)\bigr) = (1-t^2)^2.$$

If g is equal to $(1\ 2\ 3)$ or conjugate to it, then

$$\det\bigl(I-t\rho_V(g)\bigr) = (1-t^3)(1-t).$$

There is only one element of A_4 conjugate to e, namely e itself. There are 3 elements of A_4 conjugate to $(1\ 2)(3\ 4)$, and 8 elements of A_4 conjugate to $(1\ 2\ 3)$. By Molien's formula we have

$$\begin{aligned}H(K[V]^G,t) &= \frac{1}{|G|}\sum_{g\in G}\frac{1}{\det(I-t\rho_V(g))}\\ &= \frac{1}{(1-t)^4}+\frac{3}{(1-t^2)^2}+\frac{8}{(1-t^3)(1-t)} \quad = \frac{1-t^2+t^4}{(1-t)(1-t^2)^2(1-t^3)}\\ &= \frac{1+t^6}{(1-t)(1-t^2)(1-t^3)(1-t^4)}.\end{aligned}$$

The elementary symmetric functions

$$\begin{aligned} e_1 &= x_1 + x_2 + x_3 + x_4, \\ e_2 &= x_1x_2 + x_1x_3 + x_1x_4 + x_2x_3 + x_2x_4 + x_3x_4, \\ e_3 &= x_1x_2x_3 + x_1x_2x_4 + x_1x_3x_4 + x_2x_3x_4, \\ e_4 &= x_1x_2x_3x_4 \end{aligned}$$

form a homogeneous system of parameters. Now

$$H(K[e_1, e_2, e_3, e_4], t) = \frac{1}{(1-t)(1-t^2)(1-t^3)(1-t^4)}.$$

We see that

$$H(K[V]^G, t) = (1+t^6)H(K[e_1, e_2, e_3, e_4], t). \tag{4.4}$$

Since e_1, e_2, e_3, e_4 is a homogeneous system of parameters, and $K[V]^G$ is Cohen–Macaulay, we have that $K[V]^G$ is a free $K[e_1, e_2, e_3, e_4]$-module. From (4.4) it follows that $K[V]^G$ is a free module over $K[e_1, e_2, e_3, e_4]$ with generators 1 and h where h is some invariant of degree 6. In fact we can take

$$h = (x_1 - x_2)(x_1 - x_3)(x_1 - x_4)(x_2 - x_3)(x_2 - x_4)(x_3 - x_4).$$

Suppose now that $G = K^\star$ is the multiplicative group and $\rho_V : K^\star \to V$ is a representation. The action of G on V is diagonalizable. For a diagonal action of $K^\star$, the Hilbert series counts the number of invariant monomials in each degree. The Hilbert series of the invariant ring does not really depend on the base field we are working over. For the field $K = \mathbb{C}$ we have the following formula for the Hilbert series of the invariant ring.

THEOREM 4.5.

$$H(K[V]^G, t) = \frac{1}{2\pi i}\int_{S^1} \frac{1}{\det(I - t\rho_V(z))}\frac{dz}{z} \tag{4.5}$$

where $S^1 \subset \mathbb{C}$ is the unit circle $\{z; |z| = 1\}$.

PROOF. The action of a torus is diagonalizable. We may assume that with respect to some basis of V, $\rho_V(z)$ has the diagonal matrix

$$\begin{pmatrix} z^{a_1} & & & \\ & z^{a_2} & & \\ & & \ddots & \\ & & & z^{a_n} \end{pmatrix}.$$

The coefficient of t^d in the power series of

$$\frac{1}{(1-tx_1)\cdots(1-tx_n)} \tag{4.6}$$

is equal to the sum of all monomials in $x_1, x_2, \ldots, x_n$ of degree d. Note that $z^{-1} \cdot x_i = z^{a_i}x_i$ for all i. We apply the action of $z \in K^\star$ to (4.6). Then the coefficient of t^d in the power series of

$$\frac{1}{(1-tz^{a_1}x_1)(1-tz^{a_2}x_2)\cdots(1-tz^{a_n}x_n)} \tag{4.7}$$

is equal to the sum of all $z^{-1} \cdot m$ where m is a monomial of degree d. The coefficient of $t^d z^0$ in (4.7) is equal to the sum over all invariant monomials of degree d. If we now put $x_1 = x_2 = \cdots = x_n = 1$ then we see that the coefficient of $t^d z^0$ in

$$\frac{1}{(1 - tz^{a_1}) \cdots (1 - tz^{a_n})} \tag{4.8}$$

is exactly the number of invariant monomials of degree d, which is $\dim K[V]^G_d$. In particular we see now that the coefficient of z^0 in (4.8) is exactly the Hilbert series of the invariant ring $H(K[V]^G, t)$. Note that power series of (4.8) in t is uniform convergent on $|z| = 1$, if $|t| < 1$.

Finally we note that if $f(z)$ is an absolutely convergent Laurent series in z, then

$$\frac{1}{2\pi i} \int_{S^1} f(z) \frac{dz}{z}$$

is equal to the coefficient of z^0 in $f(z)$. If we apply this to (4.8) then the theorem follows because the right-hand side of (4.5) is equal to

$$\frac{1}{2\pi i} \int_{S^1} \frac{1}{(1 - tz^{a_1})(1 - tz^{a_2}) \cdots (1 - tz^{a_n})} \frac{dz}{z}. \quad \square$$

The following example illustrates how this theorem can be used to compute Hilbert series of invariant rings for the multiplicative group using the residue theorem for complex functions.

EXAMPLE 4.6. Consider the action of $K^\star$ on K^2 by

$$\lambda \cdot (x_1, x_2) = (\lambda^{-1} x_1, \lambda^3 x_2).$$

We compute $H(K[V]^{K^\star}, t)$. By the theorem, we have

$$H(K[V]^{K^\star}, t) = \frac{1}{2\pi i} \int_{S^1} \frac{1}{(1 - z^{-1}t)(1 - z^3 t)} \frac{dz}{z}. \tag{4.9}$$

To compute the right-hand side we use the Residue Theorem. We assume that $|t| < 1$. We look for poles inside the unit circle. The only poly is $z = t$. The residue at $z = t$ is

$$\lim_{z \to t} \frac{z - t}{(1 - z^{-1}t)(1 - z^3 t) z} = \lim_{z \to t} \frac{1}{1 - z^3 t} = \frac{1}{1 - t^4}.$$

The Residue Theorem says that (4.9) is equal to the sum of the residues inside the unit circle. We conclude that

$$H(K[V]^{K^\star}, t) = \frac{1}{1 - t^4}.$$

This is indeed true, since one easily checks that $K[V]^{K^\star}$ is the polynomial ring in the degree 4 monomial $x_1^3 x_2$.

Computing the Hilbert series of the invariant ring a priori can be used to compute generators of the invariant ring. We will discuss now how one can use homogeneous systems of parameters and Hilbert series to obtain generators of the invariant ring for a finite group G.

First we search for a homogeneous system of parameters. For this we observe that $p_1, \ldots, p_r$ form a regular sequence if and only if $K[V]^G/(p_1, \ldots, p_r)$ has dimension $n - r$. This is true if and only if $K[V]/(p_1, \ldots, p_r)$ has dimension $n - r$. The latter statement can be checked using a Gröbner basis computation. Also note that if $p_1, \ldots, p_n$ is a homogeneous system of parameters for $K[V]^G$, then the denominator of $H(K[V]^G, t)$ has to divide

$$(1-t^{d_1})\cdots(1-t^{d_s})$$

(where $d_i = \deg(p_i)$). This gives us additional conditions for the degrees of $p_1, \ldots, p_n$.

After we have found the homogeneous system of parameters, we search for secondary invariants, i.e., module generators of $K[V]^G$ over $P = K[p_1, \ldots, p_n]$. We know that $K[V]^G$ is a free P-module, say

$$K[V]^G = Ph_1 \oplus \cdots \oplus Ph_l.$$

If $e_j = \deg(h_j)$ then we have

$$H(K[V]^G, t) = \frac{\sum_j t^{e_j}}{(1-t^{d_1})\cdots(1-t^{d_n})}.$$

Since we know $H(K[V]^G, t)$ and $d_1, \ldots, d_n$, we can compute $e_1, \ldots, e_l$. So we know exacty in which degrees to look for secondary invariants.

4.3. Degree bounds. We give here a brief discussion on degree bounds for generators of invariant rings. For a more extensive discussions, see [**10, 34, 35**].

Suppose that V is a representation of a linearly reductive group G. We define the following constant

$$\beta(K[V]^G) = \min\{d \mid K[V]^G \text{ is generated in degree } \leq d\}.$$

We would like to have good upper bounds for $\beta(K[V]^G)$. For finite groups we have the following bound.

THEOREM 4.7. *If the characteristic of K does not divide $|G|$, then*

$$\beta(K[V]^G) \leq |G|.$$

In characteristic 0 this was proven by Noether ([**32**]). If the characteristic does not divide the group order $|G|$, the so-called nonmodular case, then this inequality is still true. This was recently proven by Fleischmann (see [**12**]), and by Fogarty independently (cf. [**13**]).

Suppose that $G = T = (K^\star)^r$ is a torus. If $z = (z_1, \ldots, z_r) \in T$, and $b = (b_1, \ldots, b_r) \in \mathbb{Z}^r$, then z^b is an abbreviation for $z_1^{b_1} \cdots z_r^{b_r}$. If V is a representation of G then the action of G is diagonalizable. We assume that $\rho(z_1, \ldots, z_r)$ is given by the matrix

$$\begin{pmatrix} z^{a_1} & & & \\ & z^{a_2} & & \\ & & \ddots & \\ & & & z^{a_n} \end{pmatrix}$$

for certain $a_1, a_2, \ldots, a_n \in \mathbb{Z}^r$. We will assume that the action is faithfull, i.e., the vectors $a_1, \ldots, a_n$ span $\mathbb{Z}^r$. The following degree bound was proven by Wehlau (see [**41**]):

THEOREM 4.8.

$$\beta(K[V]^T) \leq \max\{n-r-1, 1\} r! \operatorname{volume}(\mathcal{C}_V)$$

where $\mathcal{C}_V$ is the convex hull of $a_1, a_2, \ldots, a_n \in \mathbb{Z}^r \subseteq \mathbb{R}^r$.

A first general degree bound for invariants of connected reductive groups was found by Popov (see [**34, 35**]). This bound was radically sharpened by the author (cf. [**8**]). Let us introduce the following constant:

$$\begin{aligned}(4.10)\qquad \gamma(K[V]^G) &= \min\{d \mid K[V]^G \text{ is finite over } K\big[K[V]^G_{\le d}\big]\} \\ &= \{d \mid \text{ invariants of degree} \le d \text{ have } \mathcal{N} \text{ as zero set}\}.\end{aligned}$$

Good bounds for $\gamma(K[V]^G)$ exists, see for example Popov (see [**34, 35**]) and Hiss (cf. [**19**]). The following bound was proven by the author:

THEOREM 4.9.

$$\beta(K[V]^G) \le \max\left\{2, \frac{3}{8}s\gamma^2(K[V]^G)\right\}$$

where $s = \dim K[V]^G$.

EXAMPLE 4.10. For the group SL_2 acting on the binary forms of degree d, one can prove that

$$\gamma(K[V_d]^{SL_2}) \le 2d^3.$$

From Theorem 4.9, it follows that

$$\beta(K[V]^G) \le \frac{3}{8}(d-2)(2d^3)^2 \le \frac{3}{2}d^7.$$

This result is slightly worse than the bound $2d^6$ found by Jordan ([**21, 22, 27**]).

References

1. William W. Adams and Phillippe Loustaunau, *An introduction to Gröbner bases*, Grad. Stud. Math. vol. 3, Amer. Math. Soc., Providence, RI, 1994.
2. David J. Benson, *Polynomial invariants of finite groups*, London Math. Soc. Lecture Note Ser., vol. 190, Cambridge Univ. Press, Cambridge, 1993.
3. Thomas Becker and Volker Weispfenning, *Gröbner bases*, Springer-Verlag, Berlin, Heidelberg, New York, 1993.
4. A. E. Brouwer and A. M. Cohen, *The Poincaré series of the polynomials invariant under* SU_2 *in its irreducible representations of degree* ≤ 17, Math. Centrum Amsterdam, ZW **134/79** (1979), 1–20.
5. Winfried Bruns and Jürgen Herzog, *Cohen–Macaulay rings*, Cambridge University Press, Cambridge, 1993.
6. David Cox, John Little, and Donal O'Shea, *Ideals, varieties, and algorithms*, Springer-Verlag, New York, Berlin, Heidelberg, 1992.
7. Harm Derksen, *Computation of invariants for reductive groups*, Adv. in Math. **141** (1999), 366–384.
8. ———, *Polynomial bounds for rings of invariants*, Proc. Amer. Math. Soc. **129** (2001), no. 4, 955–963.
9. Harm Derksen and Gregor Kemper, *Computational invariant theory*, Invariant Theory and Algebraic Transformation Groups I, Encyclopaedia Math. Sci. vol. 130, Springer-Verlag, Berlin, 2002.
10. Harm Derksen and Hanspeter Kraft, *Constructive invariant theory*, Algèbre Non Commutative, Groupes Quantiques et Invariants (Reims, 1995), Sémin. Congr. vol. 2, Soc. Math. France, Paris, 1997, pp. 221–244.
11. David Eisenbud, *Commutative algebra with a view toward algebraic geometry*, Springer-Verlag, New York, 1995.
12. Peter Fleischmann, *The Noether bound in invariant theory of finite groups*, Adv. Math. **156** (2000), no. 1, 23–32.
13. John Fogarty, *On Noether's bound for polynomial invariants of a finite group*, Electron. Res. Announc. Amer. Math. Soc. **7** (2001), 5–7.

14. Paul Gordan, *Beweis, dass jede Covariante und Invariante einer binären Form eine ganze Funktion mit numerischen Coefficienten einer endlichen Anzahl solcher Formen ist*, J. Reine Angew. Math. **69** (1868), 323–354.
15. David Hilbert, *Über die Theorie der algebraischen Formen*, Math. Ann. **36** (1890), 473–534.
16. ______, *Über die vollen Invariantensysteme*, Math. Ann. **42** (1893), 313–370.
17. ______, *Mathematische Probleme*, Archiv für Math. und Physik **1** (1901), 44–63.
18. ______, *Gesammelte Abhandlungen. Band* III: *Analysis, Grundlagen der Mathematik, Physik, Verschiedenes, Lebensgeschichte*, Springer-Verlag, Berlin-New York, 1970.
19. Karin Hiss, *Constructive invariant theory for reductive algebraic groups*, Thesis, Brandeis University, Waltham, 1996.
20. Melvin Hochster and Joel L. Roberts, *Rings of invariants of reductive groups acting on regular rings are Cohen–Macaulay*, Adv. in Math. **13** (1974), 115–175.
21. Camille Jordan, *Mémoire sur les covariants des formes binaires*, J. Math. **3** (1876), no. 2, 177–232.
22. ______, *Sur les covariants des formes binaires*, J. Math. **3** (1879), no. 5, 345–378.
23. Gregor Kemper, *Computational invariant theory*, The Curves Seminar at Queen's, vol. XII, (Kingston, 1998), Queen's Papers in Pure and Applied Math., vol. 114, Queen's Univ., Kingston, ON, 1998, pp. 5–26.
24. Hanspeter Kraft, *Geometrical methods in invariant theory*, Aspects of Mathematics, D1. Friedr. Vieweg & Sohn, Braunschweig, 1984 (German).
25. Hanspeter Kraft and Claudio Procesi, *Classical invariant theory, a primer*, preprint; `http://www.math.unibas.ch/~kraft/`.
26. Hanspeter Kraft, Peter Slodowy, and Tonny A. Springer, eds., *Algebraic transformation groups and invariant theory*, DMV Seminar **13**, Birkhäuser Verlag, Basel, 1989 (German).
27. Hanspeter Kraft and Jerzy Weyman, *Degree bounds for invariants and covariants of binary forms*, preprint; `http://www.math.unibas.ch/~kraft/`.
28. Martin Kreuzer and Lorenzo Robbiano, *Computational commutative algebra*. 1, Springer-Verlag, Berlin, 2000.
29. Peter Littelmann and Claudio Procesi, *On the Poincaré series of the invariants of binary forms*, J. Algebra **133** (1990), no. 2, 490–499.
30. D. Luna, *Slices étales*, Sur les Groupes Algébriques, Bull. Soc. Math. France, Paris, Memoire 33, Soc. Math. France, Paris, 1973, pp. 81–105.
31. Masayoshi Nagata, *On the* 14th *problem of Hilbert*, Amer. J. Math. **81** (1959), 766–772.
32. Emmy Noether, *Der Endlichkeitssatz der Invarianten endlicher Gruppen*, Math. Ann. **77** (1916), 89–92.
33. Peter Olver, *Classical invariant theory*, London Math. Soc. Stud. Texts **44**, Cambridge University Press, Cambridge, 1999.
34. Vladimir Popov, *Constructive invariant theory*, Young Tableaux and Schur Functors in Agebra and Geometry (Toruń, 1980), Astérisque, vol. 87–88, Soc. Math. France, Paris, 1981, pp. 303–334.
35. ______, *The constructive theory of invariants*, Izv. Akad. Nauk SSSR Ser. Mat. **45** (1981), no. 5, 1100–1120, 1199.
36. Vladimir L. Popov and Ernest B. Vinberg, *Invariant theory*, Algebraic Geometry IV. Linear Algebraic Groups, Invariant Theory (N. N. Parshin, I. R. Shafarevich, eds.,), Encyclopaedia Math. Sci., vol. 55, Springer-Verlag, Berlin, Heidelberg, 1994.
37. W. Smoke, *Dimension and multiplicity of graded algebras*, J. Algebra **21** (1972), 149–173.
38. Tonny Springer, *On the invariant theory of* SU_2, Nederl. Akad. Wetensch. Indag. Math. **42** (1980), no. 3, 339–345.
39. Bernd Sturmfels, *Algorithms in invariant theory*, Texts Monogr. Symbol. Comput., Springer-Verlag, Vienna, 1993.
40. Wolmer V. Vasconcelos, *Computational methods in commutative algebra and algebraic geometry*, Algorithms Comput. Math., vol. 2, Springer-Verlag, Berlin, 1998.
41. David L. Wehlau, *Constructive invariant theory for tori*, Ann. Inst. Fourier (Grenoble) **43** (1993), no. 4, 1055–1066.

Department of Mathematics, University of Michigan, East Hall 525 East University, Ann Arbor, MI 48109-1109, USA.

E-mail address: `hderksen@umich.edu`

Centre de Recherches Mathématiques
CRM Proceedings and Lecture Notes
Volume **35**, 2004

On Global Degree Bounds for Invariants

H. Derksen and Gregor Kemper

ABSTRACT. Let G be a linear algebraic group over a field K of characteristic 0. An integer m is called a global degree bound for G if for every linear representation V the invariant ring $K[V]^G$ is generated by invariants of degree at most m. We prove that if G has a global degree bound, then G must be finite. The converse is well known from Noether's degree bound.

Introduction

A classical topic in invariant theory is the question of degree bounds: Is it possible to generate an invariant ring $K[V]^G$ by homogeneous invariants of degree at most m, and can any a priori upper bound for such a number m be given? Perhaps the most prominent example of such a bound is Noether's degree bound [**8**], which states that for G finite and K of characteristic 0, every invariant ring is generated in degree at most $|G|$. Upper bounds for linearly reductive groups were given by Popov [**9, 10**] and then improved by Derksen [**2**]. It is remarkable that these bounds, in contrast to Noether's bound, do not only depend on G, but also involve properties of the representation V, such as its dimension. The same is true for an a priori bound given by Derksen and Kemper [**3**, Theorem 3.9.11] for finite groups (where the characteristic of K may divide $|G|$).

This observation leads to the following question. If G is infinite does there exist any upper bound at all which only depends on G and not on the representation? In this note we answer this question for the case that $\operatorname{char}(K) = 0$. The answer is as expected from observations: A global bound only exists if G is finite. This is stated in Theorem 2.1.

In the first section we establish the result for the case that G is linearly reductive. The second section deals with the general case of a linear algebraic group over an algebraically closed field of characteristic 0.

Let us fix some notation. Throughout the paper, G is a linear algebraic group over an algebraically closed field K. By a **G-module** we mean a finite-dimensional vector space V over K with a linear action of G given by a morphism $G \times V \to V$ of varieties. Recall that there always exists a faithful G-module (see Borel [**1**,

2000 *Mathematics Subject Classification.* Primary: 13A50; Secondary: 20G15, 14L17.
The first author was partially supported by the NSF, grant DMS 0102193.
This is the final form of the paper.

Proposition I.1.10]). If V is a G-module, then G also acts on the polynomial ring $K[V]$ on V, and the invariant ring is denoted by $K[V]^G$. The ring $K[V]^G$ is a graded algebra.

If A is any graded algebra over $K = A_0$, we write

$$\beta(A) = \min\{d \in \mathbb{N} \mid A \text{ is generated by elements of degree } \leq d\},$$

where by convention the minimum over an empty set is ∞. Moreover, define

$$\beta(G) := \sup\left\{\beta\left(K[V]^G\right) \mid V\ G\text{-module}\right\} \in \mathbb{N} \cup \{\infty\}.$$

We say that G has a **global degree bound** if $\beta(G) < \infty$, i.e., there exists an integer m such that $\beta\left(K[V]^G\right) \leq m$ for all G-modules V.

1. Linearly reductive groups

If G is linearly reductive, then $K[V]$ has a unique isotypical decomposition, i.e.,

$$K[V] = \bigoplus_{\lambda \in \Lambda} K[V]_\lambda, \tag{1.1}$$

where Λ is the set of all isomorphism classes of irreducible G-modules and $K[V]_\lambda$ is a direct sum of irreducible modules which lie in the class λ (see Springer [**14**]).

LEMMA 1.1. *Suppose that G is linearly reductive and V is a faithful G-module. Assume that only finitely many components appear in the isotypical decomposition (1.1) of $K[V]$, i.e.,*

$$|\{\lambda \in \Lambda \mid K[V]_\lambda \neq 0\}| < \infty.$$

Then $|G| < \infty$.

PROOF. For every λ, $K[V]_\lambda$ is a finitely generated module over $K[V]^G$ (see Springer [**14**, III, Satz 4.2]). If there are only finitely many λ such that $K[V]_\lambda \neq 0$, then $K[V]$ is a finitely generated $K[V]^G$-module. Let $K(V)$ be the quotient field of $K[V]$. The field of invariant rational functions $K(V)^G$ contains the quotient field of $K[V]^G$. It follows that $K(V) : K(V)^G$ is an algebraic extension. Since G acts faithfully, it follows from Galois theory that G must be finite. □

PROPOSITION 1.2. *Let G be linearly reductive and infinite. Then G has no global degree bound.*

PROOF. Let U be a faithful G-module, and let k be an arbitrary non-negative integer. We write $K[U]_i$ for the homogeneous part of degree i of the polynomial ring. By Lemma 1.1 there exists an isomorphism class λ of irreducible G-modules such that $K[U]_\lambda \neq 0$ but $(K[U]_i)_\lambda = 0$ for all $i < k$. Let m be the least integer with $(K[U]_m)_\lambda \neq 0$. Choose a representative W from λ and set $V = W \oplus U$. Then $K[V] = K[W] \otimes_K K[U]$ has a G-invariant bigrading by putting $K[V]_{i,j} = K[W]_i \otimes K[U]_j$. For the part of bidegree $(1, j)$ we have

$$K[V]^G_{1,j} = (W^* \otimes K[U]_j)^G \cong \operatorname{Hom}_G(W, K[U]_j).$$

Hence $K[V]^G_{1,j} = 0$ for $j < m$, and there exists an $f \in K[V]^G_{1,m} \setminus \{0\}$. The total degree of f is $m + 1$, and by using the bigrading we see that f cannot be written as a polynomial in invariants of smaller total degree. Hence

$$\beta(K[V]^G) \geq m + 1 > k.$$

Since k was chosen arbitrarily, there exists no global bound. □

2. The general case

Let G be a linear algebraic group. It is obvious but noteworthy that for a closed normal subgroup $N \trianglelefteq G$ we have

$$\beta(G/N) \leq \beta(G). \tag{2.1}$$

We will also use a result of Schmid [**12**, Proposition 5.1] which states that if $H \leq G$ is a subgroup of finite index, then

$$\beta(H) \leq \beta(G). \tag{2.2}$$

Schmid [**12**] only states this result for finite groups, but the proof (which works by inducing representations from H to G) only uses that the index is finite.

We can now prove the main result.

THEOREM 2.1. *Let G be a linear algebraic group over an algebraically closed field K of characteristic 0. Then G has a global degree bound if and only if it is finite.*

PROOF. If G is finite, then the Noether bound [**8**] says $\beta(G) \leq |G|$.

On the other hand, assume that G is infinite. Let $U \trianglelefteq G$ be the unipotent radical. G/U is reductive and therefore linearly reductive (this uses $\operatorname{char}(K) = 0$, see Springer [**14**, V, Satz 1.1]). If G/U is infinite, then the result follows from Proposition 1.2 and the inequality (2.1). If, on the other hand, G/U is finite, then by (2.2) it suffices to prove that $\beta(U) = \infty$. It follows from Humphreys [**6**, Corollary 17.5, Proposition 17.4, and Lemma 15.1C] that U has a closed normal subgroup N such that U/N is isomorphic to the additive group G_a. By (2.1) we are reduced to showing that $\beta(G_a) = \infty$. This is done in the following lemma. □

LEMMA 2.2. *If $G = G_a$ is the additive group over an algebraically closed field K of characteristic 0, then $\beta(G) = \infty$.*

PROOF. We use Roberts' isomorphism. This states that for an SL_2-module V (on which G_a acts by the matrices $\left(\begin{smallmatrix} 1 & t \\ 0 & 1 \end{smallmatrix}\right)$) we have an isomorphism

$$\Phi : (K[U] \otimes_K K[V])^{\mathrm{SL}_2} \xrightarrow{\sim} K[V]^{G_a}, \tag{2.3}$$

where U is the natural 2-dimensional SL_2-module. A good reference for (2.3) is Kraft [**7**, page 191] (where a more general situation is considered) for the case $K = \mathbb{C}$, and Seshadri [**13**] for general K. The isomorphism is given by $\Phi(\sum_i f_i \otimes g_i) = \sum_i f_i(v) g_i$ with $v = \left(\begin{smallmatrix} 1 \\ 0 \end{smallmatrix}\right) \in U$. We have a natural bigrading on $(K[U] \otimes K[V])^{\mathrm{SL}_2}$, and if $f \in (K[U] \otimes K[V])^{\mathrm{SL}_2}$ has bidegree (i, j), then $\Phi(f)$ is homogeneous of degree j. Also consider the epimorphism

$$K[U] \otimes K[V] \twoheadrightarrow K[V], \quad \sum_i f_i \otimes g_i \mapsto \sum_i f_i(0) g_i.$$

Since SL_2 is linearly reductive, this restricts to an epimorphism

$$\pi : (K[U] \otimes K[V])^{\mathrm{SL}_2} \twoheadrightarrow K[V]^{\mathrm{SL}_2}.$$

If $f \in (K[U] \otimes K[V])^{\mathrm{SL}_2}$ has bidegree (i, j), then $\pi(f)$ has degree j.

Now let V be an SL_2-module and set $k := \beta(K[V]^{G_a})$. We can take preimages under Φ of homogeneous generating invariants for $K[V]^{G_a}$ and decompose them into their bi-homogeneous parts. It follows that $(K[U] \otimes K[V])^{\mathrm{SL}_2}$ is generated by bi-homogeneous invariants of degrees (i, j) with $j \leq k$. By applying π we obtain that $K[V]^{\mathrm{SL}_2}$ is generated by homogeneous invariants of degree at most k, so $\beta(K[V]^{\mathrm{SL}_2}) \leq k$. This argument shows that

$$\beta(\mathrm{SL}_2) \leq \beta(G_a).$$

But $\beta(\mathrm{SL}_2) = \infty$ by Proposition 1.2. This finishes the proof. □

Unfortunately, we were unable to extend this or a similar result to positive characteristic. We conjecture the following.

CONJECTURE 2.3. *Let G be a linear algebraic group over an algebraically closed field K. Then the following are equivalent:*

(a) *G has a global degree bound;*
(b) *G is finite and* $\mathrm{char}(K)$ *does not divide the group order $|G|$.*

The implication "$(b) \Rightarrow (a)$" is given by the Noether bound, which was recently proved to hold also if $\mathrm{char}(K) < |G|$ but $\mathrm{char}(K) \nmid |G|$ independently by Fleischmann [**4**] and Fogarty [**5**]. It is also known from Richman [**11**] that a finite group with $|G|$ divisible by $\mathrm{char}(K)$ does not have a global degree bound. Both results can also be found in the book by Derksen and Kemper [**3**].

Acknowledgment

We thank Julia Hartmann for valuable comments on a draft version of this paper, and the anonymous referee for pointing out that one of our intermediate results was already contained in Barbara Schmid's paper [**12**].

References

1. Armand Borel, *Linear algebraic groups*, 2nd ed., Grad. Texts in Math., vol. 126, Springer Verlag, 1991.
2. Harm Derksen, *Polynomial bounds for rings of invariants*, Proc. Amer. Math. Soc. **129** (2001), 955–963.
3. Harm Derksen and Gregor Kemper, *Computational invariant theory*, Encyclopaedia Math. Sci., vol. 130, Springer-Verlag, Berlin, Heidelberg, New York, 2002.
4. Peter Fleischmann, *The Noether bound in invariant theory of finite groups*, Adv. in Math. **156** (2000), 23–32.
5. John Fogarty, *On Noether's bound for polynomial invariants of a finite group*, Electron. Res. Announc. Amer. Math. Soc. **7** (2001), 5–7.
6. James E. Humphreys, *Linear algebraic groups*, 2nd ed., Springer-Verlag, Berlin, Heidelberg, New York, 1981.
7. Hanspeter Kraft, *Geometrische Methoden in der Invariantentheorie*, Vieweg, Braunschweig, 1985.
8. Emmy Noether, *Der Endlichkeitssatz der Invarianten endlicher Gruppen*, Math. Ann. **77** (1916), 89–92.
9. Vladimir L. Popov, *Constructive invariant theory*, Astérique **87–88** (1981), 303–334.
10. ———, *The constructive theory of invariants*, Math. USSR Izvest. **10** (1982), 359–376.
11. David R. Richman, *Invariants of finite groups over fields of characteristic p*, Adv. in Math. **124** (1996), 25–48.
12. Barbara J. Schmid, *Finite groups and invariant theory*, Topics in Invariant Theory (P. Dubreil and M.-P. Malliavin, eds.), Lecture Notes in Math., vol. 1478, Springer-Verlag, Berlin, Heidelberg, New York, 1991, pp.1–4.
13. C. S. Seshadri, *On a theorem of Weitzenböck in invariant theory*, J. Math. Kyoto Univ. **1** (1962), 403–409.
14. Tonny A. Springer, *Aktionen reduktiver Gruppen auf Varietäten*, Algebraische Transformationsgruppen und Invariantentheorie (Hanspeter Kraft, Peter Slodowy, and Tonny A. Springer, eds.), DMV Seminar vol. 13, Birkhäuser, Basel, 1987.

DEPARTMENT OF MATHEMATICS, UNIVERSITY OF MICHIGAN, EAST HALL, 525 EAST UNIVERSITY, ANN ARBOR, MI 48109-1109, USA.
E-mail address: hderksen@umich.edu

TECHNISCHE UNIVERSITÄT MÜNCHEN, ZENTRUM MATHEMATIK—M11, BOLTZMANNSTR. 3, 85748 GARCHING, GERMANY.
E-mail address: kemper@ma.tum.de

Centre de Recherches Mathématiques
CRM Proceedings and Lecture Notes
Volume **35**, 2004

On Invariant Theory of Finite Groups

P. Fleischmann

1. Construction of invariant rings

1.1. Introduction. In each basic course on algebra one will come across the following example of a **ring of invariants**:

Let R be a commutative ring, $A := R[X_1, \ldots, X_n]$ the ring of polynomials in n variables $X_1, \ldots, X_n$ and Σ_n the symmetric group on n-letters. Then Σ_n acts "naturally" on A by "permuting the variables" following the rule

$$\sigma\big(f(X_1, \ldots, X_n)\big) = f(X_{\sigma(1)}, \ldots, X_{\sigma(n)}), \quad \forall \sigma \in \Sigma_n,\ f \in A.$$

As one can see immediately, $\sigma(fg) = \sigma(f)\sigma(g)$ for $f, g \in A$ and the action of Σ_n on A respects homogeneity of polynomials and their degree. This means that Σ_n acts on A by automorphisms of the graded R-algebra A. Those polynomials which remain unchanged by the operation of Σ_n are called **symmetric polynomials**. They form a graded subring A^{Σ_n} of A, the **ring of symmetric polynomials** in A, being our first example of a **ring of invariants**.

A central objective of constructive invariant theory is, to give a uniform and efficient description of all invariant polynomials. Again we consider the example of symmetric functions:

Let us order the power products $\underline{X}^\alpha := X_1^{\alpha_1} X_2^{\alpha_1} \ldots X_n^{\alpha_n}$ by the "degree—lexicographical" order saying that $\underline{X}^\alpha < \underline{X}^\beta$ if and only if $\deg \underline{X}^\alpha < \deg \underline{X}^\beta$ or the degrees coincide and

$$\alpha_i < \beta_i \text{ for } i = \min\{j \le n \mid \alpha_j \ne \beta_j\}.$$

Then each orbit $\{\sigma(\underline{X}^\alpha) \mid \sigma \in \Sigma_n\}$ contains a unique maximal element which we call $\underline{X}^\alpha_{\max}$ and which satisfies $\alpha_1 \ge \alpha_2 \ge \cdots$. Obviously Σ_n permutes the power products, hence each symmetric polynomial will be a unique R-linear combination of **orbit-sums**

$$\operatorname{orb}_{\Sigma_n}(\underline{X}^\alpha) := \sum_{\{\underline{X}^\beta = \sigma(\underline{X}^\alpha) \mid \sigma \in \Sigma_n\}} \underline{X}^\beta.$$

Let $f \in A^{\Sigma_n}$ be an arbitrary (nonconstant) symmetric polynomial, and let

$$\underline{X}^\alpha_{\max}(f) = X_1^{\alpha_1} X_2^{\alpha_1} \ldots X_n^{\alpha_n}$$

2000 *Mathematics Subject Classification.* Primary: 13A50; Secondary: 16G99.
This is the final version of the paper.

be the largest power product appearing in f with a nonzero coefficient r, say. Then $\alpha_1 \geq \alpha_2 \geq \cdots \geq \alpha_n$. For $k \in \mathbb{N}$ let

$$e_k := \mathrm{orb}_{\Sigma_n}(X_1 X_2 \ldots X_k) = \sum_{1 \leq j_1 < j_2 < \cdots < j_k \leq n} X_{j_1} X_{j_2} \ldots X_{j_k}$$

be the kth **elementary symmetric polynomial**. Then $\underline{X}^{\alpha}_{\max}(f)$ is also the unique maximal power product appearing in the product $e := e_1^{\alpha_1-\alpha_2} e_2^{\alpha_2-\alpha_3} \ldots e_n^{\alpha_n}$. Now it is easy to see that all power products appearing in $f - re$ are strictly smaller than $\underline{X}^{\alpha}(f)$ in the lexicographical order. Of course we can iterate the argument replacing f by the symmetric polynomial $f - re$. In each iteration the "new" $\underline{X}^{\alpha}_{\max}$ will strictly decrease and the whole process must end with a constant function, because there are no infinite sequences

$$\cdots < \underline{X}^{\gamma} < \underline{X}^{\rho} < \underline{X}^{\alpha}.$$

In the summary this argument can be made into an algorithm to produce a polynomial $\Psi(Y_1, \ldots, Y_n)$ with $f = \Psi(e_1, \ldots, e_n)$. To show that Ψ is unique, it suffices to observe that the power products of the form $\underline{e}^{\beta} := e_1^{\beta_1} \ldots e_n^{\beta_n}$ are linearly independent, which in turn follows from the fact that these power products have pairwise different leading monomials of the form $\underline{X}^{\alpha}_{\max}(\underline{e}^{\beta}) = X_1^{\beta_1+\cdots+\beta_n} X_2^{\beta_2+\cdots+\beta_n} \ldots X_n^{\beta_n}$.

This proves the **main theorem of symmetric polynomials**, stating that every $f \in A^{\Sigma_n}$ can be rewritten in a unique way as a polynomial in the elementary symmetric polynomials e_i, which themselves are algebraically independent.

This result is remarkable in the following sense:

1. it gives an overview over all symmetric polynomials and how to construct them, using the finite set of homogeneous generators $e_1, \ldots, e_n$ of degrees bounded by n;
2. it describes explicitly the ring theoretic structure of $A^{\Sigma_n} = R[e_1, \ldots, e_n]$ as being isomorphic to a polynomial ring $R[Y_1, \ldots, Y_n]$.

These two *constructive* and *structural* aspects guide the investigation of more general invariant rings. Firstly we replace Σ_n by a general finite group G, secondly we skip the requirement that G acts on A by permuting variables. In most cases we will only assume that G acts via graded R-algebra automorphisms.[1] The following are natural questions in this context:

1. How can we explicitly find a finite set of (homogeneous) invariants that generate A^G as an R-algebra? Is there an a priori bound for their degrees?
2. If A is a polynomial ring over a field $R = \mathbb{F}$, say. How "far away" is A^G from being a polynomial ring?

The second question is deliberately vague at this stage and the answer certainly depends on how to define the structure of a ring to be "close to" a polynomial ring. This will be made more precise later on.

FURTHER READING. Quite recently there have appeared or are about to appear several very good monographs on the subject [**1, 6, 24, 36**]. They differ in style and focus and supplement each other in providing a full encyclopedia about all theoretical and constructive aspects of invariant theory of finite groups. Therefore I strongly recommended to read these for any deeper study in this area.

---

[1] Sometimes even this is not needed and A can be replaced by some ungraded finitely generated R-algebra, on which G acts by R-algebra automorphisms.

2. Degree bounds

Let R be a commutative ring, $A := R[a_1, a_2, \ldots, a_n]$ a finitely generated R-algebra with set of generators $\boldsymbol{a} := \{a_1, \ldots, a_n\}$ and G a finite group acting on A by R-algebra automorphisms, stabilizing the R-module $\sum_{i=1}^n Ra_i$. We define the ring of G-invariants $A^G := \{a \in A \mid g(a) = a, \forall g \in G\}$.

Following the example of symmetric functions we are interested in finding a set of **fundamental invariants** $\{f_1, f_2, \ldots, f_s\} \subseteq A^G$ such that $A^G = R[f_1, f_2, \ldots, f_s]$.

HISTORIC REMARKS. The question, whether there is always a *finite* set of fundamental invariants for arbitrary groups was considered to be one of the most important problems in 19th century algebra. It was proved to be true, using explicit calculations, by P. Gordan for $R = \mathbb{C}$ and $G = \mathrm{SL}_2(\mathbb{C})$ in the 1860/70's. In 1890 David Hilbert (1862–1943) introduced new methods into invariant theory, which still today are basic tools of modern algebra (Hilbert's basis theorem). Applying these he was able to prove finite generation for the invariants of the general linear groups $\mathrm{GL}_n(\mathbb{C})$. In the year 1900, on the occasion of the international congress of mathematics in Paris, Hilbert posed the general question of finite generation of invariant rings for arbitrary groups as the 14th of the now famous "Hilbert problems".[2] In general the answer to this question is negative:

In 1958 Nagata gave a counterexample of $G \cong \mathbb{C}^+ \times \cdots \times \mathbb{C}^+$ (finite product) acting on a polynomial ring A in finitely many variables such that A^G is *not* finitely generated.

For *finite groups* Hilbert's 14th problem has been solved affirmatively: In 1926 Emmy Noether proved [**26**] that the invariant ring A^G is finitely generated, if G is a finite group and $R = \mathbb{F}$ is an arbitrary field. This proof, which will be outlined shortly, was one of the first major applications of her newly developed theory of rings and modules with ascending chain condition. Therefore it can be generalized in a straightforward way to the case where, in modern language, the ring R is Noetherian. The price one has to pay for this generality is, that the proof is not constructive and does not provide an immediate algorithm how to compute a finite set of fundamental invariants. Ten years earlier, in [**25**] Emmy Noether had considered the problem specifically for $R = \mathbb{C}$, where she was able to find constructive procedures to compute fundamental systems explicitly. We will now follow up her ideas and at the same time try to generalize them as far as possible.

Assume that $\{f_1, f_2, \ldots, f_s\}$ is a finite set of fundamental invariants; then each f_i is an R-linear combination of power products $\underline{a}^\alpha = a_1^{\alpha_1} \ldots a_n^{\alpha_n}$ and we can look at the maximal degree of an exponent function α needed to write down all the f_i's. Let us define the **Noether number** $\beta(A^G)$ to be the *minimum* of all these numbers, taken over all possible finite fundamental systems. Notice that the Noether number depends on the chosen set of generators for A; therefore we will write $\beta(A^G, \boldsymbol{a})$ whenever we need to express this dependence. So the statement $\beta(A^G) \leq m$ says that every element in A^G can be written as linear combination of products of

[2] In his original lecture this problem had been posed a little bit differently, since finiteness was thought of being solved; later an error in the proof was found.

invariants lying in in the R-module

$$\mathcal{M}_m(\boldsymbol{a}) := \sum_{\substack{\alpha\in\mathbb{N}_0^n,\\ |\alpha|\le m}} R\cdot \underline{a}^\alpha,$$

or in other words, $A^G = R[A^G \cap \mathcal{M}_m(\boldsymbol{a})]$. For example if $R = R[X_1,\dots,X_n]$ is a polynomial ring with $\beta(A^G) \le m$, then all invariants can be generated in degree less or equal to m; in particular we have seen that $\beta(R[X_1,\dots,X_n]^{\Sigma_n}) \le n$.

The example of symmetric functions also gives us an idea how to construct invariants: Let $H \le G$ be a subgroup of index m and let $\{1 = g_1, g_2, \dots, g_m\}$ denote a set of representatives for the cosets gH. Take a new variable Y and consider for any $a \in A^H$ the polynomial $f_a(Y)$ defined by:

$$f_a(Y) = \prod_{i=1}^{m}\big(Y - g_i(a)\big) = Y^m + f_{a,1}Y^{m-1} + \cdots + f_{a,m-1}Y + f_{a,m}.$$

It is easy to check that $f_a(Y)$ is independent of the particular choice of the system $\{g_1, g_2, \dots, g_m\}$ and that $f_a(Y) \in A^G[Y]$. In fact, up to sign the coefficient $f_{a,i}$ is equal to the ith elementary symmetric function $e_i\big(g_1(a), g_2(a), \dots, g_m(a)\big)$. Note that $f_a(a) = 0$, so

$$(*) \qquad a^m = -f_{a,1}a^{m-1} + \cdots - f_{a,m-1}a - f_{a,m}.$$

In particular we can take $H = 1$ and define a subring of A^G by

$$B := R[f_{a_i,j} \mid i = 1,\dots,n, j = 1,\dots,|G|].$$

Using equation (*) we see that A can be viewed as the *module* over B generated by the finite set of power products $\{\underline{a}^\alpha \mid \alpha_i < |G|, \forall i = 1,\dots,n\}$. If R is a Noetherian ring, then by Hilbert's basis theorem, so is B and hence submodules of finitely generated B-modules are again finitely generated. In particular A^G, being a B-submodule of A, is a finitely generated B-module, hence a finitely generated R-algebra. This is precisely Emmy Noether's argument for finite generation of A^G in case of finite groups and Noetherian coefficient rings. Note that the argument tells us that $A^G = R[B, b_1, b_2, \dots, b_\ell]$ with the b_i's being B-module generators, but it does not tell us how to construct these.

To attack that problem, consider the (absolute) transfer map

$$t_1^G : A \to A^G, a \mapsto \sum_{g\in G} g(a),$$

which is a homomorphism of A^G-modules and therefore of B-modules as well. It is also surjective, whenever $|G|$ is invertible in R. In that case, using equation (*) and the fact that t_1^G is A^G-linear, we see that

$$A^G = R[B, t_1^G(\underline{a}^\alpha) \mid \text{ all } \alpha_i \le |G|]$$

from which we can conclude $\beta(A^G) \le \max\{|G|, n\cdot(|G|-1)\}$.

In [**25**] Emmy Noether used a similar approach to give two different proofs for the fact that $\beta(A^G) \le |G|$ if R is a polynomial ring over the complex numbers. This is usually referred to as the **Noether bound** in invariant theory. To get from the bound above to the Noether bound, one needs some combinatorics to reduce exponents in transfer elements $t_1^G(\underline{a}^\alpha)$. The original arguments used in [**25**] only work if the *factorial* of $|G|$ is invertible. Recently different techniques have been found, which also work in the more general case that $|G|$ is invertible in R (see [**10, 15**]; for earlier work on that problem see [**12, 13, 27, 29, 35**]).

I will present a "combined version" of these methods, incorporating an essential observation by D. Benson which makes the required combinatorics very transparent.

First let us convince ourselves that the requirement $|G|$ to be invertible in R is really needed: Consider the simple example $A := A(k,2) := \mathbb{F}_2[X_1, \dots, X_k, Y_1, \dots, Y_k]$ with $G = \Sigma_2 = \langle g \rangle$ acting by swapping the "variable types" $X_i \leftrightarrow Y_i$.

LEMMA 2.1. *The invariant* $\mathfrak{X} := (X_1, \dots, X_k)^+ := X_1, \dots, X_k + Y_1, \dots, Y_k \in A^G$ *is indecomposable, i.e. cannot be written as a sum of products of invariants of smaller degree. Therefore*

$$\beta(A^G) \geq k \to \infty \text{ if } k \to \infty.$$

PROOF. We will use the fact that A and A^G are graded not just by $\mathbb{N}_0$, but by $\mathbb{N}_0^k$: For each $M = (m_{ij}) \in \mathbb{N}_0^{k\times 2}$ and power product $\mathfrak{F}_M := X_1^{m_{11}} X_2^{m_{21}}, \dots, X_k^{m_{k1}} \cdot Y_1^{m_{12}} Y_2^{m_{22}}, \dots, Y_k^{m_{k2}}$ we define a multi-degree

$$\mathrm{md}(\mathfrak{F}_M) := (m_{11} + m_{12}, m_{21} + m_{22}, \dots, m_{k1} + m_{k2}) \in \mathbb{N}_0^k.$$

Then $A = \oplus_{v \in \mathbb{N}_0^k} A_v$, with $A_v := \langle \mathfrak{F}_M \in A \mid \mathrm{md}(\mathfrak{F}_M) := v \rangle$ and $A_v \cdot A_w \subseteq A_{v+w}$. Since the G-action on A respects the multi-degree, we have as well $A^G = \oplus_{v \in \mathbb{N}_0^k} (A_v)^G$ and $\mathfrak{X}$ is "multi-homogeneous" of multi-degree $\mathbf{1} := (1, 1, \dots, 1)$. Obviously G permutes power products in A and therefore every invariant in A^G is a linear combination either of G-stable power products or of "orbit- sums" of the form $\mathfrak{F}_M^+ := \mathfrak{F}_M + g(\mathfrak{F}_M)$. Note that the power product $\mathfrak{F}_M$ is G-stable if and only if the two columns of M coincide.

Now we consider the algebra homomorphism $\pi : A(k,2) \to A(1,2) = \mathbb{F}_2[X_1, Y_1]$ induced by the specialization $X_i \mapsto 1, Y_i \mapsto 1$ for $i > 1$. Clearly π is G-equivariant and if M has two different columns, then $\pi(\mathfrak{F}_M^+) = X_1^{m_{11}} Y_1^{m_{12}} + X_1^{m_{12}} Y_1^{m_{11}}$, which is zero if and only if $m_{11} = m_{12}$. Now assume that $\mathfrak{X}$ is decomposable; then it is a sum of products of the form $\mathfrak{F}_{M_1}^+ \cdot \mathfrak{F}_{M_2}^+ \cdots \mathfrak{F}_{M_s}^+$ with $s > 1$, $\mathbf{0} \neq \mathrm{md}(\mathfrak{F}_{M_j}^+)$ for all j and $\mathrm{md}(\mathfrak{F}_{M_1}^+) + \cdots + \mathrm{md}(\mathfrak{F}_{M_s}^+) = \mathbf{1}$. In particular all occurring M_j's have different columns and in each product there must be at least one $\mathfrak{F}_{M_{j_0}}$ with multi-degree $(0, d_2, \dots, d_k)$ and $d_i \in \{0, 1\}$. Hence $\pi(\mathfrak{F}_{M_{j_0}}^+) = X_1^0 Y_1^0 + X_1^0 Y_1^0 = 1 + 1 = 0$ and since π is linear and multiplicative, we get the contradiction $X_1 + Y_1 = \pi(\mathfrak{X}) = 0$. $\square$

It is interesting, though, to observe that for $k \geq 3$:

$$\mathfrak{X} = (X_2, \dots, X_k)^+ X_1 + (X_1 X_3, \dots, X_k)^+ Y_2 - (X_, \dots, X_k)^+ X_1 Y_2,$$

i.e. $\mathfrak{X}$ decomposes in the **Hilbert-ideal** $A^{G,+}A$, generated *in* A by all invariants of positive degree. Extending the definition of the Noether number in an obvious way to cover ideal generators, one can easily see that $\beta(A^{G,+}A) = 2$ for all $k \geq 2$. In fact a generalization of this observation led to the proof of Noether's bound in [**10**].

Let $H \leq G$ be a subgroup of index m and $G := \bigcup_{i=1}^m g_i H$ the coset decomposition. Assume that A^H is known and consider the **relative transfer map** with respect to H:

$$t_H^G : A^H \to A^G, \quad a \mapsto \sum_{i=1}^m g_i(a).$$

Again this is an A^G-module homomorphism and the image $t_H^G(A^G)$ is an ideal in A^G, called the **relative transfer ideal** (w.r.t. H). The following describes a decomposition, **in the ambient ring** A, of high degree relative transfer elements:

LEMMA 2.2. *For $b, b_1, b_2, \ldots, b_m \in A^H$ we have*

$$t_H^G(bb_1, \ldots, b_m) = \sum_{\substack{I\subset \underline{m},\\ I\neq \underline{m}}} (-1)^{m-|I|+1} t_H^G\Big(b\prod_{j\in I} b_j\Big)\prod_{j\notin I} g_j(b_j).$$

PROOF. We consider the obvious equality:

$$\prod_{j=1}^{m}(g_i(b_j) - g_j(b_j)) = 0.$$

Expansion and multiplication with $g_i(b)$ for fixed i gives:

$$0 = \sum_{I\subseteq \underline{m}} (-1)^{|I|}\prod_{j\notin I} g_j(b_j)\cdot\Big(\prod_{j\in I} g_i(b_j)\Big)\cdot g_i(b).$$

Now summation over $i \in \underline{m}$ yields the claimed identity. □

The following result contains the statement of the Noether bound for $H = 1$ and $|G|$ invertible:

THEOREM 2.3. *Let $A := R[a_1, \ldots, a_n]$ as above and let H be a subgroup of G such that either $|G|$ is invertible in R or $H \triangleleft G$ a normal subgroup with index $[G : H]$ invertible in R. Then*

$$\beta(A^G, \boldsymbol{a}) \leq \beta(A^H, \boldsymbol{a})\cdot[G : H].$$

PROOF. Under both assumptions on H, the relative transfer map t_H^G is surjective. Now suppose that $\beta := \beta(A^H) = \beta(A^H, \boldsymbol{a})$ and $A^H = R[b_1, \ldots, b_k]$ with $b_i \in A^H \cap \mathcal{M}_\beta(\boldsymbol{a})$. If $H \leq G$ and $|G|$ is invertible, we have

$$\begin{aligned} t_H^G(b_\ell b_{i_1}, \ldots, b_{i_m}) &= \frac{1}{|G|} t_1^G(t_H^G(b_\ell b_{i_1}, \ldots, b_{i_m}))\\ &= \sum_{\substack{I\subset \underline{m},\\ I\neq \underline{m}}} (-1)^{m-|I|+1} t_H^G\Big(b_\ell\prod_{j\in I} b_{i_j}\Big)\frac{1}{|G|}t_1^G\Big(\prod_{j\notin I} g_j(b_{i_j})\Big) \in R[A^G\cap\mathcal{M}_{m\beta}(\boldsymbol{a})]. \end{aligned}$$

If $H \triangleleft G$, then A^H is G-invariant and the elements $g_j(b_{i_j})$ lie in A^H. If in addition $|G/H|$ is invertible, replacing $|G|$ by $|G/H|$ and t_1^G by t_H^G we conclude in a similar way that $t_H^G(b_\ell b_{i_1}, \ldots, b_{i_m}) \in R[A^G \cap \mathcal{M}_{m\cdot\beta}(\boldsymbol{a})]$. Now an iterative application of this result finishes the proof. □

3. Further results and conjectures

Firstly one might hope to remove the requirement $H \triangleleft G$ for subgroups of invertible index:

CONJECTURE 3.1. If $H \leq G$ with index $[G : H]$ invertible in R, then

$$\beta(A^G) \leq \beta(A^H)\cdot[G : H].$$

An argument similar to the original one in [**25**] can be used to show that $\beta(A^G) \le \beta(A^H) \cdot [G:H]$ whenever $[G:H]!$ is invertible (see (6.5)).

For the rest of this section let $A = \mathbb{F}[X_1, \dots, X_n]$, a polynomial ring over a field of characteristic $p > 0$. In this case, the formula in Lemma 2.2 can be used to describe a decomposition of relative transfer elements in the Hilbert-ideal $A^{G,+}A$.[3] Sample calculations done by Harm Derksen and Gregor Kemper led them to the following far reaching conjecture:

CONJECTURE 3.2. [Noether bound for Hilbert-ideals (H. Derksen/G. Kemper)] Let G be a finite group, $\mathbb{F}$ a field and $A := \mathbb{F}[X_1, \dots, X_n]$ a polynomial ring, such that G acts by graded algebra automorphisms. Then

$$\beta(A^{G,+}A) \le |G|.$$

Let P be a fixed Sylow p-group of G with normalizer $N_G(P)$ and $N := N_G(P)/P$. For subgroups $U \le H \le G$ we define the (homogeneous) **relative transfer ideal**

$$\mathcal{I}^H_{<U} := \sum_{Y<U} t^H_Y(A^Y) \lhd A^H.$$

LEMMA 3.3.

$$\beta(A^{G,+} \cdot A) \le \max\{\beta(A^Q/\mathcal{I}^Q_{<Q}) \cdot [G:Q] \mid Q \le P\}.$$

PROOF. Let $f \in A^{G,+}$ be indecomposable in $A^{G,+}A$. Since $n := [G:P]$ is invertible, f is of the form $t^G_P(h)$ for some $h \in A^{P,+}$, which itself can be written as

$$h = \sum_{Q \le P} t^P_Q(b_{Q,1}b_{Q,2}\dots b_{Q,\ell_Q})$$

with ℓ_Q factors of the form $b_{Q,j} + \mathcal{I}^Q_{<Q} \in A^Q/\mathcal{I}^Q_{<Q}$, $b_{Q,j}$ homogeneous of positive degree $\le \beta_Q := \beta(A^Q/\mathcal{I}^Q_{<Q})$. Moreover we can assume that every nonzero transfer element

$$t^G_P\big(t^P_Q(b_{Q,1}b_{Q,2}\dots b_{Q,\ell_Q})\big) = t^G_Q(b_{Q,1}b_{Q,2}\dots b_{Q,\ell_Q})$$

is indecomposable in $A^{G,+}A$ as well. But from Lemma 2.2 we see that this requires $\ell_Q \le [G:Q]$, hence

$$f \in A^{G,+} \cap \mathcal{M}_{\beta_Q[G:Q]}(\boldsymbol{X}). \quad \square$$

Hence Noether's bound for relative transfer quotients of p-groups would imply Conjecture 3.2:

COROLLARY 3.4. *If $\beta(A^Q/\mathcal{I}^Q_{<Q}) \le |Q|$ for all $Q \le P \in \mathrm{Syl}_p(G)$, then Conjecture 3.2 holds, i.e. $\beta(A^{G,+}A) \le |G|$.*

To obtain degree bounds for A^G rather than $A^{G,+}A$ one can make use of the **Brauer homomorphism** from representation theory, i.e. is the canonical homomorphism $A^G \to \overline{A^G} := A^G/\mathcal{I}^G_{<P}$. Now Mackey's formula [**22**] for the relative transfer states

$$t^G_P(b) = \sum_{g \in P:G:P} t^P_{P \cap P^g P}({}^g b),$$

[3] If A is $\mathbb{N}_0$-graded, A^+ denotes the ideal generated by homogeneous elements of positive degree.

where $P : G : P$ denotes a chosen system of double cosets of P in G. For $g \in G\backslash N_G(P)$ the summand $t^P_{P\cap {}^gP}({}^gb)$ lies in $\mathcal{I}^P_{<P}$, hence we get

$$t^G_P(b) \equiv \sum_{g\in N/P} {}^gb \equiv t^N_1(b) \mod \mathcal{I}^P_{<P}.$$

It is easy to see that $\mathcal{I}^P_{<P} \cap A^G = \mathcal{I}^G_{<P}$, hence we get

$$\overline{A^G} = (A^G + \mathcal{I}^P_{<P})/\mathcal{I}^P_{<P} = t^N_1(A^P/\mathcal{I}^P_{<P}) \cong (A^P/\mathcal{I}^P_{<P})^N.$$

Since p does not divide $|N|$, Theorem 2.3 gives

LEMMA 3.5.

$$\beta(\overline{A^G}) \leq \beta(\overline{A^P}) \cdot |N|, \quad \beta(A^G) \leq \max\{\beta(\mathcal{I}^G_{<P}),\ \beta(\overline{A^P}) \cdot |N|\}.$$

It has been conjectured that the bound in the previous section

$$\beta(A^G) \leq \max\{|G|, n(|G| - 1)\}$$

might be a "natural degree bound" for modular invariant rings in polynomial rings. Using the above technique this certainly would follow from the next two slightly sharper conjectures:

CONJECTURE 3.6. Let P be a Sylow p-group of G. Then

1. $\beta(A^G/\mathcal{I}^G_{<P}) \leq |N_G(P)|$.
2. If $A = \mathbb{F}[X_1, \ldots, X_n]$, then

$$\beta(\mathcal{I}^G_{<P}) \leq \max\{|G|, n(|G| - 1)\}.$$

All these conjectures have been verified for p-permutation modules, as we will see in the next section. We have seen that the Noether bound is not valid for modular invariant rings. In fact it has been shown that no "global degree bound" exists, that depends only on the group G and the ring or field of scalars. In fact the following has been shown:

THEOREM 3.7 (Richman [**28**]). *Let V be a finite-dimensional vector space over $\mathbb{F}$ and G a finite subgroup of* $\mathrm{GL}(V)$ *with $p = \operatorname{char}\mathbb{F} \mid |G|$. For $m \in \mathbb{N}$ let V^m denote the direct sum of m copies of V and* $\mathrm{Sym}(V^m)$ *the corresponding symmetric algebra with natural G-action. Then there is a positive number α depending only on $|G|$ and p, such that every set of $\mathbb{F}$-algebra generators of* $\mathrm{Sym}(V^m)^G$ *contains a generator of degree $\geq \alpha m$.*

4. Permutation invariants

4.1. Basic constructions. In this section we investigate the special class of permutation invariants, constructed from a group action on a polynomial ring which permutes the independent variables. For the sake of notational flexibility the variables are indexed by a finite set Ω on which G acts. Thus we consider a polynomial ring $A = A(\Omega) := R[X_\omega \mid \omega \in \Omega]$ and without loss of generality we can assume that G is a subgroup of the symmetric group Σ_Ω acting by $\sigma(X_\omega) := X_{\sigma(\omega)}$. We will use the standard symbol Ω/G to denote the set of all G-orbits on Ω.

In this situation G stabilizes the set of power products

$$\mathcal{P} := \left\{ \underline{X}^\alpha := \prod_{\omega\in\Omega} X_\omega^{\alpha_\omega} \mid \alpha \in \mathbb{N}_0^\Omega \right\}$$

and the map $\mathcal{P} \ni \underline{X}^\alpha \to \alpha \in \mathbb{N}_0^\Omega$ is an isomorphism of G-semigroups and defines a G-equivariant isomorphism between $A(\Omega)$ and the R-semigroup–ring $R[\mathbb{N}_0^\Omega]$. Note that G acts naturally on $\mathbb{N}_0^\Omega$ by the rule $g\alpha := \alpha \circ g^{-1}$. For the ease of notation, we will occasionally abuse language and identify the multiplicative semigroup of power products $\mathcal{P}$ with the additive semigroup $\mathbb{N}_0^\Omega$. Since G permutes $\mathcal{P}$ it follows that every $f \in A^G$ is an R-linear combination of power products with constant coefficients along G-orbits on $\mathcal{P}$. This shows that A^G is a free R-module of infinite rank, with basis given by the **G-orbit–sums**

$$\mathrm{orb}_G(\underline{X}^\alpha) := \sum_{\beta \in \alpha^G} \underline{X}^\beta = t^G_{G_\alpha}(\underline{X}^\alpha).$$

Here G_α denotes the stabilizer subgroup $\{g \in G \mid \alpha \circ g = \alpha\}$ of $\alpha \in \mathbb{N}_0^\Omega$. In particular, if $G_\alpha < G$, the orbit–sum $t^G_{G_\alpha}(\underline{X}^\alpha)$ lies in the ideal $\mathcal{I}^G_{<G}$. On the other hand, if $G_\alpha = G$, then α is constant on G-orbits in Ω and therefore $\underline{X}^\alpha$ is a power product of orbit–products of "norms" $\mathfrak{n}_i := \prod_{\omega \in \mathcal{O}_i} X_\omega$, where the $\mathcal{O}_i$, $i = 1, \dots, k$ run through the set of G-orbits on Ω. As different G-orbits are disjoint, the norms $\mathfrak{n}_i$ are algebraically independent and generate a polynomial subring $S \cong R[Y_1, \dots, Y_k] \leq A^G$, where k is the number of G-orbits on Ω. For each $\alpha \in \mathbb{N}_0^\Omega$ consider the greatest common divisor $g_\alpha := \gcd(|G_\alpha|/|H_\alpha| \mid H < G) \cdot 1_R$. Then for each $H < G$ and $a = \sum_\alpha c_\alpha t^H_{H_\alpha}(X^\alpha) \in A^H$ we have

$$t^G_H(a) = \sum_\alpha c_\alpha t^G_{H_\alpha}(X^\alpha) = \sum_\alpha c_\alpha t^G_{G_\alpha}\left(t^{G_\alpha}_{H_\alpha}(X^\alpha)\right) = \sum_\alpha c_\alpha \frac{|G_\alpha|}{|H_\alpha|} t^G_{G_\alpha}(X^\alpha) \in \sum_\alpha g_\alpha A^G.$$

On the other hand, if g_α is expressed as a linear combination $\sum_{H<G} |G_\alpha|/|H_\alpha| c_H$ with $c_H \in \mathbb{Z}$, then

$$\begin{aligned} g_\alpha \cdot \mathrm{orb}_G(\underline{X}^\alpha) = \sum_{H<G} \frac{|G_\alpha|}{|H_\alpha|} c_H t^G_{G_\alpha}(\underline{X}^\alpha) &= \sum_{H<G} c_H t^G_{G_\alpha} t^{G_\alpha}_{H_\alpha}(\underline{X}^\alpha) \\ &= \sum_{H<G} c_H t^G_H t^H_{H_\alpha}(\underline{X}^\alpha) \in \mathcal{I}^G_{<G}. \end{aligned}$$

So $\mathcal{I}^G_{<G} = \oplus_{\alpha \in \mathbb{N}_0^\Omega / G} R g_\alpha \cdot \mathrm{orb}_G(\underline{X}^\alpha)$ and $A^G = \mathcal{I}^G_{<G} + S$ with $\mathcal{I}^G_{<G} \cap S = g \cdot S$ with $g := \gcd(|G|/|H| \mid H < G) \cdot 1_R$. Hence

$$A^G/\mathcal{I}^G_{<G} \cong S/gS \cong R/gR\,[Y_1, \dots Y_k],$$

where $Y_i = \mathfrak{n}_i$ is homogeneous of degree $|\mathcal{O}_i| \leq |G|$. In the case of "p-groups in characteristic p" we get:

LEMMA 4.1. *If R has prime characteristic p and P is a p-group acting on Ω, then*

$$A(\Omega)^P = \mathcal{I}^P_{<P} \oplus S \text{ with } A^P/\mathcal{I}^P_{<P} \cong S \cong R[Y_1, \dots, Y_k],$$

where k is the number of P-orbits on Ω and each Y_i is homogeneous of degree $\leq |P|$. In particular

$$\beta\left(A(\Omega)^P/\mathcal{I}^P_{<P}\right) \leq |P|.$$

We can use this result to prove some of the conjectures in the previous section, at least for the special type of "p-permutation invariants". Let $R := \mathbb{F}$ be a field of characteristic p dividing $|G|$ and V a finitely generated $\mathbb{F}G$-module. Then V is called a **p-permutation module** or **trivial source module** if its restriction to any Sylow p-subgroup of G is an ordinary permutation module, or in other words, if a basis $\mathfrak{b}$ of V can be found, which is permuted by some Sylow p-group. As any two Sylow p-groups are conjugate in G, the property of being a p- permutation module does not depend on the choice of the Sylow group. Note also that V is a p-permutation module if and only if so is the dual V^*. It is a known fact in modular representation theory that V is a p-permutation module if and only if it is a direct summand of a permutation module for G (see for example [**37**, Proposition 27.3]). Now Lemma 4.1 together with Lemma 3.3 gives:

PROPOSITION 4.1. *Let $R := \mathbb{F}$ be a field of characteristic $p \mid |G|$, V a finitely generated $\mathbb{F}G$-module and $A := \mathrm{Sym}(V^*)$ the symmetric algebra of the dual module V^*. If V is a p-permutation module and P is a Sylow p-group of G, then $\beta(A^Q/\mathcal{I}^Q_{<Q}) \le |Q|$ for each subgroup $Q \le P$. In particular Conjecture 3.2 holds for $A = \mathrm{Sym}(V^*)$.*

To show that the second degree bound of Conjecture 3.6 holds in this case, we can argue in a similar way as in the proof of Lemma 3.3:

Let $f \in \mathcal{I}^{G,+}_{<P}$ be indecomposable. By the transitivity of relative transfers[4] we can assume that f is of the form $t^G_Q(h)$ with $Q < P$ and $h \in A^Q$ is a power product in norms $\mathfrak{n}_i$ corresponding to the Q-orbits on a chosen P-permutation basis of V. Each polynomial $F[T] := \prod_{g \in G:Q}(T - g(\mathfrak{n}_i)) \in A^G[T]$ has $\mathfrak{n}_i$ as a zero. Expansion of $F[T]$ shows that the power $\mathfrak{n}_i^{[G:Q]}$ lies in the A^G-span of the $\mathfrak{n}_i^j$ with $j < [G:Q]$. This allows for reductions of exponents in h and since the operator t^G_Q is A^G-linear, we can assume that these exponents do not exceed $|G:Q| - 1$. Hence the total degree of f can be assumed to be less or equal to $s \cdot (|G:Q| - 1)$, where s is the number of Q-orbits. Since $s \le d = \dim(V)$ and $\deg \mathfrak{n}_i \le |Q|$ we conclude that $\beta(\mathcal{I}^G_{<P}) \le \max\{|G|, d \cdot (|G| - 1)\}$. Since V is a p-permutation module (or a trivial source module) if and only if V is a direct summand of a permutation module, we get

PROPOSITION 4.2. *If the $\mathbb{F}G$-module V is a trivial source module, then*

$$\beta(\mathrm{Sym}(V^*)^G) \le \max\{|G|, \dim(V) \cdot (|G| - 1)\}.$$

In 1994 M. Goebel developed a general reduction technique for permutation invariants, which produces generators for $A(\Omega)^G$ of degree less or equal to $\frac{1}{2}|\Omega|(|\Omega| - 1)$ (see [**16**]). A very good description including all details of Goebel's algorithm and theorem can be found in [**6**, 3.10.2]. Therefore I will only briefly sketch Goebel's result here.

DEFINITION 4.2. A function $\alpha \in \mathbb{N}_0^\Omega$ will be called **special**, if its image is an interval in $\mathbb{N}_0$ which includes 0, i.e. if $\alpha(\Omega) = [0, m_\alpha] = \{0, 1, 2, \ldots, m_\alpha\}$.

Note that if α is special, then

$$|\alpha| = \sum_\omega \alpha(\omega) \le 0 + 1 + 2 + \cdots + (|\Omega| - 1) = \frac{1}{2}|\Omega|(|\Omega| - 1).$$

Let $e_i := \sum_{T \subseteq \Omega, |T| = i} \prod_{\omega \in T} X_\omega$ be the ith elementary symmetric polynomial in $A(\Omega)$.

---

[4] $t^H_{U\,|A^U} = (t^H_Y \circ t^Y_U)_{|A^U}$ for $U \le Y \le H$.

THEOREM 4.3 (M. Goebel). *Let $\mathcal{S} \subset \mathbb{N}_0^\Omega$ be the subset of special functions. Then $A(\Omega)^G$ is a finitely generated $A(\Omega)^{\Sigma_\Omega}$-module with generating set*

$$\mathcal{S}^+ := \{\mathrm{orb}_G(\underline{X}^\alpha) \mid \alpha \in \mathcal{S}\}.$$

In particular

$$A(\Omega)^G = R[e_i, \mathcal{S}^+ \mid 1 \le i \le |\Omega|] \quad \text{and} \quad \beta\big(A(\Omega)^G\big) \le \binom{|\Omega|}{2}.$$

Goebel also proved a refinement of his theorem for intransitive G-actions, which leads to the bound

$$\beta\big(A(\Omega)^G\big) \le \sum_{j \in \underline{m}} \max\left\{|\Omega_j|, \binom{|\Omega_j|}{2}\right\},$$

if Ω is the disjoint union of G-sets $\Omega_1, \Omega_2, \dots, \Omega_k$.

5. The Noether homomorphism

Special types of permutation invariants can be used to construct generators of arbitrary invariant rings, using ideas from Emmy Noether's paper [**25**]. Of particular interest are the *vector invariants* of symmetric groups, which we will now introduce.

Let $A(k,n)$ be the polynomial ring $R[X_{11}, \dots, X_{k1}, \dots, X_{1n}, \dots, X_{kn}]$ in $k \times n$ variables and define the action of Σ_n on $A(k,n)$ by extending the permutation action $\sigma(X_{ij}) := X_{i\sigma(j)}$. The corresponding ring of invariants $A(k,n)^{\Sigma_n}$ is usually called the ring of (k-fold) **vector invariants**.

Let $\underline{Y} := (Y_1, \dots, Y_k)$ be a "vector of variables"; then the multivariate polynomial

$$G(\underline{X}_1, \dots, \underline{X}_n; \underline{Y}) := \prod_{j=1}^{n} \left(1 + \sum_{i=1}^{k} X_{i,j} Y_i\right) \in A(k,n)^{\Sigma_n}[Y_1, \dots, Y_k]$$

is called the **Galois-resolvent**. In Hermann Weyl's celebrated book "Classical groups" [**38**] one can find a proof of the following

THEOREM 5.1 (Weyl). *If $\mathbb{Q} \subseteq R$ then $A(k,n)^{\Sigma_n}$ is generated by the coefficients of the Galois-resolvent $G(\underline{X}_1, \dots, \underline{X}_n; \underline{Y})$. They all have total degree $\le n$, so $\beta\big(A(k,n)^{\Sigma_n}\big) \le n$.*

The analogue of Weyl's theorem is false if $R = \mathbb{Z}$ or a field of characteristic $p \le n$. This can be seen from the Σ_2-invariant $\mathfrak{X} := (X_1, \dots, X_k)^+ := X_1, \dots, X_k + Y_1, \dots, Y_k$ of Section 2.1, because $\mathfrak{X}$ is indecomposable over $\mathbb{Z}$ or $\mathbb{F}_2$ for all $k \in \mathbb{N}$. In [**27**] it was proved by D. Richman, that the analogue of Weyl's theorem holds if $n!$ is invertible in R (for a different proof, see 6). For arbitrary coefficients the following has been shown in [**8**].

THEOREM 5.2. $\beta\big(A(k,n)^{\Sigma_n}\big) \le \max\{n, k \cdot (n-1)\}$ *with equality if $n = p^s$ and* $\mathrm{char}\, R = p$ *or* $R = \mathbb{Z}$.

To make use of these results in the context of arbitrary invariant rings, consider a subgroup $H \le G$ with index n and with set of left-cosets $G/H := \{H := g_1H, g_2G, \dots, g_nH\}$. The left multiplication action of G on the set G/H gives rise to the the **Cayley–homomorphism**

$$c : G \to \Sigma_{G/H} \cong \Sigma_n, \quad g \mapsto (g_iH \mapsto g_jH := gg_iH).$$

Suppose that $A := R[a_1, \ldots, a_d]$ and $A^H = R[b_1, \ldots, b_k]$ with $b_i \in \mathcal{M}_\beta(\boldsymbol{a})$ and $\beta := \beta(A^H, \boldsymbol{a})$. Then the map $X_{si} \mapsto g_i(b_s)$ defines a unique G-equivariant homomorphism $\nu : A(k,n) \to A$ of R-algebras. In fact, ν does not depend on the choice of the g_i and

$$\nu\big(g(X_{si})\big) = \nu(X_{sj}) = g_j(b_s) = gg_ih(b_s) = gg_i(b_s) = g\nu(X_{si}),$$

because $gg_i = g_jh^{-1}$ for a suitable $h \in H$. The map ν for $H = 1$ had been used in Emmy Noether's 1916—paper to prove her degree bound in characteristic zero. It is therefore called the **Noether homomorphism**. We can use her original argument to show that the restriction $\nu_| : A(k,n)^{\Sigma_n} \to A^G$ is surjective, whenever n is invertible in R:

In that case for each $f = f(b_1, \ldots, b_k) \in A^G \leq A^H$ there is a polynomial

$$F := \frac{1}{n}\big(f(X_{11}, X_{21}, \ldots, X_{k1}) + \cdots + f(X_{1n}, X_{2n}, \ldots, X_{kn})\big) \in A(k,n)^{\Sigma_n}$$

which satisfies

$$\begin{aligned}\nu(F) &= \frac{1}{n}\big(f(g_1(b_1), g_1(b_2), \ldots, g_1(b_k)) + \cdots + f\big(g_n(b_1), g_n(b_2), \ldots, g_n(b_k)\big)\big)\\ &= \frac{1}{n}\big(g_1f(b_1, b_2, \ldots, b_k) + \cdots + g_nf(b_1, b_2, \ldots, b_k)\big) = f.\end{aligned}$$

Let $\beta(k,n) := \beta\big(A(k,n)^{\Sigma_n}\big)$ and $A(k,n)^{\Sigma_n} = R[F_1, \ldots, F_s]$ with $F_i \in \mathcal{M}_{\beta(k,n)}(\boldsymbol{X})$, then $A^G = R[\nu(F_1), \ldots, \nu(F_s)]$ with $\nu(F_i) \in \mathcal{M}_{\beta(k,n)\beta}(\boldsymbol{a}) \cap A^G$. In particular if the index $n = [G : H]$ is invertible in R, then

$$\beta(A^G, \boldsymbol{a}) \leq \beta(k,n)\beta(A^H, \boldsymbol{a}).$$

Together with (5.1) (or [**27**]) and (5.2) we get:

THEOREM 5.3. *Let A be an R-algebra as above and $H \leq G$ with $[G : H]$ invertible in R and $A^H = R[b_1, \ldots, b_k]$ with $b_i \in A^H \cap \mathcal{M}_b(\boldsymbol{a})$ and $b := \beta(A^H, \boldsymbol{a})$. Then*

$$\beta(A^G, \boldsymbol{a}) \leq \max\{[G : H], k \cdot ([G : H] - 1)\}\beta(A^H, \boldsymbol{a}).$$

In particular, if A^H is finitely generated then so is A^G. If moreover $[G : H]!$ is invertible in R, then

$$\beta(A^G, \boldsymbol{a}) \leq \ |G : H|\beta(A^H, \boldsymbol{a}).$$

If p is a prime and R is of characteristic p, then we can take $H = P$, a Sylow-p group of G. Since the index $[G : P]$ is invertible, we can apply (5.3) and construct A^G from A^P via vector invariants. So in modular invariant theory the most serious problem is to construct invariant rings of p-groups in characteristic $p > 0$. Gregor Kemper developed general construction algorithms, which work for arbitrary finite groups (see [**18**]). Current research is investigating enhanced methods for p-groups [**31, 32**]. But note that a bound for the degrees *and the numbers of generators* for A^P is needed to obtain a degree bound for A^G via (5.3).

6. Weyl's theorem on vector invariants

In this section we will give a proof of Hermann Weyl's **Main theorem on vector invariants** of the symmetric group, which is a generalized version of the theorem on symmetric polynomials. The result is beautiful and interesting in its own right, and can be used to derive "conceptually simple" fundamental systems for A^G, provided that all natural numbers $m \leq |G|$ are invertible in R. Again this approach dates back to Emmy Noether's 1916 paper [**25**], where she used it in the situation $R = \mathbb{C}$.

The following combinatorial lemma is a straightforward generalization of the identity

$$XY = \frac{1}{2}[(X+Y)^2 - X^2 - Y^2].$$

LEMMA 6.1. *For $a_1, \ldots, a_n \in A$ one has:*

$$(-1)^n n! \cdot a_1, \ldots, a_n = \sum_{I \subseteq \underline{n}} (-1)^{|I|} \Big(\sum_{i \in I} a_i \Big)^n .$$

PROOF.

$$\begin{aligned}
\sum_{I\subseteq \underline{n}} (-1)^{|I|} \Big(\sum_{i\in I} a_i\Big)^n &= \sum_{I\subseteq\underline{n}} (-1)^{|I|} \sum_{\substack{\alpha\in\mathbb{N}_0^I,\\ |\alpha|=n}} \binom{n}{\alpha_{i_1}, \ldots, \alpha_{i_{|I|}}} \underline{a}^\alpha \\
&= \sum_{\substack{\beta\in\mathbb{N}_0^n,\\ |\beta|=n}} \binom{n}{\beta_1,\ldots,\beta_n} \underline{a}^\beta \sum_{\substack{I\subseteq\underline{n},\\ \mathrm{supp}(\beta)\subseteq I}} (-1)^{|I|} \\
&= (-1)^n \binom{n}{1,\ldots,1} \underline{a}^{\mathbf{1}} = (-1)^n n! \cdot a_1 \ldots a_n.
\end{aligned}$$

The latter equation holds because $\sum_{\substack{I\subseteq\underline{n},\\ \mathrm{supp}(\beta)\subseteq I}} (-1)^{|I|} = (-1)^n$, if $\mathrm{supp}(\beta) = \underline{n}$ and $= 0$ otherwise. □

DEFINITION 6.2 (Polarization). For $f \in R[X_1, \ldots, X_n]$ and $\mu \in \mathbb{N}_0^k$ we define

$$\mathrm{Pol}(f) := f(X_{11} + \cdots + X_{k1}, \quad X_{12} + \cdots + X_{k2}, \ldots, X_{1n} + \cdots + X_{kn}) \in A(k,n)$$

and $\mathrm{Pol}(f)_\mu$ to be the homogeneous component of multi degree μ.

Obviously $f \in R[X_1, X_2, \ldots, X_n]^{\Sigma_n}$ implies $\mathrm{Pol}(f)_\mu \in A(k,n)^{\Sigma_n}$. If f is homogeneous of degree m, then $\mathrm{Pol}(f)_\mu$ is homogeneous of total degree m and $\mathrm{Pol}(f)_\mu = 0$ if $|\mu| \neq m$. It is straightforward to see that the polarized elementary symmetric polynomials $\mathrm{Pol}(e_t)_\mu$ are precisely the coefficients of the Galois resolvent

$$G(\underline{X}_1, \ldots, \underline{X}_n; \underline{Y}) = \prod_{i=1}^n \Big(1 + \sum_{s=1}^k X_{si} Y_s \Big).$$

For $t \in \mathbb{N}$ let $e_t^{(n)}$ denote the tth elementary symmetric polynomial in n-variables, i.e.

$$e_t^{(n)} := \sum_{\substack{I\subseteq\underline{n},\\ |I|=t}} \prod_{i\in I} X_i.$$

Then we have the recursion formula $e_t^{(n)} = e_t^{(n-1)} + e_{t-1}^{(n-1)} X_n$
(*e.g. for* $n = 4$):

$$\begin{aligned}
e_1^{(4)} &= X_1 + \cdots + X_4 = e_1^{(3)} + X_4,\\
e_2^{(4)} &= X_1X_2 + \cdots + X_3X_4 = e_2^{(3)} + X_4 e_1^{(3)},\\
e_3^{(4)} &= X_1X_2X_3 + \cdots + X_2X_3X_4 = e_3^{(3)} + X_4 e_2^{(3)},\\
e_4^{(4)} &= X_1X_2X_3X_4 = X_4 e_3^{(3)}.
\end{aligned}$$

Now polarization yields:

$$\begin{aligned}\mathrm{Pol}(e_t^{(n)})_\mu &= \mathrm{Pol}(e_t^{(n-1)})_\mu + [\mathrm{Pol}(e_{t-1}^{(n-1)})(X_{1n} + X_{2n} + \cdots + X_{kn})]_\mu \\ &= \mathrm{Pol}(e_t^{(n-1)})_\mu + \sum_{s=1}^{k} X_{sn} \cdot \mathrm{Pol}(e_{t-1}^{(n-1)})_{\mu-\delta_s},^{5}\end{aligned}$$

with $\mathrm{Pol}(e_t^{(n-1)})_\mu = 0$ if $|\mu| > t$. Using an obvious induction argument, we observe

LEMMA 6.3.

$$\mathrm{Pol}(e_t^{(n-1)})_\mu \in R[\mathrm{Pol}(e_s^{(n)})_\nu, X_{in} \mid s \in \underline{n}, i \in \underline{k}, \nu \in \mathbb{N}_0^k].$$

Now we can formulate Weyl's theorem on vector invariants of Σ_n. Our proof is a generalization of the one in Weyl's book "Classical groups" [**38**]. A different one can be found in [**27**], but Theorem 2.3 facilitates the proof substantially.

THEOREM 6.4. *Let $n!$ be invertible in R. Then the invariant ring*

$$R[X_{ij} \mid i \in \underline{k}, j \in \underline{n}]^{\Sigma_n}$$

is generated by the polarized elementary symmetric polynomials $\mathrm{Pol}(e_m)_\mu$. *In particular $\beta(R[X_{ij} \mid i \in \underline{n},\ j \in \underline{k}]^{\Sigma_n}) = n$.*

PROOF. The proof is by induction on n. The induction hypothesis, applied to Σ_{n-1} yields $A(k,n)^{\Sigma_{n-1}} = R[\mathrm{Pol}(e_m^{(n-1)})_\mu, X_{jn} \mid m \in \underline{n-1}, \mu \in \mathbb{N}_0^k, j \in \underline{k}]$. Using (6.3) we see $A(k,n)^{\Sigma_{n-1}} = R[\mathrm{Pol}(e_m^{(n)})_\mu, X_{jn} \mid m \in \underline{n}, j \in \underline{k}, \mu \in \mathbb{N}_0^k]$. Since $n! = |\Sigma_n|$ is invertible in R, we can apply Theorem 2.3. Its proof shows that

$$A(k,n)^{\Sigma_n} = R[t_{\Sigma_{n-1}}^{\Sigma_n}(\underline{b}^\alpha) \mid 0 \neq \alpha \in \mathbb{N}_0^{\mathbb{N}_0}, |\alpha| \leq n],$$

where $b_i \in \{\mathrm{Pol}(e_m^{(n)})_\mu, X_{jn}, \dots\}$. Since the $\mathrm{Pol}(e_m^{(n)})_\mu$'s are Σ_n-invariant, we have

$$A(k,n)^{\Sigma_n} = R[\mathrm{Pol}(e_m^{(n)})_\mu, t_{\Sigma_{n-1}}^{\Sigma_n}(\underline{X}_{*,n}^\alpha) \mid 0 \neq \alpha \in \mathbb{N}_0^k, |\alpha| \leq n],$$

and it remains to show that the elements $t_{\Sigma_{n-1}}^{\Sigma_n}(\underline{X}_{*,n}^\alpha)$ with $|\alpha| \leq n$ lie in $\mathcal{P} := R[\mathrm{Pol}(e_m^{(n)})_\mu \mid m, \mu \dots]$.

For that purpose we consider the power sums

$$p_m := X_1^m + X_2^m + \cdots + X_n^m \in R[X_1, \dots, X_n]^{\Sigma_n}.$$

We have

$$\begin{aligned}\mathrm{Pol}(p_{|\alpha|})_\alpha = \left(\sum_{i=1}^n (X_{1i} + \cdots + X_{ki})^{|\alpha|}\right)_\alpha &= \sum_{i=1}^n \left(\sum_{\beta \in \mathbb{N}_0^k, |\beta| = |\alpha|} \binom{|\alpha|}{\beta_1, \dots, \beta_k} \underline{X}_{*,i}^\beta \right)_\alpha \\ &= \binom{|\alpha|}{\alpha_1, \dots, \alpha_k} t_{\Sigma_{n-1}}^{\Sigma_n}(\underline{X}_{*,n}^\alpha).\end{aligned}$$

Since $|\alpha| \leq n$ and $n! \in R^*$, the multinomial coefficient

$$\binom{|\alpha|}{\alpha_1, \dots, \alpha_k} = \frac{|\alpha|!}{\alpha_1! \dots \alpha_k!}$$

is invertible in R. By the main theorem on symmetric functions, the power sums p_m can be expressed over $\mathbb{Z}$ by elementary symmetric functions. Polarization of the

[5] $\delta_s := (0, \dots, 0, 1, 0, \dots, 0) \in \mathbb{N}_0^k$ with 1 in position s.

corresponding formulae gives polynomial expressions for $\mathrm{Pol}(p_{|\alpha|})_\alpha$ in terms of the $\mathrm{Pol}(e_m)_\mu$'s. Finally we see that $t^{\Sigma_n}_{\Sigma_{n-1}}(\underline{X}^\alpha_{*,n}) \in \mathcal{P}$ and the proof is finished. □

Now (6.4), together with the Noether homomorphism from (5) gives the following

COROLLARY 6.5 (Richman [**27**]). *Let $H \leq G$ such that $n := [G : H]$ with $n!$ invertible in R. Let $g_1, \ldots, g_n$ be a set of representatives for the left H-cosets in G and assume that $A^H = R[b_1, b_2, \ldots, b_k]$. Then A^G is generated as an R-algebra by the coefficients of the specialized Galois resolvent*

$$\phi(A^G)(\underline{Y}) := \prod_{i=1}^{n}\Big(1 + \sum_{\ell=1}^{k} g_i(b_\ell)Y_\ell\Big).$$

In particular for $A := R[a_1, \ldots, a_d]$ one has

$$\beta(A^G, \boldsymbol{a}) \leq [G : H]\beta(A^H, \boldsymbol{a}).$$

7. On the structure of invariant rings

In this section we want to take up the question from (1.1).

> *Let A be a polynomial ring over a field $R = \mathbb{F}$, say, and G a finite group acting on A via graded $\mathbb{F}$-algebra automorphisms. How "far away" is A^G from being a polynomial ring?*

The fact that not all invariant rings are polynomial rings can be seen from the example $G = \langle g \rangle \cong C_2$ acting on $A := \mathbb{F}[X, Y]$ by $g(X) = -X$ and $g(Y) = -Y$ (with $\mathbb{F}$ of characteristic $\neq 2$). It is easy to see that $A^G = \mathbb{F}[X^2, Y^2, XY]$, which is not isomorphic to a polynomial ring, because it is not factorial. Indeed X^2, Y^2 and XY are different irreducibles in $\mathbb{F}[X, Y]^G$ giving a nonunique factorization $X^2Y^2 = (XY)^2$. However, A^G contains a polynomial subalgebra $\mathfrak{P} = \mathbb{F}[X^2, Y^2]$ such that $A^G = \mathfrak{P} \cdot 1 \oplus \mathfrak{P} \cdot XY$ is free of rank 2 as $\mathfrak{P}$-module. On the other hand the map $X_1 \mapsto X^2$, $X_2 \mapsto Y^2$, $X_3 \mapsto XY$ extends to a surjective homomorphism of algebras $\mathbb{F}[X_1, X_2, X_3] \to A^G$ which shows that

$$A^G \cong \mathbb{F}[X_1, X_2, X_3]/(X_1X_2 - X_3^2).$$

Let us start with some special geometric features that are due to the fact that G is a **finite group**.

8. Geometric aspects

Historically a major motivation for invariant theory comes from the analysis of symmetries on affine algebraic varieties. One way to set up the scene goes as follows: Let A be an **affine algebra**, i.e. a finitely generated commutative $\mathbb{F}$-algebra without nonzero nilpotent elements. Let $\overline{\mathbb{F}}$ be the algebraic closure and $\hat{A} := A \otimes_{\mathbb{F}} \overline{\mathbb{F}}$; then using Hilbert's Nullstellensatz, the set $\widehat{Y}$ of maximal ideals in $\hat{A}$ can be viewed as an affine algebraic variety in the usual sense of algebraic geometry, with $\hat{A} := \mathcal{O}(\widehat{Y})$ as corresponding algebra of regular functions. In a similar way one can view the set Y of maximal ideals of A as an $\mathbb{F}$-variety and, even more generally, consider the "affine scheme" $X := \mathrm{spec}(A)$ consisting of all prime ideals $\mathrm{p} \triangleleft A$, $\mathrm{p} \neq A$. Every $\mathbb{F}$-algebra automorphism of A induces a (dual) automorphism of X and for a group G of automorphisms of A one can define a **categorical quotient** X/G. Then a natural question is, whether X/G is again an affine algebraic variety. A necessary condition for this is, that the invariant ring

$$A^G := \{a \in A \mid g(a) = a \ \forall g \in G\}$$

is a finitely generated $\mathbb{F}$-algebra. In this case the prime ideal spectrum spec A^G is a natural candidate for X/G and one might hope that X/G coincides with the set–theoretic quotient, namely the orbit space X/G. For an arbitrary group neither does A^G have to be finitely generated, nor does the categorical quotient X/G have to coincide with the "geometric quotient" X/G, but if G is a finite group, then the situation is much better (see e.g. [**1**, 1.4.4]):

From now on let G be a finite group. Note that A is integral and finite over A^G and A as well as A^G are integrally closed in their quotient fields Quot(A) and Quot$(A)^G$ respectively. In particular A^G and A have the same Krull-dimension[6], which is equal to the dimension of the algebraic variety $\widehat{Y}$. Moreover Quot$(A) \geq$ Quot$(A)^G$ is a Galois extension with Galois group G. The dual i^* of the inclusion map $i : A^G \hookrightarrow A$ is the map spec $A \to$ spec A^G, $\mathfrak{p} \mapsto \mathfrak{p} \cap A^G$. In this situation the classical "lying over", "going up", "going down" and transitivity theorems of commutative algebra apply (see [**1**, Theorem 1.4.4, p. 7]). We have $A^G \otimes_{\mathbb{F}} \overline{\mathbb{F}} = \hat{A}^G$ and the above statements also hold for $\hat{A}$ instead of A. Hence max-spec $\hat{A}^G$ can be identified with the set $\widehat{Y}/G$ of G-orbits on $\widehat{Y}$ and $\hat{i}^*$, the dual of the inclusion $\hat{A}^G \hookrightarrow \hat{A}$ can be identified with the orbit map $y \mapsto y^G$. Thus a strict geometric quotient $\widehat{Y}/G$ exists and the invariant ring $\hat{A}^G$ can be interpreted as the ring of regular functions $\mathcal{O}(\widehat{Y}/G)$ of the quotient variety (e.g. see [**15**, Theorem 5.52, p. 187] or [**1**, p. 8]). Notice that all G-orbits are closed in the Zariski topology since G is finite. The orbit map $\hat{i}^*$, being the comorphism of a finite embedding, is surjective and closed, i.e. sends closed irreducible subvarieties of $\widehat{Y}$ to closed irreducible subvarieties of $\widehat{Y}/G$.

Now let us return to our usual (linear) setup. To avoid trivialities, we assume that $G \leq \mathrm{GL}(V)$ with V a finite-dimensional $\mathbb{F}$-vector space with dual module $V^* = \oplus_{i=1}^n \mathbb{F}X_i$ and $A = \mathbb{F}[V] = \mathrm{Sym}(V^*) = \mathbb{F}[X_1, \ldots, X_n]$. Since G is finite, we know that A^G is a finitely generated $\mathbb{N}_0$-graded $\mathbb{F}$-algebra with degree zero component isomorphic to $\mathbb{F}$. Such an algebra is called **graded connected over** $\mathbb{F}$. Many arguments in invariant theory work in this wider context of graded connected algebras. In the following we will, unless explicitly stated otherwise, use A to denote Sym(V^*) and B to denote a general finitely generated graded connected sub algebra such that the B-module A is finitely generated. Typically we have in mind $B = A^G$.

9. Hilbert-series

Let M be a finitely generated $\mathbb{N}_0$-graded B-module. Recall that the **Hilbert-series** (or Poincaré–series) of M is defined to be the power series

$$H(M,t) := \sum_{i \geq 0} \dim_{\mathbb{F}}(M_i) t^i,$$

where M_i denotes the ith homogeneous component of M. From the geometric series $1 + t + t^2 + \cdots = 1/1 - t$ and the rule $H(M \otimes_{\mathbb{F}} N, t) = H(M,t) \cdot H(N,t)$ one can easily see that $H({}_A A, t) = 1/(1-t)^n$. In general for $B = \mathbb{F}[b_1, \ldots, b_m]$

[6]recall that the Krull-dimension Dim(A) is the maximal length of a properly ascending chain of prime ideals of A.

one has $H({}_BB,t) = f/\prod_{i=1}^m(1-t^{s_i})$, with $f \in \mathbb{Z}[t]$ and s_i being the degrees of the homogeneous generators b_i.

There is a beautiful theorem about the Hilbert-series of nonmodular invariant rings and modular permutation invariants. Assume that $\mathrm{char}(\mathbb{F})$ does not divide $|G|$ and let V be a finitely generated $\mathbb{F}G$-module. It is known from representation theory of finite groups, that there is a "Brauer lift" $\hat{V}$ to zero characteristic: in brief terms, this is an $\mathcal{O}G$-module which is free as an $\mathcal{O}$-module, where $\mathcal{O}$ is a suitable discrete valuation ring with quotient field K of characteristic zero and maximal ideal $\mathfrak{P} \lhd \mathcal{O}$ such that $\mathcal{O}/\mathfrak{P} \cong \mathbb{F}$ and $\widehat{V} \otimes_{\mathcal{O}} \mathbb{F} \cong V$. In particular for each $g \in G$ there is a "lift" of $g_{|V}$ to $g_{|\widehat{V}} \in \mathrm{GL}(\widehat{V})$.

THEOREM 9.1 (Molien). *If* $\mathrm{char}(\mathbb{F})$ *does not divide* $|G|$, *then for* $A = \mathbb{F}[V]$ *one has*

$$H(A^G,t) = \frac{1}{|G|}\sum_{g\in G}\frac{1}{\det(1-g_{|\widehat{V}^*}t)}.$$

If $\mathbb{F}$ *is arbitrary and* $G \leq \Sigma_\Omega$, *then*

$$H(\mathbb{F}[\mathbb{F}^\Omega]^G,t) = H(\mathbb{C}[\mathbb{C}^\Omega]^G,t).$$

PROOF. Assume first that $\mathrm{char}(\mathbb{F}) = 0$. The dimensions of homogeneous fixed point spaces do not change with extension of scalars, so we can assume that $\mathbb{F}$ is algebraically closed and therefore we can use character theory: the dimension of A_m^G is equal to the multiplicity of the trivial G-module in the $\mathbb{F}G$-module A_m, so it is the inner product $\langle\chi_{A_m}, 1_G\rangle$. Our assumptions on $\mathbb{F}$ imply that all elements $g \in G$ are diagonalisable on V and hence as well on A_m. If $\lambda_1,\dots,\lambda_n$ are the eigenvalues of g on V (with multiplicities), then the products $\underline{\lambda}^\alpha$ with $\alpha \in \mathbb{N}_0^n$ and $|\alpha| = m$ are exactly the eigenvalues of g. We conclude

$$\begin{aligned} H(A^G,t) = \sum_{m=0}^\infty \langle\chi_{A_m},1_G\rangle t^m &= \frac{1}{|G|}\sum_{m=0}^\infty\sum_{g\in G}\sum_{\substack{\alpha\in\mathbb{N}_0^n,\\ |\alpha|=m}}\underline{\lambda}^\alpha(g)t^m \\ &= \frac{1}{|G|}\sum_{g\in G}\prod_{i=1}^n(1+\lambda_i t+\lambda_i^2t^2+\cdots) \\ &= \frac{1}{|G|}\sum_{g\in G}\prod_{i=1}^n\frac{1}{1-\lambda_i t}. \end{aligned}$$

If $\mathbb{F}$ has positive characteristic not dividing $|G|$, we replace V^* by a Brauer lift $\widehat{V}^*$. Then $\hat{A} := \mathrm{Sym}(\widehat{V}^*)$ is a Brauer lift of A. Moreover the group order $|G|$ is invertible in $\mathcal{O}$, hence we get

$$A_m^G = t_1^G(A_m) = t_1^G(\hat{A}_m \otimes_{\mathcal{O}} \mathbb{F}) = t_1^G(\hat{A}_m)\otimes_{\mathcal{O}}\mathbb{F} \cong (\hat{A}_m)^G\otimes_{\mathcal{O}}\mathbb{F}.$$

Similarly we see for $A_K := \hat{A}\otimes_{\mathcal{O}} K$: $(A_K)_m^G \cong (\hat{A}_m)^G \otimes_{\mathcal{O}} K$. Hence $\dim_{\mathbb{F}}(A_m^G) = \dim_K\big((A_K)_m^G\big)$, which shows that the Hilbert-series of A^G and $(A_K)^G$ coincide and are given by Molien's formula.

Now let $\mathbb{F}$ be arbitrary but $V = \mathbb{F}^\Omega$ a G-permutation module. Then each A_m is a permutation module as well and the dimension of A_m^G is equal to the number of G-orbits on the set $\{\alpha \in \mathbb{N}_0^\Omega \mid |\alpha| = m\}$. This number is obviously independent of the characteristic of $\mathbb{F}$. □

It has been shown by Gregor Kemper that for p-permutation modules one also has $H(\mathbb{F}[V]^G, t) = H(\mathbb{F}[\widehat{V}]^G, t)$ with Brauer lift $\widehat{V}$ [**21**]. Hence the Hilbert-series of direct summands of permutation modules can be determined using Molien's formula.

10. Homogeneous systems of parameters

There are various ways to measure how close B comes to be a polynomial algebra. A very intuitive way of comparison is to exhibit a maximal polynomial subalgebra $\mathfrak{P} \le B$ such that the module ${}_{\mathfrak{P}}B$ is finitely generated. This is always possible by Noether's normalization lemma and it can be shown that the transcendence degree of $\mathfrak{P}$ is equal to the Krull-dimension of B, i.e. $\mathfrak{P} = \mathbb{F}[h_1, \dots, h_n]$ with algebraically independent elements $h_1, \dots, h_n$ that can be chosen to be homogeneous. They form a **homogeneous system of parameters** (hsop). In fact a set $\{h_1, \dots, h_n\}$ of homogeneous elements of positive degree forms an hsop, if and only if $B/(h_1, \dots, h_n)B$ is a finite-dimensional $\mathbb{F}$-vector space. Let $\{h_1, \dots, h_n\}$ be an hsop with $d_i := \deg h_i$ and let $s_{n+1}, \dots, s_k$ be generators for the $\mathfrak{P}$-module B. Then the h_i are also called **primary generators** of B with corresponding **secondary generators** s_j.

Note that, since A is finite over B, an hsop for B is also an hsop for A. If $\mathfrak{P} \le B \le A = \mathbb{F}[V]$, then the **$\mathfrak{P}$-module** A is free of finite rank, hence ${}_{\mathfrak{P}}A \cong \mathfrak{P} \otimes_{\mathbb{F}} W$ with a finite dimensional graded $\mathbb{F}$-vector space W of dimension d, say, which is also the degree of the quotient field extension $[\mathrm{Quot}(A) : \mathrm{Quot}(\mathfrak{P})]$. We then get for the Hilbert-series, $H(A,t) = H(\mathfrak{P}, t) \cdot H(W,t)$ and we conclude

$$H(W,t) = \frac{\prod_{i=1}^{n}(t^{d_i} - 1)}{(t-1)^n} = \prod_{i=1}^{n}(1 + t + \cdots + t^{d_i - 1}),$$

hence $d := \dim_{\mathbb{F}}(W) = \prod_{i=1}^{n} d_i$. Let $\ell := [\mathrm{Quot}(B) : \mathrm{Quot}(\mathfrak{P})]$, then $\ell = d/m$ with $m = [\mathrm{Quot}(A) : \mathrm{Quot}(B)]$ and the module ${}_{\mathfrak{P}}B$ is finitely generated *free* if and only if it can be generated by ℓ homogeneous elements. Since $\mathrm{Quot}(A)$ is Galois over $\mathrm{Quot}(A^G)$ with Galois-group G, we get:

LEMMA 10.1. *Let $\mathfrak{P} = \mathbb{F}[h_1, \dots, h_n] \le A^G \le A$ with hsop $h_1, \dots, h_n$ of A^G of degrees $d_1, \dots, d_n$. Then the degree of the extension of quotient fields $\mathrm{Quot}(\mathfrak{P}) \hookrightarrow \mathrm{Quot}(A)$ is equal to the product $d := \prod_{i=1}^{n} d_i$. In particular $|G|$ divides d. Moreover A^G is a finitely generated free $\mathfrak{P}$-module[7] if and only if ${}_{\mathfrak{P}}A^G$ is generated by $\ell = d/|G|$ homogeneous elements.*

Using this lemma and the fact that polynomial algebras are integrally closed, one can now prove the following criterion for invariant rings to be polynomial algebras:

THEOREM 10.2. *Let $A := \mathbb{F}[V]$ with $\mathbb{F}G$-module V of dimension n. Then the following are equivalent:*

1. *A^G is a polynomial ring.*
2. *There is an* hsop *$\{h_1, \dots, h_n\}$ of A^G with $|G| = \prod_{i=1}^{n} \deg h_i$.*

In particular $A^G = \mathbb{F}[h_1, \dots, h_n]$ implies $|G| = \prod_{i=1}^{n} \deg h_i$.

---

[7]this is in fact equivalent to A^G being a Cohen–Macaulay algebra.

PROOF. Let $\{h_1, \ldots, h_n\} \subseteq A^G$ be an hsop with $\deg h_i = d_i$ and $\mathfrak{P} := \mathbb{F}[h_1, \ldots, h_n]$. Then $[\mathrm{Quot}(A) : \mathrm{Quot}(\mathfrak{P})] = d := \prod_{i=1}^n d_i$. Now the implication "1. ⇒ 2." follows from (10.1). To prove "2. ⇒ 1." assume that $d = |G|$. Then

$$d = [\mathrm{Quot}(A) : \mathrm{Quot}(\mathfrak{P})] = |G| \cdot [\mathrm{Quot}(A)^G : \mathrm{Quot}(\mathfrak{P})]$$

implies $\mathrm{Quot}(\mathfrak{P}) = \mathrm{Quot}(A)^G$. The polynomial ring $\mathfrak{P}$ is factorial and therefore integrally closed, hence

$$A^G = \mathrm{Quot}(A)^G \cap A = \mathrm{Quot}(\mathfrak{P}) \cap A = \mathfrak{P}. \quad \square$$

Consider the elementary symmetric polynomials $e_i \in \mathbb{F}[X_1, \ldots, X_n]$. The product of their degrees is $n! = |\Sigma_n|$, hence (10.2) contains an algebraic proof for the fact that $\mathbb{F}[X_1, \ldots, X_n]^{\Sigma_n} = \mathbb{F}[e_1, \ldots, e_n]$. In particular the elementary symmetric functions form a "generic hsop" for each ring of permutation invariants $\mathbb{F}[X_1, \ldots, X_n]^G$ with $G \leq \Sigma_n$.

If $\mathbb{F} = \mathbb{F}_q$ is finite of order q and $\widetilde{G} := \mathrm{GL}(V)$, then

$$\mathbb{F}[V]^{\widetilde{G}} = \mathbb{F}[d_{i,n} \mid i = 1, \ldots, n]$$

where the **Dickson-invariants** $d_{i,n}$ are defined by

$$F_n(T) := \prod_{v \in V^*} (T - v) = \sum_{i=0}^{n} d_{i,n} T^{q^{n-i}}$$

with $d_{0,n} = 1$. Of course it has to be shown that $F_n(T)$ is a sparse polynomial as described. An elementary way to see this has been suggested by C. Wilkerson [**39**]: consider the determinant

$$\Delta_n(T) = \begin{vmatrix} X_1 & \cdots & X_n & T \\ X_1^q & \cdots & X_n^q & T^q \\ \cdots & \cdots & \cdots & \cdots \\ X_1^{q^n} & \cdots & X_n^{q^n} & T^{q^n} \end{vmatrix}.$$

It is clear that $\Delta(v) = 0$ for each $v \in V$; comparing the coefficients at T^{q^n} we get $\Delta_n(T) = \Delta_{n-1}(X_n) \cdot F_n(T)$. The needed fact that the "constant" $\Delta_{n-1}(X_n)$ is nonzero follows for example from the normal basis theorem for the finite field extension $V \cong \mathbb{F}_{q^n} \geq \mathbb{F}_q$. Now for $i > 0$, $\deg d_{n-i,n} = q^n - q^i$, hence $\prod_{i=0}^{n-1} \deg d_{n-i,n} = |\mathrm{GL}(V)|$ and clearly $A = \mathbb{F}[V]$ is finite over $\mathbb{F}[d_{i,n} \mid i = 1, \ldots, n]$. Hence the $d_{i,n} \in \mathbb{F}[V]^{\mathrm{GL}(V)}$ form an hsop and (10.2) shows that

$$\mathbb{F}[V]^{\mathrm{GL}(V)} = \mathbb{F}[d_{i,n} \mid i = 1, \ldots, n].$$

For any subgroup $G \leq \mathrm{GL}(V)$, $\mathbb{F}[V]^G$ is a finite extension of $\mathbb{F}[V]^{\mathrm{GL}(V)}$, therefore the $d_{i,n}$ form "generic hsops" for invariant rings of finite groups over finite fields.

The arguments above contain a general recipe to prove that an invariant ring is polynomial. Following C. Wilkerson we condense them into the following three steps:

1. Guess a set of n homogeneous algebra generators $a_1, \ldots, a_n \in A^G$.
2. Show that A is finite over $\mathbb{F}[a_1, \ldots, a_n]$.
3. Show that the degree of $\mathrm{Quot}(A)$ over $\mathbb{F}(a_1, \ldots, a_n)$ equals $|G|$, e.g. by showing that $\prod_{i=1}^n \deg a_i = |G|$. [8]

[8] It has been shown by Kemper [**18**] that it suffices to verify that the a_i are algebraically independent.

Besides the cases Σ_n and $\mathrm{GL}(V)$ one can for example determine the invariant rings $\mathbb{F}[V]^{\mathrm{SL}(V)}$ and $\mathbb{F}[V]^U$ in that way, with U being the full upper triangular subgroup in $\mathrm{GL}(V)$. To find hsops in the general case, one can use a construction of E. Dade, which always gives an hsop whose elements have degrees not exceeding $|G|$. Dade's original construction works over an algebraically closed field, but can be generalized with a "scalar extension" argument.

11. Cohen–Macaulay property

One can expect to measure the "distance" of A^G from being polynomial by the complexity of the module-structure of ${}_{\mathfrak{P}}A^G$, which of course is free of rank one in case $A^G = \mathfrak{P}$ is a polynomial ring. But the choice of hsop's is far from unique, as can be seen by expressing $\mathbb{F}[X_1, \ldots, X_n]$ as a finitely generated module over the polynomial subalgebra $\mathbb{F}[X_1^{s_1}, \ldots, X_n^{s_n}]$ for $s_i \in \mathbb{N}$. Hence we look for more "intrinsic" ring theoretic properties to compare $\mathbb{F}$-algebras.

For each choice of hsop the module ${}_{\mathfrak{P}}\mathbb{F}[X_1, \ldots, X_n]$ is free of finite rank over $\mathfrak{P}$. In the context of graded connected algebras, this can be used as a defining property of a **Cohen–Macaulay algebra**. More precisely, if a graded connected algebra B is free over some hsop, it is so over every hsop.

A different definition of the Cohen–Macaulay property, which is equivalent in the case of connected algebras, uses the notion of **regular sequences** and **depth**. If B is graded connected and M is a finitely generated $\mathbb{N}_0$-graded B-module, then a sequence $\underline{b} := (b_1, b_2, \ldots, b_k)$ of homogeneous elements in B^+ is called **regular on** M, or an M-sequence for short, if multiplication with b_i on $M/(b_1, b_2, \ldots, b_{i-1})M$ is injective for each $i \leq k$. If $I \leq B^+$ is a homogeneous ideal, then all maximal M-sequences in I have the same length, which is called $\mathrm{grade}(I, M)$, or $\mathrm{depth}\, M$ if $I = B^+$. In general $\mathrm{depth}\, M \leq \mathrm{Dim}\, M := \mathrm{Dim}(B/\mathrm{ann}\, M)$ [9] and one calls M a **Cohen–Macaulay module**, if equality holds. It turns out that B is a Cohen–Macaulay algebra if and only if the regular module ${}_BB$ is a Cohen–Macaulay module. Clearly a polynomial algebra $\mathbb{F}[X_1, \ldots, X_n]$ is a Cohen–Macaulay algebra (CM-algebra) with maximal regular sequence $(X_1, X_2, \ldots, X_n)$. From (10.1) we know that A^G is Cohen–Macaulay, if and only if it has an hsop of degrees $d_1, \ldots, d_n$ with exactly $d_1 d_2 \ldots d_n/|G|$ many secondary generators. The following results give a "CM-criterion" that does not require information on hsops and generators:

Theorem 11.1. *Let $B \leq A$ be such that ${}_BA$ is a finitely generated B-module with ${}_BB$ as direct summand. Then B is a Cohen–Macaulay algebra.*

Proof. Let $\mathfrak{P} \leq B \leq A$ with $\mathfrak{P}$ generated by an hsop of B. Since A is Cohen–Macaulay, ${}_{\mathfrak{P}}A$ is finitely generated and free, hence ${}_{\mathfrak{P}}B$, as a summand of ${}_{\mathfrak{P}}A$ is projective. But for a graded connected algebra like $\mathfrak{P}$ all f.g. projective graded modules are free (see [**1**, Lemma 4.1.1]) and we conclude that ${}_{\mathfrak{P}}B$ is free. Hence B is Cohen–Macaulay. □

If $|G|$ is invertible in $\mathbb{F}$, then A^G is a direct summand of A as A^G-module and we immediately obtain the theorem of Eagon and Hochster:

Corollary 11.2. *If* $\mathrm{char}\,\mathbb{F}$ *does not divide* $|G|$, *then* A^G *is a Cohen–Macaulay algebra.*

[9] $\mathrm{ann}\, M := \{b \in B \mid bm = 0, \forall m \in M\}$.

The CM-property plays a significant role in the description of an algebra by "generators and syzygies": Let $\{h_1, \ldots, h_n\}$ be an hsop of B; then B is finitely generated, by $s_{n+1}, \ldots, s_{n+\ell}$ say, as $\mathfrak{P}$-module. This gives rise to an obvious epimorphism of $\mathfrak{P}$-modules $\chi : \mathfrak{P}^\ell \to B$. On the other hand one can pick a subset $\{a_{n+1}, \ldots, a_\mu\} \subseteq \{s_{n+1}, \ldots, s_{n+\ell}\}$ such that $B = \mathbb{F}[h_1, \ldots, h_n, a_{n+1}, \ldots, a_\mu]$. Then there is an obvious ring epimorphism $\phi : \mathbb{F}[X_1, \ldots, X_\mu] \to B$, giving B another structure as module over some polynomial ring. To find out what is "intrinsic" about these module structures, let $\mathcal{F}$ be some polynomial algebra, $\theta : \mathcal{F} \to B$ an algebra homomorphism such that B is a finitely generated $\mathcal{F}$-module via θ, and let $\theta_0 : \mathcal{F}^{r_0} \to B$ be an $\mathcal{F}$- module epimorphism. Since $\mathcal{F}$ is Noetherian, the kernel K of θ_0 is finitely generated and we can find an $\mathcal{F}$-module epimorphism $\theta_1 : \mathcal{F}^{r_1} \to K$. The kernel of θ_1 is a finitely generated submodule of $\mathcal{F}^{r_1}$ and we can iterate the process, which leads to a **free resolution** $\mathcal{F}^*$ of B as $\mathcal{F}$-module, i.e. to an exact sequence

$$\cdots \to \mathcal{F}^{r_t} \to \mathcal{F}^{r_{t-1}} \to \ldots \mathcal{F}^{r_1} \to \mathcal{F}^{r_0} \to B \to 0.$$

From Hilbert's syzygy-theorem it is known that the global dimension of $\mathcal{F}$ is equal to its Krull-dimension, which means that there is a "minimal" free resolution

$$0 \to \mathcal{F}^{r_t} \to \mathcal{F}^{r_{t-1}} \to \ldots \mathcal{F}^{r_1} \to \mathcal{F}^{r_0} \to B \to 0$$

with $t \leq \operatorname{Dim} \mathcal{F}$. In fact the minimal number $t_{\min}$ is the **projective dimension** proj. $\dim_{\mathcal{F}} B$ of B as an $\mathcal{F}$-module. By the Auslander–Buchsbaum formula we get:

$$\operatorname{Dim} \mathcal{F} - \text{proj. } \dim_{\mathcal{F}} B = \operatorname{depth} B.$$

Thus the "cohomological codimension" $\operatorname{Dim} \mathcal{F} - \text{proj. } \dim_{\mathcal{F}} B$ turns out to be an intrinsic property of B, namely the maximal length of a regular sequence in B^+. In other words, the **CM-defect** $\operatorname{def} B := \operatorname{Dim} B - \operatorname{depth} B$ is equal to the projective dimension proj. $\dim_{\mathfrak{P}} B$ of B over any hsop. This of course is zero if and only if B is Cohen–Macaulay.

12. Non-CM invariant rings

In the modular situation, i.e. if the characteristic of $\mathbb{F}$ divides $|G|$, then A^G will in general not be a Cohen–Macaulay algebra. Following [**5**], this can be seen in the following way.

Assume that $N \lhd G$ is a normal subgroup such that $G/N = \langle \bar{g} \rangle \cong C_p$ with $p = \operatorname{char} \mathbb{F}$. Then for $1 \leq i \leq p-1$ there is $j \in \mathbb{N}$ with $ij \equiv 1 \mod p$ and we have

$$(\bar{g}^i - 1)A^N = (\bar{g}-1)(1+\bar{g}+\cdots+\bar{g}^{i-1})A^N \subseteq (\bar{g}-1)A^N = \left((\bar{g}^i)^j - 1\right)A^N \subseteq (\bar{g}^i - 1)A^N,$$

hence the ideal $\mathcal{J} := (\bar{g} - 1)A^N \cap A^G \lhd A^G$ is independent of the choice of the generator $\bar{g} \in G/N$. Moreover $\mathcal{J}$ is nonzero. To see this, note that by Galois theory $\operatorname{Quot}(A)^G < \operatorname{Quot}(A)^N$, hence $A^G < A^N$ and A^G is the kernel of the operator $\bar{g} - 1$ on A^N. In particular there is $a \in A^N$ with $(\bar{g} - 1)a \neq 0$. Since $(\bar{g} - 1)^p a = (\bar{g}^p - 1)a = 0$, there is some $i \in \mathbb{N}$ with $(\bar{g} - 1)^{i+1} a = 0 \neq (\bar{g} - 1)^i a \in \mathcal{J}$.

LEMMA 12.1 ([**5**]).

$\operatorname{grade}(\mathcal{J}, A^G) = \min\{2, \operatorname{ht}(\mathcal{J})\}$, *where* $\operatorname{ht}(\mathcal{J})$ *denotes the height of the ideal* $\mathcal{J}$.

PROOF. By contradiction we assume that $f_1, f_2, f_3 \in \mathcal{J}$ form a regular sequence. Then there are $a_1, a_2, a_3 \in A^N$ with $\bar{g}(a_i) - a_i = f_i$ for $i = 1, 2, 3$ and we have

$$0 = \begin{vmatrix} f_1 & f_2 & f_3 \\ f_1 & f_2 & f_3 \\ a_1 & a_2 & a_3 \end{vmatrix} = u_{23} f_1 + u_{31} f_2 + u_{12} f_3 \quad \text{with} \quad u_{ij} := \begin{vmatrix} f_i & f_j \\ a_i & a_j \end{vmatrix}.$$

It is easy to see that $\bar{g}(u_{ij}) = u_{ij} \in A^G$ and the regularity of f_1, f_2, f_3 implies that $u_{12} \in (f_1, f_2)A^G$, so $a_2 f_1 - a_1 f_2 = u_{12} = r f_1 + s f_2$ with $r, s \in A^G$ which gives $(a_2 - r) f_1 = (a_1 + s) f_2$. Since the polynomial ring A is Cohen–Macaulay and finite over A^G, f_1, f_2 are part of a system of parameters for A and therefore regular. It is easy to see that this implies that f_1 and f_2 are coprime as polynomials. Hence f_1 divides $a_1 + s$ in A, i.e. $a_1 = q f_1 - s$. From this we get $f_1 = \bar{g}(a_1) - a_1 = f_1(\bar{g} - 1)(q)$, hence $(\bar{g} - 1)(q) = 1$ which is impossible. This shows that $\text{grade}(\mathcal{J}, A^G) \leq 2$. In general one knows that $\text{grade}(\mathcal{J}, A^G) \leq \text{ht}(\mathcal{J})$. If $\text{ht}(\mathcal{J}) = 1$, then $\mathcal{J}$ contains a regular element of A^G and $\text{grade}(\mathcal{J}, A^G) = 1$. So assume that $\text{ht}(\mathcal{J}) \geq 2$. Then there are homogeneous elements $f_1, f_2 \in \mathcal{J}$ with $\text{ht}(f_1, f_2)A^G = 2 = \text{ht}(f_1, f_2)A$ (the equality of heights, because A is finite over A^G). This implies that f_1 and f_2 are coprime in A, otherwise $f_i = d\tilde{f}_i$, $i = 1, 2$, with greatest common divisor $d \in A$, hence $(f_1, f_2)A \subseteq (d)A$ which would imply the contradiction $\text{ht}(f_1, f_2)A^G \leq 1$. Now it follows that $(f_1, f_2) \subseteq A^G$ is a regular sequence: indeed a relation $r f_1 + s f_2 = 0$ in A^G implies that f_1 divides s in A but also in A^G. Hence $\text{grade}(\mathcal{J}, A^G) \geq 2$ and the proof is complete. □

As an application we get the following beautiful result

THEOREM 12.2 (Campbell, Hughes, Kemper, Shank, Wehlau [**5**]). *Let* $\text{char}\,\mathbb{F} = p > 0$ *and let* N *be a normal subgroup of* G *with cyclic factor group* $G/N \cong C_p$. *Then for every* $\mathbb{F}G$*-module* V *the ring of* m*-fold vector–invariants* $\mathbb{F}[V^m]^G$ *is not Cohen–Macaulay for* $m \geq 3$.

PROOF. Using the notation above, the discussion in the beginning of this section shows that there is $a \in \mathbb{F}[V]^N$ with $0 \neq \bar{g}a - a \in \mathbb{F}[V]^G$. Let $a_1, \ldots, a_m$ be the elements in $\mathbb{F}[V^m]^N$ corresponding to a, with $f_i := \bar{g}a_i - a_i$. Then the sequence $(f_1, \ldots, f_m)$ is a partial hsop of $\mathbb{F}[V^m]^G$ whose elements lie in $\mathcal{J}$. From (12.1) we know that $(f_1, \ldots, f_m)$ cannot be regular if $m \geq 3$. Since in a CM-algebra each partial hsop forms a regular sequence, we conclude that $\mathbb{F}[V^m]^G$ is not CM for $m \geq 3$. □

It has been shown by Gregor Kemper that for every finite group G of order divisible by $p = \text{char}\,\mathbb{F}$ and for every finite-dimensional $\mathbb{F}G$-module V there is some $m \in \mathbb{N}$ such that the ring of m-fold vector invariants $\text{Sym}\big((V^*)^m\big)^G$ is not Cohen–Macaulay (see [**20**, Corollary 2.4]).

The following "Three Copies Conjecture" is still open:

CONJECTURE 12.3. If V is a faithful $\mathbb{F}G$-module and $\text{char}\,\mathbb{F}$ divides $|G|$, then for W the direct sum of three copies of V, the invariant ring $\mathbb{F}[W]^G$ is not Cohen–Macaulay.

13. On the depth of invariant rings

We have seen that modular invariant rings are in general not Cohen–Macaulay. Therefore we will refine our initial question to the following one

- How "far away" is A^G from being a Cohen–Macaulay algebra?

It is known that the length of a maximal regular sequence in $A^{G,+}$ is less or equal to the Krull-dimension with equality if and only if A^G is CM. Hence the parameters $\mathrm{depth}(A^G) = \mathrm{grade}(A^{+G}, A^G)$ or $\mathrm{def}(A^G) := \mathrm{Dim}(A^G) - \mathrm{depth}(A^G)$ provide a natural measure for the distance of A^G to being CM. The following is a generalization of (11.1):

LEMMA 13.1. *Let $C \leq B$ be graded connected over $\mathbb{F}$ such that ${}_C B$ is finite. Assume $\theta : B \to C$ is a* ***Reynolds-operator****, i.e. a C-module homomorphism which is a left inverse to the embedding $C \hookrightarrow B$. Then*

$$\mathrm{depth}(B) \leq \mathrm{depth}(C).$$

COROLLARY 13.2. *Let $p = \mathrm{char}\,\mathbb{F}$ and P a Sylow p-group. Then the following hold:*

1. *If $H \leq G$ with $[G : H]1_{\mathbb{F}} \neq 0$, then $\mathrm{depth}\, A^H \leq \mathrm{depth}\, A^G$ (Kemper [**18**]).*
2. *If A^P is CM, so is A^G (Campbell–Hughes–Pollack [**4**]).*

PROOF. : 1. follows from (13.1) with $\theta := t_H^G : A^H \to A^G$. Now 2. follows from the fact that, since G is finite, the extensions $A^G \leq A^P \leq A$ are finite and therefore all of equal dimension. □

It is known that in 1. equality is not true in general: for $G = \Sigma_5$, $P = \langle(12345)\rangle$, $p = 5$ and $V = \mathbb{F}^5$ with natural permutation action, one knows that $\mathbb{F}[V]^P$ is not CM, whereas $\mathbb{F}[V]^{\Sigma_5}$ is even a polynomial ring.

In 1980 G. Ellingsrud and T. Skjelbred proved the celebrated result that, if P is a Sylow p-group of G with fixed point space V^P, one has

$$\mathrm{depth}\,\mathrm{Sym}(V^*)^G \geq 2 + \dim(V^P) \tag{13.1}$$

if $\dim(V) \geq \dim(V^P)+2$, with equality if G is a cyclic p-group (see [**7**]). For almost two decades this has been the only general result on the depth of modular invariant rings, which remains to be one of their most interesting, but difficult to determine parameters. In particular the classification of modular Cohen–Macaulay–invariant rings is an open problem. The result of Ellingsrud and Skjelbred was achieved using homological algebra, in particular a Grothendieck spectral sequence. During the last five years or so, these techniques have been revitalized (see e.g. [**20, 23, 34**]), most notably by Gregor Kemper who was able to classify all groups, whose modular regular representation has a Cohen–Macaulay ring of invariants:

THEOREM 13.3 (G. Kemper [**19**]). *For a finite group G with group algebra $\Lambda := \mathbb{F}G$ the following are equivalent:*

1. *$\mathrm{Sym}(\Lambda)^G$ is Cohen–Macaulay;*
2. *$\mathrm{char}\,\mathbb{F}$ does not divide $|G|$ or $G \in \{\mathbb{Z}_2, \mathbb{Z}_3, \mathbb{Z}_2 \times \mathbb{Z}_2\}$.*

Due to constraints in length and time we have not mentioned the use of Steenrod-operations in modular invariant rings. An extensive account can be found in Section 8 of [**24**]. Without doubt the most striking success of these techniques in modular invariant theory is the proof by Bourguiba and Zarati of an earlier conjecture of Landweber and Stong:

THEOREM 13.4 (Bourguiba–Zarati [2] and [17]). *Let* $A = \mathbb{F}_q[V]$ *with* $\mathbb{F}_qG$*-module* V. *Then one can use the Dickson invariants (see (10)) as a test sequence to measure the depth of* A^G. *More precisely the depth of* A^G *is the largest* ℓ *such that the sequence of Dickson–invariants* $d_{1,n}, \dots, d_{\ell,n}$ *is* A^G*-regular.*

14. Relative transfer and depth

With regard to the techniques laid out in the previous sections of these notes, some recent results show that the relative transfer ideal $\mathcal{I}^G_{<P}$ and its radical ideal $\sqrt{\mathcal{I}^G_{<P}}$ [10] shed some new light on the problem of determining the depth of modular invariant rings. Let $P \leq G$ be a fixed Sylow p-group of G and assume that $\mathbb{F}$ is algebraically closed of characteristic $p > 0$. This allows us to consider $A := \mathrm{Sym}(V^*)$ as the algebra of polynomial functions on V, and A^G as the algebra of polynomial functions on the orbit space V/G. Hence for an ideal $\mathcal{I} \triangleleft A^G$, the variety $\mathcal{V}(\mathcal{I})$ consists of all orbits v^G in V such that $f(x) = 0$ for every $f \in \mathcal{I}$ and $x \in v^G$. On the other hand for each subset $S \subseteq V$ there is the ideal

$$\mathcal{I}(S) := \{f \in A \mid f(s) = 0 \forall s \in S\}$$

and $\mathcal{I}^G(S) := \mathcal{I}(S) \cap A^G \triangleleft A^G$. In [9] relative transfer ideals have been investigated geometrically, which led to the following description of $\sqrt{\mathcal{I}^G_{<P}}$ in terms of its variety in the orbit space V/G:

THEOREM 14.1 ([9]).

1. $\mathcal{V}(\mathcal{I}^G_{<P}) = \{v^G \in V/G \mid p \not| |v^G|\}$;
2. $\sqrt{\mathcal{I}^G_{<P}} = \mathcal{I}^G(V^P)$ *is a prime ideal of height* $\mathrm{codim}_{\mathbb{F}}(V^P)$, *where* V^P *denotes the space of* P*-fixed points in* V.

Note that for a permutation module of a p–group Q, the previous result $A^Q = \mathcal{I}^Q_{<Q} \oplus \mathbb{F}[\mathfrak{n}_1, \dots, \mathfrak{n}_s]$ implies that $\sqrt{\mathcal{I}^Q_{<Q}} = \mathcal{I}^Q_{<Q}$ with $A^Q/\mathcal{I}^Q_{<Q} \cong \mathbb{F}[\mathfrak{n}_1, \dots, \mathfrak{n}_s]$ being a polynomial ring. In general the quotient $A^G/\mathcal{I}^G(V^P)$ is still Cohen–Macaulay, which was proved in [9] for p-permutation modules. A second proof in that paper, designed to work under slightly more general hypotheses, contains a gap and I am indebted to Jim Shank for pointing that out to me. I take this occasion to present a proof that works without any additional hypotheses. We will make use of a very useful technique to extend regular sequences, which was developed by Gregor Kemper [20]:

PROPOSITION 14.1 (G. Kemper). *Let* $W \leq V^*$ *be an* $\mathbb{F}G$*-submodule such that the kernel* N *of the* G*-action on the quotient* V^*/W *has an index coprime to* p. *Then there exist homogeneous* $g_1, g_2, \dots, g_m \in A^G$ *with the following two properties:*

(a) *The images* $\overline{g_1}, \overline{g_2}, \dots, \overline{g_m}$ *form an* hsop *in the quotient ring* $A^G/[(W)A \cap A^G]$.

(b) *For each* i, *multiplication with* g_i *has a left inverse in the endomorphism ring* $\mathrm{End}_{A_iG}(A)$, *where* A_i *is the subalgebra*

$$A_i := \mathrm{Sym}(W)^G[g_1, \dots, g_{i-1}, g_{i+1}, \dots, g_m] \subseteq A^G,$$

generated by $\mathrm{Sym}(W)^G$ *and all* g_j *with* $j \neq i$.

---

[10] for an ideal $\mathcal{I}$ in A: $\sqrt{\mathcal{I}} := \{a \in A \mid a^m \in I,$ for some $m \in \mathbb{N}\}$.

Any $g_1, g_2, \ldots, g_m$ *having property* (a) *also have property* (b) *and* $m = \operatorname{codim}(W)$.

Let $W := (V^G)^\perp = \{f \in V^* \mid f(V^G) = 0\}$ with $\mathbb{F}$-basis $\{y_1, \ldots, y_r\}$, let $\mathcal{I} := \mathcal{I}(V^G) = ((V^G)^\perp)A$ and $\mathfrak{i} := \mathcal{I} \cap A^G = \mathcal{I}^G(V^G)$. Note that G acts trivially on the quotient V^*/W, hence we can take $N = G$ in Proposition 14.1. The proof in [**20**] shows, that there always exists a sequence $g_1, g_2, \ldots, g_m \in A^G$, $m = \dim(V^G)$, with $\overline{g_1}, \overline{g_2}, \ldots, \overline{g_m}$ being an hsop in $\overline{A^G} := A^G/\mathfrak{i}$, such that

1. for each i there is a map $\Psi_i : A \to A$, commuting with the G $(= N)$-action and with elements in $\mathbb{F}[g_1, \ldots, g_{i-1}, g_{i+1}, \ldots, g_m, y_1, y_2, \ldots, y_r]$;
2. $\Psi_i(g_i a) = a$ for every $a \in A$.

Hence Ψ_i is a left-inverse to the multiplication–map $\mu_{g_i} : A \to A$, $a \mapsto g_i a$. Moreover we see that $\Psi_i(\mathcal{I}) \subseteq \mathcal{I}$ and $\Psi_i(\mathfrak{i}) \subseteq \mathfrak{i}$, therefore the Ψ_i's induce left-inverses in $\operatorname{End}_{\mathbb{F}[g_1,\ldots,g_{i-1},g_{i+1},\ldots,g_m]}(\overline{A^G})$ to the multiplication–map $\mu_{\overline{g_i}}$. Now assume that $(\overline{g_1}, \overline{g_2}, \ldots, \overline{g_s})$ is a regular sequence on $\overline{A^G}$ with $s < m$, and $\overline{a} \in \overline{A^G}$ with $\overline{g_{s+1}a} \in (\overline{g_1}, \overline{g_2}, \ldots, \overline{g_s})\overline{A^G}$. Then $\overline{a} = \Psi_{s+1}(\overline{g_{s+1}a}) \in (\overline{g_1}, \overline{g_2}, \ldots, \overline{g_s})\overline{A^G}$, proving that $(\overline{g_1}, \overline{g_2}, \ldots, \overline{g_{s+1}})$ is regular as well. Hence $m = \dim(V^G) \leq \operatorname{depth}(\overline{A^G}) \leq \operatorname{Dim}(\overline{A^G})$.

THEOREM 14.2. *Let G be a finite group, P a Sylow p-subgroup of G and $A = \mathbb{F}[V]$ with $\mathbb{F}G$-module V. Then $\overline{A^G} := A^G/\mathcal{I}^G(V^P)$ is Cohen–Macaulay of Krull-dimension* $\dim_{\mathbb{F}}(V^P)$.

PROOF. Taking $U := V^P$ and $\overline{A^P} := A^P/\mathcal{I}^P(V^P)$, we get from from the above: $\operatorname{Dim}(\overline{A^P}) \geq \operatorname{depth}(\overline{A^P}) \geq \dim_{\mathbb{F}}(V^P)$. Since A and A^P are finite over A^G, the (prime) ideals $\mathcal{I}(V^P) \triangleleft A$, $\mathcal{I}^P(V^P) \triangleleft A^P$ and $\mathcal{I}^G(V^P) \triangleleft A^G$ have the same height and $\overline{A^P}$ is finite over $\overline{A^G}$ with the same Krull-dimension $\dim_{\mathbb{F}}(V^P)$. In particular we see that $\operatorname{depth}(\overline{A^P}) = \operatorname{Dim}(\overline{A^P}) = \dim_{\mathbb{F}}(V^P)$, and $\overline{A^P}$ is Cohen–Macaulay. The Mackey-formula from Section 1 shows that

$$t_P^G(\mathcal{I}^P(V^P)) = \mathcal{I}^P(V^P) \cap A^G = \mathcal{I}^G(V^P),$$

hence the relative transfer t_P^G induces a Reynolds operator $\overline{A^P} \to \overline{A^G}$. Now Lemma (13.1) yields

$$\dim_{\mathbb{F}}(V^P) = \operatorname{depth}(\overline{A^P}) \leq \operatorname{depth}(\overline{A^G}) \leq \operatorname{Dim}(\overline{A^G}) = \dim_{\mathbb{F}}(V^P). \quad \square$$

The Theorems 14.1 and 14.2 indicated that the ideal $\sqrt{\mathcal{I}^G_{<P}} = \mathcal{I}^G(V^P)$ might "measure the depth" of $\operatorname{Sym}(V^*)^G$ in general: the formula (3) of Ellingsrud–Skjelbred shows, that $\dim(V^P)$ is a lower bound for the depth of A^G, so it was conceivable that the "missing part" of the depth is provided by regular elements in the ideal $\mathcal{I}^G_{<P}$[11]. In the case of p-groups this is a consequence of [**20**, Theorem 1.5] and for arbitrary finite groups it is a consequence of the following

THEOREM 14.3 (P. Fl., R. J. Shank [**14**]).

$$\operatorname{depth} A^G = \operatorname{grade}(\mathcal{I}^G_{<P}, A^G) + \dim(V^P).$$

Moreover, if V is defined over $\mathbb{F}_q$, one can apply Theorem 13.4 and at least in principle use the "Dickson invariants" $d_{i,n} \in \operatorname{Sym}(V^*)^{\mathrm{GL}_n(q)} \leq \operatorname{Sym}(V^*)^G$ to determine the grade of $\mathcal{I}^G_{<P}$ on A^G.

[11] note that the grade of an ideal always coincides with the grade of its radical.

Theorem 14.3, in connection with Lemma and Conjecture 3.6, shows that the relative transfer ideal $\mathcal{I}^G_{<P}$ contains the clues for many important structural and constructive properties of modular invariant rings. Therefore an efficient algorithm to find minimal generating sets for the ideal $\sqrt{\mathcal{I}^G_{<P}}$ is very much needed as an important step to determine its grade and henceforth the depth of A^G.

Connecting (14.3) with the cohomological methods developed by Gregor Kemper, a substantial generalization of G. Ellingsrud and T. Skjelbred's depth formula could be achieved. Note that the following result has purely group-theoretic hypotheses: the group G is called p-nilpotent if it has a normal p-complement, i.e. a normal subgroup N of order coprime to p, such that G/N is a p-group (which then has to be isomorphic to a Sylow p-group of G.)

THEOREM 14.4 ([**11**]). *If G is p-nilpotent with cyclic Sylow p-subgroup $P \leq G$, then*

$$\operatorname{depth}(A^G) = \min\{\dim_{\mathbb{F}}(V^P) + 2, \dim_{\mathbb{F}}(V)\}.$$

References

1. D. J. Benson, *Polynomial invariants of finite groups*, London Math. Soc. Lecture Note Ser., vol. 190, Cambridge University Press, Cambridge, 1993.
2. D. Bourguiba and S. Zarati, *Depth and the Steenrod algebra*, Invent. Math. **128** (1997), 589–602.
3. Winfried Bruns and Jürgen Herzog, *Cohen–Macaulay rings*, Cambridge Stud. Adv. Math., vol. 39, Cambridge University Press, Cambridge, 1993.
4. H. E. A. Campbell, I. Hughes, and R. D. Pollack, *Rings of invariants and p-Sylow subgroups*, Canad. Math. Bull **34** (1991), 42–47.
5. H. E. A. Campbell et al, *Depth of modular invariant rings*, Transform. Groups **5** (2000), no. 1, 21–34.
6. H. Derksen and G. Kemper, *Computational invariant theory*, Encyclopaedia Math. Sci., Springer-Verlag, Berlin Heidelberg, New York, 2002,
7. G. Ellingsrud and T. Skjelbred, *Profondeur d'anneaux d'invariants en caractéristique p*, Compos. Math. **41** (1980), 233–244.
8. P. Fleischmann, *A new degree bound for vector invariants of symmetric groups*, Trans. Amer. Math. Soc. **350** (1998), no. 4, 1703–1712.
9. ______, *Relative trace ideals and Cohen–Macaulay quotients of modular invariant rings*, Computational Methods for Representations of Groups and Algebras (P. Dräxler, G. O. Michler, and C. M. Ringel, eds.) (April, 1997), Euroconference in Essen, Progress in Mathematics, vol. 173, Birkhäuser, Basel, 1999.
10. ______, *The Noether bound in invariant theory of finite groups*, Adv. Math. **156** (2000), 23–32.
11. P. Fleischmann, G. Kemper, and J. Shank, Work in progress.
12. P. Fleischmann and W. Lempken, *On generators of modular invariant rings of finite groups*, Bull. London Math. Soc. **29** (1997), 585–591.
13. ______, *On degree bounds for invariant rings of finite groups over finite fields*, Contemp. Mathematics **225** (1999), 31–41.
14. P. Fleischmann and J. Shank, *The relative trace ideal and the depth of modular rings of invariants*, Report UKC/IMS/01/30, IMS, University of Kent, UK., 2001; Archiv der Mathematik **80** (2003), no. 3, 347–353.
15. J. Fogarty, *On Noether's bound for polynomial invariants of a finite group*, Electron. Res. Announc. Amer. Math. Soc. **7** (2001), 5–7.
16. M. Göbel, *Computing bases for rings of permutation-invariant polynomials*, J. Symbolic Comput. **19** (1995), 285–291.
17. H.-W. Henn, *A variant of the proof of the Landweber–Stong conjecture*, Proc. Sympos. Pure Math., vol. 63, 1998, pp. 271–275.
18. G. Kemper, *Calculating invariant rings of finite groups over arbitrary fields*, J. Symbolic Comput. **21** (1996), 351–366.

19. ———, *Die Cohen–Macaulay–Eigenschaft in der modularen Invariantentheorie*, Universität Heidelberg, Report, Habilitationsschrift, 1999.
20. ———, *On the Cohen–Macaulay property of modular invariant rings*, J. of Algebra **215** (1999), 330–351.
21. ———, *The depth of invariant rings and cohomology, with an appendix by Kay Magaard*, J. of Algebra **245** (2001), 463–531.
22. P. Landrock, *Finite group algebras and their modules*, London Math. Soc. Lecture Note Ser., vol. 84, Cambridge University Press, Cambridge, 1983.
23. M. Lorenz and J. Pathak, *On Cohen–Macaulay rings of invariants*, J. of Algebra **245** (2001), 247–264.
24. M. Neusel and L. Smith, *Invariant theory of finite groups*, Math. Surveys Monogr., vol. 94, Amer. Math. Soc., Providence, RI, 2002.
25. E. Noether, *Der Endlichkeitssatz der Invarianten endlicher Gruppen*, Math. Ann. **77** (1916), 89–92.
26. ———, *Der Endlichkeitssatz der Invarianten endlicher linearer Gruppen der Charakteristik p*, Nachr. Ges. Wiss. Göttingen (1926), 28–35; Collected Papers, Springer Verlag, Berlin, 1983, pp. 485–492.
27. D. Richman, *Explicit generators of the invariants of finite groups*, Adv. Math. **124** (1996), 49–76.
28. ———, *Invariants of finite groups over fields of characteristic p*, Adv. Math. **124** (1996), 25–48.
29. B. J. Schmid, *Generating invariants of finite groups*, C. R. Acad. Sci. Paris **308** (1989), 1–6.
30. R. J. Shank, S.A.G.B.I. *bases for rings of formal modular seminvariants [semi-invariants]*, Comment. Math. Helv. **73** (1998), no. 4, 548–565.
31. R. J. Shank and D. L. Wehlau, *Noether numbers for subrepresentations of cyclic groups of prime order*, Bull. London Math. Soc. **4** (2002), 438–450.
32. ———, *Computing modular invariants of p-groups*, J. Symbolic Comput. **34** (2002), no. 5, 307–327.
33. L. Smith, *Polynomial invariants of finite groups. A survey of recent developments*, Bull. Amer. Math. Soc. (N.S.) **34** (1997), no. 3, 211–250.
34. ———, *Homological codimension of modular rings of invariants and the Koszul complex*, J. Math. Kyoto Univ. **38** (1998), 727–747.
35. ———, *E. Noether's bound in the invariant theory of finite groups*, Arch. Math. (Basel) **66** (1996), no. 2, 89–92.
36. ———, *Polynomial invariants of finite groups*, Res. Notes Math., vol. 6, A. K. Peters, Ltd., Wellesley, MA, 1995.
37. J. Thévenaz, *G-algebras and modular representation theory*, Oxford Math. Monogr., Oxford Science Publications, The Clarendon Press, Oxford University Press, New York, 1995.
38. H. Weyl, *The classical groups*, Princeton University Press, Princeton, N.J., 1953.
39. C. W. Wilkerson, *A primer on the Dickson invariants*, Proc. of the Northwestern Homotopy Theory Conference (Evanston, 1982), Contemp. Math., vol. 19, Amer. Math. Soc., Providence, RI, 1983, pp. 421–434.

Institut of Mathematics and Statistics, University of Kent, Canterbury, CT2 7NF, UK.

E-mail address: `P.Fleischmann@ukc.ac.uk`

Centre de Recherches Mathématiques
CRM Proceedings and Lecture Notes
Volume **35**, 2004

Combinatorics Related to Orbit Closures of Symmetric Subgroups in Flag Varieties

Aloysius G. Helminck

ABSTRACT. Let G be a connected reductive group over an algebraically closed field k of characteristic not 2; let $\theta \in \operatorname{Aut}(G)$ be an involution and $K = G^\theta \subseteq G$ the fixed point group of θ and let $P \subseteq G$ be a parabolic subgroup. The set $K\backslash G/P$ of (K, P)-double cosets in G plays an important role in the study of Harish Chandra modules.

In [**2**] we gave a description of the orbits of symmetric subgroups in a flag variety G/P mainly using geometric arguments. For general P, it is difficult to describe the combinatorics of the decomposition of the closure of a double coset in terms of $K \times P$ double cosets. However, in some special cases one can describe the combinatorics of the closures of the double cosets in more detail. This paper discusses the special case that P contains a θ-stable Levi factor L and the set of roots of the connected center S of L is a root system with Weyl group $W(S) = N_G(S)/Z_G(S)$. Here $N_G(S)$ (resp. $Z_G(S)$) is the normalizer (resp. centralizer) of S in G. In this case the combinatorics of the Weyl Group can be used to describe the closures of the double cosets of a part of the double coset space which includes the open and closed orbits and we get a number of results similar to the case that $P = B$ a Borel subgroup. This root system condition on P is satisfied in many cases. For example in the case that P is a minimal parabolic k_0-subgroup of G or a minimal θ-split parabolic subgroup of G or a minimal (θ, k_0)-split parabolic k_0-subgroup of G. Here $k_0 \subseteq k$ is a subfield of k and G, θ are defined over k_0.

Introduction

Let G be a connected reductive group defined over an algebraically closed field k of characteristic not 2 and let $B \subseteq G$ be a Borel subgroup. A closed subgroup $K \subseteq G$ is called *spherical* if the number of K-orbits in the flag variety G/B is finite; equivalently, the set of (K, B)-double cosets $K\backslash G/B$ is finite. If H and K are spherical subgroups, we may also consider the double coset space $H\backslash G/K$. Results on double coset decompositions are important in representation theory (see, e.g., [**1, 5, 6, 18, 23**]), for the orbit method of Kirillov and Kostant [**17**] and in the study of Harish Chandra modules (see [**24**]). In the later case the group K is a

2000 *Mathematics Subject Classification.* 14M15, 20G05.
Author is partially supported by N.S.F. Grant DMS-9977392.
This is the final form of the paper.

symmetric subgroup of G, that is, K consists of all fixed points of an involutive automorphism θ of G, and $H = P$ a parabolic subgroup of G.

These double cosets have been studied for several types of spherical subgroups, but best known are the case that both groups are parabolic subgroups or the case that one or both of the subgroups is a symmetric subgroup. One usually distinguishes the finite from the infinite case. Results in the infinite case include a detailed description of these double cosets and their invariants in the case that both H and K are symmetric subgroups (see [**13, 15**]), as well as some results for general spherical subgroups [**14**].

Several of the finite cases have been studied in detail as well. In the finite case a lot of the geometry related to these double cosets, for example their closures, can be described by combinatorial properties of the set of double cosets. The best example of this is the case when $H = P$ is a parabolic subgroup and $K = B$ is a Borel subgroup. Then by the Bruhat decomposition, each (B, P)-double coset intersects the Weyl group W of a maximal torus into a unique coset of W_P, the Weyl group of P. This decomposition is one of the fundamental tools in the study of algebraic groups and their representations and in this case the double cosets and their closures can be completely described in terms of the Bruhat order of the Weyl group. Another important case is the case that K is a symmetric subgroup and $H = B$ is a Borel subgroup. In [**21**] Springer gave several characterizations of these double cosets, including one similar to the Bruhat decomposition of the group. In this case also much is know about the closures of the (K, B)-double cosets in G. For example, Richardson and Springer [**20**] gave a combinatorial description of a Bruhat type order on, which enables one to compute the double cosets contained in a closure of the (K, B)-double coset in G using combinatorial methods (see [**9**]).

When $H = P$ a parabolic subgroup and K a symmetric subgroup the set $K\backslash G/P$ of (K, P)-double cosets in G was studied extensively in [**2**]. Again there are several descriptions of the double cosets, similar in nature to the previous case where $H = B$ is a Borel subgroup. However, for general parabolic subgroups P the combinatorics of the decomposition of the closure of a double coset in terms of $K \times P$ double cosets is much more complicated than in the case that $P = B$ a Borel subgroup. In some special cases one can give a more detailed description of these double cosets. In this paper we consider the special case that P contains a θ-stable Levi factor L and the set of roots of the connected center S of L is a root system with Weyl group $W(S) = N_G(S)/Z_G(S)$. This root system structure can be used to obtain a detailed description of the combinatorics of the closures of the double cosets, which can then be used to describe a short resolution of singularities of closures of double cosets (see Remark 4). In this paper we will give some of these new results on the combinatorics of these orbits. There are many classes of parabolic subgroups for which there is a natural root system and Weyl group associated with the Levi subgroups of a parabolic subgroup. Examples are the case that P is a minimal parabolic k_0-subgroup of G or a minimal θ-split parabolic subgroup of G or a minimal (θ, k_0)-split parabolic k_0-subgroup of G (see [**12, 16, 19**]). Here $k_0 \subseteq k$ is a subfield of k and G, θ are defined over k_0.

A brief summary of the contents is as follows. In Section 1 we first introduce the notation and briefly review the description of the double cosets $K\backslash G/P$ from [**2**]. After that we restrict ourselves to the case that P contains a θ-stable Levi factor L and give a characterization of the double cosets $V^S := K\backslash G^P/P$, where

G^P is the set of all $g \in G$ such that gPg^{-1} contains a θ-stable Levi subgroup. The set V^S contains a significant part of the $K \times P$ double cosets in G, including the open and closed orbits. Moreover in the case that G, K are defined over a subfield $k_0 \subseteq k$ of k and P is a minimal parabolic k_0-subgroup of G the set V^S contains all the double cosets KgP with $g \in G(k_0)$. Here $G(k_0)$ is the set of k_0-rational points of G. Let $S = Z(L)^0$ be the connected center of L; let $\Phi(S) = \Phi(G, S)$ be the set of the roots of S in G, let $N_G(S)$ (resp. $Z_G(S)$) be the normalizer (resp. centralizer) of S in G and let $W(S) = W(G, S) = N_G(S)/Z_G(S)$ the Weyl group of S. In the case that $\Phi(S)$ is a root system with Weyl group $W(S)$ one has a natural map from V^S into the set $\mathcal{I} = \mathcal{I}_\theta = \{w \in W(S) \mid \theta(w) = w^{-1}\}$ of twisted involutions of the Weyl group $W(S)$. In the remainder of Section 1 we analyze this map and in Section 2 we analyze the combinatorics of the root system related to the set of orbits V^S. Let $\Phi^+ = \Phi(P, S)$ be the set of positive roots of $\Phi(S)$ related to P and Δ the corresponding basis of $\Phi(S)$. By a result of Steinberg [**22**] any Borel subgroup admits a θ-stable conjugate, hence if $P = B$ one may assume that Δ is θ-stable. In [**2**] it was shown that this does not extend to arbitrary parabolic subgroups. In particular G does not need to contain a θ-stable conjugate of P and so we cannot assume that Δ is θ-stable. This condition was used by Springer [**21**] in the case that $P = B$ is a Borel subgroup and this enables one to give a detailed description of the twisted involutions in the Weyl group. To enable us to use this characterization of the twisted involutions we switch to an involution $\theta' = \theta w_0$, which stabilizes Δ. Here $w_0 \in W(S)$. This involution is conjugate to θ if and only if the subset G^P/P of G/P contains a θ-stable conjugate of P. In Proposition 9 it is shown that the set G^P/P actually contains a θ-stable conjugate of P if $\Phi(S)$ is a root system with Weyl group $W(S)$. In the remainder of this section we analyze the relation between the orbit decompositions for the involutions θ' and θ and use these results to obtain some dimension formulas.

In Section 3 we show that the open and closed orbits are contained in V^S and give several characterizations of these orbits using the combinatorial structure developed in Section 2. Finally in Section 4 we study the decomposition of the closures of the double cosets KgP with $g \in G^P$ in terms of $K \times P$ double cosets and give a combinatorial description of these orbit closures using the combinatorial description of the related twisted involution in the Weyl group.

The author would like to thank Michel Brion and the referee for their very useful comments.

1. Preliminaries and recollection of results on double cosets

In this section, we recall some results about the double cosets and discuss some properties of the parabolic subgroups containing a θ-stable Levi factor.

1.1. Notation and recollection. Throughout the paper, the ground field k is algebraically closed of characteristic $\neq 2$. We denote by G a connected reductive group, by B a Borel subgroup of G, and by T a maximal torus of B. The unipotent part of B is denoted by U. We denote by P a parabolic subgroup of G containing B, and by L the Levi subgroup of P which contains T. If S is a subset of G and H a closed subgroup of G, then we write $N_H(S)$ (resp. $Z_H(S)$) for the normalizer (resp. centralizer) of S in H and we write $\mathrm{Aut}(G, S)$ for the set of automorphisms $\phi \in \mathrm{Aut}(G)$ satisfying $\phi(S) \subset S$. We write $Z(G)$ for the center of G. The commutator subgroup of G is denoted by $D(G)$ or $[G, G]$.

For a torus A of G we denote by $X^*(A)$ (resp. $X_*(A)$) the group of characters of A (resp. one-parameter subgroups of A) and by $\Phi(A) = \Phi(G, A)$ the set of the roots of A in G. If $\Phi(A)$ is a root system, then we let $\Delta(A)$ denote a set of simple roots of $\Phi(A)$. For a closed subgroup H of G we denote the Weyl group of H relative to A by $W_H(A) = N_H(A)/Z_H(A)$. If $H = G$, then we will also write $W(A) = W(G, A) = N_G(A)/Z_G(A)$.

For each $\alpha \in \Phi(T)$, let $U_\alpha \subset G$ be the corresponding root subgroup. If $\Delta(T)$ is the basis of $\Phi(T)$ corresponding to B, then each simple root $\alpha \in \Delta(T)$ defines a parabolic subgroup P_α of semisimple rank one, generated by B and $U_{-\alpha}$. We denote by L_α the Levi subgroup of P_α which contains T, and by G_α the quotient of L_α by its center; then G_α is isomorphic to PSL(2). We shall identify U_α and $U_{-\alpha}$ with their images in G_α, and we denote by T_α the image of T. Any parabolic subgroup $P \supseteq B$ is generated by the P_α's that it contains. We write $P = P_\Pi$ where Π is the set of all $\alpha \in \Delta$ such that $P_\alpha \subseteq P$. We denote by Φ_Π the sub-root system of Φ generated by Π, and by W_Π its Weyl group.

Let $\mathfrak{g}$, $\mathfrak{b}$, $\mathfrak{t}, \ldots$ be the Lie algebras of G, B, $T, \ldots$. For a subgroup H of G will also write $L(H)$ for the Lie algebra of H. We have the decomposition $\mathfrak{g} = \mathfrak{t} \oplus \bigoplus_{\alpha \in \Phi(T)} \mathfrak{g}_\alpha$; for each $\alpha \in \Phi(T)$, we choose a non-zero root vector $X_\alpha \in \mathfrak{g}_\alpha$.

Let θ be an automorphism of order 2 of G; let $K \subseteq G$ be the fixed point subgroup. Then K is reductive by [**22**, §8]; let $K^0 = G^{\theta,0}$ be its connected component containing 1. For the θ-action on $\mathfrak{g}$, the fixed point subspace $\mathfrak{g}^\theta$ is the Lie algebra of K by [**3**, Corollary 9.2]. Given g, $x \in G$, the *twisted action* associated to θ is given by $(g, x) \mapsto g * x = g^{-1} x\, \theta(g)$. This is also called the twisted θ-action. Let Q denote the set given by $Q = \{g \in G \mid \theta(g) = g^{-1}\}$. Then Q is a finite union of orbits under the twisted θ-action and $Q^0 = \{g^{-1}\theta(g) \mid g \in G\}$ is the connected component of Q containing the identity. Let $\tau : G \to G$ be the map $g \mapsto g^{-1}\theta(g)$; observe that $\theta(x) = x^{-1}$ for all $x \in \tau(G)$ and $\tau(G) = Q^0$.

If $A \subset G$ is a torus and $\phi \in \mathrm{Aut}(G, A)$ an involution, then we write $A_\phi^+ = (A \cap G^\phi)^0$ and $A_\phi^- = \{x \in A \mid \phi(x) = x^{-1}\}^0$. It is easy to verify that the product map

$$\mu : A_\phi^+ \times A_\phi^- \to A, \quad \mu(t_1, t_2) = t_1 t_2$$

is a separable isogeny. In particular $A = A_\phi^+ A_\phi^-$ and $A_\phi^+ \cap A_\phi^-$ is a finite group. (In fact it is an elementary abelian 2-group.) If $\phi = \theta$ we reserve the notation A^+ and A^- for A_θ^+ and A_θ^- respectively. For other involutions of A, we shall keep the subscript. The involution ϕ of (G, A) induces an automorphism of $W(A)$, also denoted by ϕ, which is given by

$$\phi(w) = \phi \circ w \circ \phi, \quad w \in W(A).$$

The action of $\Phi(A)$ induced by ϕ will also be denoted by ϕ or $\phi|\Phi(A)$.

1.1.1. *Characterization of double cosets.* For $g \in G$, define an automorphism ζ_g of G by

$$\zeta_g := \mathrm{Int}(g^{-1}) \circ \theta \tag{1.1}$$

and similar as in [**2**] define the automorphism

$$\psi_g := \zeta_{g^{-1}\theta(g)} = \mathrm{Int}(g^{-1}) \circ \theta \circ \mathrm{Int}(g). \tag{1.2}$$

This is an involution of G. If $x \in G$, then $\theta(gxg^{-1}) = g\zeta_{g^{-1}\theta(g)}(x)g^{-1} = g\psi_g(x)g^{-1}$. Hence for a subgroup M of G we have the condition that

$$(gMg^{-1})^\theta = gM^{\zeta_{g^{-1}\theta(g)}}g^{-1} = gM^{\psi_g}g^{-1}.$$

So in particular, when g normalizes M we have

$$M^{\psi_g} = g^{-1}M^\theta g. \tag{1.3}$$

From now on we assume that T is a θ-stable maximal torus of G; then its normalizer N is θ-stable, too. Set $\mathcal{V} = \{g \in G \mid T \text{ is } \psi_g\text{-stable}\} = \{g \in G \mid g^{-1}\theta(g) \in N\}$. Then $\mathcal{V}$ is stable under left multiplication by K and right multiplication by N and by [**21**] and [**16**], any (K, B)-double coset in G meets $\mathcal{V}$, along a unique (K,T)-double coset. For parabolic subgroups we obtained in [**2**] a similar parametrization of the (K, P)-double cosets in G. Set $\mathcal{V}^P := \{g \in \mathcal{V} \mid KgB \text{ is open in } KgP\}$. Then by [**2**, Proposition 2] any (K, P)-double coset in G meets $\mathcal{V}^P$, along a unique (K,T)-double coset.

1.1.2. *θ-stable Levi subgroups.* In the following, we assume that P contains a θ-stable Levi subgroup L. By [**2**, Lemma 5] any θ-stable Levi subgroup of P is conjugate to L in $R_u(P)^\theta$. Let G^P be the set of all $g \in G$ such that gPg^{-1} contains a θ-stable Levi subgroup. Clearly, G^P is a union of (K, P)-double cosets, but it can happen that $G^P \subsetneq G$. In the case that $P = B$ we have $G^B = G$ (see [**22**]). Let $S = Z(L)^0$ denote the connected center of L, and $N_G(S)$ resp. $Z_G(S)$ the normalizer, resp. centralizer of S in G. Then $L = Z_G(S)$, $N_G(L) = N_G(S)$, and these groups are θ-stable. Let $\mathcal{V}^S = \{g \in G \mid g^{-1}\theta(g) \in N_G(S)\}$, a union of $(K, N_G(S))$-double cosets contained in G^P. Finally, let $\mathcal{V}^{S,P} = \mathcal{V}^S \cap \mathcal{V}^P$. Then by [**2**, Proposition 3] any (K, P)-double coset in G^P meets $\mathcal{V}^S$ along a unique (K, L)-double coset, which itself meets $\mathcal{V}^{S,P}$ along a unique (K,T)-double coset.

Set $V^S := K\backslash\mathcal{V}^S/L$; then we have $V^S = K\backslash G^P/P = K\backslash\mathcal{V}^{S,P}/T$. The action of $N_G(S)$ on $\mathcal{V}^S$ by right multiplication induces an action of the Weyl group $W(S) := N_G(S)/Z_G(S)$ on V^S. We denote this action of $W(S)$ on V^S by $(w, v) \to w \cdot v$ $(w \in W(S), v \in V^S)$. We interpret the orbit set $V^S/W(S)$ in terms of certain conjugacy classes of θ-stable tori, as follows.

Let $\mathcal{S}$ be the set of all conjugates of S by elements of G. This is an affine variety, isomorphic to $G/N_G(S)$, on which θ acts. Let $\mathcal{S}^\theta$ be the fixed point set of θ, then $\mathcal{S}^\theta$ is the set of conjugates of S by elements of $\mathcal{V}^S$. It is an affine variety, on which K acts by conjugation. The bijective map $\mathcal{V}^S/N_G(S) \to \mathcal{S}^\theta : gN_G(S) \mapsto gSg^{-1}$ is K-equivariant; thus, the induced map $\eta : V^S/W(S) \to \mathcal{S}^\theta/K$ is bijective. In the case that $P = B$ this was observed in [**20**, Proposition 2.7].

1.2. Orbits and twisted involutions. In the remainder of this section we assume that $\Phi(S)$ is a root system with Weyl group $W(S) = N_G(S)/Z_G(S)$. Similar as in the case of $K\backslash G/B$-double cosets there exists a natural map from the orbit set into the set of twisted involutions in the Weyl group $W(S)$. First we give a characterization of this Weyl group.

1.2.1. *The group $W(S)$.* Since S is θ-stable its centralizer $Z_G(S)$ is θ-stable as well. By [**22**, 7.5] $Z_G(S)$ contains a θ-stable maximal torus $T \supset S$. Let $X_0(S) = \{\chi \in X(T) \mid \chi(S) = e\}$ and $\Phi_0(S) = \Phi(T) \cap X_0(S)$. This is a closed subsystem of $\Phi(T)$. Denote the Weyl group of $\Phi_0(S)$ by $W_0(S)$ and identify it with the subgroup of $W(T)$ generated by the reflections s_α, $\alpha \in \Phi_0(S)$. Let $W_1 = \{w \in W(T) \mid w(X_0(S)) = X_0(S)\}$. We note that $W_0(S)$ is a normal subgroup of W_1. Put

$X = X(T)/X_0(S)$, π the natural projection from $X(T)$ to X and let $\overline{\Phi} = \pi\big(\Phi(T) - \Phi_0(S)\big)$ denote the set of *restricted roots of* Φ *relative to* $X_0(S)$. Using similar arguments as in [**11**, Proposition 4.11] we easily get the following characterization of $\Phi(S)$ and $W(S)$:

LEMMA 1. $\overline{\Phi} = \Phi(S)$ *and* $W(S) \cong W_1/W_0(S)$.

In this case we also have the following characterization of $\mathcal{V}^{S,P}$:

LEMMA 2. *Let* $v \in \mathcal{V}^P$. *Then the following are equivalent:*

(i) $v \in \mathcal{V}^{S,P}$.
(ii) S *is* ψ_v*-stable.*
(iii) $\psi_v\big(X_0(S)\big) = X_0(S)$.

1.2.2. *Twisted involutions.* Recall that an element $w \in W(S)$ is a *twisted involution* if $\theta(w) = w^{-1}$ (see [**21**, §3] or [**16**, §7]). Let $\mathcal{I} = \mathcal{I}_\theta = \{w \in W(S) \mid \theta(w) = w^{-1}\}$ be the set of twisted involutions in $W(S)$. If $v \in V^S$ with representative $x(v)$, then $\varphi(v) := \tau\big(x(v)\big)Z_G(S) \in W(S)$ is a twisted involution. The element $\varphi(v) \in \mathcal{I}_\theta$ is independent of the choice of representative $x(v) \in \mathcal{V}^S$ for v. So this defines a map $\varphi : V^S \to \mathcal{I}_\theta$. In the case that $P = B$ a Borel subgroup the map φ plays an important role in the study of the Bruhat order on V. For more details, see [**20**].

The Weyl group $W(S)$ acts also on $\mathcal{I}_\theta$. This action comes from the *twisted action* of $W(S)$ on (the set) $W(S)$, which is defined as follows: if $w, w_1 \in W(S)$, then $w * w_1 = ww_1\theta(w)^{-1}$. If $w_1 \in W(S)$, then $W(S) * w_1 = \{w * w_1 \mid w \in W(S)\}$ is the *twisted* $W(S)$*-orbit* of w_1. Now $\mathcal{I}_\theta$ is stable under the twisted action, so that we get a twisted action of $W(S)$ on $\mathcal{I}_\theta$. The image of φ in $\mathcal{I}_\theta$ is a union of twisted $W(S)$-orbits, as follows from the following result:

LEMMA 3. *Let* $w \in W(S)$ *and* $v \in V^S$. *Then* $\varphi(w \cdot v) = w * \varphi(v)$.

From this result it follows now that the map $\varphi : V^S \to \mathcal{I}_\theta$ is equivariant with respect to the action of $W(S)$ on V^S and the twisted action of $W(S)$ on $\mathcal{I}_\theta$. So there is a natural orbit map $\phi : V^S/W(S) \to \mathcal{I}_\theta/W(S)$. In the case that $P = B$ a Borel subgroup Richardson and Springer [**20**] showed that this map is in fact injective. Moreover in this case one can easily proves then the following properties of the maps φ and ϕ (see [**16**] and [**20**]).

PROPOSITION 1. *Let* G, φ *and* ϕ *be as above and assume* S *is a maximal torus. Then we have the following:*

(i) $\phi : V^S/W(S) \to \mathcal{I}_\theta/W(S)$ *is injective.*
(ii) *There is a bijection from* $\varphi(V^S)/W(S)$ *onto* $\mathcal{S}^\theta/K$.

REMARK 1. It is an open question if the above result holds when S not a maximal torus but still satisfies the condition that $\Phi(S)$ is a root system with Weyl group $W(S)$. The statement (ii) is immediate from (i). To prove (i) it suffices to show that if $\varphi(v_1) = \varphi(v_2) = \mathrm{id} \in W(S)$, then $v_1 = w \cdot v_2$ for some $w \in W(S)$.

2. Twisted involutions and dimension formulas

In this section we first recall and establish some results on twisted involutions in a Weyl group. In the last subsection we apply these results to obtain some dimension formulas.

2.1. Preliminaries about twisted involutions. For the remainder of this paper we assume that P is a parabolic subgroup of G containing a θ-stable Levi factor L, $S = Z(L)^0$ and that $\Phi(S)$ is a root system with Weyl group $W(S)$. To simplify notation we will write $X = X^*(S)$, $\Phi = \Phi(S)$ and $W = W(S)$. For $\alpha \in \Phi$ denote the corresponding reflection in W by s_α. Let $\Phi^+ = \Phi(P, S)$ be the set of positive roots of Φ related to P, Δ the corresponding basis of Φ and $\Sigma = \{s_\alpha \mid \alpha \in \Delta\}$. The Weyl group W is generated by Σ. In the sequel, the length function ℓ and Bruhat order $\leq$ on W are defined relative to Σ.

For a subset Λ of Δ denote the subset of Φ consisting of the integral combinations of Λ by Φ_Λ. Then Φ_Λ is a closed subsystem of Φ with Weyl group W_Λ. We identify W_Λ with the finite subgroup $W(\Phi_\Lambda)$ of W generated by the s_α for $\alpha \in \Phi_\Lambda$. Let w_Λ^0 denote the longest element of W_Λ with respect to Λ.

2.1.1. The roots of Φ can be divided into three subsets, according to the action of θ, as follows:

(a) $\theta(\alpha) \neq \pm\alpha$. Then α is called *complex* (relative to θ).
(b) $\theta(\alpha) = -\alpha$. Then α is called *real* (relative to θ).
(c) $\theta(\alpha) = \alpha$. Then α is called *imaginary* (relative to θ).

These definitions of real, complex and imaginary roots carry over to the Weyl group as follows. Let $\mathcal{I} = \mathcal{I}_\theta = \{w \in W \mid \theta(w) = w^{-1}\}$ denote the set of twisted involutions. Given $w \in \mathcal{I}_\theta$, an element $\alpha \in \Phi$ is called *complex* (resp. *real, imaginary*) relative to w if $w\theta\alpha \neq \pm\alpha$ (resp. $w\theta\alpha = -\alpha$, $w\theta\alpha = \alpha$). We use the following notation:

$$C'(w,\theta) = \{\alpha \in \Phi^+ \mid -\alpha \neq w\theta\alpha < 0\}, \qquad R(w,\theta) = \{\alpha \in \Phi^+ \mid -\alpha = w\theta\alpha\},$$
$$C''(w,\theta) = \{\alpha \in \Phi^+ \mid \alpha \neq w\theta\alpha > 0\}, \qquad I(w,\theta) = \{\alpha \in \Phi^+ \mid \alpha = w\theta\alpha\}.$$

We will omit θ from this notation if there is no ambiguity as to which involution we consider. These sets are related to the following subsets of Φ^+. For $\sigma \in \mathrm{Aut}(X, \Phi)$ let $\Psi(\sigma)$ (resp. $\Psi^-(\sigma)$) denote the set given by $\Psi(\sigma) = \{\alpha \in \Phi^+ \mid \sigma^{-1}\alpha > 0\}$ (resp. $\Psi^-(\sigma) = \Phi^+ - \Psi(\sigma)$).

LEMMA 4. *Let $w \in \mathcal{I}_\theta$. Then we have the following conditions*:

(i) $\Psi(w\theta) = \{\alpha \in \Phi^+ \mid w\theta\alpha > 0\}$ *and* $\Psi^-(w\theta) = \{\alpha \in \Phi^+ \mid w\theta\alpha < 0\}$.
(ii) $\Psi^-(w\theta) = C'(w) \cup R(w)$ *and* $\Psi(w\theta) = C''(w) \cup I(w)$ *are disjoint unions.*
(iii) $\Psi^-(w\theta)$, $C'(w)$ *are* $-w\theta$ *stable and* $-w\theta \mid R(w) = 1$.
(iv) $\Psi(w\theta)$, $C''(w)$ *are* $w\theta$ *stable and* $w\theta \mid I(w) = 1$.

PROOF. Since $\theta(w) = w^{-1}$, $(w\theta)^2 = 1$ the first statement is clear. Since $\Psi(w\theta)$ is $w\theta$ stable and $\Psi^-(w\theta))$ is $-w\theta$ stable, the other statements are clear. □

2.2. Switch to an involution fixing Φ^+. The discussion on twisted involutions in [**21**] in the case that $P = B$ depends on the fact that $\theta(\Phi^+) = \Phi^+$. This is equivalent to the condition that $\theta(B) = B$. This in its turn is equivalent to the condition that the orbit KB is closed (see [**21**] or [**16**]). In general θ does not need to fix Φ^+, but we can modify the involution θ so that this is the case. Let $w_0 \in W$ be the unique element such that

$$\theta(\Phi^+) = w_0(\Phi^+) \tag{2.1}$$

and let $\theta' = \theta w_0$. Then w_0 and θ' satisfy the following conditions:

PROPOSITION 2. *Let Φ, Φ^+, θ, w_0 and θ' be as above, $w \in \mathcal{I}_\theta$ and $w' = ww_0$. Then we have the following properties:*

(i) $w_0 \in \mathcal{I}_\theta$ *and* $\theta w_0 = w_0^{-1}\theta$.
(ii) $\theta'(\Phi^+) = \Phi^+$.
(iii) θ' *is an involution of* Φ.
(iv) $\mathcal{I}_{\theta'} = \mathcal{I}_\theta \cdot w_0$.
(v) $w'\theta' = w\theta$.
(vi) $I(w,\theta) = I(w',\theta')$, $R(w,\theta) = R(w',\theta')$, $C'(w,\theta) = C'(w',\theta')$, $C''(w,\theta) = C''(w',\theta')$.

PROOF.

(i) Since $\theta(\Phi^+) = w_0(\Phi^+)$ it follows that $w_0^{-1}\theta(\Phi^+) = \Phi^+$ and consequently $\theta(w_0^{-1})(\Phi^+) = \theta w_0^{-1}\theta(\Phi^+) = \theta(\Phi^+) = w_0(\Phi^+)$. So $\theta(w_0^{-1}) = w_0$, which proves (i).
(ii) Follows from the fact that $\theta w_0 = w_0^{-1}\theta$.
(iii) From (i) it follows that $\theta(w_0) = \theta w_0 \theta = w_0^{-1}$, hence $(\theta')^2 = \theta w_0 \theta w_0 = \mathrm{id}$.
(iv) Let $w \in \mathcal{I}_{\theta'}$ and consider ww_0^{-1}. Since $\theta w_0 = w_0^{-1}\theta$ we have

$$\theta(ww_0^{-1}) = \theta w w_0^{-1}\theta = \theta w \theta w_0 = w_0 w_0^{-1}\theta w \theta w_0 = w_0\theta' w\theta' = w_0 w^{-1}.$$

So $\mathcal{I}_{\theta'} \subset \mathcal{I}_\theta \cdot w_0$. A similar argument shows the opposite inclusion, which proves the result.
(v) Let $w \in \mathcal{I}_\theta$ and $w' = ww_0$. Then $w'\theta' = ww_0\theta w_0 = ww_0\theta w_0\theta\theta = ww_0\theta(w_0)\theta = ww_0w_0^{-1}\theta = w\theta$, what proves the result.

Finally (vi) is immediate from (v).

□

2.2.1. *Relation between θ and θ'.* On the level of the root system and Weyl group we can now work with θ' and w' instead of θ and w. The question remains if θ' can be lifted to an involution of G, and if so, whether that involution is conjugate to θ and what is the relation between the orbits for θ and θ'. For $P = B$ these questions were studied in [**10**]. In the following we generalize this to the parabolic subgroups as in this section. First we address the question of lifting.

Let $w \in W(S)$ and $\xi = w\theta$. Then ξ is an involution of Φ if and only if $w \in \mathcal{I} = \mathcal{I}_\theta$. Moreover ξ can be lifted to an involution of G if and only if w has a representative $n \in N_G(S)$ satisfying $\theta(n)n \in Z(G)$. The question remains which twisted involutions in $w \in \mathcal{I}$ have such a representative and when this lifted involution is conjugate to θ. This comes down to twisted involutions in $\varphi(V^S)$:

LEMMA 5. *Let $\mathcal{V}^S$, V^S and $\varphi : V^S \to \mathcal{I}_\theta$ be as in 1.2.2 and let $w \in \mathcal{I}$. Then the following are equivalent:*

(i) *There exists a representative $n \in N_G(S)$ for w, such that $\xi = \mathrm{Int}(n)\theta$ is an involution of G conjugate to θ.*
(ii) $w \in \varphi(V^S) \subset \mathcal{I}$.

PROOF. Assume first that $\xi = \mathrm{Int}(n)\theta$ is an involution of G conjugate to θ. Let $g \in G$ such that $\mathrm{Int}(g)^{-1}\theta\,\mathrm{Int}(g) = \mathrm{Int}(n)\theta$. Then $g^{-1}\theta(g) = nz$ for some $z \in Z(G)$. It follows that $g \in \mathcal{V}^S$ and $w \in \varphi(V^S)$.

Conversely if $w \in \varphi(V^S)$, take $v \in V^S$ with $w = \varphi(v)$, let $x = x(v) \in \mathcal{V}^S$ be a representative and set $n = x^{-1}\theta(x) \in N_G(S)$; then n represents w, and $\xi = \mathrm{Int}(n)\theta = \mathrm{Int}(x)\theta\,\mathrm{Int}(x^{-1}) = \psi_x$. So the involution ξ is conjugate to θ. □

The question whether θ' as in (2.1) can be lifted to an involution conjugate to θ reduces now to the question if $w_0 \in \varphi(V^S)$. The latter condition is equivalent to the following:

PROPOSITION 3. *Let $\mathcal{V}^S$, V^S, $\varphi : V^S \to \mathcal{I}$ and w_0 be as above. Then $w_0 \in \varphi(V^S)$ if and only if there exists $g \in G^P$ such that gPg^{-1} is θ-stable.*

PROOF. Assume first $w_0 \in \varphi(V^S)$. Let $v_0 \in V^S$ be such that $\varphi(v_0) = w_0$ and let $x_0 = x(v_0) \in \mathcal{V}^S$ be a representative of v_0. Then $\tau(x_0)$ is a representative of w_0 in $N_G(S)$. Since $\theta(\Phi^+) = w_0(\Phi^+)$, we have $\theta(P) = \tau(x_0)P\tau(x_0^{-1})$. But then $P_1 = \theta(x_0^{-1})P\theta(x_0)$ is a θ-stable parabolic subgroup of G.

Conversely, assume $P_0 \subset G$ is a θ-stable conjugate of P. By [**2**, Proposition 3] there exists $x \in \mathcal{V}^S$ such that $P_0 = xPx^{-1}$. Since $\theta(P_0) = P_0$ it follows that

$$\theta(P) = \theta(x)^{-1}xPx^{-1}\theta(x) = \tau\big(\theta(x)^{-1}\big)P\tau\big(\theta(x)\big).$$

Now $\tau\big(\theta(x)^{-1}\big) \in N_G(S)$. Let $w \in W$ be the corresponding Weyl group element. Then $\theta(\Phi^+) = w(\Phi^+)$, so $w = w_0 \in \varphi(V^S)$. □

REMARK 2. For arbitrary parabolic subgroups G/P does not need to contain a θ-stable parabolic subgroup (see [**2**]).

In Proposition 9 we will show that $G^P/P \subset G/P$ contains a θ-stable conjugate of P. Since this result does not depend on the results in this section we can combine it with the above result to obtain the following:

COROLLARY 1. *Let w_0, θ' be as above. There exists a representative $n \in N_G(S)$ of w_0, such that $\xi = \mathrm{Int}(n)\theta$ is an involution of G conjugate to θ satisfying $\xi|\Phi = \theta'$.*

2.2.2. Since by Corollary 1 the involutions θ and θ' are conjugate, the orbit decompositions of the corresponding symmetric varieties under the action of a parabolic subgroup are similar. In the remainder of this section we discuss the relation between these double coset decompositions.

Let P, S, $\mathcal{V}^S$, V^S, $\mathcal{I}$ and $\varphi : V^S \to \mathcal{I}$ be as above. Take $w_0 \in W(S)$ such that $\theta\big(\Phi(S)^+\big) = w_0\big(\Phi(S)^+\big)$, $n_0 = x_0^{-1}\theta(x_0) \in N_G(S) \cap \tau(G)$ a representative of w_0^{-1}, $\theta' = \mathrm{Int}(n_0)\theta$ and $K' = G^{\theta'} = x_0 G^\theta x_0^{-1}$. Denote the actions of θ and θ' on $\Phi(S)$ also by θ and θ'. Then $\theta' = \theta w_0 = w_0^{-1}\theta$. As for θ let $\tau' : G \to G$ be the map defined by $\tau'(x) = x^{-1}\theta'(x)$, $\mathcal{V}^{S'} = \{x \in G \mid \tau'(x) \in N_G(S)\}$, $V^{S'}$ the set of $\big(Z_G(S) \times K'\big)$-orbits in $\mathcal{V}^{S'}$, $\mathcal{I}_{\theta'}$ the set of twisted involutions of $W(S)$ with respect to θ' and $\varphi' : V^{S'} \to \mathcal{I}_{\theta'}$ as in 1.2.2. We have the following relations between the sets $\mathcal{V}^S$, $\mathcal{V}^{S'}$, $\mathcal{I}$ and $\mathcal{I}_{\theta'}$.

LEMMA 6. *Let $\mathcal{V}^S$, $\mathcal{V}^{S'}$, $\mathcal{I}$, $\mathcal{I}_{\theta'}$, x_0, n_0 and w_0 be as above. Then we have the following*:

(i) $\mathcal{V}^{S'} = \mathcal{V}^S \cdot x_0^{-1}$.

(ii) $\tau'(\mathcal{V}^{S'}) = \tau(\mathcal{V}^S) \cdot n_0^{-1}$.

As in 1.2.2 there is also an action of the Weyl group $W(S)$ on $\mathcal{I}_{\theta'}$. Namely if $w \in W(S)$ and $a' \in \mathcal{I}_{\theta'}$, then define an action $w *' a' = wa'\theta'(w)^{-1}$. Since $\theta' = \theta w_0$ and $a' = aw_0$ for some $a \in \mathcal{I}_\theta$, we get

$$\begin{aligned} w *' a' &= waw_0\theta' w^{-1}\theta' = waw_0w_0^{-1}\theta w^{-1}\theta w_0 \\ &= wa\theta w^{-1}\theta w_0 = (w * a)w_0. \end{aligned} \tag{2.2}$$

This means that right translation by w_0 gives an isomorphism $\iota_{w_0} : \mathcal{I} \to \mathcal{I}_{\theta'}$, which is equivariant with respect to the actions $*$ resp. $*'$ of $W(S)$ on $\mathcal{I}$ resp. $\mathcal{I}_{\theta'}$. On the other hand, the map $g \to gx_0^{-1}$ from $\mathcal{V}^S$ to $\mathcal{V}^{S'}$ induces a map $\delta : V^S \to V^{S'}$. Then $\varphi' \circ \delta = \iota_{w_0} \circ \varphi$. We summarize the above analysis in the following result.

PROPOSITION 4. *Let $\mathcal{V}^S$, $\mathcal{V}^{S'}$, $\mathcal{I}$, $\mathcal{I}_{\theta'}$, n_0 and w_0 be as above. Then we have the following*:

(i) *The map $\iota_{w_0} : \mathcal{I} \to \mathcal{I}_{\theta'}$ induces an isomorphism between $\mathcal{I}/W(S)$ and $\mathcal{I}_{\theta'}/W(S)$.*
(ii) $V^{S'}/W(S') \simeq V^S/W(S) \simeq \varphi'(V^{S'})/W(S) \simeq \varphi(V^S)/W(S)$.

2.3. Dimension formulas. Using the description of the twisted involutions as in the previous subsections we can refine the description of the group $P^\theta = P \cap K$ as given in [**2**]. This leads to some dimension formulas which are needed for our discussion on closures of double cosets. First we need some notation. Given a quasi-closed subset ψ of $\Phi(S)$, we write U_ψ for the unipotent subgroup generated by the product of the U_α with $\alpha \in \psi$.

PROPOSITION 5. *Let $g \in G^P$, $M = gPg^{-1}$ and $v \in \mathcal{V}^S$ such that $M = vPv^{-1}$ (see 1.1.2). Let $n = v^{-1}\theta(v) \in N_G(S)$, w the image of n in $W(S)$, $\Psi(w\theta) = \Phi(S)^+ \cap w\theta\Phi(S)^+$ as in 2.1 and $\mathfrak{g}_w = \bigoplus_{\alpha \in C''(w)} \mathfrak{g}_\alpha$. Then we have the following*:

(i) $M^\theta \simeq Z_G(S)^{\psi_v} \ltimes U^{\psi_v}_{\Psi(w\theta)}$.
(ii) $\dim(M^\theta) = \dim\big(Z_G(S)^{\psi_v}\big) + \dim(U^{\psi_v}_{I(w)}) + \frac{1}{2}\dim(\mathfrak{g}_w)$.

PROOF.

(i) Since $M^\theta = v(P \cap n\theta(P)n^{-1})v^{-1}$ it follows that $M^\theta = v\big(Z_G(S) \ltimes U_{\Psi(w\theta)}\big)v^{-1}$. The groups $vZ_G(S)v^{-1}$ and $vU_{\Psi(w\theta)}v^{-1}$ are θ-stable. Hence by (1.3), $M^\theta = v\big(Z_G(S)^{\psi_v} \ltimes U^{\psi_v}_{\Psi(w\theta)}\big)v^{-1}$. Now (i) is obvious.
(ii) By Lemma 4 (i), $\Psi(w\theta) = I(w) \cup C''(w)$ is a disjoint union. The set $I(w)$ is a closed subset of $\Phi(S)^+$ and $L(U_{I(w)}) = \bigoplus_{\alpha \in I(w)} \mathfrak{g}_\alpha$. It follows that $L(U_{\Psi(w\theta)}) = L(U_{I(w)}) \oplus \mathfrak{g}_w$. Note that $\psi_v\alpha = w\theta\alpha$, $\alpha \in \Phi(S)$. Since $I(w)$ and $C''(w)$ are $w\theta$-stable by Lemma 4, $L(U_{\Psi(w\theta)})^{\psi_v} = L(U_{I(w)})^{\psi_v} \oplus \mathfrak{g}_w^{\psi_v}$. Then we have that $\dim(U^{\psi_v}_{\Psi(w\theta)}) = \dim(U^{\psi_v}_{I(w)}) + \dim(\mathfrak{g}_w^{\psi_v})$. It remains to show that $\dim(\mathfrak{g}_w^{\psi_v}) = \frac{1}{2}\dim(\mathfrak{g}_w)$. From the definition of $C''(w)$ and the fact that $\psi_v = w\theta$ on $\Phi(S)$, $\psi_v\alpha \neq \alpha$ for $\alpha \in C''(w)$. Hence there exists a subset J of $C''(w)$ such that $C''(w) = J \cup \psi_v(J)$ is a disjoint union. Set $V = \bigoplus_{\alpha \in J} \mathfrak{g}_\alpha$. Then $\mathfrak{g}_w = V \oplus \psi_v(V)$. It is now easy to see that $\mathfrak{g}_w^{\psi_v} \simeq V$ and $\dim(\mathfrak{g}_w^{\psi_v}) = \frac{1}{2}\dim(\mathfrak{g}_w)$.

□

COROLLARY 2. *Let $v \in \mathcal{V}^S$ and $v_1 = vm$ with $m \in N_G(S)$. Let $n = v^{-1}\theta(v)$, $n_1 = v_1^{-1}\theta(v_1) = m^{-1}n\theta(m)$ and w, w_1, w_2 the images of n, n_1, m in $W(S)$ respectively. If $w_2\big(I(w)\big) = I(w_1)$, then*

$$\dim\big((v_1Pv_1^{-1})^\theta\big) - \dim\big((vPv^{-1})^\theta\big) = \tfrac{1}{2}\big(\dim(\mathfrak{g}_{w_1}) - \dim(\mathfrak{g}_w)\big).$$

PROOF. From Proposition 5, we have the identities

$$\begin{aligned} \dim\big((v_1Pv_1^{-1})^\theta\big) &= \dim\big(Z_G(S)^{\psi_{v_1}}\big) + \dim(U_{I(w_1)}^{\psi_{v_1}}) + \tfrac{1}{2}\dim(\mathfrak{g}_{w_1}), \\ \dim\big((vPv^{-1})^\theta\big) &= \dim\big(Z_G(S)^{\psi_v}\big) + \dim(U_{I(w)}^{\psi_v}) + \tfrac{1}{2}\dim(\mathfrak{g}_w). \end{aligned} \tag{2.3}$$

On the other hand from (1.3) it follows that $Z_G(S)^{\psi_{v_1}} = m^{-1}v^{-1}\big(Z_G(S)^\theta\big)vm = m^{-1}\big(Z_G(S)^{\psi_v}\big)m$ and $U_{I(w_1)}^{\psi_{v_1}} = m^{-1}(U_{I(w)}^{\psi_v})m$. Since $w_2\big(I(w)\big) = I(w_1)$ we have $m^{-1}U_{I(w)}m = U_{I(w_1)}$, what proves the result. □

2.3.1. The group $\theta(P)$ is also a parabolic subgroup of G with $Z_G(S)$ as a Levi factor. Since $\Phi(S)$ is a root system with Weyl group $W(S)$, there exists $n_0 \in N_G(S)$ with $n_0Pn_0^{-1} = \theta(P)$. Let w_0 denote the image of n_0 in $W(S)$. By our choice, $\theta\big(\Phi(S)^+\big) = w_0\big(\Phi(S)^+\big)$. Similarly as in 2.2 set $\theta' = \theta w_0$ and for $w \in W(S)$ set $w' = ww_0$. For $\alpha \in \Phi(S)$ let P_α as in 3.2.1 be the parabolic subgroup of G containing P with $\Phi(P_\alpha, S) = \big(\mathbb{Z}\alpha \cap \Phi(S)\big) \cup \Phi(S)^+$. Then we have:

PROPOSITION 6. *Let $v \in \mathcal{V}^S$, $n = v^{-1}\theta(v)$, w the image of n in $W(S)$ and $w' = ww_0$. Let $\alpha \in \Delta(S)$, s_α in $W(S)$ the reflection defined by α, $n_\alpha \in N_G(S)$ a representative of s_α, $v_1 = vn_\alpha$ and $n_1 = v_1^{-1}\theta(v_1) = n_\alpha^{-1}n\theta(n_\alpha)$. If $\ell\big(s_\alpha w'\theta'(s_\alpha)\big) = \ell(w') + 2$, then we have the following*:

(i) $\dim\big((v_1Pv_1^{-1})^\theta\big) = \dim\big((vPv^{-1})^\theta\big) + \dim P - \dim P_\alpha$.
(ii) $\dim(P * n_1) = \dim(P * n) + \dim P_\alpha - \dim P$.

PROOF.

(i) By Proposition 2 (6) $w'\theta' = w\theta$, $I(w,\theta) = I(w',\theta')$ and $C''(w,\theta) = C''(w',\theta')$. Since $s_\alpha w'\theta'(s_\alpha)\theta' = s_\alpha w\theta(s_\alpha)\theta$ we also have $I\big(s_\alpha w\theta(s_\alpha),\theta\big) = I\big(s_\alpha w'\theta'(s_\alpha),\theta'\big)$, and $C''\big(s_\alpha w\theta(s_\alpha),\theta\big) = C''\big(s_\alpha w'\theta'(s_\alpha),\theta'\big)$. From [**16**, Lemma 7.8] it follows that

$$I\big(s_\alpha w\theta(s_\alpha)\big) = s_\alpha I(w), \tag{2.4}$$

$$C''\big(s_\alpha w\theta(s_\alpha)\big) = s_\alpha\big(C''(w) - Y\big), \tag{2.5}$$

where $Y = \Phi(S)^+ \cap (\mathbb{Z}\alpha \cup \mathbb{Z}w\theta\alpha)$. But then by (2.4) and Corollary 2,

$$\dim\big((v_1Pv_1^{-1})^\theta\big) - \dim\big((vPv^{-1})^\theta\big) = \tfrac{1}{2}\big(\dim(\mathfrak{g}_{s_\alpha w\theta(s_\alpha)}) - \dim(\mathfrak{g}_w)\big).$$

By (2.5) this number coincides with $\dim\big(\bigoplus_{\gamma\in\Phi(S)^+\cap\mathbb{Z}\alpha}\mathfrak{g}_\alpha\big)$. Now (i) follows from (3.1).

(ii) By (i), we have that $\dim\big((v_1Pv_1^{-1})^\theta\big) = \dim\big((vPv^{-1})^\theta\big) + \dim P - \dim P_\alpha$. Since $\dim(P*n) = \dim(P) - \dim\big((vPv^{-1})^\theta\big)$ and $\dim(P*n_1) = \dim(P) - \dim\big((v_1Pv_1^{-1})^\theta\big)$, the result follows.

□

3. Open and closed orbits

In this subsection we give a characterization of the open and closed orbits in V and show that they are actually contained in V^S. We first set the notation and prove some preliminary results.

3.1. Preliminaries and facets. Let P, L, S, etc. be as in Section 2. By chambers, facets of $X_*(S) \otimes_{\mathbb{Z}} \mathbb{R}$, we mean those with respect to the hyperplanes $\ker(\alpha)$, $\alpha \in \Phi(G,S)$. In the case that $\Phi(S)$ is a root system with Weyl group $W(S) = N_G(S)/Z_G(S)$ the parabolic subgroups of G which contain a parabolic subgroup with $Z_G(S)$ as a Levi factor are in bijective correspondence with the facets of $X_*(S) \otimes_{\mathbb{Z}} \mathbb{R}$. Given a facet F, the corresponding parabolic subgroup $P(F)$ of G containing S is determined by

$$\Phi\big(P(F), S\big) = \{\alpha \in \Phi(G,S) \mid \langle x, \alpha\rangle \geq 0, \quad x \in F\}.$$

For $\lambda \in X_*(S)$, let $F(\lambda)$ denote the facet containing λ. For simplicity, we write $P(\lambda)$ for the parabolic subgroup $P\big(F(\lambda)\big)$ of G containing S.

LEMMA 7. *Assume S is as above, $\Phi(S)$ is a root system with Weyl group $W(S)$, P a parabolic subgroup containing $Z_G(S)$, F the facet with $P = P(F)$ and $L = Z_G(F)$ the corresponding Levi factor of P. If P is θ-stable, then we have the following*:

(i) $\theta(F) = F$.
(ii) *There is $\lambda \in X_*(S^+)$ such that $P = P(\lambda)$ and $L = Z_G(\lambda)$.*

PROOF.

(i) If P is θ-stable, then $P(F) = \theta\big(P(F)\big) = P\big(\theta(F)\big)$. It follows that $\theta(F) = F$.
(ii) There exists $\mu \in X_*(S) \cap F$. By (i), $\theta(\mu) \in F$ and $\lambda = \mu + \theta(\mu) \in X_*(S^+) \cap F$. Then λ has the desired properties.

□

For θ-split parabolic subgroups of G we have a similar characterization. First we define the following:

DEFINITION 1. A parabolic subgroup P_1 of G will be called $(\theta, \mathcal{S}^\theta)$-split if P_1 is θ-split and there exists $S_1 \in \mathcal{S}^\theta$ such that $S_1 \subset P_1$ and $P_1 \cap \theta(P_1) = Z_G(S_1^-)$.

Similar as in [**16**, §4] we have the following result.

LEMMA 8. *Let P_1 be a $(\theta, \mathcal{S}^\theta)$-split parabolic subgroup of G and $S_1 \in \mathcal{S}^\theta$ such that $S_1 \subset P_1$. Then there exists $\lambda \in X_*(S_1^-)$ such that $P_1 = P(\lambda)$ and $P_1 \cap \theta(P_1) = Z_G(\lambda)$.*

PROOF. Since S_1 is θ-stable, $S_1 \subset \theta(P_1)$. Let F be the facet of $X_*(S_1) \otimes_{\mathbb{Z}} \mathbb{R}$ such that $P_1 = P(F)$. Note that $\theta(P_1) = P(-F)$. Hence $\theta(F) = -F$. Now choose $\tau \in X_*(S_1) \cap F$. Since τ, $-\theta(\tau) \in F$ and F is a convex cone, $\lambda = \tau - \theta(\tau) \in F$. Then λ has the desired property. □

3.2. θ-singular roots. In this subsection we prove some results about θ-singular roots which are needed for the description of the open and closed orbits.

3.2.1. Similarly as for Borel subgroups we can characterize the parabolic subgroups containing P by subsets of the roots in $\Phi(S)$. Let $\Phi(S) = \Phi(G,S)$ denote the set of roots of S in G, $\Phi(S)^+ = \Phi(P,S)$ and $\Delta(S)$ the corresponding basis. For $\lambda \in \Phi(S)$, let $\mathfrak{g}_\lambda$ be the root subspace of the Lie algebra $\mathfrak{g}$ of G corresponding to λ. We have the decomposition $\mathfrak{g} = L\big(Z_G(S)\big) \bigoplus_{\lambda \in \Phi(S)} \mathfrak{g}_\lambda$. Given $\lambda \in \Delta$, let P_λ denote the parabolic subgroup of G containing P such that $\Phi(P_\lambda, S) = \big(\mathbb{Z}\lambda \cap \Phi(S)\big) \cup \Phi(S)^+$. It is easy to see that

$$\dim(P_\lambda) = \dim(P) + \dim\left(\bigoplus_{\beta\in\mathbb{Z}\lambda\cap\Phi(S)^+} \mathfrak{g}_\beta\right). \tag{3.1}$$

If $\Lambda \subset \Delta(S)$, then similar as in 2.1 we denote by $\Phi(S)_\Lambda$ the subsystem of $\Phi(S)$ consisting of integral combinations of Λ and by P_Λ the parabolic subgroup of G containing P generated by all the P_λ with $\lambda \in \Lambda$. Note that $\Phi(P_\Lambda, S) = \Phi(S)_\Lambda \cup \Phi(S)^+$. We denote by L_Λ the Levi subgroup of P_Λ which contains S, by G_Λ the quotient of L_Λ by its center, and by S_Λ the image of S. In the case that $\Lambda = \{\lambda\}$ we will also write L_λ for L_Λ, G_λ for G_Λ and S_λ for S_Λ. We shall identify U_λ and $U_{-\lambda}$ with their images in G_λ.

DEFINITION 2. A root $\lambda \in \Phi(S)$ with $\theta(\lambda) = \pm\lambda$ is called *θ-singular* (resp. *θ-compact*) if $G_\lambda \not\subset K$ (resp. $G_\lambda \subset K$). A root $\lambda \in \Phi$ with $\theta(\lambda) = \lambda$ is called *θ-compact imaginary* if $G_\lambda \subset K$. A root $\lambda \in \Phi$ with $\theta(\lambda) = \lambda$ and $G_\lambda \not\subset K$ is called *imaginary θ-singular*. This definition is a refinement of 2.1.1.

In the case that S is a maximal torus of G all real roots are θ-singular (see [**7**] or [**21**]). For S not a maximal torus we can show the following:

PROPOSITION 7. *Let $S \in \mathcal{S}^\theta$. Then we have the following*:

(i) *If $\lambda \in \Phi(S)$ is real, then there exist a $x \in G_\lambda$ such that $S_1 = xSx^{-1}$ is θ-stable and* $\dim S^- > \dim S_1^-$.
(ii) *If $\lambda \in \Phi(S)$ is θ-singular imaginary, then there exist a $x \in G_\lambda$ such that $S_1 = xSx^{-1}$ is θ-stable and* $\dim S^- < \dim S_1^-$.

PROOF. We may assume that $G = G_\lambda$ and S is θ-split. Let $T \supset S$ be a θ-stable maximal torus with T^+ a maximal torus of $Z_G(S) \cap K$. If T^+ is not a maximal torus of K, then there exists $\alpha \in \Phi(T)$, $\theta(\alpha) = -\alpha$. If $\alpha \in \Phi(S) \cap \Phi(T)$, then by [**8**, 4.6] α is θ-singular and there exists $g \in Z_G\big((\ker\alpha)^0\big)$ such that $T_1 = gTg^{-1}$ is θ-stable with $\dim T_1^+ = \dim T^+ + 1$. Then also $S_1 = gSg^{-1}$ is θ-stable and $\dim S_1^+ = \dim S^+ + 1$, what proves the result. So assume $\alpha \notin \Phi(S)$. Since T^+ is maximal in $Z_G(S) \cap K$, we get $\alpha|S \neq 0$, so $\pi(\alpha) \in \Phi(S)$. Moreover reducing to $Z_G(T^+)$ we may assume $T = T^-$ is θ-split. Since $\alpha \notin \Phi(S)$ there exists $\beta \in \Phi(T)$ with $\beta \neq \pm\alpha$. Then also $\beta|S \neq 0$, so $\pi(\alpha) \in \Phi(S)$. Since $\Phi(S) = \{\pm\lambda, \pm 2\lambda\}$ or $\Phi(S) = \{\pm\lambda, \pm\frac{1}{2}\lambda\}$ it follows that α and β are perpendicular and span a root system of type $A_1 + A_1$. Repeating the same argument it follows that $\Phi(T)$ is of type $A_1 + \cdots + A_1$ (n copies of A_1). Let $\alpha_1, \ldots, \alpha_n$ be a basis of $\Phi(T)$. Since $\alpha_1, \ldots, \alpha_n$ are real roots, it follows from [**8**, 4.6] that they are θ-singular strongly orthogonal roots and there exists $g \in G$ such that $T_1 = gTg^{-1}$ is θ-stable with $\dim T_1^+ = \dim T^+ + n$. Since $gSg^{-1} \subset T_1^+$ the result follows.

Assume next that T^+ is a maximal torus of K. Let $w_0 \in \Phi(T)$ be a maximal θ-singular involution of $W(T)$ (see [**8**]). Then $T_{w_0}^- \subset T^+$ and $T_{w_0}^+ \supset S$. We may restrict to $Z_G(T_{w_0}^-)$. So assume that $\Phi(T)$ has no θ-singular roots. Then the θ-index of (G, θ) must be one of the following:

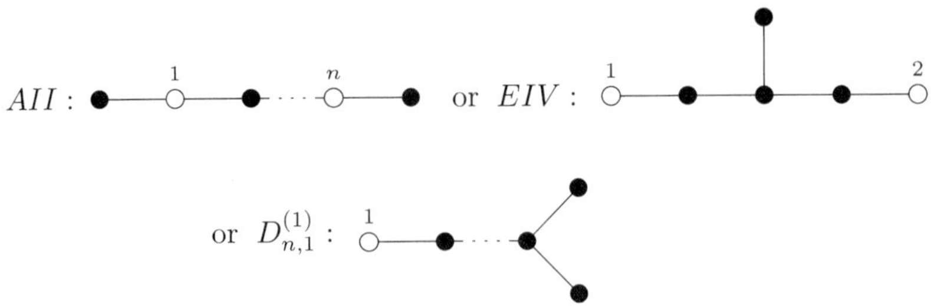

In each of these cases one easily shows that there exists $w_1 \in W(T_1)$ such that $w_1(T_1^-) \subset T_1^+$. Let $n \in N_G(T_1)$ be a representative of w_1. Then, since $S \subset T_1^-$, we get $nSn^{-1} \subset nT_1^- n^{-1} \subset T_1^+$, what proves (i).

(ii) follows with a similar argument as in (i).

□

DEFINITION 3. Let $S \in \mathcal{S}^\theta$. We will say that S^+ is maximal if $\Phi(S)$ has no real roots. Similarly we will say that S^- is maximal if $\Phi(S)$ has no θ-singular imaginary roots.

3.3. Closed orbits. In [**2**, Proposition 8] we gave a description of the closed (K, P)-orbits in G. Whether an orbit KgP is closed in G can also be seen from the order induced by P on the restricted root system $\Phi(S)$. For this we need to recall the notion of θ-order from [**8**]. In the following let θ be an involution in $\operatorname{Aut}(X, \Phi)$. Let $X_0(-\theta) = \{\chi \in X \mid \theta(\chi) = -\chi\}$, $\Phi_0(-\theta) = \Phi \cap X_0(-\theta)$ and π^- the natural projection from X to $X/X_0(-\theta)$. As in [**8**] we define a θ^--order on Φ by choosing orders on $X_0(-\theta)$ and $X/X_0(-\theta)$:

DEFINITION 4. Let $\succ$ be a linear order on X. The order $\succ$ is called a θ^--order if it has the following property:

$$\text{if } \chi \in X,\ \chi \succ 0, \quad \text{and} \quad \chi \notin X_0(-\theta), \text{ then } \theta(\chi) \succ 0. \tag{3.2}$$

A basis Δ of Φ with respect to a θ^--order on X will be called a θ^--basis of Φ. If Δ is a basis of Φ with respect to a θ^--order on X, then we write $\Delta_0(-\theta) = \Delta \cap \Phi_0(-\theta)$ and $\bar{\Delta}_{-\theta} = \pi^-\big(\Delta - \Delta_0(-\theta)\big)$. Clearly $\Delta_0(-\theta)$ is a basis of $\Phi_0(-\theta)$. A similar property holds for $\bar{\Delta}_{-\theta}$ (see [**7**, 2.4]).

3.3.1. *A characterization of θ on a θ^--basis of Φ.* Let Δ be a θ^--basis of Φ. Then $\theta = \theta^* \cdot w_0(-\theta)$, where $w_0(-\theta)$ is the longest element of $W\big(\Phi_0(-\theta)\big)$ with respect to $\Delta_0(-\theta)$ and $\theta^* \in \operatorname{Aut}\big(X, \Phi, \Delta, \Delta_0(-\theta)\big) = \{\phi \in \operatorname{Aut}(X, \Phi) \mid \phi(\Delta) = \Delta$ and $\phi\big(\Delta_0(-\theta)\big) = \Delta_0(-\theta)\}$, $(\theta^*)^2 = \mathrm{id}$. This is called a characterization of θ on its (-1)-eigenspace.

For closed orbits we have now the following:

PROPOSITION 8. *Let P be a parabolic subgroup containing a θ-stable Levi subgroup L of P. Let $S = Z(L)^0$ and assume $\Phi(S)$ is a root system with Weyl group $W(S)$. Then P is θ-stable if and only if*

(i) *$\Phi(S)$ has no real roots.*
(ii) *P induces a θ^--order on $\Phi(S)$.*

PROOF. Assume P is θ-stable. Let Δ be the basis of Φ corresponding to the order induced by P. If $\alpha \in \Phi(S)^+$ is a real root then $\theta(\alpha) = -\alpha \in \Phi(S)^-$, what contradicts the fact that P is θ-stable. Similarly, it follows from the fact that P is θ-stable that $\theta(\Phi^+) = \Phi^+$ and consequently also $\theta(\Delta) = \Delta$. The result now follows from 3.3.1.

Conversely assume (i) and (ii) hold. Let Δ be the basis of $\Phi(S)$ related to the order induced by P. Since $\Phi(S)$ has no real roots, it follows from 3.3.1 that $\theta(\Delta) = \Delta$, what proves the result. □

PROPOSITION 9. *Let P be a parabolic subgroup containing a θ-stable Levi subgroup L of P. Let $S = Z(L)^0$ and assume $\Phi(S)$ is a root system with Weyl group $W(S)$. Then $G^P/P \subset G/P$ contains a θ-stable conjugate of P.*

PROOF. Let L be a θ-stable Levi factor of P and $S = Z(L)^0$. Let $\Phi_1 = \{\alpha \in \Phi(S) \mid \theta(\alpha) = -\alpha\}$ be the set of real roots, $\Delta_1 \subset \Phi_1$ a basis and w_1 the longest involution of $W(\Phi_1)$ with respect to Δ_1. By [**8**, 2.7] there exist $\alpha_1, \ldots, \alpha_n \in \Phi_1$ strongly orthogonal roots, such that $w_1 = s_{\alpha_1}, \ldots, s_{\alpha_n}$. Since each of the α_i, $(i = 1, \ldots, n)$ is θ-singular, it follows from Proposition 7, with an easy induction, that there exists $x \in Z_G(S^-_{w_1})$ such that $S_1 = xSx^{-1}$ is θ-stable, $\dim S_1^- = \dim S - n$ and S_1 has no real roots. From Proposition 8 and Lemma 7 it follows that there exists $\lambda \in X_*(S_1)$ such that $P(\lambda)$ is θ-stable. □

REMARK 3. If P is θ-stable, then KP is closed in G and any closed orbit is of the form KxP with xPx^{-1} a θ-stable conjugate of P.

3.4. Open orbit. For $g \in \mathcal{V}^S$ let $\Delta_{c,g}(S)$ be the set of all ψ_g-compact simple roots. From Lemma 2 it follows that

$$\mathcal{V}^{S,P} = \{v \in G \mid S \text{ and } T \text{ are } \psi_v\text{-stable and } \Pi \subset \Delta_{c,v} \cup \psi_v(\Phi^-)\}.$$

Combined with [**2**, Proposition 5] we get:

COROLLARY 3. *Let $\mathcal{O}$ be the unique open orbit of (K, P) in G. Then the following are equivalent:*

(i) *$\mathcal{O} \subset G^P$.*
(ii) *There exists $v \in \mathcal{V}^{S,P}$ such that $\mathcal{O} = KvP$.*
(iii) *There exists $v \in \mathcal{V}^{S,P}$ such that $\Delta \subset \Delta_{c,v} \cup \psi_v(\Phi^-)$.*
(iv) *There exists $v \in \mathcal{O}$ such that S and T are ψ_v-stable and $\Delta \subset \Delta_{c,v} \cup \psi_v(\Phi^-)$.*

Using the restricted root system $\Phi(S)$ we can actually show that the open orbit is contained in G^P. For this we first characterize the conjugates of P containing a $S \in \mathcal{S}^\theta$ with S^- maximal. These are related to $(\theta, \mathcal{S}^\theta)$-split parabolic subgroups.

PROPOSITION 10. *Let P_1 be a $(\theta, \mathcal{S}^\theta)$-split parabolic subgroup of G and $S \in \mathcal{S}^\theta$ such that $S \subset P_1$ and $P_1 \cap \theta(P_1) = Z_G(S^-)$. Then the following are equivalent:*

(i) *P_1 is a minimal $(\theta, \mathcal{S}^\theta)$-split parabolic subgroup of G.*
(ii) *S^- is maximal.*

PROOF.

(i) ⇒ (ii) Assume S^- is not maximal. Let $\lambda \in \Phi(S)$ be a θ-singular imaginary root. Then $S^- \subset \ker(\lambda)^0$. Let $T \supset S$ be a θ-stable maximal torus with T^- maximal θ-split in $P_1 \cap \theta(P_1) = Z_G(S^-)$. From Proposition 7 it follows that there exist $x \in Z_G(T^-.\ker(\lambda)^0)$ such that $S_1 = xSx^{-1}$ is θ-stable with $\dim S_1^- = \dim S^- + 1$.

Since $Z_G(T^-.\ker(\lambda)^0) \subset Z_G(S^-)$ it follows that $S_1 \subset P_1 \cap \theta(P_1) = Z_G(S^-)$. Let δ be a nontrivial element of $X_*(S_1^-)$ and $P(\delta)$ the parabolic subgroup of G containing S_1 given by $\Phi(P(\delta), S_1) = \{\alpha \in \Phi(S_1) \mid \langle \delta, \alpha \rangle \geq 0\}$. Then $P(\delta)$ is a proper parabolic subgroup of P_1 and since $\theta(\delta) = -\delta$ it follows that $P(\delta)$ is $(\theta, \mathcal{S}^\theta)$-split. Clearly this is a contradiction. So S^- is maximal.

(ii) $\Rightarrow$ (i) Assume P_0 a $(\theta, \mathcal{S}^\theta)$-split parabolic subgroup with $P_0 \subset P_1$. Then $P_0 \cap \theta(P_0) \subset P_1 \cap \theta(P_1) = Z_G(S^-)$. By Lemma 8 there exists $\delta \in X_*(S^-)$ such that $P_0 = P(\delta)$ and $P_0 \cap \theta(P_0) = Z_G(\delta)$. Since clearly $Z_G(S^-) \subset Z_G(\delta)$ it follows that $P_0 \cap \theta(P_0) = P_1 \cap \theta(P_1) = Z_G(S^-)$. So $P_0 = P_1$. □

Proposition 11. *Let P_1 be a minimal $(\theta, \mathcal{S}^\theta)$-split parabolic subgroup of G containing a conjugate P_0 of P. Then $K^0 P_1 = K^0 P_0$ is open in G.*

Proof. Since $\Phi(Z_G(S^-), S)$ has no θ-singular roots it follows that $K^0 Z_G(S^-) = K^0 Z_G(S)$. Since $P_1 = Z_G(S^-) P_0$ we get $K^0 P_1 = K^0 P_0$, which by [**25**, Theorem 1.3.] is open. □

Proposition 12. *All minimal $(\theta, \mathcal{S}^\theta)$-split parabolic subgroups of G are conjugate under K^0.*

Proof. Let P_1, P_2 be minimal $(\theta, \mathcal{S}^\theta)$-split parabolic subgroups of G and let $P_0 \subset P_1$, $P_0' \subset P_2$ be parabolic subgroups of G conjugate to P. Let $g \in G$ such that $gP_0'g^{-1} = P_0$. By Proposition 11 both $K^0 P_0'$ and $K^0 P_0 = K^0 g P_0' g^{-1}$ are open in G. This yields that $K^0 P_0'$ and $K^0 g P_0' g^{-1}$ are the same open orbit of P_0' in $K^0 \backslash G$. Hence $g \in K^0 P_0'$. Write $g = h.p$, $h \in K^0$, $p \in P_0'$. Then $hP_0'h^{-1} = P_0$. So we may assume that $P_0 \subset P_1 \cap hP_2h^{-1}$.

Let $S_0 \in \mathcal{S}^\theta$ such that $S_0 \subset P_0$ and δ_1, $\delta_2 \in X_*(S_0^-)$ such that $P_1 = P(\lambda_1)$, $P(\lambda_2) = hP_2h^{-1}$. Since $P(\lambda_1) \cap P(\lambda_2) \supset P_0$ it follows with the same argument as in [**25**, 1.2. Prop. 4] that $P_1 \cap hP_2h^{-1} = P(\lambda_1 + \lambda_2)$. Clearly $P(\lambda_1 + \lambda_2)$ is a $(\theta, \mathcal{S}^\theta)$-split parabolic subgroup of G. By the minimality condition of P_1 we get $P_1 = P(\lambda_1 + \lambda_2) \subset hP_2h^{-1}$. By symmetry, we also have that $h^{-1}P_1h \subset P_2$. Thus $hP_2h^{-1} = P_1$. □

For the tori $S \in \mathcal{S}^\theta$ with S^- maximal we can prove a similar result:

Corollary 4. *Let $S_1, S_2 \in \mathcal{S}^\theta$ with S_1^- and S_2^- maximal. Then S_1 and S_2 are conjugate under K.*

Proof. Let $P_1 \supset S_1$ and $P_2 \supset S_2$ be minimal $(\theta, \mathcal{S}^\theta)$-split parabolic subgroups of G. By Proposition 12 There exists $h \in K^0$ such that $hP_2h^{-1} = P_1$. So we may assume $P_1 = P_2$. Let $P_0 \supset S_1$ and $P_0' \supset S_2$ be parabolic subgroups, conjugate to P, contained in P_1. Since $K^0 P_0 = K^0 P_0'$ open in G, it follows that P_0 and P_0' are K^0-conjugate. So we may assume $P_0 = P_0'$ contains S_1 and S_2. But then by [**2**, Lemma 5] S_1 and S_2 are conjugate under $K \cap R_u(P_0)$. □

We conclude this subsection by giving a characterization of the open orbit in terms of $\Phi(S)$ and $W(S)$. Let $\mathcal{O}$ be the unique open orbit of (K, P) in G and let $v \in \mathcal{V}^S$ such that $\mathcal{O} = KvP$. If $n = v^{-1}\theta(v)$, then the orbit $P * n$ is the unique open orbit of P in Q^0, which is also called the big cell. Similar as in 2.3.1 let w_0 be the element in the Weyl group $W(S)$ with $w_0(\Phi(S)^+) = \theta(\Phi(S)^+)$ and let $\theta' = \theta w_0$. For $w \in W(S)$ let $w' = ww_0$.

THEOREM 1. *Let* $v \in \mathcal{V}^S$, $n = v^{-1}\theta(v)$, *w the image of n in* $W(S)$, $w' = ww_0$ *and* $\Lambda = I(w) \cap \Delta(S)$. *The following conditions are equivalent*:

(i) $P * n$ *is open in* $\tau(G) = Q^0$.

(ii) $C''(w) \cap \Delta(S) = \emptyset$ *and* $\Lambda \subset \Delta_{c,v}(S)$.

(iii) $w' = w^0_\Lambda w^0_{\Delta(S)}$ *and* $\Lambda \subset \Delta_{c,v}(S)$.

(iv) $vP_\Lambda v^{-1}$ *is a minimal* $(\theta, \mathcal{S}^\theta)$*-split parabolic subgroup of G, containing P.*

(v) *There exists a minimal* $(\theta, \mathcal{S}^\theta)$*-split parabolic subgroup of G, containing* $v^{-1}Pv$.

PROOF.

(i) $\Rightarrow$ (ii) By Proposition 6 (ii), $C''(w) \cap \Delta(S) = \emptyset$. By Proposition 2 (vi) $w'\theta' = w\theta$. Since $C''(w) \cap \Delta(S) = \emptyset$ it follows from [**21**, Lemma 3.2 (ii)] that $s_\alpha w'\theta'(s_\alpha) = w'$ for all $\alpha \in \Delta(S)$ with $s_\alpha w' > w'$. Then from [**16**, Proposition 7.11] it follows that $w' = w^0_\Lambda w^0_{\Delta(S)}$. Observe that $vP_\Lambda v^{-1} \cap \theta(vP_\Lambda v^{-1}) = v(P_\Lambda \cap n\theta(P_\Lambda)n^{-1})v^{-1}$. From [**16**, Proposition 7.11] we get $\big(\Phi(S)_\Lambda \cup \Phi(S)^+\big) \cap w\theta\big(\Phi(S)_\Lambda \cup \Phi(S)^+\big) = \Phi(S)_\Lambda \cup \Phi(S)^+_\Lambda = \Phi(S)_\Lambda$. It follows that $vP_\Lambda v^{-1} \cap \theta(vP_\Lambda v^{-1}) = vG_{\Phi(S)_\Lambda} v^{-1}$.

Now let $P_0 = P \cap G_{\Phi(S)_\Lambda}$. A simple dimension argument shows that $P_0 * n$ is open in $G_{\Phi(S)_\Lambda} * n$. It follows that $P_0 G^{\psi_v}_{\Phi(S)_\Lambda}$ is open in $G_{\Phi(S)_\Lambda}$. However $\psi_v | \Phi(S)_\Lambda = 1$ and P_0 is ψ_v-stable. Now by [**2**, Proposition 8], $G_{\Phi(S)_\Lambda} = P_0 G^{\psi_v}_{\Phi(S)_\Lambda}$ and by [**16**, Lemma 1.8], $\Lambda \subset \Delta_{c,v}(S)$.

(ii) $\Rightarrow$ (iii) Using the same argument as in (i) $\Rightarrow$ (ii), the condition $C''(w) \cap \Delta(S) = \emptyset$ yields that $w' = w^0_\Lambda w^0_{\Delta(S)}$.

(iii) $\Rightarrow$ (iv) Using the same argument as in (i) $\Rightarrow$ (ii), the condition $w' = w^0_\Lambda w^0_{\Delta(S)}$ yields that $vP_\Lambda v^{-1} \cap \theta(vP_\Lambda v^{-1}) = vG_{\Phi(S)_\Lambda} v^{-1}$. Consequently $vP_\Lambda v^{-1}$ is a θ-split parabolic subgroup of G. Let $S_1 = vSv^{-1}$. Since $\Lambda \subset \Delta_{c,v}(S)$, $S^-_{\psi_v}$ is maximal with respect to ψ_v. But then S^-_1 is maximal with respect to θ. By Proposition 10, $vP_\Lambda v^{-1}$ is a minimal θ-split parabolic subgroup of G.

(iv) $\Rightarrow$ (v) is trivial.

(v) $\Rightarrow$ (vi) By Proposition 11, $KvPv^{-1}$ is open in G. Hence KvP is open in G and (i) is immediate. □

4. Orbit closures

In this section we study the decomposition of the closures in G of double cosets in G^P in terms of $K \times P$ double cosets. These results are very similar to those obtained in [**16**] for minimal parabolic k_0-subgroups acting on the symmetric k_0-variety G_{k_0}/H_{k_0}, where $k_0 \subseteq k$ is a subfield of k and G, θ are defined over k_0.

For $g \in G^P$, consider the double coset KgP. Let $v \in \mathcal{V}^S$ such that $KvP = KgP$, $n = v^{-1}\theta(v)$, w the image of n in $W(S)$, $\theta' = \theta w_0$ and $w' = ww_0$. Let $\Phi^+ = \Phi(P, S)$ be the set of positive roots related to P and let $\Delta(S)$ be the corresponding basis of Φ. By [**16**, Proposition 7.9] we can write

$$w' = s_{\alpha_1}, \dots, s_{\alpha_h} w^0_\Lambda \theta'(s_{\alpha_h}), \dots, \theta'(s_{\alpha_1}),$$

with Λ a θ'-stable subset of $\Delta(S)$ satisfying $w^0_\Lambda \theta' \alpha = -\alpha$ for $\alpha \in \Phi(S)_\Lambda$, $\alpha_1, \dots, \alpha_h \in \Delta(S)$ and $\ell(w') = 2h + \ell(w^0_\Lambda)$. Choose $n_1, \dots, n_h \in N_G(S)_k$ with images $s_{\alpha_1}, \dots, s_{\alpha_h}$ in $W(S)$ respectively. Set $u = n_h^{-1} \dots n_1^{-1} v$ and $m = u\theta(u)^{-1}$. First we look at P_Λ:

LEMMA 9. *Let* $v \in \mathcal{V}^S$, $n = v^{-1}\theta(v)$, *w the image of n in* $W(S)$ *and* $w' = ww_0$. *Let* $\Lambda = \Delta(S) \cap R(w) = \{\alpha \in \Delta(S) \mid w\theta\alpha = -\alpha\}$, P_Λ *the parabolic subgroup of G* *containing P defined by Λ and $P' = vP_\Lambda v^{-1}$. Assume $w' = w^0_\Lambda$. Then we have the following conditions:*

(i) *If $w' = w^0_\Lambda$, then the group $P' = vP_\Lambda v^{-1}$ is θ-stable.*
(ii) *$\tau(P')$ is a closed irreducible and smooth subvariety of $\tau(G) = Q^0$.*
(iii) *$\tau(vPv^{-1})$ is dense in $\tau(P')$.*

PROOF.

(i) First note that $w\theta = w'\theta'$. From the condition $w' = w^0_\Lambda$, $w'\theta'\big(\Phi(S)_\Lambda \cup \Phi(S)^+\big) = \Phi(S)_\Lambda \cup \Phi(S)^+$. By [**4**, 3.22], we have that $P' \cap \theta(P') = vP_\Psi v^{-1}$ with $\Psi = \big(\Phi(S)_\Lambda \cup \Phi(S)^+\big) \cap w\theta\big(\Phi(S)_\Lambda \cup \Phi(S)^+\big) = \Phi(S)_\Lambda \cup \Phi(S)^+$. It follows that P' is θ-stable.
(ii) Since P' is θ-stable, by [**2**, Proposition 8], KP' is a closed subvariety of G. It follows that $\tau(P')$ is a closed irreducible subvariety of Q. As $\tau(P')$ is homogeneous, it is smooth.
(iii) Note that $vG_{\Phi(S)_\Lambda}v^{-1}$ is θ-stable. By the assumption on Λ, $vG_{\Phi(S)^+_\Lambda}v^{-1}$ is a θ-split parabolic subgroup of $vG_{\Phi(S)_\Lambda}v^{-1}$. It follows from Proposition 11 that $KvG_{\Phi(S)^+_\Lambda}v^{-1}$ is dense in $KvG_{\Phi(S)_\Lambda}v^{-1}$. Hence $KvPv^{-1}$ is dense in KP'. Now the assertion is obvious.

□

For a subset Y of G denote the Zariski closure of Y in G by $\mathrm{cl}(Y)$.

THEOREM 2. *Let $v \in \mathcal{V}^S$, $n = v^{-1}\theta(v)$, w, etc. be as above. Then we have the following:*

(i) $\mathrm{cl}(P * n) = P_{\alpha_1} * \cdots * P_{\alpha_h} * P_\Lambda * m$.
(ii) $\dim(P * n) = \Sigma_{i=1}^h\big(\dim(P_{\alpha_i}) - \dim(P)\big) + \dim(P_\Lambda * m)$.

PROOF. We prove the assertion by induction on h, starting $h = 0$. When $h = 0$, the result is immediate from Lemma 9. Set $v_1 = n_1^{-1}v = n_2 \dots n_h u$. We may assume that

$$\mathrm{cl}\big(P * v_1^{-1}\theta(v_1)\big) = P_{\alpha_2} * \cdots * P_{\alpha_h} * P_\Lambda * m.$$

By Proposition 6 (ii), we have that

$$\dim(P * n) = \dim\big(P * v_1^{-1}\theta(v_1)\big) + \dim P_{\alpha_1} - \dim P. \tag{4.1}$$

Clearly $\dim\big(P_{\alpha_1} * P * v_1^{-1}\theta(v_1)\big) \le \dim\big(P * v_1^{-1}\theta(v_1)\big) + \dim P_{\alpha_1} - \dim P$. Hence we have the relation that $\dim\big(P_{\alpha_1} * P * v_1^{-1}\theta(v_1)\big) \le \dim(P * n)$. However $P_{\alpha_1} * P * v_1^{-1}\theta(v_1)$ contains $P * n$. This yields that

$$\dim\big(P_{\alpha_1} * P * v_1^{-1}\theta(v_1)\big) = \dim(P * n). \tag{4.2}$$

Since $P_{\alpha_1} * \mathrm{cl}\big(P * v_1^{-1}\theta(v_1)\big)$ is closed and irreducible, we conclude that $\mathrm{cl}(P * n) = \mathrm{cl}\big(P_{\alpha_1} * P * v_1^{-1}\theta(v_1)\big) = P_{\alpha_1} * \mathrm{cl}\big(P * v_1^{-1}\theta(v_1)\big) = P_{\alpha_1} * \cdots * P_{\alpha_h} * P_\Lambda * m$. Finally (ii) follows from (4.1) by an easy induction. □

When $P = B$ a Borel subgroup, the above result is due to Springer [**21**, Theorem 6.5]. In the case that G, θ are defined over a subfield k_0 of k and P is a minimal parabolic k_0-subgroup of G this result can be found in [**16**].

REMARK 4. Since $P_\Lambda * m$ closed in Q^0 as well it follows from Theorem 2 that the product map defines a proper morphism from $P_{\alpha_1} \times^P P_{\alpha_2} \times^P \cdots \times^P P_\Lambda * m$ to the closure of $P * m$. From Theorem 2 (ii) it follows that its general fibers are finite. One can show that this morphism is in fact birational and that we have a resolution of singularities of orbit closures. We intend to discuss this in more detail in future work.

References

1. J.-L. Brylinski and P. Delorme, *Vecteurs distributions H-invariants pour les séries principales généralisées d'espaces symétriques réductifs et prolongement méromorphe d'intégrales d'Eisenstein*, Invent. Math. **109** (1992), no. 3, 619–664.
2. M. Brion and A. G. Helminck, *On orbit closures of symmetric subgroups in flag varieties*, Canad. J. Math. **52** (2000), no. 2, 265–292.
3. A. Borel, *Linear algebraic groups*, 2nd enlarged edition ed., Graduate texts in mathematics, vol. 126, Springer Verlag, New York, 1991.
4. A. Borel and J. Tits, *Groupes réductifs*, Inst. Hautes Études Sci. Publ. Math. **27** (1965), 55–152.
5. P. Delorme, *Formule de Plancherel pour les espaces symétriques réductifs*, Ann. of Math. (2) **147** (1998), no. 2, 417–452.
6. M. Flensted-Jensen, *Discrete series for semisimple symmetric spaces*, Annals of Math. **111** (1980), 253–311.
7. A. G. Helminck, *Algebraic groups with a commuting pair of involutions and semisimple symmetric spaces*, Adv. in Math. **71** (1988), 21–91.
8. ———, *Tori invariant under an involutorial automorphism* I, Adv. in Math. **85** (1991), 1–38.
9. ———, *Computing B-orbits on G/H*, J. Symbolic Comput. **21** (1996), 169–209.
10. ———, *Tori invariant under an involutorial automorphism* II, Adv. in Math. **131** (1997), no. 1, 1–92.
11. ———, *On the classification of k-involutions* I, Adv. in Math. **153** (2000), no. 1, 1–117.
12. A. G. Helminck and G. F. Helminck, *A class of parabolic k-subgroups associated with symmetric k-varieties*, Trans. Amer. Math. Soc. **350** (1998), 4669–4691.
13. A. G. Helminck and G. W. Schwarz, *Orbits and invariants associated with a pair of commuting involutions*, Duke Math. J. **106** (2001), no. 2, 237–279.
14. ———, *Orbits and invariants associated with a pair of spherical varieties: some examples*, The 2000 Twente Conference on Lie Groups (Enschede), Acta Appl. Math. **73** (2002), no. 1-2, 103–113.
15. ———, *Smoothness of quotients associated with a pair of commuting involutions*, Canad. J. Math. (2003) (to appear).
16. A. G. Helminck and S. P. Wang, *On rationality properties of involutions of reductive groups*, Adv. in Math. **99** (1993), 26–96.
17. A. Kirillov, *The orbit method. I. Geometric quantization*, Representation theory of groups and algebras (Providence, RI), Contemp. Math., vol. 145, Amer. Math. Soc., Providence, RI, 1993, pp. 1–32.
18. T. Oshima and J. Sekiguchi, *Eigenspaces of invariant differential operators in an affine symmetric space*, Invent. Math. **57** (1980), 1–81.
19. R. W. Richardson, *Orbits, invariants and representations associated to involutions of reductive groups*, Invent. Math. **66** (1982), 287–312.
20. R. W. Richardson and T. A. Springer, *The Bruhat order on symmetric varieties*, Geom. Dedicata **35** (1990), no. 1-3, 389–436.
21. T. A. Springer, *Some results on algebraic groups with involutions*, Algebraic Groups and Related Topics, Adv. Stud. in Pure Math., vol. 6, Academic Press, Orlando, FL, 1984, pp. 525–543.
22. R. Steinberg, *Endomorphisms of linear algebraic groups*, Mem. Amer. Math. Soc., vol. 80, Amer. Math. Soc., Providence, RI, 1968.
23. E. P. van den Ban and H. Schlichtkrull, *The most continuous part of the Plancherel decomposition for a reductive symmetric space* I, Ann. Math. **145** (1997), 267–364.

24. D. A. Vogan, *Irreducible characters of semi-simple Lie groups* III. *Proof of the Kazhdan–Lusztig conjectures in the integral case*, Invent. Math. **71** (1983), 381–417.
25. T. Vust, *Opération de groupes reductifs dans un type de cônes presque homogènes*, Bull. Soc. Math. France **102** (1974), 317–334.

Department of Mathematics, North Carolina State University, Raleigh, NC 27695, USA.

E-mail address: loek@math.ncsu.edu

Centre de Recherches Mathématiques
CRM Proceedings and Lecture Notes
Volume **35**, 2004

Deformation of Symmetric Functions and the Rational Steenrod Algebra

F. Hivert and N. M. Thiéry

ABSTRACT. In 1999, Reg Wood conjectured that the quotient of $\mathbb{Q}[x_1, \dots, x_n]$ by the action of the rational Steenrod algebra is a graded regular representation of the symmetric group $\mathfrak{S}_n$. As pointed out by Reg Wood, the analog of this statement is a well known result when the rational Steenrod algebra is replaced by the ring of symmetric functions; actually, much more is known about the structure of the quotient in this case.

We introduce a non-commutative q-deformation of the ring of symmetric functions, which specializes at $q = 1$ to the rational Steenrod algebra. We use this formalism to obtain some partial results. Finally, we describe several conjectures based on an extensive computer exploration. In particular, we extend Reg Wood's conjecture to q formal and to any $q \in \mathbb{C}$ not of the form $-a/b$, with $a \in \{1, \dots, n\}$ and $b \in \mathbb{N}$.

1. Introduction

The *rational Steenrod algebra* is the subalgebra $\mathcal{A}$ of the Weyl algebra generated by the *rational Steenrod squares* $D_k := \sum_i x_i^{k+1}\partial_i$ [**18**]. The Steenrod squares are derivations, and satisfy the Lie relations $[D_k, D_l] = (l-k)D_{k+l}$. In particular, the Steenrod algebra is generated by just D_1 and D_2.

Let $\mathbb{C}[X_n] := \mathbb{C}[x_1, \dots, x_n]$ be a ring of polynomials, and set $\pi := x_1 \cdots x_n$. Consider the action of the Steenrod algebra on $\mathbb{C}[X_n]$ through the algebraic Thom map, so that for $f \in \mathcal{A}$ and $p \in \mathbb{C}[X_n]$,

$$\rho(f) \cdot p := \frac{1}{\pi} f \pi \cdot p. \tag{1.1}$$

This action seems to share properties with the natural action of the ring of symmetric polynomials on polynomials by multiplication.

CONJECTURE 1 (Rational hit conjecture of Reg Wood [**16, 17**]). The quotient $\mathbb{C}[X_n]_{/\mathcal{A}^+\mathbb{C}[X_n]}$ is a graded regular representation of the symmetric group $\mathfrak{S}_n$.

In this article, we present preliminary results from research in progress around this conjecture. We refer to [**16–18**] for motivations.

2000 *Mathematics Subject Classification.* Primary: 20G42; Secondary: 13N10, 55S10.
This is the final version of the paper.

It is actually easier to get rid of the algebraic Thom map, by conjugating once for all the Steenrod algebra by this map. So, we consider instead the isomorphic algebra denoted Steen generated by the Weyl operators

$$P_k := \frac{1}{\pi} D_k \pi = \sum_i x_i^k (1 + x_i \partial_i). \tag{1.2}$$

Note that those operators are no longer derivations.

Now, P_k appears clearly as a usual symmetric power-sum, with a deformation. Since the goal is to transfer properties of the ring of symmetric polynomials to the Steenrod algebra, this suggests that we interpolate continuously between the two. Hence, we introduce the *q-Steenrod algebra* denoted $\mathrm{Steen_q}$ which is generated by the Weyl operators

$$P_{q,k} := \sum_i x_i^k (1 + q x_i \partial_i). \tag{1.3}$$

The 0-Steenrod algebra is the ring of symmetric polynomials, while the 1-Steenrod algebra is the rational Steenrod algebra conjugated by the algebraic Thom map.

In the following, q can be either a given complex number q_0, or a formal parameter. In the later case we take $\mathbb{K} := \mathbb{C}(q)$ as base field, or $\mathbb{K} := \mathbb{C}[q]$, when we just need a base ring. Roughly speaking, all the statements we consider are of algebraic nature, and hold for q formal if, and only if, they hold for q generic (that is in most cases for all but a countable set).

1.1. Summary of the results. In [**4**], a concrete realization Harm of the quotient $\mathbb{C}[X_n]/_{\mathrm{Sym}^+ \mathbb{C}[X_n]}$ is constructed as a space of so-called *harmonics*. Following a similar approach, we construct a concrete realization $\mathrm{Harm_q}$ of the quotient $\mathbb{C}[X_n]/_{\mathrm{Steen}_q^+ \mathbb{C}[X_n]}$.

Both our theoretical results, and our computer exploration leads us to extend Reg Wood's conjecture as follows:

CONJECTURE 2. Let q be formal, or a complex number not of the form $-a/b$, with $a \in \{1, \ldots, n\}$, $b \in \mathbb{N}$ and $a \leq b$. Then, for any n, $\mathrm{Harm_q}$ is isomorphic to Harm as a graded $\mathfrak{S}_n$-module. In particular, it is isomorphic to the regular representation of $\mathfrak{S}_n$.

The main results of this article are summarized in the following two propositions. Recall that a monomial $x^K = x_1^{k_1} \cdots x_n^{k_n}$ in $\mathbb{K}[X_n]$ is called a *staircase monomial* if $k_i \leq n - i$, for all i.

PROPOSITION 1. *The following are equivalent:*

(a) *Conjecture* 2 *holds for* q *formal;*
(b) *For any* n, $\dim \mathrm{Harm_q} \geq \dim \mathrm{Harm}$;
(c1) *For any* n, *and any polynomial* $p \in \mathrm{Harm_q}$, *all the variables have degree at most* $n - 1$ *in* p;
(c2) *For any* n, *the staircase monomials are the leading terms of the polynomials in* $\mathrm{Harm_q}$ *for the lexicographic order;*
(d) *Any basis of* $\mathbb{C}[X_n]/_{\mathrm{Sym}^+ \mathbb{C}[X_n]}$ *is a basis of* $\mathbb{K}[X_n]/_{\mathrm{Steen}_q^+ \mathbb{K}[X_n]}$.

PROPOSITION 2. *Assume that conjecture* 2 *holds for* q *formal. Then, for* $q_0 \in \mathbb{C}$, *the following are equivalent:*

(a) *Conjecture* 2 *holds for* q_0;

(b) *For any* n, $\dim \mathrm{Harm}_q[q_0] \leq \dim \mathrm{Harm}$;
(c) *For any* n, *the non-staircase monomials are the leading terms of the polynomials in* Hit_q *for the lexicographic order.*

1.2. Organization of the paper. After a background section, we define the q-deformed Steenrod algebra Steen_q (Definition 5), together with q-hit and q-harmonic polynomials. We compute the dimension and the structure of Steen_q (Theorems 3.1 and 3.2).

In the Section 4, we describe the relations between the generic setting (when q is a formal parameter), the complex setting (when $q = q_0$ is a complex number) and the classical setting (when $q = 0$).

Then, we analyze the space of polynomials that can be hit from harmonics by successive applications of at most n Steenrod operators. This provides information in small degrees. In particular we show that the submodules of degree smaller than n of $\mathbb{K}[X_n]/\mathrm{Sym}^+$ and $\mathbb{K}[X_n]/\mathrm{Steen}_q^+$ are isomorphic (Corollary 1).

In the next section, we investigate the relations between the q-harmonics in n variables and in $n+1$ variables. We construct a trick to go from n to $n+1$: the so-called *strings*. The main result of this part is that the conjecture is equivalent to the fact that the harmonics in n-variables are of degree smaller than n in each variable (Proposition 17). We also analyze completely the simple cases of 1 and 2 variables.

In the next to last section we describe some attempts to prove conjecture 2 by looking for q-analogs of elementary symmetric function (Subsection 7.1) and q-analogs of Schubert polynomials (Subsection 7.2), and we explain why they fail.

Finally, the last section is devoted to a computer exploration which motivates our conjectures. In particular, we give, for small numbers of variables and small degrees, a tight superset of the values of q for which Conjecture 2 does not hold. Most computations where realized with the open source library `MuPAD-Combinat` [**11**] [1] for the Computer Algebra System `MuPAD` [**12**]. The extra tools we developed at this occasion are available upon request, and will eventually be integrated into `MuPAD-Combinat`, after appropriate cleanup and documentation.

2. Preliminaries

The main topic of this paper is the study of the q-deformation of a module. Hence we will work in two different settings: either $q := q_0$ is a complex number, in which case the base field is $\mathbb{K} := \mathbb{C}$, or q is a formal complex number, and $\mathbb{K} := \mathbb{C}(q)$. Many of the results are true in both settings; in this case, we use $\mathbb{K}$ as base field without precision.

For an alphabet X of commuting variables, we denote respectively by $\mathbb{K}[X]$, $\mathbb{K}(X)$, and $\mathrm{Sym}(X)$ the ring of polynomials, the field of fractions, and the ring of symmetric functions in the variables in X. We denote by X_n the alphabet of the n commuting variables $\{x_1, x_2, \ldots, x_n\}$. Unless explicitly stated, $X := X_\infty$ is the alphabet in the countably many variables $\{x_1, \ldots, x_n, \ldots\}$, and we use the shorthand $\mathrm{Sym} = \mathrm{Sym}(X_\infty)$.

2.1. Graded rings and Hilbert series. In the following, most of the vector spaces we consider are graded, with finite dimensional homogeneous components, and comparing their dimensions will be one of our main tools. The *Hilbert series*

[1] `http://mupad-combinat.sf.net`

of a graded vector space $A := \bigoplus_{d=0}^{\infty} A_d$ is the generating series of the dimensions of the homogeneous components of A:

$$\mathrm{Hilb}_t(A) := \sum_{d=0}^{\infty} \dim A_d t^d. \tag{2.1}$$

Given two formal series $H(t) := \sum_d h_d t^d$ and $L(t) := \sum_d l_d t^d$, we say that $H(t)$ is *dominated* by $L(t)$, and write $H(t) \leq L(t)$, whenever $h_d \leq l_d$ for all d. The relation of domination is preserved by multiplication by series with nonnegative coefficients, such as Hilbert series and generating functions.

Let us recall some basic Hilbert series.

PROPOSITION 3. *The Hilbert series of* $\mathbb{K}[X_n]$ *is*

$$\mathrm{Hilb}_t(\mathbb{K}[X_n]) = \frac{1}{(1-t)^n}. \tag{2.2}$$

The Hilbert series of the algebra $\mathrm{Sym} := \mathrm{Sym}(X_\infty)$ *of symmetric functions on countably many variables is*

$$\mathrm{Hilb}_t(\mathrm{Sym}) = \prod_{i>0} \frac{1}{1-t^i}. \tag{2.3}$$

The Hilbert series of the algebra $\mathrm{Sym}(X_n)$ *of symmetric polynomials on n variables is*

$$\mathrm{Hilb}_t\big(\mathrm{Sym}(X_n)\big) = \prod_{i=1}^{n} \frac{1}{1-t^i}. \tag{2.4}$$

2.2. Representations of the symmetric group on polynomials. The representation theory of the symmetric groups is a well known combinatorial subject. The goal of this subsection is to recall several basic useful facts. For more details, we refer the reader to [**3, 4, 13**] or, in French, to [**6**].

The irreducible representations V_λ of the symmetric group $\mathfrak{S}_n$ are naturally indexed by partitions $\lambda := (\lambda_1 \geq \lambda_2 \geq \cdots \geq \lambda_k > 0)$ of sum n (denoted by $\lambda \vdash n$). There is a particular occurrence of V_λ, called the Specht module, in the natural representation of $\mathfrak{S}_n$ on polynomials. This module is spanned by the so-called Specht polynomials. Recall that the partition λ is depicted (in the French way) as a diagram of boxes putting λ_i boxes in the i-th row, starting from the bottom. A filling of the boxes by the numbers $(1, 2 \ldots, n)$ is called a *standard filling* F of shape λ. A standard filling which is increasing from left to right along rows and from bottom to top along columns is called a *standard tableau.*

For example here is the diagram of the partition $\lambda := (5, 3, 2)$ together with a standard filling and a standard tableau of shape λ

8	2			
9	1	5		
10	4	3	7	6

6	9			
3	5	7		
1	2	4	8	10

Recall that the *Vandermonde determinant* of a finite alphabet of variables $\{y_1, \ldots, y_k\}$ is defined by

$$\Delta(y_1, \ldots, y_k) := \begin{vmatrix} 1 & y_1 & y_1^2 & \cdots & y_1^{k-1} \\ 1 & y_2 & y_2^2 & \cdots & y_2^{k-1} \\ \vdots & \vdots & \vdots & \ddots & \vdots \\ 1 & y_k & y_k^2 & \cdots & y_k^{k-1} \end{vmatrix} = \prod_{1 \leq i < j \leq k} (y_j - y_i). \tag{2.5}$$

For each standard filling F, the *Specht polynomial* Sp_F is defined as the product of the Vandermonde determinants of the variables indexed by the columns of F.

For example the Specht polynomial associated with the previous filling is

$$\begin{aligned} \mathrm{Sp}_F &:= \Delta(x_{10}, x_9, x_8) \cdot \Delta(x_4, x_1, x_2) \cdot \Delta(x_3, x_5) \cdot \Delta(x_7) \cdot \Delta(x_6) \\ &= (x_9 - x_{10})(x_8 - x_{10})(x_8 - x_9) \cdot (x_1 - x_4)(x_2 - x_4)(x_2 - x_1) \cdot (x_5 - x_3) \cdot 1 \cdot 1. \end{aligned}$$

It is easy to see that, for $\sigma \in \mathfrak{S}_n$, and F a filling, $\sigma \,\mathrm{Sp}_F = \mathrm{Sp}_{\sigma F}$. Hence, the span of $\{\mathrm{Sp}_F\}$, where F runs through the standard fillings of a given shape $\lambda \vdash n$ is a representation V_λ of $\mathfrak{S}_n$, which is called *Specht module* indexed by λ.

PROPOSITION 4. *The Specht modules $(V_\lambda)_{\lambda \vdash n}$ form a complete family of irreducible representations of $\mathfrak{S}_n$.*

The Specht polynomials indexed by fillings are not linearly independent. A basis of the Specht module V_λ is given by $\{\mathrm{Sp}_F\}$, where F runs through the standard tableaux of shape λ.

Note that the polynomials Sp_F are homogeneous. Furthermore, each Specht module is the lowest degree occurrence of the corresponding irreducible representation; more precisely:

PROPOSITION 5. *Let d be the degree of the Specht module V_λ. The multiplicity of the irreducible representation V_λ in $\mathbb{K}[X_n]$ in degree d is 1; Any occurrence of a representation isomorphic to V_λ in $\mathbb{K}[X_n]$ occurs in degree at least d.*

There is a construction for higher degree occurrences of representations isomorphic to V_λ by **[15]** using so-called *higher Specht polynomials*. They form an explicit basis for the quotient module $\mathbb{K}[X_n]_{/\,\mathrm{Sym}^+(X_n)\mathbb{K}[X_n]}$.

Let W a representation of $\mathfrak{S}_n$. Recall that the direct sum of all representations isomorphic to some V_λ is a sub-representation called the *isotypic component* indexed by λ of W and denoted W_λ. Then, W decomposes canonically as $W = \bigoplus_{\lambda \vdash n} W_\lambda$. Moreover, in the group ring $\mathbb{K}\mathfrak{S}_n$ of $\mathfrak{S}_n$, there are special elements e_λ called *central orthogonal idempotents* such that $e_\lambda W = W_\lambda$ and $e_i e_j = \delta_{i,j}$.

Consider a non zero polynomial f in V_λ and a representation W isomorphic to V_λ. By Maschke's theorem, the set of the images of f by all Sym_n-morphisms from V_λ to W is a line. Moreover, for two linearly independent polynomials f_1 and f_2, these lines are in direct sum.

DEFINITION 1. Let $f \in V_\lambda$ be a non zero polynomial, and W be a $\mathfrak{S}_n$-representation. The set of the images of f by all $\mathfrak{S}_n$-morphisms from V_λ to W is called the *Garnir component* of W associated to f denoted by $W_\lambda(f)$.

The Garnir component $W_\lambda(f)$ has a natural structure of vector space isomorphic to $\hom(V_\lambda, W)$ of which it is a sort of concrete realization.

PROPOSITION 6. *Let W be a representation of $\mathfrak{S}_n$. Then*

$$W = \bigoplus_{\lambda \vdash n} W_\lambda = \bigoplus_{\lambda \vdash n} \left(\bigoplus_{\mathrm{Shape}(T)=\lambda} W_\lambda(\mathrm{Sp}_T) \right), \tag{2.6}$$

where the last sum is on standard tableaux of shape λ.

Note that this direct sum decomposition is highly dependent on the chosen basis of V_λ (here the Specht basis).

2.3. The Weyl algebra. The following brief introduction to the Weyl algebra is essentially taken from [**18**]. For a variable x_i define the partial derivative

$$\partial_i := \frac{\partial}{\partial x_i}.$$

Following standard practice, we write abbreviated expressions for monomials:

$$x^K := \prod x_i^{k_i}, \quad \partial^L := \partial_i^{l_i},$$

where $K := (k_1, \ldots, k_n, \ldots)$ and $L := (l_1, \ldots, l_n, \ldots)$ are exponent vectors. The *degree* of x^K is $\deg(x^K) := \sum_i k_i$, and the *order* of ∂^L is $\mathrm{ord}(x^L) := \sum_i l_i$.

Weyl monomials are monomials in the variables x_i and the partial derivatives ∂_i. They act by multiplication and derivation on $\mathbb{K}[X_n]$, and form an algebra under addition and composition. For example,

$$\partial_1 \cdot x_1^4 = 4x_1^3, \quad \partial_1 \cdot x_2^2 = 0, \quad \partial_1\partial_2 \cdot p = \partial_2\partial_1 \cdot p,$$
$$x_1x_2 \cdot (1 + x_2x_4) = x_2x_1(1 + x_2x_4) = x_1x_2 + 2x_1x_2^2x_4,$$
$$\partial_1 x_1 \cdot p = \partial_1 \cdot (x_1p) = (\partial_1 \cdot x_1)p + x_1(\partial_1 \cdot p) = (1 + x_1\partial_1) \cdot p.$$

This is formulated more precisely as follows:

DEFINITION 2. The Weyl algebra Weyl := Weyl(X) is the associative algebra, with unit, generated by x_i, ∂_i, subject to the relations

$$[x_i, x_j] = 0, \quad [\partial_i, \partial_j] = 0, \quad [x_i, \partial_j] = \delta_{ij},$$

where square brackets denotes the Lie product and δ_{ij} the Kronecker symbol.

The Weyl algebras Weyl(X_n) are defined for each n in a similar way by restricting to the finite set of variables $x_1, \ldots, x_n$ and the corresponding partial derivatives.

By repeated differentiation, any element of the Weyl algebra can be expressed as a linear combination of Weyl monomials where the polynomial part is on the left and the derivatives on the right. For example,

$$\partial_1 x_1 = x_1\partial_1 + 1 \tag{2.7}$$
$$\partial_1^k x_1 = x_1\partial_1^k + k\partial_1^{k-1} \tag{2.8}$$
$$\partial_1\partial_2 x_1x_2 = x_1x_2\partial_1\partial_2 + x_1\partial_1 + x_2\partial_2 + 1. \tag{2.9}$$

We shall refer to such monomials as the *standard monomials.* As an Abelian group, Weyl is freely generated by the standard monomials $x^K\partial^L$, as K and L range over exponent vectors.

An element f in the Weyl algebra lies in *filtration* k if the maximum order of a term of f in standard form does not exceed k. Then, Weyl(X_n) and Weyl are filtered algebras in the sense that the composition product of two elements, in filtration m and k respectively lies in filtration $m+k$. The Weyl algebras are graded by assigning grading $\deg(x^K) - \operatorname{ord}(\partial^L)$ to $x^K\partial^L$.

Given a polynomial or a Weyl operator p, we denote by $\mathrm{o}(p)$ the *orbit sum* of p, that is the sum of all the polynomials in the orbit of m under the action of the symmetric group. Typically, if λ is a partition, then $m_\lambda := \mathrm{o}(x^\lambda)$ is the *monomial symmetric function* indexed by λ. Note that we have to be a little bit careful, since $\mathrm{o}(p)$ might be undefined for an infinite alphabet X, as for example for $\mathrm{o}(x_1+x_2)$; however, for a monomial or Weyl monomial m, $\mathrm{o}(m)$ is always defined. We will consider differential operators such as

$$P_k := \sum_i x_i^k(1+x_i\partial_i) := \mathrm{o}\big(x_1^k(1+x_1\partial_1)\big) \tag{2.10}$$

which are obtained by such symmetrization over an infinite alphabet. To be precise, such operators do not belong to the Weyl algebra, because they involve infinite sums. They do not even necessarily act properly on $\mathbb{K}[X]$; for example $P_k \cdot 1 = x_1^k + \cdots + x_n^k + \ldots$ involves an infinite sum. In general, an infinite sum makes sense as an operator on $\mathbb{K}[X]$ providing that, for each n, all but a finite number of the monomials $x^K\partial^L$ annihilate $\mathbb{K}[X_n]$.

In practice, this is not an issue; such symmetric operators can be manipulated in the same classical manner as, for example, symmetric functions on an infinite number of variables. They restrict to elements of the Weyl algebra $\mathrm{Weyl}(X_n)$ for each n, and the notions of order, filtration and grading carry over properly. By extension, we also call them Weyl operators.

The *wedge product* (or *formal product*) of two differential operators is defined on standard monomials of the Weyl algebra by

$$x^K\partial^L \wedge x^M\partial^N := x^{K+L}\partial^{M+N}.$$

In other words, the partial derivatives are made to commute with all the variables, and Weyl becomes polynomial in both sets of variables x_i, ∂_i under the wedge product. In working out the composition product of two standard forms in the Weyl algebra, the wedge product is the term of top filtration. The wedge product is extended to infinite sums by linearity.

2.4. Duality in the Weyl algebra and scalar product on polynomials.

DEFINITION 3. The *dual* f^* of a Weyl operator f is defined by sesquilinearity on the standard monomials:

$$(x^K\partial^L)^* := x^L\partial^K.$$

For example,

$$(2x_1x_2^2\partial_2^2\partial_3^3 + 3ix_2\partial_1^4)^* = 2x_2^2x_3^3\partial_1\partial_2^2 - 3ix_1^4\partial_2.$$

Note that, in the literature, the dual of a polynomial p is often written $p(\partial)$ instead of p^* [**4**].

REMARK 1. The dual is an anti-morphism with respect to composition:

$$(fg)^* = g^*f^*. \tag{2.11}$$

Denote by $p(0)$ the constant term of a polynomial p. For two polynomials p and q, set

$$\langle p \mid q\rangle := (\overline{p}^* \cdot q)(0). \tag{2.12}$$

It is easy to check that this defines a scalar product which makes the monomials $\{x^K\}$ into an orthogonal basis. In fact, we have

$$\langle x^K \mid x^L \rangle = \begin{cases} 0 & \text{if } K \neq L \\ K! & \text{if } K = L \end{cases}, \tag{2.13}$$

where $K! = k_1!k_2! \cdots k_n!$.

PROPOSITION 7. *Let f be a Weyl operator, seen as a linear endomorphism of $\mathbb{K}[X]$. Then, the adjoint of f with respect to the scalar product $\langle \cdot \mid \cdot \rangle$ is its dual Weyl operator f^*.*

PROOF. Note first that, for a Weyl operator f and a polynomial p, we have

$$(f^* \cdot p)(0) = \big((f \cdot 1)^* \cdot p\big)(0); \tag{2.14}$$

indeed, $(f - f \cdot 1)$ is a pure differential operator, hence $(f - f \cdot 1)^* \cdot p$ is a pure polynomial with no constant term. Then,

$$\begin{aligned} \langle f \cdot p \mid q \rangle = \big((f \cdot p)^* \cdot q\big)(0) &= \big((fp \cdot 1)^* \cdot q\big)(0) \\ &= \big((fp)^* \cdot q\big)(0) = (p^* f^* \cdot q)(0) \\ &= \big(p^* \cdot (f^* . q)\big)(0) = \langle p \mid f^* \cdot q \rangle, \end{aligned} \tag{2.15}$$

as desired. □

2.5. Hit and harmonic polynomials. In this subsection, we fix a set S of homogeneous Weyl operators, and denote by S^+ the operators in S of positive degree.

DEFINITION 4. A polynomial is called *S-hit* if there exists some polynomials $p_1, \dots, p_k$ and operators $f_1, \dots, f_k \in S^+$ such that

$$p = f_1 \cdot p_1 + \cdots + f_k \cdot p_k. \tag{2.16}$$

A polynomial p is called *S-harmonic* if it satisfies the differential equations

$$f^* \cdot p = 0, \quad \text{for all } f \in S^+. \tag{2.17}$$

The subspaces of S-hit and S-harmonic polynomials are denoted respectively by Hit and Harm.

For example, if $S := \{x_1^2 + \cdots + x_n^2\}$, the harmonics in $\mathbb{K}[X_n]$ are the polynomials p which satisfy

$$\partial_1^2 p + \cdots + \partial_n^2 p = 0;$$

hence the name *harmonic* [**4**]. If S consists only of polynomials, as in the previous example, then the space of S-hits is simply the ideal generated by S^+ in $\mathbb{K}[X_n]$.

REMARK 2. Consider an operator $f \in S^+$, any Weyl operator g, and a polynomial p. Then, $fg \cdot p = f \cdot (g \cdot p)$ is hit. Dually, if p is harmonic then $gf^* \cdot p = g \cdot (f^* \cdot p) = g \cdot 0 = 0$: it is killed by any operator in the left ideal generated by S^{+*} in the Weyl algebra. Hence, S and the right ideal $S^+ \operatorname{Weyl}(X_n)$ generated by S^+ in the Weyl algebra have the same hit and harmonic polynomials.

Furthermore, if A^+ is any subspace of the Weyl algebra such that $S^+ \subset A^+ \subset S^+ \operatorname{Weyl}(X_n)$, then Hit is the image $A \cdot \mathbb{K}[X_n]$ of $A^+ \otimes \mathbb{K}[X_n]$ by the bilinear evaluation map $(f, p) \mapsto f \cdot p$.

PROPOSITION 8. *A polynomial p is S-harmonic if, and only if, it is orthogonal to any S-hit polynomial q. That is, the space of S-harmonic polynomials is the orthogonal complement of the space of S-hit polynomials.*

PROOF. This is a simple variation on the fact that the image of a linear morphism is the orthogonal complement of the kernel of its adjoint. Let p be a polynomial. By duality, for any polynomial q and operator $f \in S^+$, we have

$$\langle f^* \cdot p \mid q \rangle = \langle p \mid f \cdot q \rangle.$$

Assume that p is S-harmonic; then it is obviously orthogonal to any $f \cdot q$ and by linearity to any S-hit. Reciprocally, assume that p is orthogonal to any hit polynomial, and consider an operator $f \in S^+$; then $f^* \cdot p = 0$ because it is orthogonal to all polynomials. □

PROPOSITION 9. *Let A be the graded algebra generated by S^+, acting on $\mathbb{K}[X_n]$. Then,* Harm *generates $\mathbb{K}[X_n]$ as an A-module. In particular,*

$$\mathrm{Hilb}_t(A) \cdot \mathrm{Hilb}_t(\mathrm{Harm}) \geq \mathrm{Hilb}_t(\mathbb{K}[X_n]), \tag{2.18}$$

with equality if, and only if, $\mathbb{K}[X_n]$ is a free A-module.

PROOF. It is sufficient to verify, by induction on the degree, that any homogeneous polynomial can be written in the form $p = 1 \cdot p_0 + f_1 \cdot p_1 + \cdots + f_k \cdot p_k$ where the p_i's are harmonic. Then, $\mathbb{K}[X_n] = A \cdot \mathrm{Harm}$ is a graded quotient of $A \otimes \mathrm{Harm}$, which yields the desired inequality on the Hilbert series. □

PROPOSITION 10. *Assume that the operators in S are symmetric. Then, any Specht polynomial Sp_F is S-harmonic. Hence, the space of S-harmonics contains at least one copy of each irreducible representation of the symmetric group $\mathfrak{S}_n$.*

PROOF. Consider an operator $f \in S^+$. Its dual f^* is an $\mathfrak{S}_n$-morphism of negative degree. Hence, $f^* \cdot \mathrm{Sp}_F$ lies in the same isotypic component as Sp_F, and $\deg(f^* \cdot \mathrm{Sp}_F) < \deg(\mathrm{Sp}_F)$. By minimality of the degree of Sp_F, $f^* \cdot \mathrm{Sp}_F = 0$. □

As a final example, take for S the n symmetric power-sums on n variables $(p_k := x_1^k + \cdots + x_n^k)_{k=1,\ldots,n}$. Then, the space Hit of hit polynomials is the ideal generated by the symmetric polynomials. There exists an explicit Gröbner basis for the lexicographic term order, which shows that the leading terms of Hit are the non-staircase monomials (see e.g. [**14**]).

The harmonics are the polynomials p which satisfy the generalized harmonic equations

$$p_k^* \cdot p = \partial_1^k \cdot p + \cdots + \partial_n^k \cdot p = 0, \quad \text{for } k = 1, \ldots, n. \tag{2.19}$$

The Vandermonde $\Delta := \prod(x_j - x_i)$ is the Specht polynomial $\mathrm{Sp}_{(1,\ldots,1)}$, and as such is harmonic. Moreover, any partial derivative ∂_i commutes with the p_k^*'s, and it follows that it stabilizes the space Harm of harmonic polynomials. Actually, any harmonic is of the form $p^* \cdot \Delta$, for some polynomial p, and if a set B of $n!$ polynomials is a basis of $\mathbb{K}[X_n]/\,\mathrm{Sym} \cdot \mathbb{K}[X_n]$, then $(p^* \cdot \Delta)_{p \in B}$ is a basis of Harm. Finally, $\mathbb{K}[X_n]$ is a free $\mathrm{Sym}(X_n)$-module, and the Hilbert series of $\mathrm{Harm}(X_n)$ is given by

$$\text{Hilb}_t\big(\text{Harm}(X_n)\big) = \frac{\text{Hilb}_t(\mathbb{K}[X_n])}{\text{Hilb}_t(\text{Sym}(X_n))} \tag{2.20}$$
$$= \prod_{i=1}^{n} \frac{1-t^i}{1-t} = (1+t)(1+t+t^2)\cdots(1+t+\cdots+t^{n-1}).$$

For further details, we refer to [**4**].

2.6. Specialization lemmas. In this subsection, U is a $\mathbb{C}$-vector space with a distinguished basis B. The goal here is to deal with q-deformations of finite dimensional subspaces of $\mathbb{C}(q) \otimes U$ and their specializations.

LEMMA 2.1. *Let $\big(u_1(q), \dots, u_k(q)\big)$ be a family of vectors of $\mathbb{C}(q) \otimes U$. For all but a finite set of values $q_0 \in \mathbb{C}$, namely the roots of the denominators of the coefficients of the u_i's on the basis B, the specialization $q = q_0$ is well defined. In this case,*

$$\text{Rank}_{\mathbb{C}}\big(u_1(q_0), \dots, u_k(q_0)\big) \leq \text{Rank}_{\mathbb{C}(q)}\big(u_1(q), \dots, u_k(q)\big). \tag{2.21}$$

Moreover, there is only a finite number of values $q_0 \in C$ such that the inequality above is strict, namely the roots of the non-trivial principal minors.

Let $V(q)$ be a subspace of $\mathbb{C}(q) \otimes U$ of finite dimension. For any $q_0 \in \mathbb{C}$, let $V(q_0)$ be its specialization, that is the vector space of all meaningful specializations of elements of $V(q)$. Note that this definition is highly dependent on the basis B.

LEMMA 2.2. *There exists a basis $B'(q) = \big(v_1(q), \dots, v_k(q)\big)$ of $V(q)$ such that the coefficients of the elements of $B'(q)$ in the basis B are polynomials in q, and for any specialization $q = q_0$, the family $B'(q_0)$ is a basis of $V(q_0)$ (in particular, $\dim_{\mathbb{C}} V(q_0) = \dim_{\mathbb{C}(q)} V(q)$).*

This is a classical result, but let us give a short effective proof.

PROOF. First remark that, by the preceding lemma, the vector space $V(q_0)$ of all meaningful specializations of elements of $V(q)$ is of $\mathbb{C}$-dimension at most the $\mathbb{C}(q)$-dimension of $V(q)$. Consequently we just have to construct a basis of $V(q)$ without denominators such that all specializations are linearly independent. Let us construct such a basis.

Choose a $\mathbb{C}(q)$-basis $A(q)$ of $V(q)$ and expand the vectors of $A(q)$ over B. By a suitable multiplication, we can remove the denominators and then suppose that the coefficients of the expansion are in $\mathbb{C}[q]$. Now we apply the fraction-free Gram–Schmidt orthogonalization to $A(q)$, and get a basis $A'(q) = \big(a_1(q), \dots, a_n(q)\big)$. By dividing each $a_i(q)$ by the greatest common divisor of its coefficient, (this is often called the *content* of $a_i(q)$) we get a new basis $B'(q) = \big(b_i(q)\big)$ of $V(q)$.

This basis is orthogonal on $\mathbb{C}[q]$ and thus remains orthogonal for any specialization at $q = q_0$. Furthermore, an element $b_i(q)$ of $B'(q)$ is content-free, so $b_i(q_0)$ never vanishes. We conclude that, for any q_0, the family $B'(q_0)$ is a $\mathbb{C}$-basis for $V(q_0)$. □

Moreover, since the character table of $\mathfrak{S}_n$ is with integer coefficients, the action of central idempotents commutes with specializations. Thus if $V(q)$ is an $\mathfrak{S}_n$-module then

1. $V(q)$ is isomorphic as $\mathfrak{S}_n$-module to $\mathbb{C}(q) \otimes V(q_0)$ for all q_0;

2. $\mathbb{C}(q_0)$ and $\mathbb{C}(q_1)$ are isomorphic as $\mathfrak{S}_n$-modules for all q_0, q_1.

3. The q-Steenrod algebra

The similarities between Sym and Steen leads us to define a new algebra to interpolate between the two.

DEFINITION 5. Let $P_{q,k}$ be the following q-analogue of the P_k operator:

$$P_{q,k} := \sum_i x_i^k(1 + qx_i\partial_i).$$

The *q-Steenrod algebra* $\mathrm{Steen_q}$ is the algebra over $\mathbb{C}(q)$ generated by the $P_{q,k}$, $k > 0$.

Note that the specializations at $q = 0$ and $q = 1$ yield respectively the symmetric functions and the rational Steenrod algebra (conjugated by the Thom map):

1. $P_{0,k}$ is the usual operator of multiplication by the symmetric power-sum p_k.
2. $P_{1,k} = P_k$.

REMARK 3. It is tempting to actually introduce two parameters q and q':

$$P_{q,q',k} := \sum_i x_i^k(q' + qx_i\partial_i). \tag{3.1}$$

Now, the rational Steenrod algebra itself is also a special case of the (q, q')-Steenrod algebra, with $q' = 0$ and $q = 1$: working in projective space versus q removes the singularity at $q = \infty$. The price to pay is that $\mathbb{C}[q, q']$ is not anymore a principal ideal domain; this renders computations, and especially linear algebra, far less practical. Since we are mainly interested in the cases $q = 0$ and $q = 1$, we stick in the following to a single parameter.

3.1. Structure and dimension of the q-Steenrod algebra. The q-Steenrod squares satisfy the following commutation rule:

PROPOSITION 11. *For $k, l \in \mathbb{N}$ and q complex or formal,*

$$[P_{q,k}, P_{q,l}] = q(l - k)P_{q,k+l}. \tag{3.2}$$

For a composition $\mu := (\mu_1, \ldots, \mu_k)$, define $P_{q,\mu} := P_{q,\mu_1} \cdots P_{q,\mu_k}$. Using the commutation rules, it is obvious that any $P_{q,\mu}$ can be rewritten as a linear combination of $P_{q,\lambda}$'s indexed by partitions. For example,

$$P_{q,(1,2)} = qP_{q,(3)} + P_{q,(2,1)}.$$

THEOREM 3.1. *Let q be formal or a complex number. Then,*

(a) *The operators $P_{q,\lambda}$, where λ runs through all integer partitions form a vector space basis of* $\mathrm{Steen_q}$;

(b) *The Hilbert series of the q-Steenrod algebra is given by*

$$\mathrm{Hilb}_t(\mathrm{Steen_q}) = \mathrm{Hilb}_t(\mathrm{Sym}) = \prod_{i>0} \frac{1}{1 - t^i}; \tag{3.3}$$

(c) *The ideal of relations between the $P_{q,k}$'s is generated by the commutation relations*

$$[P_{q,k}, P_{q,l}] - q(l - k)P_{q,k+l}. \tag{3.4}$$

(d) $\mathrm{Steen_q}$ *is the enveloping algebra of the Lie algebra spanned by the $P_{q,k}$.*

This is a straightforward generalization of results in [**18**] about the rational Steenrod algebra, and follows from an easy triangularity property with respect to monomial functions. Namely, consider the expansion of

$$P_{q,(2,1)} = \mathrm{o}(x_1^2x_2) + (q+1)\,\mathrm{o}(x_1^3) + q\,\mathrm{o}(x_1^3x_2\partial_1) + 2q(q+1)\,\mathrm{o}(x_1^4\partial_1) \\ + q\,\mathrm{o}(x_1^2x_2^2\partial_1) + q^2\,\mathrm{o}(x_1^5\partial_1^2) + q^2\,\mathrm{o}(x_1^3x_2^2\partial_1\partial_2).$$

Its polynomial part is a symmetric function, and can be expressed in terms of monomial symmetric function: $P_{q,(2,1)} \cdot 1 = m_{(2,1)} + (q+1)m_{(3)}$. Note that $m_{(2,1)}$ has coefficient 1, while m_3 is indexed by a shorter partition. This generalizes immediately:

LEMMA 3.1. *The polynomial part $P_{q,\lambda} \cdot 1$ of $P_{q,\lambda}$ is a symmetric function of the form*

$$P_{q,\lambda} \cdot 1 = m_\lambda + \sum_\mu c_\mu m_\mu,$$

where μ runs through partitions of length $\ell(\mu) < \ell(\lambda)$.

It follows that, except for $q = 0$, all the q-Steenrod algebras on infinite alphabets are isomorphic independently of q; only their actions on polynomials differ:

THEOREM 3.2. *Let q be formal (respectively a non zero complex number). Then, the q-Steenrod algebra* $\mathrm{Steen_q}$ *is isomorphic as a graded algebra to $\mathbb{C}(q) \otimes$* Steen (*respectively* Steen).

PROOF. By renormalizing the q-Steenrod squares $\widetilde{P}_{q,k} := 1/qP_{q,k}$, we obtain the commutation relations

$$[\widetilde{P}_{q,k}, \widetilde{P}_{q,l}] = (l-k)\widetilde{P}_{q,k+l}. \tag{3.5}$$

Hence the algebra $\mathrm{Steen_q}$ has the same presentation with generators and relations as $\mathbb{C}(q) \otimes$ Steen. □

We consider now the q-Steenrod algebra over a finite alphabet.

PROPOSITION 12. *For any finite alphabet X_n,*

$$\prod_{i=1}^{n} \frac{1}{1-t^i} = \mathrm{Hilb}_t\big(\mathrm{Sym}(X_n)\big) \leq \mathrm{Hilb}_t\big(\mathrm{Steen_q}(X_n)\big) \leq \frac{1}{(1-t)^n} \prod_{i=1}^{n} \frac{1}{1-t^i}.$$

PROOF. The first inequality follows again from Lemma 3.1. Note that, for $k \geq 1$, any Weyl monomial $x^K\partial^L$ appearing in $P_{q,k}$ satisfies $k_i \geq 2l_i$, for all i. This property is preserved by multiplication, and it follows that $P_{q,\lambda}$ can be written as a linear combination of orbit-sums $\mathrm{o}(x^{2K+L}\partial^L)$, where $\sum_i k_i + l_i = n$, and the k_i's are decreasing. The second inequality follows. □

It appears from computations that the first inequality is strict, starting from degree $n+1$. Namely, for $n \leq 5$, we noticed that the polynomials $\big(P_{q,\lambda}(X_n) \cdot x_1\big)_{\lambda \vdash n+1}$ are always linearly independent (see Sections A.2 and 7.1), whereas there is a linear relation between the symmetric functions $\big(p_\lambda(X_n)\big)_{\lambda \vdash n+1}$.

The second inequality is very coarse, and could certainly be refined. For example, one can easily check that

$$\mathrm{Hilb}_t\big(\mathrm{Steen_q}(X_1)\big) = \frac{1}{1-t} \ll \frac{1}{(1-t)^2}. \tag{3.6}$$

Still, this inequality implies that $\mathrm{Hilb}_t(\mathrm{Steen_q}(X_n)) < \mathrm{Hilb}_t(\mathrm{Steen_q}(X))$; indeed the dimension of $\mathrm{Steen}_{\mathrm{q}_d}$ is counted by the number $p(d)$ of partitions whose asymptotic [**5**] is:

$$p(d) \approx \frac{1}{4\sqrt{3n}} \exp \pi \sqrt{\frac{2n}{3}}. \tag{3.7}$$

3.2. Action on polynomials; hit and harmonic polynomials. We consider now the natural action of $\mathrm{Steen_q}$ on $\mathbb{K}[X_n]$. Following Subsection 2.5 with $S = \mathrm{Steen}_{\mathrm{q}}^{+}$, we define the spaces $\mathrm{Hit_q}$ of *q-hit polynomials* and $\mathrm{Harm_q}$ of *q-harmonic* polynomials. For $q \neq 0$, the q-Steenrod algebra is generated by $P_{q,1}$ and $P_{q,2}$, and it follows that the q-harmonics can be described by just two differential equations:

$$P_{q,1}^{*} \cdot p = 0 \quad \text{and} \quad P_{q,2} \cdot p = 0. \tag{3.8}$$

Furthermore, it follows immediately from the triangularity property described in Lemma 3.1 that any symmetric function is q-hit.

PROPOSITION 13. *Let q be formal or a complex number, and X be a commutative alphabet. Then,* $\mathrm{Steen_q}(X) \cdot 1 = \mathrm{Sym}(X)$.

4. Specializations

The action of the q-Steenrod algebra behaves properly with respect to specialization. That is, if a polynomial $p(q)$ specializes properly to $p(q_0)$, then $P_{q,k} \cdot p(q)$ and $P_{q,k}^{*} \cdot p(q)$ specializes properly to $P_{q_0,k} \cdot p(q_0)$ and $P_{q_0,k}^{*} \cdot p(q_0)$ respectively. Hence, we can apply the specialization lemmas to obtain some properties of the q-harmonics and q-hits with respect to specialization. In particular we obtain the implication (b) $\Rightarrow$ (a) of Proposition 1:

PROPOSITION 14. *Let $q_0 \in \mathbb{C}$. Then,*

(a) *Let $S(q)$ be a set of homogeneous polynomials in $\mathbb{C}(q)[X_n]$, which specializes properly at $q = q_0$. Then, the* $\mathrm{Steen_q}$*-module (resp.* $\mathrm{Steen}_{\mathrm{q}}^{+}$*-module) $M(q)$ generated by S in $\mathbb{C}(q)[X_n]$ specializes to the* $\mathrm{Steen}_{\mathrm{q_0}}$*-module (resp.* $\mathrm{Steen}_{\mathrm{q_0}}^{+}$*-module) $M(q_0)$ generated by $S(q_0)$;*
(b) *The Hilbert series of $M(q)$ dominates the Hilbert series of $M(q_0)$;*
(c) *Let $p \in \mathbb{C}[X_n]$. The Hilbert series of the* $\mathrm{Steen_q}$*-module generated by p dominates the Hilbert series of the corresponding* $\mathrm{Sym}(X_n)$*-module;*
(d) *Let B be a* $\mathrm{Sym}(X_n)$*-module basis of* $\mathbb{C}[X_n]$*, e.g. the Schubert polynomials or the staircase monomials. Then B spans $\mathbb{C}(q)[X_n]$ as a* $\mathrm{Steen_q}$*-module; note that this module is not necessarily free;*
(e) $\mathrm{Hit_q}$ *specializes at $q = q_0$ to* $\mathrm{Hit}_{\mathrm{q_0}}$*; in particular the Hilbert series of* $\mathrm{Hit_q}$ *dominates the Hilbert series of* $\mathrm{Hit}_{\mathrm{q_0}}$*; furthermore, for all but a countable number of $q_0 \in \mathbb{C}$, there is equality;*
(f) $\mathrm{Hit_q} + \mathrm{Harm} = \mathbb{C}(q)[X_n]$.

PROOF. (a) Fix a degree d. Extract from $S(q)$ a finite subset $A(q)$ which generates linearly all the other elements of $S(q)$ of degree less than d, and specializes properly at $q = q_0$. Consider the set $B(q)$ of all $P_{q,\lambda} \cdot p(q)$, where p belongs to $A(q)$ and λ is a partition of $d - \deg(p)$. Then, $B(q)$ generates $M(q)_d$, and specializes to $B(q_0)$ which generates $M(q_0)_d$. Hence, $M(q)_d$ specializes to $M(q_0)_d$.

The others are direct applications of (a). □

Similarly, the following proposition provides the implication (b) $\Rightarrow$ (a) of Proposition 2, and more:

PROPOSITION 15. (a) *If $p(q)$ is q-harmonic and specializes properly at $q = q_0$, then $p(q_0)$ is harmonic;*
(b) *There exists a basis of* Harm_q *which specializes properly for any q_0;*
(c) Harm_q *specializes to a subspace* $\mathrm{Harm}_\mathrm{q}(q_0)$ *of* $\mathrm{Harm}_\mathrm{q}[q_0]$;
(d) *The Hilbert series of* Harm_q *is dominated by the Hilbert series of* Harm; *furthermore, for all but a countable number of $q_0 \in \mathbb{C}$, there is equality;*
(d) Hit *and* Harm_q *are in direct sum;*
(e) *The orthogonal projector* $\mathbb{C}(q)[X_n] \mapsto \mathrm{Harm} \otimes \mathbb{C}(q)$ *is injective on* Harm_q.

PROOF. (a) For any k, $P^*_{q,k} \cdot p(q)$ specializes to $P^*_{q_0,k} \cdot p(q)$ for all k; hence the latter is zero whenever the former is;
(b) Direct application of the specialization Lemma 2.2;
(c) Consequence of (a) and (b);
(d) Take $f \in \mathrm{Hit} \cap \mathrm{Harm}_\mathrm{q}$. Then, for all k, $\sum_i (1 + qx_i\partial_i)\partial_i^k f = 0$.
Write $f = q^{\mathrm{val}(f)}(f_0 + qf_1)$, with $0 \neq f_0 \in \mathbb{C}[X_n]$ and $f_1 \in \mathbb{C}(q)[X_n]$ such that f_1 has no $1/q$ coefficient. Then, $f_0 + qf_1$ is also q-harmonic, and it follows that f_0 is harmonic. This is in contradiction with f_0 being hit. □

5. The low degree and truncated cases

One of the main difficulties is that the q-Steenrod algebra on a finite alphabet X_n is much bigger than $\mathrm{Sym}(X_n)$. In this section, we get around this difficulty by truncating the q-Steenrod algebra to make it share the dimension of $\mathrm{Sym}(X_n)$. No truncation occurs in low degree, so all the results of this section also apply to the full q-Steenrod algebra in degree $d \leq n$.

The *truncated q-Steenrod algebra* $\mathrm{TSteen}_\mathrm{q}(X_n)$ is the subspace of $\mathrm{Steen}_\mathrm{q}(X_n)$ spanned by $(P_{q,\lambda})_{\ell(\lambda)\leq n}$. Note that $\mathrm{Steen}_\mathrm{q}(X_n)$ and $\mathrm{TSteen}_\mathrm{q}(X_n)$ coincide in degree $d \leq n$; however $\mathrm{TSteen}_\mathrm{q}(X_n)$ is not an algebra.

Here also, applying Lemma 3.1, yields $\mathrm{TSteen}_\mathrm{q}(X_n) \cdot 1 = \mathrm{Sym}$, and it follows that $\mathrm{Hilb}_t(\mathrm{TSteen}_\mathrm{q}(X_n)) = \mathrm{Hilb}_t(\mathrm{Sym}(X_n))$, for q formal or a complex number.

PROBLEM 1. For which polynomials p do $\mathrm{TSteen}_\mathrm{q} \cdot p$ and $\mathrm{Steen}_\mathrm{q} \cdot p$ coincide?
For such a polynomial p, find a straightening algorithm for rewriting an element of $\mathrm{Steen}_\mathrm{q} \cdot p$ as element of $\mathrm{TSteen}_\mathrm{q}$.

The proper analogue of $\mathrm{Hit}_\mathrm{q}(X_n)$ is the space of *truncated q-hit polynomials* $\mathrm{THit}_\mathrm{q}(X_n)$ which is spanned by the $f \cdot p$, where $f \in \mathrm{TSteen}^+_\mathrm{q}(X_n)$, and $p \in \mathrm{Harm}(X_n)$. The polynomials h in the orthogonal $\mathrm{THarm}_\mathrm{q}(X_n) := \mathrm{THit}_\mathrm{q}(X_n)^\perp$ are called *truncated q-harmonics.* Note that they do not necessarily satisfy $P^*_{q,1} \cdot h = 0$ and $P^*_{q,2} \cdot h = 0$.

Reg Wood's conjecture holds for the truncated q-Steenrod algebra with q formal:

THEOREM 5.1. *Let q be formal. Then,*
(a) $\mathrm{Hilb}_t(\mathrm{THarm}_\mathrm{q}(X_n)) = \mathrm{Hilb}_t(\mathrm{Harm}(X_n))$;
(b) $\mathrm{THarm}_\mathrm{q}(X_n)$ *is a graded regular representation of the symmetric group;*
(c) $\mathbb{K}[X_n] = \mathrm{THarm}_\mathrm{q}(X_n) \bigoplus \mathrm{Hit}(X_n) = \mathrm{Harm}(X_n) \bigoplus \mathrm{THit}_\mathrm{q}(X_n)$.

Furthermore, similar statements hold in the isotypic and Garnir components of $\mathbb{K}[X_n]$.

PROOF. The specialization $q = 0$ of TSteen_q is Sym. Then THit_q specializes to Hit. Hence, by Proposition 15, $\mathrm{Hilb}_t(\mathrm{THarm}_q) \leq \mathrm{Hilb}_t(\mathrm{Harm})$.

Furthermore, $\mathbb{K}[X_n]$ is a free $\mathrm{Sym}(X_n)$ module, so applying Proposition 9 yields

$$\begin{aligned}\mathrm{Hilb}_t(\mathbb{C}[X_n]) &\leq \mathrm{Hilb}_t\big(\mathrm{TSteen}_q(X_n)\big)\,\mathrm{Hilb}_t\big(\mathrm{THarm}_q(X_n)\big)\\ &\leq \mathrm{Hilb}_t\big(\mathrm{Sym}(X_n)\big)\,\mathrm{Hilb}_t(\mathrm{Harm}) = \mathrm{Hilb}_t(\mathbb{C}[X_n]).\end{aligned} \tag{5.1}$$

It follows that equality must hold, and that $\mathrm{Hilb}_t\big(\mathrm{THarm}_q(X_n)\big) = \mathrm{Hilb}_t\big(\mathrm{Harm}(X_n)\big)$. The same reasoning can be carried over in any isotypic or Garnir component, and all the statements of the theorem follow. □

COROLLARY 1. *Let* $\mathrm{Hilb}_{t\leq n}(A)$ *denote the truncation of the Hilbert series of* A *up to degree* d

1. *If* q *is formal, then* $\mathrm{Hilb}_{t\leq n}\big(\mathrm{Steen}_q(X_n)\big) = \mathrm{Hilb}_{t\leq n}\big(\mathrm{Sym}(X_n)\big)$;
2. *If* q *is a complex number, then* $\mathrm{Hilb}_{t\leq n}\big(\mathrm{Steen}_q(X_n)\big) \leq \mathrm{Hilb}_{t\leq n}\big(\mathrm{Sym}(X_n)\big)$.

PROOF. This is a reformulation of the preceding theorem by the fact that for $d \leq n$ the algebra TSteen_q and Steen_q coincide. □

6. Changing the number of variables

In the preceding section, we proved that Conjecture 2 holds when the number of the variable is greater than the degree. The goal here is to relate q-harmonics on a large number of variable with q-harmonics on fewer variables, trying to take down as much information as possible. In particular we prove the equivalence (a) $\iff$ (c1) of Proposition 1.

PROPOSITION 16. *Let* X *be a set of variables and* $y \notin X$ *be another variable. Let us denote by* $\mathrm{Harm}_q(X,y)_{\deg(y)\leq d}$ *the subspace of* $\mathrm{Harm}_q(X,y)$ *of polynomials of degree at most* d *in the variable* y. *The isomorphism*

$$\pi_{y^d} : \begin{cases} \mathbb{K}(X,y)_{\deg(y)\leq d}/\mathbb{K}(X,y)_{\deg(y)\leq d-1} & \hookrightarrow \mathbb{K}(X)\\ f & \longmapsto \mathrm{Coeff}(f, y^d)\end{cases} \tag{6.1}$$

defines an injective mapping:

$$\tilde{\pi}_{y^d} : \mathrm{Harm}_q(X,y)_{\deg(y)\leq d}/\,\mathrm{Harm}_q(X,y)_{\deg(y)\leq d-1} \hookrightarrow \mathrm{Harm}_q(X). \tag{6.2}$$

PROOF. The injectivity of $\tilde{\pi}_{y^d}$ follows from the definition. The only thing to prove is that the image $\tilde{\pi}_{y^d}$ of a q-harmonic polynomial of degree d in $\mathbb{K}[X,y]$ is in $\mathrm{Harm}_q(X)$. It comes from the fact that the $P^*_{q,k}$ are primitive, that is

$$P_{q,k}(X,y) = P_{q,k}(X) + P_{q,k}(y)\left(\sum_{x\in X}(1+qx\partial_x)\partial_x^k + (1+qy\partial_y)\partial_y^k\right)\cdot f = 0. \tag{6.3}$$

Write $f = f_0 y^d + f_1$ with $\deg_y(f_1) < d$. The identification of the homogeneous components of degree d in y in the preceding equation proves that

$$\left(\sum_{x\in X}(1+qx\partial_x)\partial_x^k\right)\cdot f_0 = 0. \tag{6.4}$$

This ends the proof. □

The following proposition yields as corollary the equivalence (a) $\Rightarrow$ (c1) and by an easy induction the implication (a) $\Rightarrow$ (c2) of Proposition 2.

PROPOSITION 17. *Let q be generic. Then, the following are equivalent*

(a) *Conjecture 2 holds;*
(b) *For all n, any q-harmonic on n variables has degree at most $n-1$ on each variable;*
(c) *For all n, the mapping*

$$\tilde{\pi}_{y^d} : \operatorname{Harm}_q(X_n, y)_{\deg(y)\leq d} / \operatorname{Harm}_q(X_n, y)_{\deg(y)\leq d-1} \hookrightarrow \operatorname{Harm}_q(X_n), \tag{6.5}$$

is an isomorphism for $d \leq n$. For $d > n$, $\operatorname{Harm}_q(X_n, y)_{\deg(y)=d} = \{0\}$.

PROOF. (c) $\Rightarrow$ (b) is obvious.

(a) $\Rightarrow$ (c): For any degree d, the injectivity of $\tilde{\pi}_{y^d}$ implies that

$$\operatorname{Hilb}_t\big(\operatorname{Harm}_q(X_n, y)_{\deg(y)\leq d} / \operatorname{Harm}_q(X_n, y)_{\deg(y)\leq d-1}\big) \leq t^d \operatorname{Hilb}_t\big(\operatorname{Harm}_q(X_n)\big), \tag{6.6}$$

with equality if, and only if, $\tilde{\pi}_{y^d}$ is an isomorphism. On the other hand, we assumed that Conjecture 2 holds. Identifying y with x_{n+1}, it follows that

$$\operatorname{Hilb}_t\big(\operatorname{Harm}_q(X_{n+1})\big) = (1 + t + \cdots + t^n) \operatorname{Hilb}_t\big(\operatorname{Harm}_q(X_n)\big). \tag{6.7}$$

If for some $d \leq n-1$, the inequality in Equation 6.6 above is strict, the terms of degree d in Equation 6.7 will differ. Then, by a similar reasoning, the existence of a non-trivial component $\operatorname{Harm}_q(X_n, y)_{\deg(y)\leq d} / \operatorname{Harm}_q(X_n, y)_{\deg(y)\leq d-1})$ for $d \geq n$ would contradict Equation 6.7.

(b) $\Rightarrow$ (a): By the assumption, and the injectivity of $\tilde{\pi}_{x_n^d}$, for any n,

$$\operatorname{Hilb}_t\big(\operatorname{Harm}_q(X_{n+1})\big) \leq (1 + t + \cdots + t^n) \operatorname{Hilb}_t\big(\operatorname{Harm}_q(X_n)\big). \tag{6.8}$$

By induction

$$\operatorname{Hilb}_t\big(\operatorname{Harm}_q(X_{n+1})\big) \leq (1+t)(1+t+t^2)\cdots(1+t+\cdots+t^n) = \operatorname{Hilb}_t(\operatorname{Harm}). \tag{6.9}$$

Suppose now that, for a given n_0, this inequality is strict:

$$\operatorname{Hilb}_t\big(\operatorname{Harm}_q(X_{n_0+1})\big) < (1+t)(1+t+t^2)\cdots(1+t+\cdots+t^{n_0}). \tag{6.10}$$

Let d be the valuation of the difference. Applying again induction, we obtain that

$$\operatorname{Hilb}_t\big(\operatorname{Harm}_q(X_{d+1})\big) < (1+t)(1+t+t^2)\cdots(1+t+\cdots+t^d), \tag{6.11}$$

with the inequality still appearing in degree d. However, the number of variables now exceeds d, and this is in contradiction with Corollary 1. □

6.1. Strings. We introduce a little bit more formalism to analyze when q-harmonic polynomials in $\mathbb{K}[X_{n+1}]$ have variables of degree at most n.

PROPOSITION 18. *Consider $f \in \mathbb{K}[X, y]$, and write its expansion on the variable y as*

$$f(X, y) = \sum_{i\geq 0} \frac{1}{i!} f_i(X) y^i. \tag{6.12}$$

Then, f is q-harmonic if, and only if, for any $i \geq 0$ and $k \geq 1$,

$$P^*_{q,k} \cdot f_i = -(1 + qi) f_{i+k}. \tag{6.13}$$

PROOF. Since f is q-harmonic, $P^*_{q,k} \cdot f = 0$. Hence,

$$\sum_{i\geq 0} \frac{1}{i!}(P^*_{q,k} \cdot f_i)y^i + \sum_{i\geq 0} \frac{1}{i!} f_i(P^*_{q,k} \cdot y^i) = 0, \tag{6.14}$$

and then

$$\sum_{i\geq 0} \frac{1}{i!}(P^*_{q,k} \cdot f_i)y^i + \sum_{i\geq k} \frac{1}{(i-k)!} f_i\big(1 + q(i-k)\big)y^{i-k} = 0. \tag{6.15}$$

Setting $j := i - k$ in the second sum, it follows that

$$\sum_{i\geq 0} \frac{1}{i!}(P^*_{q,k} \cdot f_i)y^i + \sum_{j\geq 0} \frac{1}{j!} f_{j+k}(1 + qj)y^j = 0. \tag{6.16}$$

Equating the coefficients of y yields the desired equality, for all i. □

A remarkable consequence of Proposition 18 is that a q-harmonic polynomial f can be reconstructed from its tail f_0. This motivates the following definition:

DEFINITION 6. Let $g \in \mathbb{K}(X)$ be a homogeneous polynomial. The *string* generated by g is the sequence $F = \text{String}(g) := (f_i)_{i\geq 0}$ defined by

$$f_0 := g, \quad \text{and for all } i > 0, \quad f_i := -P^*_{q,i} \cdot g. \tag{6.17}$$

The *length* of the string $F = (f_i)$ is defined by

$$\ell(F) := 1 + \max\{i \mid f_i \neq 0\}. \tag{6.18}$$

The *head* of the string is $f_{\ell(f)}$.

The length of a string is nothing but the degree in the variable y of the associated polynomial $f(X, y) := \sum 1/i! f_i(X)y^i$. Obviously, if g is of degree d then $f_i = 0$ for all $i > d$, consequently $\ell\big(\text{String}(g)\big) < \deg(g)$ and the length is well defined.

DEFINITION 7. A string $F = (f_i)_i$ is called *harmonic* if

$$P^*_{q,k} \cdot f_i = -(1 + qi)f_{i+k}, \quad \text{for all } i, k. \tag{6.19}$$

Thanks to the relations in the q-Steenrod algebra, it is sufficient in the above definition to check that $P^*_{q,1}$ acts properly:

PROPOSITION 19. *Let $F = (f_i)_i$ a string. Then F is harmonic if, and only if,*

$$P^*_{q,1} \cdot f_i = -(1 + qi)f_{i+1}, \quad \textit{for all } i. \tag{6.20}$$

PROOF. This is a simple consequence of the commutation relation. Namely, we have

$$\begin{aligned} P^*_{q,k} \cdot f_l = P^*_{q,k}P^*_{q,l} \cdot f_0 &= q(l-k)P^*_{q,(k+l)} \cdot f_0 - P^*_{q,l}P^*_{q,k} \cdot f_0 \\ &= -q(l-k)f_{k+l} + P^*_{q,l}f_k. \end{aligned} \tag{6.21}$$

The proposition follows by an easy induction. □

For example,

$$P^*_{q,2} \cdot f_1 = -P^*_{q,2}P^*_{q,1} \cdot f_0 = -qP^*_{q,3} \cdot f_0 - P^*_{q,1}P^*_{q,2} \cdot f_0 = qf_3 - (1+2q) \cdot f_3 = -(1+q)f_3.$$

Here are some basic properties of harmonic strings:

PROPOSITION 20. *Let $F = (f_i)_i$ be an harmonic string. Then,*

(a) *The head $f_{\ell(F)}$ of the string is harmonic;*
(b) *$f_i \neq 0$ whenever $0 \leq i \leq \ell(F)$ and $q \neq -1/i$.*

PROOF. These are obvious consequences of the definitions: for $k > 0$, one has $P^*_{q,k}(f_{\ell(F)}) = f_{\ell(F)+k} = 0$ and $P^*_{q,\ell(F)-i}(f_i) = -(1+qi)f_{\ell(F)} \neq 0$. □

6.2. Length of the strings. Proving that no q-harmonic polynomial in $\mathbb{K}[X_{n+1}]$ has variable degree $\leq n$ is obviously equivalent to proving that all strings on $\mathbb{K}[X_n]$ have length $\leq n-1$. The rest of this section aims at some partial results in this direction.

PROPOSITION 21. (a) *Let $q = 0$, and F be a 0-harmonic string on $\mathbb{K}[X_n]$. Then, the length of F is at most n;*
(b) *Let q be formal, and F be a q-harmonic string on $\mathbb{K}[X_n]$ of length d, so that f_0 is of the form $f_d \cdot p$, where p is symmetric. Then, the length of F is at most n.*

(a) is a classical result, and (b) is a little extension to certain q-harmonic strings using a similar proof. Note that a q-harmonic string F for q generic will specialize at q_0 to a string $F(q_0)$ of same or smaller length. In particular, the above proposition shows that the string $F(0)$ will be of length at most $n-1$, so all the $f_i(q), i \geq n$ are divisible by q.

PROOF. (a) Let F be a 0-harmonic string on $\mathbb{K}[X_n]$ of length $k > n$ and degree $d = n+1$. On n variables, there is an explicit relation between symmetric polynomials:

$$\sum_{|\lambda|=d} (-1)^{\ell(\lambda)} c_\lambda P^*_{0,\lambda} = 0 \tag{6.22}$$

where the c_λ are positive constants (cf. [**10**]). This comes from the fact that, on n variables, the n first power sums generate $\mathrm{Sym}(X_n)$ as a $\mathbb{Q}$-algebra. Applying $P^*_{0,\lambda}$ to f_{k-d} yields $(-1)^{\ell(\lambda)} d_\lambda f_k$ where d_λ are polynomials in k. Hence a contradiction.

(b) Let F be a string. Thanks to the string relations,

$$P^*_{q,\lambda}(f_{k-|\lambda|}) = (-1)^{\ell(\lambda)} R_\lambda(q) f_k \tag{6.23}$$

for some polynomial $R_\lambda(q)$ with positive coefficients and constant term 1.

Assume now that d is the length of F, and that f_0 is of the form $f_d p$ with p symmetric. Write the matrix $\left(P^*_{q,\lambda}(m_\mu f_d)\right)_{|\lambda|=|\mu|=n+1, \ell(\mu)\leq n}$. This matrix has one more column than rows. Furthermore, for $q = 0$ it has maximal rank. Hence, for q generic it is also of maximal rank, and there exists a unique linear combination of the rows, up to a scalar coefficient in $\mathbb{C}(q)$. We pick up such a linear combination whose coefficients C_λ are polynomials in $\mathbb{C}[q]$ which are relatively prime in their set:

$$\forall \mu, \quad \sum_\lambda (-1)^{\ell(\lambda)} C_\lambda(q) P^*_{q,\lambda}(m_\mu f_d) = 0. \tag{6.24}$$

At $q = 0$, this relation specializes, up to a coefficient $c \neq 0$, to the relation between the p_λ's described in Equation 6.22. We cancel out this coefficient, so that $C_\lambda(0) = c_\lambda$.

Applying this relation on f_{k-d} yields

$$0 = \sum_{\lambda}(-1)^{\ell(\lambda)} C_\lambda P^*_{q,\lambda}(f_0) = \Big(\sum_{\lambda} C(q)_\lambda R_\lambda(q)\Big) f_d. \tag{6.25}$$

However,

$$\sum_{\lambda} C_\lambda(0) R_\lambda(0) = \sum_{\lambda} c_\lambda \neq 0. \tag{6.26}$$

Hence, $f_d = 0$, which is a contradiction. □

6.3. Complete analysis of the 2 variables case. We work over the alphabet $X = \{x\}$, and make use of the strings machinery. For simplicity, let us use the notation $x^{[d]} := x^d/d!$. The Steenrod operators act by the rule:

$$P^*_{q,k} \cdot x^{[d]} = \big(1 + q(d-k)\big) x^{[d-1]}. \tag{6.27}$$

Let (f_i) be a harmonic string. Then,

$$-P^*_{q,1} \cdot f_k = -P^*_{q,1} \cdot (-P^*_{q,k} \cdot f_0) = -(1+kq) P^*_{q,k+1} \cdot f_0. \tag{6.28}$$

Define A_k so that

$$A_k(f_0) = \big(P^*_{q,1} 1 P^*_{q,k} + (1+kq) P^*_{q,k+1}\big) \cdot f_0 = 0, \tag{6.29}$$

and suppose that $f_0 = x^{[d]}$. Then, $A_k(f_0)$ is of the form $a_k x^{[d-k]}$, with

$$a_k = \big(1 + q(d-k-1)\big)\big(1 + q(d-k) + 1 + kq\big). \tag{6.30}$$

Hence, $F = \mathrm{String}(x^{[d]})$ is harmonic if, and only if,

$$\text{for all } 0 < k < d \quad (2+qd)\big(1 + q(d-k-1)\big) = 0. \tag{6.31}$$

So, there are two possible solutions, namely $d < 2$ and $q = -2/d$.

Going back from strings to harmonic polynomials, we get the proposition:

PROPOSITION 22. *Let q be formal or a complex number. Then, the q-harmonic polynomials in $\mathbb{K}[x, y]$ are spanned by*

$$1 \quad \textit{and} \quad (x-y) \tag{6.32}$$

if q is not of the form $q = -2/d$ for d integer, $d \geq 2$, and by

$$1, \quad (x-y), \quad \textit{and} \quad (x-y)(x+y)^{d-1} \tag{6.33}$$

otherwise.

In particular, if $n \geq 2$ and $d > n(n-1)/2$ the polynomial $f := (x_1 - x_2)(x_1 + x_2)^{d-1}$ is q_0-harmonic for $q_0 := -2/d$. However, the highest degree of an harmonic polynomial in n variables is $n(n-1)/2$. This shows that Conjecture 2 would not hold without condition on q.

PROPOSITION 23. *Take $d > 2$, and $n >= 2$ such that $n(n-1)/2 < d$. Then, the statement of Conjecture 2 would not hold for $q_0 := -2/d$.*

7. Why is the q case more difficult than the $q = 0$ case?

The difficulties when trying to prove Conjecture 2 arises from the fact that the algebra $\mathrm{Steen_q}(X_n)$ is much bigger than $\mathrm{Sym}(X_n)$. We cite here some consequences of this; in this section we will expand more on the last two items

1. $\mathbb{K}[X_n]$ is not a free $\mathrm{Steen_q}(X_n)$ module;
2. The $\mathrm{Steen_q}$-modules generated by two polynomials p_1 and p_2 may be non-isomorphic, even when p_1 and p_2 are of the same degree, or in the same isotypic or Garnir component;
3. There are no proper q-analogs of elementary symmetric functions e_k which vanishes on X_n when $k > n$;
4. Their are very few operators which commute with the action of $\mathrm{Steen_q}$.

With this in view, Conjecture 2 almost comes as a surprise: why is not the increased size of $\mathrm{Steen_q}(X_n)$ counterbalanced by fewer q-harmonics?

7.1. Elementary symmetric functions do not exist. As mentioned in Section 2.5, one way to treat the classical case $q = 0$ is to construct an explicit Gröbner basis of the ideal Hit. This construction relies on the existence of the elementary e_k and complete h_k symmetric functions which satisfy for any k

$$\sum_{i=1}^{k} e_i h_{k-i} = 0, \quad e_k(X+Y) = \sum_{i=1}^{k} e_i(X) e_{k-i}(Y), \tag{7.1}$$

and, last but not least,

$$e_k(X) = 0, \quad \text{whenever } k > |X|. \tag{7.2}$$

Finding q-analogs of the elementary and complete symmetric functions which satisfy the first two conditions is feasible, for example by considering $\mathrm{Steen_q}$ as an appropriate quotient of the ring of non commutative symmetric functions. On the other hand, the third condition does not transfer properly. Indeed, it appears from our computations that, for $n \leq 5$, the operators $\big(P_{q,\lambda}(X_n)\big)_{\lambda \vdash n+1}$ are linearly independent (see Sections A.2). Hence, no linear combination of them vanishes on X_n, and can serve as a reasonable q-analog of e_{n+1}.

7.2. Schubert polynomials do not exist. In the classical case, the quotient $\mathbb{K}[X]/\mathrm{Sym}^+(X)$ is not only isomorphic to the regular representation of $\mathfrak{S}_n$ but also to many deformations of it; among them one can find Hecke algebras. The more general construction has been described by Lascoux and Schützenberger in [**8**], with a 5-parameters deformation.

Recall that the symmetric group is generated by the elementary transpositions σ_i which exchange i and $i+1$ for $i < n$. A presentation is given by the relations

$$\begin{aligned}
\sigma_i^2 &= 1, && \text{for } 1 \leq i \leq n-1,\\
\sigma_i\sigma_j &= \sigma_j\sigma_i, && \text{for } |i-j| > 1,\\
\sigma_i\sigma_{i+1}\sigma_i &= \sigma_{i+1}\sigma_i\sigma_{i+1}, && \text{for } 1 \leq i \leq n-2.
\end{aligned}$$

The last two relations are called the *braid relations*, since they give a presentation of the braid group. A decomposition $\sigma = \sigma_{i_1}\sigma_{i_2}\cdots\sigma_{i_k}$ of minimal length is called a *reduced word* for σ.

The Hecke algebras $H_n(v_1, v_2)$ are a 2-parameters family of quotients of dimension $n!$ of the braid algebra, generated by T_i with the relations

$$(T_i - v_1)(T_i - v_2) = 0, \quad \text{for } 1 \le i \le n-1, \tag{7.3}$$

$$T_iT_j = T_jT_i, \quad \text{for } |i-j| > 1, \tag{7.4}$$

$$T_iT_{i+1}T_i = T_{i+1}T_iT_{i+1}, \quad \text{for } 1 \le i \le n-2. \tag{7.5}$$

Note that, as for the (q, q')-Steenrod algebra, the structure of the algebra depends only on the quotient v_1/v_2.

Here, the useful special case is $v_1 = v_2 = 0$. Indeed, $H_n(0,0)$ acts on polynomials by the *divided differences operators*

$$d_i(f) := \frac{f - \sigma_i f}{x_i - x_{i+1}}. \tag{7.6}$$

To avoid confusion with the derivations we use d_i instead of the usual notation ∂_i. Note that the d_i's are homogeneous operators of degree -1 on polynomials.

Given a reduced word $\sigma_{i_1}\sigma_{i_2}\cdots\sigma_{i_k}$ for a permutation σ, define $d_\sigma := d_{i_1}d_{i_2}\dots d_{i_k}$. Thanks to the braid relations, d_σ is independent of the choice of the reduced word for σ. The family $(d_\sigma)_{\sigma\in\mathfrak{S}_n}$ happens to be a basis of $H_n(0,0)$.

The key point here is the following proposition:

PROPOSITION 24. *The divided differences commute with the action of symmetric functions: for any $p \in \mathrm{Sym}(X_n)$, any $i < n$ and any polynomial f,*

$$d_i(pf) = pd_i(f). \tag{7.7}$$

The quotient $\mathbb{K}[X_n]/\mathrm{Sym}^+(X_n)$ is isomorphic to the regular representation of the divided differences algebra $H_n(0,0)$.

This construction was first defined in geometry, where the quotient $\mathbb{K}[X_n]/\mathrm{Sym}^+(X_n)$ is interpreted as the cohomology ring $H^*(F, \mathbb{Z})$ of the flag manifold F [**1, 2**].

Using the divided differences, one can construct a basis of $\mathbb{K}[X_n]/\mathrm{Sym}^+(X_n)$ composed of the so-called Schubert polynomials:

DEFINITION 8. Let $\rho := (n-1, n-2, \dots, 1, 0)$, so that $X^\rho = x_1^{n-1}x_2^{n-2}\dots x_{n-1}$. For each permutation $\sigma \in \mathfrak{S}_n$ the *Schubert polynomial* S_σ is defined by

$$\mathrm{S}_\sigma := d_{\sigma\omega}(x^\rho), \tag{7.8}$$

where, for technical reasons, ω has been chosen to be the maximal permutation of $\mathfrak{S}_n$.

For more details about Schubert polynomials, we refer to [**9**]. Note that they are denoted by $\mathfrak{S}_\sigma$ in [**9**], and by X_σ in Lascoux and Schützenberger work [**7, 8**].

The main theorem is

THEOREM 7.1 (Lascoux–Schützenberger, 1982). *The family of Schubert polynomials $(\mathrm{S}_\sigma)_{\sigma\in\mathfrak{S}_n}$ is a basis of $\mathbb{K}[X_n]$ as a free* Sym*-module. In particular, the family of Schubert polynomial $(\mathrm{S}_\sigma)_{\sigma\in\mathfrak{S}_n}$ is a basis of $\mathbb{K}[X_n]/\mathrm{Sym}^+$.*

Moreover the divided difference machinery provides a very efficient algorithm to decompose a polynomial in this basis.

This machinery relies heavily on the commutation property of Proposition 24. One strategy to prove Conjecture 2 is to search for a family of linear operators of degree -1 which commute with the action of $\mathrm{Steen_q}$, hoping that they will generate a q-analog of the divided differences algebra. Unfortunately, we obtained by a computation that no such operator exists for, for example, $n = 1, 2, 3, 4$.

Maybe this machinery could still be made to work with some weaker skew commutation condition. A precise statement of this might be:

PROBLEM 2. Does there exist an operator T of degree -1 on polynomials such that for any $f \in \mathrm{Steen_q}$, there exists $g \in \mathrm{Steen_q}$ such that

$$fT = Tg. \tag{7.9}$$

What is the structure of the algebra generated by such operators?

For a similar reason, we could not generalize the fact that Harm is spanned by the derivatives of the Vandermonde. Indeed, there are no reasonable q-analogs of the derivations which commute with the operators $P^*_{q,k}$'s.

Appendix A. Tables

We conclude by a series of tables of computer results which we found interesting during our exploration.

A.1. Tables of q-harmonics. The following listing describes the q-harmonics for $n = 2, 3, 4, 5$. Inside each irreducible representation, we choose only one polynomial. For example, in 4 variables, degree 2 the q-harmonics are of dimension 5. There is an irreducible representation of type $(3, 1)$ generated by the polynomial

$$(x_1 - x_2)\big((q + 1)(x_3 + x_4) - x_2 - x_1\big),$$

and an irreducible representation of type $(2, 2)$ generated by the polynomial

$$(x_1 - x_2)(x_3 - x_4),$$

which is actually a Specht polynomial.

A.2. Tables of cone dimensions. The following tables present, for various polynomials p, the graded dimension of the *cone* $\mathrm{Steen_q} \cdot p$ generated by p, for q generic, as well as for all the specializations at singular q_0's which yield a cone of strictly smaller dimension. For example, Table 3 shows that for $p := x_1 \in \mathbb{K}[X_2]$, and in degree $2 + \deg p$ the dimension of the cone generated by p is 1 for $q_0 = -1$, and 2 for all over values of q.

Some comments are in order:

1. As expected, 0 is never singular in degree $d \leq n$;
2. As expected, there is no singular value for the cone generated by 1;
3. For the cone generated by x_1^k in $\mathbb{K}[X_n]$, the singular values seem to be $-1/k, 0, 1/k, \dots, (n-2)/k$, appearing respectively in degree $n, n+1, \dots, n$; at this degree, the dimension falls by exactly 1. Aside for the scaling of the singular values by $1/k$, the dimensions seems to be independent of k.
4. For the cone generated by the VanderMonde of all the variables in $\mathbb{K}[X_n]$, $n \geq 2$, the singular values seem to be all the rational number of the form $-2/k$, with k even or $k > n$. The singular values for $k = 2n-2, 2n-1, \dots$ appear respectively in degree $1, 2, \dots$; at this degree, the dimension falls by exactly 1.

5. For $q = 0$, the cones generated by any two polynomials are obviously always isomorphic as Sym-module. As is shown by Tables 18 and 19, this does not carry over to $q \neq 0$, even for two polynomials of same degree lying in the same isotypic component. Some further computations actually show that, in a Garnir component indexed by the partition $[3, 1, 1]$, and in degree 5, the higher Specht polynomial indexed by the tableau $[[5], [2], [1, 3, 4]]$ is the unique polynomial p, up to a scalar coefficient, such that -1 is singular for the cone in degree $2 + \deg p$.

TABLE 1. Graded dimension of the cone generated by $p := 1 \in \mathbb{K}[X_5]$.

d	q
1	1
2	2
3	3
4	5
5	7
6	10
7	13
8	18

TABLE 2. Graded dimension of the cone generated by $p := x_1 \in \mathbb{K}[X_1]$.

d	q	-1
1	1	0
2	1	0
3	1	0
4	1	0
5	1	0
6	1	0
7	1	0
8	1	0
9	1	0
10	1	0

TABLE 3. Graded dimension of the cone generated by $p := x_1 \in \mathbb{K}[X_2]$.

d	q	-1	0
1	1		
2	2	1	
3	3	1	2
4	4	1	3
5	5	1	3
6	6	1	4
7	7	1	4
8	8	1	5
9	9	1	5
10	10	1	6

TABLE 4. Graded dimension of the cone generated by $p := x_1 \in \mathbb{K}[X_3]$.

d	q	-1	0	1
1	1			
2	2			
3	3	2		
4	5	3	4	
5	7	3	5	6
6	10	4	7	9
7	13	4	8	11
8	17	5	10	15
9	21	5	12	18

TABLE 5. Graded dimension of the cone generated by $p := x_1 \in \mathbb{K}[X_4]$.

d	q	-1	0	1	2
1	1				
2	2				
3	3				
4	5	4			
5	7	5	6		
6	11	7	9	10	
7	15	8	11	13	14
8	21	10	15	18	20
9	28	12	18	23	26

TABLE 6. Graded dimension of the cone generated by $p := x_1 \in \mathbb{K}[X_5]$.

d	q	-1	0	1	2
1	1				
2	2				
3	3				
4	5				
5	7	6			
6	11	9	10		
7	15	11	13	14	
8	22	15	18	20	21

TABLE 7. Graded dimension of the cone generated by $p := x_1^2 \in \mathbb{K}[X_4]$.

d	q	$-\frac{1}{2}$	0	$\frac{1}{2}$	1
1	1				
2	2				
3	3				
4	5	4			
5	7	5	6		
6	11	7	9	10	
7	15	8	11	13	14
8	21	10	15	18	20

TABLE 8. Graded dimension of the cone generated by $p := x_1^3 \in \mathbb{K}[X_4]$.

d	q	$-\frac{1}{3}$	0	$\frac{1}{3}$	$\frac{2}{3}$
1	1				
2	2				
3	3				
4	5	4			
5	7	5	6		
6	11	7	9	10	
7	15	8	11	13	14
8	21	10	15	18	20

TABLE 9. Graded dimension of the cone generated by $p := x_1 - x_2 \in \mathbb{K}[X_2]$.

d	q	-1	$-\frac{2}{3}$	$-\frac{2}{4}$	$-\frac{2}{5}$	$-\frac{2}{6}$	$-\frac{2}{7}$	$-\frac{2}{8}$	$-\frac{2}{9}$	$-\frac{2}{10}$	$-\frac{2}{11}$
1	1	0									
2	2	1	1								
3	2	1		1							
4	3	1			2						
5	3	1				2					
6	4	1					3				
7	4	1						3			
8	5	1							4		
9	5	1								4	
10	6	1									5

TABLE 10. Graded dimension of the cone generated by $p := x_1 - x_2 \in \mathbb{K}[X_3]$.

d	q	-1	$-\frac{1}{2}$	$-\frac{1}{3}$	0	1
1	1					
2	2	1				
3	3	2				
4	5	3			4	
5	7	3	6		5	6
6	10	4			7	9
7	13	4			8	11
8	16	5		15	10	14
9	20	5			12	17

TABLE 11. Graded dimension of the cone generated by $p := x_1 - x_2 \in \mathbb{K}[X_4]$.

d	q	-1	$-\frac{1}{2}$	0	1	2
1	1					
2	2					
3	3	2				
4	5	4				
5	7	5		6		
6	11	7		9	10	
7	15	8	14	11	13	14
8	21	10		15	18	20

TABLE 12. Graded dimension of the cone generated by $p := x_1 - x_2 \in \mathbb{K}[X_5]$.

d	q	-1	0	1	2
1	1				
2	2				
3	3				
4	5	4			
5	7	6			
6	11	9	10		
7	15	11	13	14	
8	22	15	18	20	21

TABLE 13. Graded dimension of the cone generated by $p := x_1 - x_2 \in \mathbb{K}[X_6]$.

d	q	-1	0
1	1		
2	2		
3	3		
4	5		
5	7	6	
6	11	10	
7	15	13	14

TABLE 14. Graded dimension of the cone generated by $p := \Delta(x_1, x_2, x_3) \in \mathbb{K}[X_3]$.

d	q	-1	$-\frac{2}{4}$	$-\frac{2}{5}$	$-\frac{2}{6}$	$-\frac{2}{7}$	$-\frac{2}{8}$	$-\frac{2}{9}$	$-\frac{2}{10}$	$-\frac{2}{11}$
1	1		0							
2	2	1	1	1						
3	3	2	1	2	2					
4	4	2	2	3	3	3				
5	5	3	2	4	4	4	4			
6	7	3	3	6	6	6	6	6		
7	8	4	3	7	7	7	7	7	7	
8	10	4	4	9	9	9	9	9	9	9

TABLE 15. Graded dimension of the cone generated by $p := \Delta(x_1, x_2, x_3, x_4) \in \mathbb{K}[X_4]$.

d	q	-1	$-\frac{2}{4}$	$-\frac{2}{5}$	$-\frac{2}{6}$	$-\frac{2}{7}$	$-\frac{2}{8}$	$-\frac{2}{9}$	$-\frac{2}{10}$	$-\frac{2}{11}$
1	1				0					
2	2		1		1	1				
3	3	2	2		1	2	2			
4	5	4	3	4	2	3	4	4		
5	6	4	4		2	4	4	5	5	
6	9	6	5		4	6	7	7	8	8

TABLE 16. Graded dimension of the cone generated by $p := \Delta(x_1, x_2, x_3, x_4, x_5) \in \mathbb{K}[X_5]$.

d	q	-1	$-\frac{2}{4}$	$-\frac{2}{6}$	$-\frac{2}{7}$	$-\frac{2}{8}$	$-\frac{2}{9}$	$-\frac{2}{10}$	$-\frac{2}{11}$	$-\frac{2}{12}$
1	1					0				
2	2			1		1	1			
3	3		2	2		1	2	2		
4	5	4	4	3	4	2	3	4	4	
5	7	6	5	4	6	2	4	5	6	6

TABLE 17. Graded dimension of the cone generated by $p := \Delta(x_1, x_2, x_3, x_4, x_5, x_6) \in \mathbb{K}[X_6]$.

d	q	-1	$-\frac{2}{4}$	$-\frac{2}{6}$	$-\frac{2}{8}$	$-\frac{2}{9}$	$-\frac{2}{10}$	$-\frac{2}{11}$	$-\frac{2}{12}$	$-\frac{2}{13}$	$-\frac{2}{14}$
1	1						0				
2	2				1		1	1			
3	3			2	2		1	2	2		
4	5		4	4	3	4	2	3	4	4	
5	7	6	6	5	4	6	2	4	5	6	6

TABLE 18. Graded dimension of the cone generated by the higher Specht polynomial indexed by the tableau $[[4], [3], [1, 2, 5]]$ in the canonical Garnir component, namely $p := 2(x_2 - x_1) \times (x_3 - x_1)(x_3 - x_2)(x_4^2 + x_5^2) \in \mathbb{K}[X_5]$.

d	q	$-\frac{1}{2}$
1	1	
2	2	
3	3	2
4	5	4

TABLE 19. Graded dimension of the cone generated by the higher Specht polynomial indexed by the tableau $[[5], [2], [1, 3, 4]]$ in the canonical Garnir component, namely $p := 2(x_2 - x_1) \times (x_3 - x_1)(x_3 - x_2)x_5x_4 \in \mathbb{K}[X_5]$.

d	q	-1	$-\frac{1}{2}$
1	1		
2	2	1	
3	3	2	2
4	5	2	4

A.3. Tables of singular values of q. Let $d > 0$ and λ be a partition. A complex number q_0 is said to be a *singular value* in the isotypic component V_λ in degree d if the dimension of the q_0-harmonic polynomials in V_λ in degree d exceeds the corresponding dimension for q-harmonic polynomials. The following tables list, for each degree and isotypic component, a set which contains all singular values.

They were computed as follows. Start from a generating family M of the Sym-module $\mathbb{K}[x_1, \ldots, x_n]$, for example the staircase monomials. In the classical case the family $(p_\lambda \cdot m)_{\ell(\lambda)\leq n, m\in M}$ is a basis of $\mathbb{K}[x_1, \ldots, x_n]$. The idea is to compute the determinant $D_{q,n,d}$ of the q-deformation $(P_{q,\lambda} \cdot m)_{\ell(\lambda)\leq n, m\in M}$ in a given degree d. If, for a given $q = q_0$, the determinant does not vanish, then

$$\dim(\mathrm{Hit}_{q_0,n,d}) \geq \dim(\mathrm{Hit}_{n,d}).$$

From the computations of the Tables A.1, we also know that 0 is not singular; hence:

$$\dim(\mathrm{Hit}_{q_0,n,d}) \geq \dim(\mathrm{Hit}_{n,d}) = \dim(\mathrm{Hit}_{q,n,d}).$$

Note that the determinant may vanish even for some non-singular q_0's, because for practical reasons we only used a subset of the Steenrod algebra $(P_{q,\lambda})_{\ell(\lambda)\leq n}$. The complete family $(P_{q,\lambda} \cdot m)_{\lambda, m\in M}$ is not free any more for q generic, and finding the values q_0 which reduces the rank of this family is more complicated than computing a single determinant. Actually, what we did here was to consider only one of the maximal minors, instead of all of them.

We can refine these computation by working inside a given isotypic component of the representation $\mathfrak{S}_n$ or better inside one of its Garnir sub-component. This also drastically reduces the size of the considered families and determinants. Here, instead of computing the determinant of the family, we actually compute the determinant of the Gram matrix. The following tables show for λ and d the vanishing points of the determinant for the basis of the Garnir component composed of Higher Specht Polynomials (see [**15**]).

Note that the homogeneous component of lowest degree of a Garnir component V_λ is generated by its Specht polynomial S_λ; this polynomial is always q_0-harmonic, and q_0 is never singular at this degree. Consider for example the Garnir component $V_{(1,1,1)}$. The Specht polynomial $S_{(1,1,1)}$ is the Vandermonde $(x_1 - x_2)(x_1 - x_3)(x_2 - x_3)$ of degree 3, and there are no polynomials in $V_{(1,1,1)}$ of lower degree. The first singular value is $-\frac{1}{2}$ in degree 4; it corresponds to the polynomial $(x_1 + x_2 + x_3)(x_1 - x_2)(x_1 - x_3)(x_2 - x_3)$.

Consider now the Garnir component $V_{(3,1)}$ in degree 7. The determinant factorizes (upto a huge integer constant) in linear terms:

$$D_{q,(3,1),7} = (q+1)^{34}\,(2q+1)^6\,(3q+1)^2\,(3q+2)^6\,(4q+3)^2\,(5q+2)^4 \\ (5q+3)^2\,(7q+2)^2\,(7q+3)^2\,(7q+4)^2.$$

As far as we have computed, all the determinants factored similarly.

TABLE 20. Singular q_0's for $n = 3$, $d \leq 10$.

λ	d	{Singular values} $\subset$
(3)	$1 \ldots \infty$	$\emptyset$
(2, 1)	1	$\emptyset$
(2, 1)	2	$\{-1\}$
(2, 1)	3	$\{-1, -\frac{2}{3}\}$
(2, 1)	4	$\{-1, -\frac{3}{4}, -\frac{2}{3}, -\frac{1}{2}\}$
(2, 1)	5	$\{-1, -\frac{2}{3}, -\frac{3}{5}, -\frac{1}{2}, -\frac{2}{5}\}$
(2, 1)	6	$\{-1, -\frac{2}{3}, -\frac{1}{2}, -\frac{2}{5}, -\frac{1}{3}\}$
(2, 1)	7	$\{-1, -\frac{2}{3}, -\frac{1}{2}, -\frac{3}{7}, -\frac{2}{5}, -\frac{1}{3}, -\frac{2}{7}\}$
(2, 1)	8	$\{-1, -\frac{2}{3}, -\frac{1}{2}, -\frac{2}{5}, -\frac{3}{8}, -\frac{1}{3}, -\frac{2}{7}, -\frac{1}{4}\}$
(2, 1)	9	$\{-1, -\frac{2}{3}, -\frac{1}{2}, -\frac{2}{5}, -\frac{1}{3}, -\frac{2}{7}, -\frac{1}{4}, -\frac{2}{9}\}$
(2, 1)	10	$\{-1, -\frac{2}{3}, -\frac{1}{2}, -\frac{2}{5}, -\frac{1}{3}, -\frac{3}{10}, -\frac{2}{7}, -\frac{1}{4}, -\frac{2}{9}, -\frac{1}{5}\}$
(1, 1, 1)	3	$\emptyset$
(1, 1, 1)	4	$\{-\frac{1}{2}\}$
(1, 1, 1)	5	$\{-1, -\frac{1}{2}, -\frac{2}{5}\}$
(1, 1, 1)	6	$\{-1, -\frac{1}{2}, -\frac{2}{5}, -\frac{1}{3}\}$
(1, 1, 1)	7	$\{-1, -\frac{1}{2}, -\frac{2}{5}, -\frac{1}{3}, -\frac{2}{7}\}$
(1, 1, 1)	8	$\{-1, -\frac{1}{2}, -\frac{2}{5}, -\frac{1}{3}, -\frac{2}{7}, -\frac{1}{4}\}$
(1, 1, 1)	9	$\{-1, -\frac{1}{2}, -\frac{2}{5}, -\frac{1}{3}, -\frac{2}{7}, -\frac{1}{4}, -\frac{2}{9}\}$
(1, 1, 1)	10	$\{-1, -\frac{1}{2}, -\frac{2}{5}, -\frac{1}{3}, -\frac{2}{7}, -\frac{1}{4}, -\frac{2}{9}, -\frac{1}{5}\}$

TABLE 21. Singular q_0's for $n = 4$, $d \leq 9$

λ	d	{Singular q_0's} $\subset$
(4)	$1 \ldots \infty$	$\emptyset$
$(3,1)$	1	$\emptyset$
$(3,1)$	2	$\{-1\}$
$(3,1)$	3	$\{-1, -\frac{2}{3}\}$
$(3,1)$	4	$\{-1, -\frac{3}{4}, -\frac{2}{3}, -\frac{1}{2}\}$
$(3,1)$	5	$\{-1, -\frac{4}{5}, -\frac{3}{4}, -\frac{2}{3}, -\frac{3}{5}, -\frac{1}{2}, -\frac{2}{5}\}$
$(3,1)$	6	$\{-1, -\frac{3}{4}, -\frac{2}{3}, -\frac{3}{5}, -\frac{1}{2}, -\frac{2}{5}, -\frac{1}{3}\}$
$(3,1)$	7	$\{-1, -\frac{3}{4}, -\frac{2}{3}, -\frac{3}{5}, -\frac{4}{7}, -\frac{1}{2}, -\frac{3}{7}, -\frac{2}{5}, -\frac{1}{3}, -\frac{2}{7}\}$
$(3,1)$	8	$\{-1, -\frac{3}{4}, -\frac{2}{3}, -\frac{3}{5}, -\frac{1}{2}, -\frac{3}{7}, -\frac{2}{5}, -\frac{3}{8}, -\frac{1}{3}, -\frac{2}{7}, -\frac{1}{4}\}$
$(3,1)$	9	$\{-1, -\frac{3}{4}, -\frac{2}{3}, -\frac{3}{5}, -\frac{1}{2}, -\frac{4}{9}, -\frac{3}{7}, -\frac{2}{5}, -\frac{3}{8}, -\frac{1}{3}, -\frac{2}{7}, -\frac{1}{4}, -\frac{2}{9}\}$
$(3,1)$	10	$\{-1, -\frac{3}{4}, -\frac{2}{3}, -\frac{3}{5}, -\frac{1}{2}, -\frac{3}{7}, -\frac{2}{5}, -\frac{3}{8}, -\frac{1}{3}, -\frac{3}{10}, -\frac{2}{7}, -\frac{1}{4}, -\frac{2}{9}, -\frac{1}{5}\}$
$(2,2)$	2	$\emptyset$
$(2,2)$	3	$\{-1\}$
$(2,2)$	4	$\{-1\}$
$(2,2)$	5	$\{-1, -\frac{3}{5}, -\frac{1}{2}\}$
$(2,2)$	6	$\{-1, -\frac{2}{3}, -\frac{3}{5}, -\frac{1}{2}, -\frac{2}{5}\}$
$(2,2)$	7	$\{-1, -\frac{3}{5}, -\frac{1}{2}, -\frac{3}{7}, -\frac{2}{5}, -\frac{1}{3}\}$
$(2,2)$	8	$\{-1, -\frac{3}{5}, -\frac{1}{2}, -\frac{3}{7}, -\frac{2}{5}, -\frac{3}{8}, -\frac{1}{3}, -\frac{2}{7}\}$
$(2,2)$	9	$\{-1, -\frac{3}{5}, -\frac{1}{2}, -\frac{3}{7}, -\frac{2}{5}, -\frac{3}{8}, -\frac{1}{3}, -\frac{2}{7}, -\frac{1}{4}\}$
$(2,2)$	10	$\{-1, -\frac{3}{5}, -\frac{1}{2}, -\frac{3}{7}, -\frac{2}{5}, -\frac{3}{8}, -\frac{1}{3}, -\frac{3}{10}, -\frac{2}{7}, -\frac{1}{4}, -\frac{2}{9}\}$
$(2,1,1)$	3	$\emptyset$
$(2,1,1)$	4	$\{-\frac{1}{2}\}$
$(2,1,1)$	5	$\{-1, -\frac{1}{2}, -\frac{2}{5}\}$
$(2,1,1)$	6	$\{-1, -\frac{1}{2}, -\frac{2}{5}, -\frac{1}{3}\}$
$(2,1,1)$	7	$\{-1, -\frac{4}{7}, -\frac{1}{2}, -\frac{3}{7}, -\frac{2}{5}, -\frac{1}{3}, -\frac{2}{7}\}$
$(2,1,1)$	8	$\{-1, -\frac{1}{2}, -\frac{3}{7}, -\frac{2}{5}, -\frac{3}{8}, -\frac{1}{3}, -\frac{2}{7}, -\frac{1}{4}\}$
$(2,1,1)$	9	$\{-1, -\frac{1}{2}, -\frac{4}{9}, -\frac{3}{7}, -\frac{2}{5}, -\frac{3}{8}, -\frac{1}{3}, -\frac{2}{7}, -\frac{1}{4}, -\frac{2}{9}\}$
$(2,1,1)$	10	$\{-1, -\frac{1}{2}, -\frac{3}{7}, -\frac{2}{5}, -\frac{3}{8}, -\frac{1}{3}, -\frac{3}{10}, -\frac{2}{7}, -\frac{1}{4}, -\frac{2}{9}, -\frac{1}{5}\}$
$(1,1,1,1)$	6	$\emptyset$
$(1,1,1,1)$	7	$\{-\frac{1}{3}\}$
$(1,1,1,1)$	8	$\{-\frac{1}{2}, -\frac{1}{3}, -\frac{2}{7}\}$
$(1,1,1,1)$	9	$\{-1, -\frac{1}{2}, -\frac{1}{3}, -\frac{2}{7}, -\frac{1}{4}\}$
$(1,1,1,1)$	10	$\{-1, -\frac{1}{2}, -\frac{2}{5}, -\frac{1}{3}, -\frac{2}{7}, -\frac{1}{4}, -\frac{2}{9}\}$

TABLE 22. Singular q_0's for $n = 5$, $d \leq 8$.

λ	d	$\{\text{Singular } q_0\text{'s}\} \subset$
(5)	$1 \ldots \infty$	$\emptyset$
$(4,1)$	1	$\emptyset$
$(4,1)$	2	$\{-1\}$
$(4,1)$	3	$\{-1, -\frac{2}{3}\}$
$(4,1)$	4	$\{-1, -\frac{3}{4}, -\frac{2}{3}, -\frac{1}{2}\}$
$(4,1)$	5	$\{-1, -\frac{4}{5}, -\frac{3}{4}, -\frac{2}{3}, -\frac{3}{5}, -\frac{1}{2}, -\frac{2}{5}\}$
$(4,1)$	6	$\{-1, -\frac{5}{6}, -\frac{4}{5}, -\frac{3}{4}, -\frac{2}{3}, -\frac{3}{5}, -\frac{1}{2}, -\frac{2}{5}, -\frac{1}{3}\}$
$(4,1)$	7	$\{-1, -\frac{4}{5}, -\frac{3}{4}, -\frac{5}{7}, -\frac{2}{3}, -\frac{3}{5}, -\frac{4}{7}, -\frac{1}{2}, -\frac{3}{7}, -\frac{2}{5}, -\frac{1}{3}, -\frac{2}{7}\}$
$(4,1)$	8	$\{-1, -\frac{4}{5}, -\frac{3}{4}, -\frac{2}{3}, -\frac{5}{8}, -\frac{3}{5}, -\frac{4}{7}, -\frac{1}{2}, -\frac{3}{7}, -\frac{2}{5}, -\frac{3}{8}, -\frac{1}{3}, -\frac{2}{7}, -\frac{1}{4}\}$
$(3,2)$	2	$\emptyset$
$(3,2)$	3	$\{-1\}$
$(3,2)$	4	$\{-1\}$
$(3,2)$	5	$\{-1, -\frac{3}{5}, -\frac{1}{2}\}$
$(3,2)$	6	$\{-1, -\frac{2}{3}, -\frac{3}{5}, -\frac{1}{2}, -\frac{2}{5}\}$
$(3,2)$	7	$\{-1, -\frac{5}{7}, -\frac{2}{3}, -\frac{3}{5}, -\frac{4}{7}, -\frac{1}{2}, -\frac{3}{7}, -\frac{2}{5}, -\frac{1}{3}\}$
$(3,2)$	8	$\{-1, -\frac{2}{3}, -\frac{5}{8}, -\frac{3}{5}, -\frac{4}{7}, -\frac{1}{2}, -\frac{3}{7}, -\frac{2}{5}, -\frac{3}{8}, -\frac{1}{3}, -\frac{2}{7}\}$
$(3,1,1)$	3	$\emptyset$
$(3,1,1)$	4	$\{-\frac{1}{2}\}$
$(3,1,1)$	5	$\{-1, -\frac{1}{2}, -\frac{2}{5}\}$
$(3,1,1)$	6	$\{-1, -\frac{1}{2}, -\frac{2}{5}, -\frac{1}{3}\}$
$(3,1,1)$	7	$\{-1, -\frac{4}{7}, -\frac{1}{2}, -\frac{3}{7}, -\frac{2}{5}, -\frac{1}{3}, -\frac{2}{7}\}$
$(3,1,1)$	8	$\{-1, -\frac{5}{8}, -\frac{4}{7}, -\frac{1}{2}, -\frac{3}{7}, -\frac{2}{5}, -\frac{3}{8}, -\frac{1}{3}, -\frac{2}{7}, -\frac{1}{4}\}$
$(2,2,1)$	4	$\emptyset$
$(2,2,1)$	5	$\{-\frac{1}{2}\}$
$(2,2,1)$	6	$\{-1, -\frac{1}{2}, -\frac{2}{5}\}$
$(2,2,1)$	7	$\{-1, -\frac{1}{2}, -\frac{3}{7}, -\frac{2}{5}, -\frac{1}{3}\}$
$(2,2,1)$	8	$\{-1, -\frac{1}{2}, -\frac{3}{7}, -\frac{2}{5}, -\frac{3}{8}, -\frac{1}{3}, -\frac{2}{7}\}$
$(2,1,1,1)$	6	$\emptyset$
$(2,1,1,1)$	7	$\{-\frac{1}{3}\}$
$(2,1,1,1)$	8	$\{-\frac{1}{2}, -\frac{1}{3}, -\frac{2}{7}\}$

TABLE 23. Singular q_0's for $n = 6$, $d \leq 8$.

λ	d	$\{$Singular q_0's$\} \subset$
(6)	$1 \dots \infty$	$\emptyset$
$(5,1)$	1	$\emptyset$
$(5,1)$	2	$\{-1\}$
$(5,1)$	3	$\{-1, -\frac{2}{3}\}$
$(5,1)$	4	$\{-1, -\frac{3}{4}, -\frac{2}{3}, -\frac{1}{2}\}$
$(5,1)$	5	$\{-1, -\frac{4}{5}, -\frac{3}{4}, -\frac{2}{3}, -\frac{3}{5}, -\frac{1}{2}, -\frac{2}{5}\}$
$(5,1)$	6	$\{-1, -\frac{5}{6}, -\frac{4}{5}, -\frac{3}{4}, -\frac{2}{3}, -\frac{3}{5}, -\frac{1}{2}, -\frac{2}{5}, -\frac{1}{3}\}$
$(5,1)$	7	$\{-1, -\frac{6}{7}, -\frac{5}{6}, -\frac{4}{5}, -\frac{3}{4}, -\frac{5}{7}, -\frac{2}{3}, -\frac{3}{5}, -\frac{4}{7}, -\frac{1}{2}, -\frac{3}{7}, -\frac{2}{5}, -\frac{1}{3}, -\frac{2}{7}\}$
$(5,1)$	8	$\{-1, -\frac{5}{6}, -\frac{4}{5}, -\frac{3}{4}, -\frac{5}{7}, -\frac{2}{3}, -\frac{5}{8}, -\frac{3}{5}, -\frac{4}{7}, -\frac{1}{2}, -\frac{3}{7}, -\frac{2}{5}, -\frac{3}{8}, -\frac{1}{3}, -\frac{2}{7}, -\frac{1}{4}\}$
$(4,2)$	2	$\emptyset$
$(4,2)$	3	$\{-1\}$
$(4,2)$	4	$\{-1\}$
$(4,2)$	5	$\{-1, -\frac{3}{5}, -\frac{1}{2}\}$
$(4,2)$	6	$\{-1, -\frac{2}{3}, -\frac{3}{5}, -\frac{1}{2}, -\frac{2}{5}\}$
$(4,2)$	7	$\{-1, -\frac{5}{7}, -\frac{2}{3}, -\frac{3}{5}, -\frac{4}{7}, -\frac{1}{2}, -\frac{3}{7}, -\frac{2}{5}, -\frac{1}{3}\}$
$(4,2)$	8	$\{-1, -\frac{3}{4}, -\frac{5}{7}, -\frac{2}{3}, -\frac{5}{8}, -\frac{3}{5}, -\frac{4}{7}, -\frac{1}{2}, -\frac{3}{7}, -\frac{2}{5}, -\frac{3}{8}, -\frac{1}{3}, -\frac{2}{7}\}$
$(4,1,1)$	3	$\emptyset$
$(4,1,1)$	4	$\{-\frac{1}{2}\}$
$(4,1,1)$	5	$\{-1, -\frac{1}{2}, -\frac{2}{5}\}$
$(4,1,1)$	6	$\{-1, -\frac{1}{2}, -\frac{2}{5}, -\frac{1}{3}\}$
$(4,1,1)$	7	$\{-1, -\frac{4}{7}, -\frac{1}{2}, -\frac{3}{7}, -\frac{2}{5}, -\frac{1}{3}, -\frac{2}{7}\}$
$(4,1,1)$	8	$\{-1, -\frac{5}{8}, -\frac{4}{7}, -\frac{1}{2}, -\frac{3}{7}, -\frac{2}{5}, -\frac{3}{8}, -\frac{1}{3}, -\frac{2}{7}, -\frac{1}{4}\}$
$(3,3)$	3	$\emptyset$
$(3,3)$	4	$\{-1\}$
$(3,3)$	5	$\{-1\}$
$(3,3)$	6	$\{-1, -\frac{1}{2}\}$
$(3,3)$	7	$\{-1, -\frac{4}{7}, -\frac{1}{2}, -\frac{2}{5}\}$
$(3,3)$	8	$\{-1, -\frac{5}{8}, -\frac{4}{7}, -\frac{1}{2}, -\frac{3}{7}, -\frac{2}{5}, -\frac{1}{3}\}$
$(3,2,1)$	4	$\emptyset$
$(3,2,1)$	5	$\{-\frac{1}{2}\}$
$(3,2,1)$	6	$\{-1, -\frac{1}{2}, -\frac{2}{5}\}$
$(3,2,1)$	7	$\{-1, -\frac{1}{2}, -\frac{3}{7}, -\frac{2}{5}, -\frac{1}{3}\}$
$(3,2,1)$	8	$\{-1, -\frac{1}{2}, -\frac{3}{7}, -\frac{2}{5}, -\frac{3}{8}, -\frac{1}{3}, -\frac{2}{7}\}$
$(3,1,1,1)$	6	$\emptyset$
$(3,1,1,1)$	7	$\{-\frac{1}{3}\}$
$(3,1,1,1)$	8	$\{-\frac{1}{2}, -\frac{1}{3}, -\frac{2}{7}\}$
$(2,2,2)$	6	$\emptyset$
$(2,2,2)$	7	$\{-\frac{1}{2}\}$
$(2,2,2)$	8	$\{-\frac{1}{2}\}$
$(2,2,1,1)$	7	$\emptyset$
$(2,2,1,1)$	8	$\{-\frac{1}{3}\}$

TABLE 24. Singular q_0's for $n = 7$, $d \leq 7$.

λ	d	$\{$Singular q_0's$\} \subset$
(7)	$1 \ldots \infty$	$\emptyset$
$(6,1)$	1	$\emptyset$
$(6,1)$	2	$\{-1\}$
$(6,1)$	3	$\{-1, -\frac{2}{3}\}$
$(6,1)$	4	$\{-1, -\frac{3}{4}, -\frac{2}{3}, -\frac{1}{2}\}$
$(6,1)$	5	$\{-1, -\frac{4}{5}, -\frac{3}{4}, -\frac{2}{3}, -\frac{3}{5}, -\frac{1}{2}, -\frac{2}{5}\}$
$(6,1)$	6	$\{-1, -\frac{5}{6}, -\frac{4}{5}, -\frac{3}{4}, -\frac{2}{3}, -\frac{3}{5}, -\frac{1}{2}, -\frac{2}{5}, -\frac{1}{3}\}$
$(6,1)$	7	$\{-1, -\frac{6}{7}, -\frac{5}{6}, -\frac{4}{5}, -\frac{3}{4}, -\frac{5}{7}, -\frac{2}{3}, -\frac{3}{5}, -\frac{4}{7}, -\frac{1}{2}, -\frac{3}{7}, -\frac{2}{5}, -\frac{1}{3}, -\frac{2}{7}\}$
$(5,2)$	2	$\emptyset$
$(5,2)$	3	$\{-1\}$
$(5,2)$	4	$\{-1\}$
$(5,2)$	5	$\{-1, -\frac{3}{5}, -\frac{1}{2}\}$
$(5,2)$	6	$\{-1, -\frac{2}{3}, -\frac{3}{5}, -\frac{1}{2}, -\frac{2}{5}\}$
$(5,2)$	7	$\{-1, -\frac{5}{7}, -\frac{2}{3}, -\frac{3}{5}, -\frac{4}{7}, -\frac{1}{2}, -\frac{3}{7}, -\frac{2}{5}, -\frac{1}{3}\}$
$(5,1,1)$	3	$\emptyset$
$(5,1,1)$	4	$\{-\frac{1}{2}\}$
$(5,1,1)$	5	$\{-1, -\frac{1}{2}, -\frac{2}{5}\}$
$(5,1,1)$	6	$\{-1, -\frac{1}{2}, -\frac{2}{5}, -\frac{1}{3}\}$
$(5,1,1)$	7	$\{-1, -\frac{4}{7}, -\frac{1}{2}, -\frac{3}{7}, -\frac{2}{5}, -\frac{1}{3}, -\frac{2}{7}\}$
$(4,3)$	3	$\emptyset$
$(4,3)$	4	$\{-1\}$
$(4,3)$	5	$\{-1\}$
$(4,3)$	6	$\{-1, -\frac{1}{2}\}$
$(4,3)$	7	$\{-1, -\frac{4}{7}, -\frac{1}{2}, -\frac{2}{5}\}$
$(4,2,1)$	4	$\emptyset$
$(4,2,1)$	5	$\{-\frac{1}{2}\}$
$(4,2,1)$	6	$\{-1, -\frac{1}{2}, -\frac{2}{5}\}$
$(4,2,1)$	7	$\{-1, -\frac{1}{2}, -\frac{3}{7}, -\frac{2}{5}, -\frac{1}{3}\}$
$(4,1,1,1)$	6	$\emptyset$
$(4,1,1,1)$	7	$\{-\frac{1}{3}\}$
$(3,3,1)$	5	$\emptyset$
$(3,3,1)$	6	$\{-\frac{1}{2}\}$
$(3,3,1)$	7	$\{-1, -\frac{1}{2}, -\frac{2}{5}\}$
$(3,2,2)$	6	$\emptyset$
$(3,2,2)$	7	$\{-\frac{1}{2}\}$
$(3,2,1,1)$	7	$\emptyset$

TABLE 25. Singular q_0's for $n = 8$, $d \leq 7$.

λ	d	{Singular q_0's} $\subset$
(8)	$1 \ldots \infty$	$\emptyset$
$(7,1)$	1	$\emptyset$
$(7,1)$	2	$\{-1\}$
$(7,1)$	3	$\{-1, -\frac{2}{3}\}$
$(7,1)$	4	$\{-1, -\frac{3}{4}, -\frac{2}{3}, -\frac{1}{2}\}$
$(7,1)$	5	$\{-1, -\frac{4}{5}, -\frac{3}{4}, -\frac{2}{3}, -\frac{3}{5}, -\frac{1}{2}, -\frac{2}{5}\}$
$(7,1)$	6	$\{-1, -\frac{5}{6}, -\frac{4}{5}, -\frac{3}{4}, -\frac{2}{3}, -\frac{3}{5}, -\frac{1}{2}, -\frac{2}{5}, -\frac{1}{3}\}$
$(7,1)$	7	$\{-1, -\frac{6}{7}, -\frac{5}{6}, -\frac{4}{5}, -\frac{3}{4}, -\frac{5}{7}, -\frac{2}{3}, -\frac{3}{5}, -\frac{4}{7}, -\frac{1}{2}, -\frac{3}{7}, -\frac{2}{5}, -\frac{1}{3}, -\frac{2}{7}\}$
$(6,2)$	2	$\emptyset$
$(6,2)$	3	$\{-1\}$
$(6,2)$	4	$\{-1\}$
$(6,2)$	5	$\{-1, -\frac{3}{5}, -\frac{1}{2}\}$
$(6,2)$	6	$\{-1, -\frac{2}{3}, -\frac{3}{5}, -\frac{1}{2}, -\frac{2}{5}\}$
$(6,2)$	7	$\{-1, -\frac{5}{7}, -\frac{2}{3}, -\frac{3}{5}, -\frac{4}{7}, -\frac{1}{2}, -\frac{3}{7}, -\frac{2}{5}, -\frac{1}{3}\}$
$(6,1,1)$	3	$\emptyset$
$(6,1,1)$	4	$\{-\frac{1}{2}\}$
$(6,1,1)$	5	$\{-1, -\frac{1}{2}, -\frac{2}{5}\}$
$(6,1,1)$	6	$\{-1, -\frac{1}{2}, -\frac{2}{5}, -\frac{1}{3}\}$
$(6,1,1)$	7	$\{-1, -\frac{4}{7}, -\frac{1}{2}, -\frac{3}{7}, -\frac{2}{5}, -\frac{1}{3}, -\frac{2}{7}\}$
$(5,3)$	3	$\emptyset$
$(5,3)$	4	$\{-1\}$
$(5,3)$	5	$\{-1\}$
$(5,3)$	6	$\{-1, -\frac{1}{2}\}$
$(5,3)$	7	$\{-1, -\frac{4}{7}, -\frac{1}{2}, -\frac{2}{5}\}$
$(5,2,1)$	4	$\emptyset$
$(5,2,1)$	5	$\{-\frac{1}{2}\}$
$(5,2,1)$	6	$\{-1, -\frac{1}{2}, -\frac{2}{5}\}$
$(5,2,1)$	7	$\{-1, -\frac{1}{2}, -\frac{3}{7}, -\frac{2}{5}, -\frac{1}{3}\}$
$(5,1,1,1)$	6	$\emptyset$
$(5,1,1,1)$	7	$\{-\frac{1}{3}\}$
$(4,4)$	4	$\emptyset$
$(4,4)$	5	$\{-1\}$
$(4,4)$	6	$\{-1\}$
$(4,4)$	7	$\{-1, -\frac{1}{2}\}$
$(4,3,1)$	5	$\emptyset$
$(4,3,1)$	6	$\{-\frac{1}{2}\}$
$(4,3,1)$	7	$\{-1, -\frac{1}{2}, -\frac{2}{5}\}$
$(4,2,2)$	6	$\emptyset$
$(4,2,2)$	7	$\{-\frac{1}{2}\}$
$(4,2,1,1)$	7	$\emptyset$
$(3,3,2)$	7	$\emptyset$

TABLE 26. Singular q_0's for $n = 9$, $d \leq 6$.

λ	d	{Singular q_0's} $\subset$
(9)	$1 \ldots \infty$	$\emptyset$
$(8,1)$	1	$\emptyset$
$(8,1)$	2	$\{-1\}$
$(8,1)$	3	$\{-1, -\frac{2}{3}\}$
$(8,1)$	4	$\{-1, -\frac{3}{4}, -\frac{2}{3}, -\frac{1}{2}\}$
$(8,1)$	5	$\{-1, -\frac{4}{5}, -\frac{3}{4}, -\frac{2}{3}, -\frac{3}{5}, -\frac{1}{2}, -\frac{2}{5}\}$
$(8,1)$	6	$\{-1, -\frac{5}{6}, -\frac{4}{5}, -\frac{3}{4}, -\frac{2}{3}, -\frac{3}{5}, -\frac{1}{2}, -\frac{2}{5}, -\frac{1}{3}\}$
$(7,2)$	2	$\emptyset$
$(7,2)$	3	$\{-1\}$
$(7,2)$	4	$\{-1\}$
$(7,2)$	5	$\{-1, -\frac{3}{5}, -\frac{1}{2}\}$
$(7,2)$	6	$\{-1, -\frac{2}{3}, -\frac{3}{5}, -\frac{1}{2}, -\frac{2}{5}\}$
$(7,1,1)$	3	$\emptyset$
$(7,1,1)$	4	$\{-\frac{1}{2}\}$
$(7,1,1)$	5	$\{-1, -\frac{1}{2}, -\frac{2}{5}\}$
$(7,1,1)$	6	$\{-1, -\frac{1}{2}, -\frac{2}{5}, -\frac{1}{3}\}$
$(6,3)$	3	$\emptyset$
$(6,3)$	4	$\{-1\}$
$(6,3)$	5	$\{-1\}$
$(6,3)$	6	$\{-1, -\frac{1}{2}\}$
$(6,2,1)$	4	$\emptyset$
$(6,2,1)$	5	$\{-\frac{1}{2}\}$
$(6,2,1)$	6	$\{-1, -\frac{1}{2}, -\frac{2}{5}\}$
$(6,1,1,1)$	6	$\emptyset$
$(5,4)$	4	$\emptyset$
$(5,4)$	5	$\{-1\}$
$(5,4)$	6	$\{-1\}$
$(5,3,1)$	5	$\emptyset$
$(5,3,1)$	6	$\{-\frac{1}{2}\}$
$(5,2,2)$	6	$\emptyset$
$(4,4,1)$	6	$\emptyset$

References

1. I. N. Bernšteĭn, I. M. Gel′fand, and S. I. Gel′fand, *Schubert cells, and the cohomology of the spaces G/P*, Russian Math. Survey **28** (1973), no. 3(171), 3–26.
2. M. Demazure, *Désingularisation des variétés de Schubert généralisées*, Collection of articles dedicated to Henri Cartan on the occasion of his 70th birthday, I, Ann. Sci. École Norm. Sup. (4) **7** (1974), 53–88.
3. W. Fulton and J. Harris, *Representation theory*, A first course, Grad. Texts in Math., vol. 129, Readings in Mathematics, Springer-Verlag, New York, 1991.
4. A. M. Garsia and M. Hamain, *Orbit harmonics and graded representations* (S. Brlek, ed.), Research Monograph, Lab. de Comb. et Informatique Mathématique, Université du Québec à Montréal; J. Algebraic Combin. **3** (1994), 17–76.
5. G. H. Hardy and S. Ramajuan, *Asymptotic formulae in combinatory analysis*, Proc. London Math. Soc. **17** (1918), no. 2, 75–115.
6. D. Krob, *Éléments de combinatoire*, Notes de cours, Magistère 1ère année, École Normale Supérieure, 1995.
7. A. Lascoux and M.-P. Schützenberger, *Polynômes de Schubert*, C. R. Acad. Sci. Paris Sér. I Math. **294** (1982), no. 13, 447–450.
8. ———, *Symmetrization operators in polynomial rings*, Funktsional. Anal. i Prilozhen. **21** (1987), no. 4, 77–78. (Russian)
9. I. G. Macdonald, *Schubert polynomials*, Surveys in Combinatorics (Guildford, 1991), London Math. Soc. Lecture Note Ser., vol. 166, Cambridge Univ. Press, Cambridge, 1991, pp. 73–99.
10. ———, *Symmetric functions and Hall polynomials*, 2nd ed., Oxford Mathematical Monographs, vol. 166, Clarendon, 1995.
11. F. Hivert and N. Thiéry, et al., `MuPAD`-*Combinat—a library for Algebraic Combinatorics in* `MuPAD`, `http://mupad-combinat.sf.net/`.
12. The `MuPAD` group, B. Fuchssteiner et al., `MuPAD` *user's manual*—`MuPAD` *version* 1.2.2, 1st ed., John Wiley and sons, Chichester, New York, 1996.
13. B. E. Sagan, *The symmetric group*, Representations, Combinatorial Algorithms, and Symmetric Functions, The Wadsworth & Brooks/Cole Mathematics Series, Wadsworth & Brooks/Cole Advanced Books & Software, Pacific Grove, CA, 1991.
14. B. Sturmfels, *Algorithms in invariant theory*, Springer-Verlag, Vienna, 1993.
15. T. Terasoma and H. Yamada, *Higher specht polynomials for the symmetric group*, Proc. Japan Acad. Ser. A **69** (1993), 41-44.
16. R. M. W. Wood, *The hit problems and the Steenrod algebra*, Proceedings of the summer school Interactions Between Algebraic Topology and Invariant Theory (Greece, 2000), University of Ioannina reports, 2001.
17. ———, *Problems in the Steenrod algebra*, Bull. London Math. Soc. **30**, (1998), no. 5, 449–517.
18. ———, *Differential operators and the Steenrod algebra*, Proc. London Math. Soc. (3) **75** (1997), no. 1, 194–220.

INSTITUT GASPARD MONGE, UNIVERSITÉ DE MARNE-LA-VALLÉE, 5, BD DESCARTES, CHAMPS SUR MARNE, 77 454 MARNE LA VALLÉE CEDEX 2, FRANCE.

E-mail address: `Florent.Hivert@univ-mlv.fr`

LAPCS, UNIVERSITÉ CLAUDE BERNARD LYON I, BÂTIMENT RECHERCHE [B] 50, AVENUE TONY-GARNIER, DOMAINE DE GERLAND, 69366 LYON CEDEX 07, FRANCE.

E-mail address: `nthiery@users.sourceforge.net`

Centre de Recherches Mathématiques
CRM Proceedings and Lecture Notes
Volume **35**, 2004

Cohomology with Grosshans Graded Coefficients

W. van der Kallen

ABSTRACT. Let the reductive group G act on the finitely generated commutative k-algebra A. We ask if the finite generation property of the ring of invariants A^G extends to the full cohomology ring $H^*(G, A)$. We confirm this for $G = \mathrm{SL}_2$ and also when the action on A is replaced by the "contracted" action on the Grosshans graded ring $\operatorname{gr} A$, provided the characteristic of k is large.

1. Introduction

Consider a linear algebraic group or group scheme G defined over an algebraically closed field k of positive characteristic p. Let A be a finitely generated commutative k-algebra on which G acts rationally by k-algebra automorphisms. So G acts on $\operatorname{Spec}(A)$. One may then ask if the cohomology ring $H^*(G, A)$ is finitely generated as a k-algebra.

We think of this as a question in invariant theory. Our question is not what combinations of G and A yield a finitely generated $H^*(G, A)$. Instead we are interested in finding those G for which every A as above will give a finitely generated $H^*(G, A)$. In particular, G should be such that for all these A the ring of invariants $A^G = H^0(G, A)$ is finitely generated. This is why we must restrict attention to geometrically reductive G. We believe no further restriction is needed, and our aim is to present some evidence for this. In characteristic zero there would be nothing to do. Indeed, suppose G is a reductive linear algebraic group over $\mathbb{C}$. We are concerned with representations of the algebraic group, the so-called rational representations. The group is linearly reductive, meaning that all extensions of representations split (even when the representations are infinite dimensional). So there is no higher rational cohomology and $H^*(G, A) = A^G$. Thus our problem asks nothing new in this case. That is why we now return to characteristic p. Notice that in characteristic p there is one type of module or algebra for which the invariants, including the higher invariants known as cohomology, behave as in characteristic zero. They are the modules/algebras with good filtration. They will serve as a natural tool in the sequel.

2000 *Mathematics Subject Classification.* Primary: 20G10; Secondary: 14L24.
This is the final version of the paper.

Our proofs combine arguments and results from several earlier works. As our question concerns "higher invariant theory", it is clear that invariant theory will play its part. On the other hand our work is a direct descendent of the work of Evens for the case that G is a finite group, and the work of Friedlander and Suslin for the case of finite group schemes. We try to merge this strand with invariant theory and emphasize that in both cases G happens to be geometrically reductive.

Thus say G is geometrically reductive. Then we know by Nagata that at least the ring of invariants $A^G = H^0(G, A)$ is finitely generated. (As explained in [**4**], Nagata's proof [**21**] extends to group schemes.)

By Waterhouse [**28**] a finite group scheme is geometrically reductive. If G is a finite group scheme, then A is a finite module over A^G and hence the cohomology ring $H^*(G, A)$ is indeed finitely generated by Friedlander and Suslin [**11**]. (If G is finite reduced, see Evens [**8**, Thm. 8.1]. If G is finite and connected, take $C = A^G$ in [**11**, Theorem 1.5, 1.5.1]. If G is not connected, one finishes the argument by following [**8**] [**11**, pages 220–221].)

Note that if the geometrically reductive G is a subgroup of GL_n, then GL_n/G is affine, $\mathrm{ind}_G^{\mathrm{GL}_n} A$ is finitely generated, and $H^*(G, A) = H^*(\mathrm{GL}_n, \mathrm{ind}_G^{\mathrm{GL}_n} A)$. (Compare [**21, 23**], [**13**, Ch. II], [**15**, I 4.6, I 5.13].) Therefore let us now assume $G = \mathrm{GL}_n$, or rather $G = \mathrm{SL}_n$ to keep it semi-simple. (The SL_n case suffices, as $H^*(\mathrm{GL}_n, A) = H^*(\mathrm{SL}_n, A)^{\mathbb{G}_m}$ for a GL_n-algebra A.)

Grosshans [**12**] has introduced a filtration on A. The associated graded ring $\mathrm{gr}\, A$ is finitely generated [**12**, Lemma 14]. There is a flat family with general fibre A and special fibre $\mathrm{gr}\, A$ [**12**, Theorem 13]. Its counterpart in characteristic zero was introduced by Popov [**22**]. Our first main result says

THEOREM 1.1. *If $n < 6$ or $p > 2^n$, then $H^*(\mathrm{SL}_n, \mathrm{gr}\, A)$ is finitely generated as a k-algebra.*

REMARK 1.2. In Problems 3.10 and 3.11 we discuss how one could try to remove the gr in the conclusion of the theorem.

Our second main result concerns the case $n = 2$, where we succeed in removing the gr, using a family of universal cohomology classes which behaves as if it is obtained by taking divided powers of the class e_1 of Friedlander and Suslin [**11**].

THEOREM 1.3 (Cohomological invariant theory in rank one). *Let A be a finitely generated commutative k-algebra on which SL_2 acts rationally by algebra automorphisms. Then $H^*(\mathrm{SL}_2, A)$ is finitely generated as a k-algebra.*

REMARK 1.4. We know very little about the size of $H^*(G, A)$, even in simple examples. For instance, let $p = 2$ and consider the second Steinberg module $V = \mathrm{St}_2$ of SL_2. The dimension of V is four. Its highest weight is three times the fundamental weight. By the theorem $H^*(\mathrm{SL}_2, S^*(V))$ is finitely generated, where $S^*(V)$ denotes the symmetric algebra. But that is all we know about the size of $H^*(\mathrm{SL}_2, S^*(V))$. Note that $S^*(V)$ does not have a good filtration [**5**, p. 71 Example], as $H^1(\mathrm{SL}_2, S^2(V)) \neq 0$.

2. Recollections

For simplicity we stay with the important case $G = \mathrm{SL}_n$ until 3.13. We choose a Borel group $B^+ = TU^+$ of upper triangular matrices and the opposite Borel group B^-. The roots of B^+ are positive. If $\lambda \in X(T)$ is dominant, then $\mathrm{ind}_{B^-}^G(\lambda)$ is the dual Weyl module $\nabla_G(\lambda)$ with highest weight λ. In a good filtration of a G-module the layers are of the form $\nabla_G(\mu)$. As in [**27**] we will actually also allow a layer to be a direct sum of any number of copies of the same $\nabla_G(\mu)$. If M is a G-module, and $m \geq -1$ is an integer so that $H^{m+1}(G, \nabla_G(\mu) \otimes M) = 0$ for all dominant μ, then we say as in [**10**] that M has *good filtration dimension* at most m. The case $m = 0$ corresponds with M having a good filtration. We say that M has good filtration dimension precisely m, notation $\dim_\nabla(M) = m$, if m is minimal so that M has good filtration dimension at most m. In that case $H^{i+1}(G, \nabla_G(\mu) \otimes M) = 0$ for all dominant μ and all $i \geq m$. In particular $H^{i+1}(G, M) = 0$ for $i \geq m$. If there is no finite m so that $\dim_\nabla(M) = m$, then we put $\dim_\nabla(M) = \infty$.

LEMMA 2.1. *Let* $0 \to M' \to M \to M'' \to 0$ *be exact. Then*

1. $\dim_\nabla(M) \leq \max(\dim_\nabla(M'), \dim_\nabla(M''))$,
2. $\dim_\nabla(M') \leq \max(\dim_\nabla(M), \dim_\nabla(M'') + 1)$,
3. $\dim_\nabla(M'') \leq \max(\dim_\nabla(M), \dim_\nabla(M') - 1)$,
4. $\dim_\nabla(M' \otimes M'') \leq \dim_\nabla(M') + \dim_\nabla(M'')$.

2.2. Filtrations. If M is a G-module, and λ is a dominant weight, then $M_{\leq\lambda}$ denotes the largest G-submodule all whose weights μ satisfy $\mu \leq \lambda$ in the usual partial order [**15**, II 1.5]. Similarly $M_{<\lambda}$ denotes the largest G-submodule all whose weights μ satisfy $\mu < \lambda$. As in [**27**], we form the $X(T)$-graded module

$$\mathrm{gr}_{X(T)} M = \bigoplus_{\lambda \in X(T)} M_{\leq\lambda}/M_{<\lambda}.$$

Each $M_{\leq\lambda}/M_{<\lambda}$, or $M_{\leq\lambda/<\lambda}$ for short, has a B^+-socle $(M_{\leq\lambda/<\lambda})^U = M_\lambda^U$ of weight λ. We view M^U as a B^--module through restriction (inflation) along the homomorphism $B^- \to T$. Then $M_{\leq\lambda/<\lambda}$ embeds naturally in its "good filtration hull" $\mathrm{hull}_\nabla(M_{\leq\lambda/<\lambda}) = \mathrm{ind}_{B^-}^G M_\lambda^U$. This good filtration hull has the same B^+-socle and is the injective hull in the category $\mathcal{C}_\lambda$ of G-modules N that satisfy $N = N_{\leq\lambda}$. Compare [**27**, 3.1.10].

Let us apply this in particular to our finitely generated commutative k-algebra with G action A. We get an $X(T)$-graded algebra $\mathrm{gr}_{X(T)} A$. We convert it to a $\mathbb{Z}$-graded algebra through an additive height function $\mathrm{ht} : X(T) \to \mathbb{Z}$, defined by $\mathrm{ht}(\gamma) = 2\sum_{\alpha>0}\langle\gamma, \widehat{\alpha}\rangle$, the sum being over the positive roots. (Our ht is twice the one used by Grosshans, because we prefer to get even degrees rather than just integer degrees.) The Grosshans graded algebra is now

$$\mathrm{gr}\, A = \bigoplus_{i \geq 0} \mathrm{gr}_i\, A,$$

with

$$\mathrm{gr}_i\, A = \bigoplus_{\mathrm{ht}(\lambda)=i} A_{\leq\lambda/<\lambda}.$$

It embeds in a good filtration hull, which Grosshans calls R, and which we call $\mathrm{hull}_\nabla(\mathrm{gr}\, A)$,

$$\mathrm{hull}_\nabla(\mathrm{gr}\, A) = \mathrm{ind}_{B^-}^G A^U = \bigoplus_i \bigoplus_{\mathrm{ht}(\lambda)=i} \mathrm{hull}_\nabla(A_{\leq\lambda}/A_{<\lambda}).$$

Grosshans shows that A^U, $\operatorname{gr} A$, $\operatorname{hull}_\nabla(\operatorname{gr} A)$ are finitely generated with $\operatorname{hull}_\nabla(\operatorname{gr} A)$ finite over $\operatorname{gr} A$. Mathieu did a little better in [**19**]. His argument shows that in fact $\operatorname{hull}_\nabla(\operatorname{gr} A)$ is a p-root closure of $\operatorname{gr} A$. That is,

LEMMA 2.3. *For every $x \in \operatorname{hull}_\nabla(\operatorname{gr} A)$, there is an integer $r \geq 0$, so that $x^{p^r} \in \operatorname{gr} A$.*

PROOF. It suffices to take $x \in \operatorname{hull}_\nabla(A_{\leq\lambda}/A_{<\lambda})$ for some λ. If $\lambda = 0$, then $\operatorname{hull}_\nabla(A_{\leq\lambda}/A_{<\lambda}) = A^G = \operatorname{gr}_0 A$. So say $\lambda > 0$ and consider the subalgebra $S = k \oplus \bigoplus_{i>0} \operatorname{hull}_\nabla(A_{\leq i\lambda}/A_{<i\lambda})$ of $\operatorname{hull}_\nabla(\operatorname{gr} A)$, with its subalgebra $S \cap \operatorname{gr} A$. Apply [**27**, Thm 4.2.3] to conclude that the p-root closure of $S \cap \operatorname{gr} A$ in S has a good filtration. As it contains all of S^U, it must be S itself. □

EXAMPLE 2.4. Consider the multicone [**16**]

$$k[G/U] = \operatorname{ind}_U^G k = \operatorname{ind}_{B^+}^G \operatorname{ind}_U^{B^+} k = \operatorname{ind}_{B^+}^G k[T] = \bigoplus_{\lambda \text{ dominant}} \nabla_G(\lambda).$$

It is its own Grosshans graded ring. Recall [**16**] that it is finitely generated by the sum of the $\nabla(\varpi_i)$, where ϖ_i denotes the ith fundamental weight. Together with the transfer principle $A^U \cong (k[G/U] \otimes A)^G$, see [**13**, Chapter 2], this gives finite generation of A^U.

3. Proof of Theorem 1.1

Choose r so big that $x^{p^r} \in \operatorname{gr} A$ for all $x \in \operatorname{hull}_\nabla(\operatorname{gr} A)$. We may view $\operatorname{gr} A$ as a finite $\operatorname{hull}_\nabla(\operatorname{gr} A)^{(r)}$-module, where the exponent (r) denotes an rth Frobenius twist [**15**, I 9.2].

3.1. Key hypothesis. We assume that for every fundamental weight ϖ_i the symmetric algebra $S^*(\nabla(\varpi_i))$ has a good filtration.

The hypothesis in Theorem 1.1 is explained by

LEMMA 3.2. *If $n < 6$ or $p > 2^n$, then the key hypothesis is satisfied.*

PROOF. We follow [**1**, Section 4]. If $p > 2^n$, then $p > \sum_i \dim(\nabla(\varpi_i))$, so $\wedge^j(\nabla(\varpi_i))$ has a good filtration for all j by [**1**], so $S^*(\nabla(\varpi_i))$ has a good filtration by [**1**]. If $n < 6$, then by symmetry of the Dynkin diagram (contragredient duality) it suffices to consider ϖ_1 and ϖ_2. But $S^*(\nabla(\varpi_1))$ has a good filtration for every n because $\wedge^j(\nabla(\varpi_1))$ has a good filtration for all j. And $S^*(\nabla(\varpi_2))$ has a good filtration for every n by Boffi [**3**]. (Thanks to J. Weyman and T. Jozefiak for pointing this out.) □

EXAMPLE 3.3 (J. Weyman). If $p = 2$ and $n \geq 6$, there is a submodule with highest weight ϖ_6 in $S^2(\nabla(\varpi_3))$. (If $n = 6$, read zero for ϖ_6, as we are working with SL_6 rather than GL_6.) With a character computation this implies that $S^2(\nabla(\varpi_3))$ does not have a good filtration. The submodule in question is generated by a highest weight vector which is a sum of ten terms $(e_1 \wedge e_{\sigma(2)} \wedge e_{\sigma(3)})(e_{\sigma(4)} \wedge e_{\sigma(5)} \wedge e_{\sigma(6)})$. One sums over the ten permutations σ of $2, 3, 4, 5, 6$ that satisfy $\sigma(2) < \sigma(3)$ and $\sigma(4) < \sigma(5) < \sigma(6)$.

We want to view $S^*(\nabla(\varpi_i))$ as a *graded polynomial G-algebra with good filtration*. Let us collect the properties that we will use in a rather artificial definition.

DEFINITION 3.4. Let D be a diagonalizable group scheme [**15**, I 2.5]. We say that P is a graded polynomial $G \times D$-algebra with good filtration, if the following holds. First of all P is a polynomial algebra over k in finitely many variables. Secondly, these variables are homogeneous of non-negative integer degree, thus making P into a graded k-algebra. There is also given an action of $G \times D$ on P by algebra automorphisms that are compatible with the grading, with $G \times D$ acting trivially on the degree zero part P_0 of P. This P_0 is thus the polynomial algebra generated by the variables of degree zero. The variables of positive degree generate their own polynomial algebra P^c, which we also assume to be $G \times D$ invariant. Thus $P = P_0 \otimes_k P^c$. And finally, P^c is an algebra with a good filtration for the action of G. Then of course, so is P.

EXAMPLE 3.5. Our key hypothesis makes that one gets a *graded polynomial $G \times D$-algebra with good filtration* by taking for D any subgroup scheme of T, with T acting on $P = S^*(\nabla(\varpi_i))$ through its natural $X(T)$-grading: On $S^j(\nabla(\varpi_i))$ we make T act with weight $j\varpi_i$.

If P_i are graded polynomial $G \times D_i$-algebras with good filtration for $i = 1, 2$, then $P_1 \otimes P_2$ is a graded polynomial $G \times (D_1 \times D_2)$-algebra with good filtration.

DEFINITION 3.6. If P is a graded polynomial $G \times D$-algebra with good filtration, then by a *finite graded P-module* M we mean a finitely generated $\mathbb{Q}$-graded module for the graded polynomial ring, together with a $G \times D$ action on M which is compatible with the grading and with the action on P. It is not required that the action is trivial on the degree zero part of M. We call M *free* if it is free as a module over the polynomial ring. We call M *extended* if there is a finitely generated $\mathbb{Q}$-graded P_0-module V with $G \times D$ action, so that $M = V \otimes_{P_0} P$ as graded P-modules.

LEMMA 3.7. *Let P be a graded polynomial $G \times D$-algebra with good filtration, and let M be a finite graded P-module*:

1. *There is a finite free resolution*

$$0 \to F_s \to F_{s-1} \to \cdots \to F_0 \to M.$$

2. *Every finite free P-module has a finite filtration whose quotients are extended.*
3. $\dim_\nabla(M) < \infty$.
4. *$H^i(G, M)$ is a finite P^G-module for every $i \geq 0$.*

PROOF. Take a finite dimensional graded $G \times D$-submodule V of M that generates M as a P-module. Then $F_0 = V \otimes_k P$ is free and it maps onto M. As M has finite projective dimension [**20**, 18C], we may repeat until the syzygy is projective. But a projective module over a polynomial ring is free by Quillen and Suslin [**17**].

Now consider a free P-module F. Let P_+ be the ideal generated by the P_i with $i > 0$. If F is free of rank t, then F/P_+F is free of rank t over P_0. Choose homogeneous elements $e_1, \ldots, e_t$ in F so that their classes form a basis of F/P_+F. Then the e_i generate F (cf. [**14**, 6.13 Lemma 5]), so they also form a basis of F. Let F_m be the component of lowest degree in F, assuming $F \neq 0$. The e_i that lie in F_m form a basis of F_m over P_0. So F_m generates a free, extended submodule PF_m of F and the quotient F/PF_m is free of lower rank.

To show that $\dim_\nabla(M) < \infty$, it suffices to consider the case of an extended module $M = V \otimes_{P_0} P$. As $V \otimes_{P_0} P = V \otimes_k P^c$, it suffices to check that V has finite good filtration dimension. But V is a finitely generated P_0-module and thus has only finitely many weights. Therefore the argument used in [**10**] to show that finite dimensional G modules have finite good filtration dimension, applies to V.

Finally let M be any finite graded P-module again. As G is reductive by Haboush [**15**, 10.7], it is well known that $H^0(G, M)$ is a finite P^G-module, because $\left(S_P^*(M)\right)^G$ is finitely generated. So we argue by dimension shift. We claim that for s sufficiently large the module $M \otimes \mathrm{St}_s \otimes \mathrm{St}_s$ has a good filtration. It suffices to check this for the extended case $M = V \otimes_{P_0} P$ and then one can use again that V has only finitely many weights, so that one may choose s so large that all weights of $V \otimes k_{-(p^s-1)\rho}$ are anti-dominant. Then $V \otimes \mathrm{St}_s = \mathrm{ind}_{B^+}^G(V \otimes k_{-(p^s-1)\rho})$ has a good filtration and so does $M \otimes \mathrm{St}_s \otimes \mathrm{St}_s$. Then $H^i(G, M)$ is the cokernel of $H^{i-1}(G, M \otimes \mathrm{St}_s \otimes \mathrm{St}_s) \to H^{i-1}(G, M \otimes \mathrm{St}_s \otimes \mathrm{St}_s / M)$ for $i \geq 1$. □

Recall that we choose r so big that $x^{p^r} \in \mathrm{gr}\, A$ for all $x \in \mathrm{hull}_\nabla(\mathrm{gr}\, A)$. Let G_r denote the rth Frobenius kernel. We will need multiplicative structure on a Hochschild–Serre spectral sequence. See for instance [**2**], translating from modules over a Hopf algebra to comodules over a Hopf algebra.

PROPOSITION 3.8. *Assume the Key hypothesis* 3.1. *With r as indicated, the spectral sequence*

$$E_2^{ij} = H^i\big(G/G_r, H^j(G_r, \mathrm{gr}\, A)\big) \Rightarrow H^{i+j}(G, \mathrm{gr}\, A)$$

stops, i.e. $E_s = E_\infty$ for some $s < \infty$.

In fact, $H^*(G_r, \mathrm{gr}\, A)^{(-r)}$ has finite good filtration dimension.

Moreover, E_2^{**} is a finite module over the even part of E_2^{0*}.

PROOF. By [**11**, Th 1.5, 1.5.1] the ring $H^*(G_r, \mathrm{gr}\, A)^{(-r)}$ is a finite module over the algebra

$$\bigotimes_{a=1}^{r} S^*\big((\mathfrak{gl}_n)^\#(2p^{a-1})\big) \otimes \mathrm{hull}_\nabla(\mathrm{gr}\, A).$$

Here $(\mathfrak{gl}_n)^\#(2p^{a-1})$ denotes the dual of $\mathfrak{gl}_n$ placed in degree $2p^{a-1}$. As $G = \mathrm{SL}_n$, it follows from [**1**, 4.3] that the algebra $\bigotimes_{a=1}^{r} S^*\big((\mathfrak{gl}_n)^\#(2p^{a-1})\big)$ is a graded polynomial $G \times \{1\}$-algebra with good filtration. So we now try to replace $\mathrm{hull}_\nabla(\mathrm{gr}\, A)$ by a similarly nice algebra. That is, we seek a graded polynomial $G \times D$-algebra P with good filtration and a surjection $P^D \to \mathrm{hull}_\nabla(\mathrm{gr}\, A)$ of graded G-algebras. This is where the key hypothesis comes in. Let v be a weight vector in $\mathrm{hull}_\nabla(\mathrm{gr}\, A)^U = A^U$. If v has weight zero, we may map x to v, where x is the variable in a polynomial ring $k[x]$ with trivial grading and trivial $G \times T$ action. If v has weight $\lambda \neq 0$, take for D the scheme theoretic kernel of λ and observe that by our hypothesis the $X(T)$-graded algebra

$$P = \bigotimes_{i=1}^{n-1} S^*\big(\nabla(\varpi_i)\big)$$

is a graded polynomial $G \times D$-algebra with good filtration, if we give $\nabla(\varpi_i)$ degree $\mathrm{ht}(\varpi_i)$. We have $P^D = \bigoplus_j P_{j\lambda}$, where $P_{j\lambda}$ denotes the summand $\bigotimes_{i=1}^{n-1} S^{m_i}(\nabla(\varpi_i))$ with $\sum_i m_i \varpi_i = j\lambda$. Choose a weight vector x of weight λ in P_λ (for the G action). The map from the polynomial ring $k[x]$ to $\mathrm{hull}_\nabla(\mathrm{gr}\, A)^U$ which sends x to v extends uniquely to a G equivariant algebra map $P^D \to \mathrm{hull}_\nabla(\mathrm{gr}\, A)$ because $P_{j\lambda} = (P_{j\lambda})_{\leq j\lambda}$. (The first subscript in the right hand side refers to a T action associated with the $X(T)$-grading, the second to the G action on P.) Compare [**27**, 4.2.4]. As A^U is finitely generated, we may combine finitely many such maps $P(i)^{D(i)} \to \mathrm{hull}_\nabla(\mathrm{gr}\, A)$ into one surjective map $P^D \to \mathrm{hull}_\nabla(\mathrm{gr}\, A)$, with D the product of the $D(i)$ and P the tensor product of the $P(i)$. If we let the linearly reductive D act on $P \otimes_{P^D} H^*(G_r, \mathrm{gr}\, A)^{(-r)}$ through its action on P, then $H^*(G_r, \mathrm{gr}\, A)^{(-r)}$ is just the direct summand, as a G-module, consisting of the D-invariants. So in order to show that $H^*(G_r, \mathrm{gr}\, A)^{(-r)}$ has finite good filtration dimension, it suffices to show that $P \otimes_{P^D} H^*(G_r, \mathrm{gr}\, A)^{(-r)}$ has finite good filtration dimension. We now view $P \otimes \bigotimes_{a=1}^r S^*\big((\mathfrak{gl}_n)^\#(2p^{a-1})\big)$ as a graded polynomial $G \times D$-algebra with good filtration. (Collect the bigrading into a single total grading.) The algebra $P \otimes_{P^D} H^*(G_r, \mathrm{gr}\, A)^{(-r)}$ is a finite graded $P \otimes \bigotimes_{a=1}^r S^*\big((\mathfrak{gl}_n)^\#(2p^{a-1})\big)$-module. (We need a $\mathbb{Q}$-grading on $P \otimes_{P^D} H^*(G_r, \mathrm{gr}\, A)^{(-r)}$ because of the twist.) By Lemma 3.7 such a P-module has finite good filtration dimension. Therefore there are only finitely many i with $E_2^{i*} \neq 0$ and the spectral sequence stops. By the same lemma $H^i\big(G, P \otimes_{P^D} H^*(G_r, \mathrm{gr}\, A)^{(-r)}\big)$ is finite over $H^0\big(G, P \otimes \bigotimes_{a=1}^r S^*\big((\mathfrak{gl}_n)^\#(2p^{a-1})\big)\big)$. Taking D-invariants again, we see that E_2^{**} is a finite module over $H^0\big(G/G_r, (P^D \otimes \bigotimes_{a=1}^r S^*\big((\mathfrak{gl}_n)^\#(2p^{a-1})\big))^{(r)}\big)$. But this ring acts by way of the the even part of E_2^{0*}. □

3.9. End of proof of Theorem 1.1. Look some more at the spectral sequence of Proposition 3.8. We argue partly as in Evens' proof of his finite generation theorem [**2**, 4.2], [**8**]. As we have already observed, it follows from [**11**] that $H^*(G_r, \mathrm{gr}\, A)$ is noetherian over its finitely generated even degree part. By the proposition E_2 is noetherian over the even degree part of E_2^{0*}, which is finitely generated because G/G_r is reductive. Or, recall from the proof of the proposition that the E_2 term is finite over the finitely generated k-algebra $H^0(G/G_r, (P^D \otimes \bigotimes_{a=1}^r S^*\big((\mathfrak{gl}_n)^\#(2p^{a-1})\big))^{(r)})$. Now the E_2 term is a differential graded algebra in characteristic p, so the pth power of an element of the even part passes to the next term in the spectral sequence. Therefore E_3 is also noetherian over its finitely generated even degree part. As the spectral sequence stops, we get by repeating this argument that E_∞ is finitely generated. But then so is the abutment.

PROBLEM 3.10. There is a spectral sequence

$$E_2^{ij} = H^{i+j}(G, \mathrm{gr}_{-i}\, A) \Rightarrow H^{i+j}(G, A),$$

and the theorem says that E_2 is finitely generated. The generators are in bidegree $(0,0)$ or (i,j) with $i+j > 0$, $i < 0$. We would like this spectral sequence to stop too. It would suffice to know that if J is a G-stable ideal in a ring like A, then the even part of $H^*(G, A/J)$ is integral over the image of the even part of $H^*(G, A)$. Of course this integrality would be implied by finite generation of $H^*\big(G, S_A^*(A/J)\big)$. But recall that Nagata proves first that $H^0(G, A/J)$ is integral over $H^0(G, A)$. So one could hope for a direct proof.

PROBLEM 3.11. If M is a vector space, let $\Gamma^*(M) = \bigoplus_m \big(S^m(M^\#)\big)^\#$ denote its divided power algebra. It is probably too much to ask for a divided power structure on the even part of the bigraded algebra $H^*\big(\mathrm{GL}_n, \Gamma^*\big((\mathfrak{gl}_n)(2)^{(1)}\big)\big)$, extending the divided power structure [**7**, Appendix A2], [**24**] on the graded algebra $H^0\big(\mathrm{GL}_n, \Gamma^*\big((\mathfrak{gl}_n)(2)^{(1)}\big)\big)$. But such a structure would be very helpful, as it would explain and enrich the supply of universal cohomology classes from [**11**]. And

$H^{i+j}(G, \mathrm{gr}_{-i} A) \Rightarrow H^{i+j}(G, A)$ would then undoubtedly stop. See our treatment of the rank one case below.

PROBLEM 3.12. To improve on the conditions of Theorem 1.1, rather than on its conclusion, one should try and prove the following. If our finitely generated A has a good filtration and M is a finite A-module on which G acts compatibly, then $H^*(G, M)$ is finite over A^G. Of course this would again be implied by finite generation of $H^*\big(G, S_A^*(M)\big)$.

REMARK 3.13. Let G be a semi-simple group defined over k, and let V be a tilting G-module [**6**] of dimension n. Choose a basis in V and assume that the representation is faithful, so that G can be identified with a subgroup of SL_n. Assume $p > n/2$. Then $S^*\big((\mathfrak{gl}_n)^\#\big)$ has a good filtration as a G-module [**1**, 4.3]. Also assume for every fundamental weight ϖ_i of G that the symmetric algebra $S^*\big(\nabla(\varpi_i)\big)$ has a good filtration. This happens for instance when p exceeds the dimensions of the fundamental representations of G. Let A be a finitely generated commutative k-algebra on which G acts rationally by k-algebra automorphisms. Then again $H^*(G, \mathrm{gr}\, A)$ is finitely generated as a k-algebra. The proof is the same as the proof of Theorem 1.1.

4. Divided powers

Let $W_2(k) = W(k)/p^2W(k)$ be the ring of Witt vectors of length two over k, see [**25**, II §6]. One has an extension of algebraic groups

$$1 \to \mathfrak{gl}_n^{(1)} \to \mathrm{GL}_n\big(W_2(k)\big) \to \mathrm{GL}_n(k) \to 1,$$

whence a cocycle class $e_1 \in H^2(\mathrm{GL}_n, \mathfrak{gl}_n^{(1)})$. We call it the Witt vector class for GL_n. Analogously there are Witt vector extensions and Witt vector classes for other groups that are originally defined over the integers, and for Frobenius kernels in them.

REMARK 4.1. Observe that if $x_\alpha : \mathbb{G}_a \to \mathrm{SL}_n$ is a root homomorphism corresponding to the root α, then the restriction of e_1 to $\mathbb{G}_a$ is non-trivial. In fact one may restrict further to the Frobenius kernel $\mathbb{G}_{a1}$. This $\mathbb{G}_{a1}$ acts trivially on $\mathfrak{gl}_n^{(1)}$, and the T-equivariant projection of $\mathfrak{gl}_n^{(1)}$ onto its $p\alpha$ weight space is thus also $\mathbb{G}_{a1}$-equivariant. Altogether we get an image β of e_1 in $H^2(\mathbb{G}_{a1}, (\mathfrak{gl}_n^{(1)})_{p\alpha})^T$. It is just the Witt vector class of $\mathbb{G}_{a1}$. It is well known that this class is non-trivial, compare [**15**, I 4.22, I 4.25], [**11**, §6]. Also note that the image of β in $H^2(\mathbb{G}_{a1}, \mathfrak{gl}_n^{(1)})$ under the map induced by the inclusion of $(\mathfrak{gl}_n^{(1)})_{p\alpha}$ into $\mathfrak{gl}_n^{(1)}$ is the same as the restriction of e_1 to $\mathbb{G}_{a1}$.

LEMMA 4.2. *The Witt vector class for* GL_n *coincides with the universal cohomology class* e_1 *of Friedlander and Suslin, up to a non-zero scalar factor.*

PROOF. Using [**11**, Remark 1.2.1, Corollary 3.13] we see it suffices to take $r = j = 1$ and $q = 2$ in [**11**, Theorem 4.5]. □

4.3. Divided powers in rank one. If R is a commutative k-algebra, and M is a finite dimensional vector space over k, then the divided power algebra $R \otimes \Gamma^*(M) = \Gamma_R^*(R \otimes M)$ is a ring with divided powers ([**7**, Appendix A2]). We write $\Delta_{i,j}$ for the component $\Gamma^{i+j}(M) \to \Gamma^i(M) \otimes \Gamma^j(M)$ of the comultiplication map $\Gamma^*(M) \to \Gamma^*(M) \otimes \Gamma^*(M)$. So $\Delta_{i,j}$ is the dual of the multiplication map

$S^i(M^\#) \otimes S^j(M^\#) \to S^{i+j}(M^\#)$. If $v \in M$ has divided powers $v^{[i]} \in \Gamma^i(M)$, then $\Delta_{i,j}(v^{[i+j]}) = v^{[i]} \otimes v^{[j]}$. Put $G = \mathrm{GL}_n$ and define T, B^+, B^- as usual. Actually n will be two, but we optimistically keep writing n. Let $r \geq 1$. As G_r acts trivially on $\mathfrak{gl}_n^{(r)}$, there is a divided power structure on $H^{\mathrm{even}}(G_r, \Gamma^*(\mathfrak{gl}_n^{(r)})) = H^{\mathrm{even}}(G_r, k) \otimes \Gamma^*(\mathfrak{gl}_n^{(r)})$, with the mth divided power operation going from $H^{2a}(G_r, \Gamma^b(\mathfrak{gl}_n^{(r)}))$ to $H^{2am}(G_r, \Gamma^{bm}(\mathfrak{gl}_n^{(r)}))$ for $b \geq 1$.

It would be nice to extend the next theorem to arbitrary n. We do not know how to put a divided power structure on $\bigoplus_m H^{2m}(\mathrm{GL}_n, \Gamma^m(\mathfrak{gl}_n^{(1)}))$. Nevertheless we feel the theorem is named appropriately.

THEOREM 4.4 (Divided powers in rank one). *Let $n = 2$. There are classes $c[m] \in H^{2m}(\mathrm{GL}_n, \Gamma^m(\mathfrak{gl}_n^{(1)}))$ so that*

1. *$c[1]$ is the Witt vector class e_1,*
2. *$(\Delta_{i,j})_*(c[i+j]) = c[i] \cup c[j]$ for $i, j \geq 1$.*

PROOF. Let α be the negative root, and let $x_\alpha : \mathbb{G}_a \to \mathrm{GL}_n$ be its root homomorphism. By Kempf vanishing $H^{2m}(\mathrm{GL}_n, \Gamma^m(\mathfrak{gl}_n^{(1)})) = H^{2m}(\mathbb{G}_a, \Gamma^m(\mathfrak{gl}_n^{(1)}))^T$, see [**15**, II 4.7c]. So we restrict to $\mathbb{G}_a$ along x_α. As $\mathbb{G}_a$ acts trivially on $\Gamma^*((\mathfrak{gl}_n^{(1)})_{p\alpha})$, we also have a divided power structure on $H^{\mathrm{even}}(\mathbb{G}_a, \Gamma^*((\mathfrak{gl}_n^{(1)})_{p\alpha})) = H^{\mathrm{even}}(\mathbb{G}_a, k) \otimes \Gamma^*((\mathfrak{gl}_n^{(1)})_{p\alpha})$. Take the mth divided power in $H^{\mathrm{even}}(\mathbb{G}_a, \Gamma^*((\mathfrak{gl}_n^{(1)})_{p\alpha}))^T$ of the Witt vector class of $\mathbb{G}_a$ and map it to $H^{2m}(\mathbb{G}_a, \Gamma^m(\mathfrak{gl}_n^{(1)}))^T$ using the inclusion of $(\mathfrak{gl}_n^{(1)})_{p\alpha}$ into $\mathfrak{gl}_n^{(1)}$. One lands in $H^{2m}(B^-, \Gamma^m(\mathfrak{gl}_n^{(1)})) \cong H^{2m}(G, \Gamma^m(\mathfrak{gl}_n^{(1)}))$. This gives the desired class $c[m]$. Indeed $(\Delta_{i,j})_*(c[i+j]) = c[i] \cup c[j]$ holds in the context of the divided power algebra $H^{\mathrm{even}}(\mathbb{G}_a, \Gamma^*((\mathfrak{gl}_n^{(1)})_{p\alpha})) = H^{\mathrm{even}}(\mathbb{G}_a, k) \otimes \Gamma^*((\mathfrak{gl}_n^{(1)})_{p\alpha}))$ and thus the result follows by functoriality of $\Delta_{i,j}$ and of cup product. □

4.5. Other universal classes. If M is a finite dimensional vector space over k and $r \geq 1$, we have a natural homomorphism between symmetric algebras $S^*(M^{\#(r)}) \to S^*(M^{\#(1)})$ induced by the map $M^{\#(r)} \to S^{p^{r-1}}(M^{\#(1)})$ which raises an element to the power p^{r-1}. It is a map of bialgebras. Dually we have the bialgebra map $\pi^{r-1} : \Gamma^{p^{r-1}*}(M(1)) \to \Gamma^*(M(r))$ whose kernel is the ideal generated by $\Gamma^1(M^{(1)})$ through $\Gamma^{p^{r-1}-1}(M^{(1)})$. So π^{r-1} maps $\Gamma^{p^{r-1}a}(M(1))$ onto $\Gamma^a(M(r))$.

NOTATION 4.6. We now introduce analogues of the classes e_r and $e_r^{(j)}$ of Friedlander and Suslin [**11**, Theorem 1.2, Remark 1.2.2]. We write $\pi_*^{r-1}(c[ap^{r-1}]) \in H^{2ap^{r-1}}(G, \Gamma^a(\mathfrak{gl}_n^{(r)}))$ as $c_r[a]$. Next we get $c_r[a]^{(j)} \in H^{2ap^{r-1}}(G, \Gamma^a(\mathfrak{gl}_n^{(r+j)}))$ by Frobenius twist.

LEMMA 4.7. *The $c_i[a]^{(r-i)}$ enjoy the following properties ($r \geq i \geq 1$)*

1. *There is a homomorphism of algebras $S^*(\mathfrak{gl}_n^{\#(r)}) \to H^{2p^{i-1}*}(G_r, k)$ given on $S^a(\mathfrak{gl}_n^{\#(r)}) = H^0(G_r, S^a(\mathfrak{gl}_n^{\#(r)}))$ by cup product with the restriction of $c_i[a]^{(r-i)}$.*
2. *For $r \geq 1$ the restriction of $c_r[1]$ to $H^{2p^{r-1}}(G_1, \mathfrak{gl}_n^{(r)})$ is nontrivial, so that $c_r[1]$ may serve as the universal class e_r in* [**11**, Thm 1.2].

PROOF.

1. By Theorem 4.4 we have $(\Delta_{a,b})_*(c_i[a+b]) = (\Delta_{a,b}\pi^{r-1})_*(c[(a+b)p^{r-1}]) = (\pi^{r-1}\otimes\pi^{r-1})_*(\Delta_{ap^{r-1},bp^{r-1}})_*(c[(a+b)p^{r-1}]) = (\pi^{r-1}\otimes\pi^{r-1})_*\,(c[ap^{r-1}]\cup c[bp^{r-1}]) = c_i[a]\cup c_i[b]$ and thus $(\Delta_{a,b})_*(c_i[a+b]^{(r-i)}) = c_i[a]^{(r-i)}\cup c_i[b]^{(r-i)}$ by pull back along a Frobenius homomorphism. Put $R = H^{\mathrm{even}}(G_r, k)$ and restrict from G to G_r. We view $H^{\mathrm{even}}\big(G_r, \Gamma^*(\mathfrak{gl}_n^{(r)})\big)$ as $\Gamma^*_R\big(R\otimes\mathfrak{gl}_n^{(r)})\big)$. Now the cup product agrees with the usual R-valued pairing between $S^*(\mathfrak{gl}_n^{\#(r)})$ and $\Gamma^*_R(R\otimes\mathfrak{gl}_n^{(r)})$. Thus if $x\in S^a(\mathfrak{gl}_n^{\#(r)})$, $y\in S^b(\mathfrak{gl}_n^{\#(r)})$, then $(xy)\cup(c_i[a+b]^{(r-i)}) = (x\otimes y)\cup\big((\Delta_{a,b})_*(c_i[a+b]^{(r-i)})\big) = (x\cup c_i[a]^{(r-i)})(y\cup c_i[b]^{(r-i)})$.

2. In fact if we restrict $c_r[1]$ as in Remark 4.1 to $H^{2p^{r-1}}\big(\mathbb{G}_{a1}, (\mathfrak{gl}_n^{(r)})_{p^r\alpha}\big) = H^{2p^{r-1}}(\mathbb{G}_{a1}, k)\otimes(\mathfrak{gl}_n^{(r)})_{p^r\alpha}$, then even that restriction is nontrivial. That is because the Witt vector class generates the polynomial ring $H^{\mathrm{even}}(\mathbb{G}_{a1}, k)$, see [**15**, I 4.26]. □

COROLLARY 4.8. *Let $n = 2$, $r\geq 1$. Further let A be a commutative k-algebra with SL_n action and J an invariant ideal in A so that the algebra A/J is finitely generated. Then $H^0\big(\mathrm{SL}_n, H^*\big((\mathrm{SL}_n)_r, A/J\big)\big)$ is a noetherian module over a finitely generated subalgebra of $H^{\mathrm{even}}(\mathrm{SL}_n, A)$.*

PROOF. We may assume A is finitely generated. Put $C = H^0\big((\mathrm{SL}_n)_r, A)\big)$. Then C contains the elements of A raised to the power p^r, so C is also a finitely generated algebra and A/J is a noetherian module over it. By [**11**, Thm 1.5] $H^*\big((\mathrm{SL}_n)_r, A/J\big)$ is a noetherian module over the finitely generated algebra

$$R = \bigotimes_{a=1}^{r} S^*\big((\mathfrak{gl}_n^{(r)})^{\#}(2p^{a-1})\big)\otimes C.$$

Then, by invariant theory [**13**, Thm. 16.9], $H^0\big(\mathrm{SL}_n, H^*\big((\mathrm{SL}_n)_r, A/J)\big)\big)$ is a noetherian module over the finitely generated algebra $H^0(\mathrm{SL}_n, R)$. By Lemma 4.7 we may take the algebra homomorphism $R\to H^*\big((\mathrm{SL}_n)_r, A/J\big)$ of [**11**] to be based on cup product with our $c_i[a]^{(r-i)}$. But then the map $H^0(\mathrm{SL}_n, R)\to H^*\big((\mathrm{SL}_n)_r, A/J\big)$ factors, as a linear map, through $H^{\mathrm{even}}(\mathrm{SL}_n, A)$. □

4.9. The contraction. Let A be a finitely generated commutative k-algebra on which SL_n acts rationally. Recall that A comes with an increasing filtration $A_{\leq i} = \sum_{\mathrm{ht}(\lambda)\leq i} A_{\leq\lambda}$ whose associated graded is the ring $\mathrm{gr}\,A$. Let $\mathcal{A}$ be the subring of the polynomial ring $A[t]$ generated by the subsets $t^iA_{\leq i}$. This ring $\mathcal{A}$, denoted D by Grosshans, is the coordinate ring of a flat family with general fibre A and special fibre $\mathrm{gr}\,A$ [**12**, Theorem 13].

4.10. The special fibre. There is a homomorphism of graded algebras $\mathcal{A}\to\mathrm{gr}\,A$ with kernel $t\mathcal{A}$, mapping $t^i\sum_{\mathrm{ht}(\lambda)=i} A_{\leq\lambda}$ onto $\sum_{\mathrm{ht}(\lambda)=i} A_{\leq\lambda}/A_{<\lambda}$ by "dropping t^i". By Corollary 4.8 the even part of E_2^{0*} in Proposition 3.8 is noetherian over a finitely generated subalgebra R of $H^{\mathrm{even}}(\mathrm{SL}_n,\mathcal{A})$. Therefore Proposition 3.8 implies in the usual way ([**11**, Lemma 1.6]) that in fact the abutment $H^*(\mathrm{SL}_n, \mathrm{gr}\,A)$ is noetherian over R.

4.11. The general fibre. One gets a homomorphism $\mathcal{A}\to A$ by substituting a nonzero scalar for t. Let us use the substitution $t\mapsto 1$. On $\mathcal{A}$ we put the filtration with $\mathcal{A}_{\leq m} = \sum_{i\leq m} t^iA_{\leq i}$. Then the associated graded $\mathrm{gr}\,\mathcal{A}$ is just $\mathcal{A}$ itself. The map $\mathcal{A} \to A$ is compatible with the filtrations, so we get a map of spectral sequences from

$$E(\mathcal{A}) : E_2^{ij}(\mathcal{A}) = H^{i+j}(G, \operatorname{gr}_{-i} \mathcal{A}) \Rightarrow H^{i+j}(G, \mathcal{A})$$

to

$$E(A) : E_2^{ij}(A) = H^{i+j}(G, \operatorname{gr}_{-i} A) \Rightarrow H^{i+j}(G, A).$$

Note that by the construction of the Grosshans filtration $H^0(G, \operatorname{gr}_{-i} A)$ vanishes for $i \neq 0$. Further $E_2^{**}(A)$ is finitely generated by Theorem 1.1 and therefore there are for each m only finitely many non-zero $E_2^{m-i,i}(A)$. This makes that, even though the spectral sequence is a second quadrant spectral sequence, the abutment will be finitely generated as soon as $E_\infty^{**}(A)$ is.

THEOREM 4.12 (Cohomological invariant theory in rank one). *Let A be a finitely generated commutative k-algebra on which* SL_2 *acts rationally by algebra automorphisms. Then $H^*(\mathrm{SL}_2, A)$ is finitely generated as a k-algebra.*

PROOF. We combine the above. The spectral sequence

$$E(\mathcal{A}) : E_2^{ij}(\mathcal{A}) = H^{i+j}(G, \operatorname{gr}_{-i} \mathcal{A}) \Rightarrow H^{i+j}(G, \mathcal{A})$$

is pleasantly boring: It does not just degenerate, even its abutment is the same as its E_2. The spectral sequence $E(A)$ is a module over it [**18**]. In particular, $E(A)$ is a module over the finitely generated subring R of $H^{\mathrm{even}}(\mathrm{SL}_n, \mathcal{A})$ over which $E_2^{**}(A)$ is noetherian by 4.10. (Yes, the homomorphism $R \to E_2^{**}(A)$ agrees with the homomorphism $E_2^{**}(\mathcal{A}) \to E_2^{**}(A)$.) So the usual argument (see Proof of [**11**, Lemma 1.6] or [**9**, Lemma 7.4.4]) shows that $E(A)$ stops and that $E_\infty^{**}(A)$ is noetherian over R. □

COROLLARY 4.13. *Let A be a finitely generated commutative k-algebra on which* SL_2 *acts rationally by algebra automorphisms. Assume that A has a good filtration. Let M be a finitely generated A module with a compatible* SL_2 *action. Then M has finite good filtration dimension.*

PROOF. Write $G = \mathrm{SL}_2$. If we tensor A with the multicone $k[G/U]$ of Example 2.4, then the result is a finitely generated G-acyclic k-algebra $A \otimes k[G/U]$ over which we have a finitely generated module $M \otimes k[G/U]$. As $H^*\big(G, S^*_{A \otimes k[G/U]}(M \otimes k[G/U])\big)$ is finitely generated by the theorem, $H^i(G, M \otimes k[G/U])$ vanishes for $i \gg 0$. But this means that M has finite good filtration dimension. □

References

1. H. H. Andersen and J.-C. Jantzen, *Cohomology of induced representations for algebraic groups*, Math. Ann. **269** (1984), 487–525.
2. D. J. Benson, *Representations and cohomology.* II. *Cohomology of groups and modules*, 2nd ed., Cambridge Stud. Adv. Math., vol. 31, Cambridge University Press, Cambridge, 1998.
3. G. Boffi, *On some plethysms*, Adv. Math. **89** (1991), 107–126.
4. H. Borsari and W. Ferrer Santos, *Geometrically reductive Hopf algebras*, J. Algebra **152** (1992), 65–77.
5. S. Donkin, *Good filtrations of rational modules for reductive groups*, Arcata Conf. on Repr. of Finite Groups, Proceedings of Symp. in Pure Math., vol. 47, 1987, pp. 69-80.
6. ———, *On tilting modules for algebraic groups*, Math. Z. **212** (1993), 39–60.
7. D. Eisenbud, *Commutative algebra, with a view toward algebraic geometry*, Grad. Texts in Math., vol. 150, Springer-Verlag, New York, 1995.
8. L. Evens, *The cohomology ring of a finite group*, Trans. Amer. Math. Soc. **101** (1961), 224–239.

9. ———, *The cohomology of groups*, Oxford Math. Monographs, Oxford Univ. Press, New York, 1991.
10. E. M. Friedlander and B. J. Parshall, *Cohomology of Lie algebras and algebraic groups*, Amer. J. Math. **108** (1986), 235–253.
11. E. M. Friedlander and A. A. Suslin, *Cohomology of finite group schemes over a field*, Invent. Math. **127** (1997), 209–270.
12. F. D. Grosshans, *Contractions of the actions of reductive algebraic groups in arbitrary characteristic*, Invent. Math. **107** (1992), 127–133.
13. ———, *Algebraic homogeneous spaces and invariant theory*, Lecture Notes in Math., vol. 1673, Springer-Verlag, Berlin and and Heidelberg GmbH & Co. KG, 1997.
14. N. Jacobson, *Basic algebra.* II, 2nd ed., W. H. Freeman and Company, New York, 1989.
15. J.-C. Jantzen, *Representations of algebraic groups*, Pure Appl. Math., vol. 131, Academic Press, Inc., Boston, MA, 1987.
16. G. Kempf and A. Ramanathan, *Multicones over Schubert varieties*, Invent. Math. **87** (1987), 353–363.
17. T.-Y. Lam, *Serre's conjecture*, Lecture Notes in Math., vol. 635, Springer, Berlin, 1978.
18. W. S. Massey, *Products in exact couples*, Ann. of Math. **59** (1954), 558–569.
19. O. Mathieu, *Filtrations of G-modules*, Ann. Sci. École Norm. Sup. **23** (1990), 625–644.
20. H. Matsumura, *Commutative algebra*, W. A. Benjamin, Inc., New York, 1970.
21. M. Nagata, *Invariants of a group in an affine ring*, J. Math. Kyoto Univ. **3** (1963–1964), 369–377.
22. V. L. Popov, *Contractions of actions of reductive algebraic groups*, (Russian), Mat. Sb. (N.S.) **130** (172) (1986), no. 3, 310–334, 431; Transl. in Math. USSR-Sb. **58** (1987), no. 2, 311–335.
23. R. W. Richardson, *Affine coset spaces of reductive algebraic groups*, Bull. London Math. Soc. **9** (1977), 38–41.
24. N. Roby, *Construction de certaines algèbres à puissances divisées*, Bull. Soc. Math. France **96** (1968), 97–113.
25. J.-P. Serre, *Local fields*, Transl. from the French by Marvin Jay Greenberg, Grad. Texts in Math., vol. 67, Springer-Verlag, New York, Berlin, 1979.
26. T. A. Springer, *Invariant theory*, Lecture Notes in Math., vol. 585, Springer-Verlag, Berlin, New York, 1977.
27. W. van der Kallen, *Lectures on Frobenius splittings and B-modules*, Notes by S. P. Inamdar, Tata Inst. Fund. Res., Bombay and Springer-Verlag, Berlin, 1993.
28. W. C. Waterhouse, *Geometrically reductive affine group schemes*, Arch. Math. **62** (1994), 306–307.

MATH. INSTITUT, UNIV. UTRECHT, P.O. BOX 80010, 3508 TA UTRECHT, THE NETHERLANDS.
E-mail address: vdkallen@math.uu.nl

Centre de Recherches Mathématiques
CRM Proceedings and Lecture Notes
Volume **35**, 2004

The Module Structure of a Group Action on a Polynomial Ring: Examples, Generalizations, and Applications

D. B. Karagueuzian and P. Symonds

ABSTRACT. The proof of a module-structure result for group actions on polynomial rings, which the authors published in a previous paper, is discussed in some detail. Also examined are possible generalizations of the result and an application to invariant theory.

1. Introduction

This article is a commentary on the authors' paper [**13**]. We address three main themes. First, the difficulties in the proof of [**13**, The main theorem (9.1)]. Second, possible generalizations of this module-structure result. Third, applications of the result to invariant theory.

The main theorem of [**13**] can be stated as follows.

Let k be a finite field of characteristic p and order q and let $U_n = U_n(k)$ denote the group of $n \times n$ upper triangular matrices over k with 1s on the diagonal. This acts on the polynomial ring in n variables $S = k[x_1, \ldots, x_n]$. The ring of invariants under this group, for which we write S^{U_n}, is known to be polynomial in generators $\{d_i : i = 1, \ldots, n\}$, where the degree of d_i is q^{i-1}. Note that any p-group P acting faithfully on S may be considered to be a subgroup of U_n after a change of variables. Our structure theorem describes the kP-module structure of S for any such P:

THEOREM 1.1. *There is an isomorphism of graded kP-modules:*

$$S \cong \bigoplus_{J \subseteq I} k[d_i : i \in I \cup \{n\} - J] \otimes_k \overline{X}_J(I),$$

where $\overline{X}_J(I)$ is a finite dimensional graded kP-module, P acts trivially on $k[d_i]$, and I is the set $\{1, 2, \ldots, n-1\}$.

This should be read as saying that S contains one copy of $\overline{X}_J(I)$ for each monomial in the d_i with $i \in I \cup \{n\} - J$.

One consequence of this theorem is a finiteness result.

2000 *Mathematics Subject Classification.* Primary: 13A50; Secondary: 20C20.
This is the final form of the paper.

COROLLARY 1.2. *For any group G (not necessarily a p-group) acting on S, only a finite number of isomorphism classes of indecomposable kG-modules appear as a summand of S.*

Section 2 presents a worked example, illustrating the proof of 1.1 and Section 3 comments on the trace lemma ([**13**, 15.1]), which is the most complicated part of the proof.

Next, in Section 4, we comment on several possible generalizations of the structure theorem or its finiteness corollary; in particular Section 5 calculates some examples of a structure theorem for $\mathrm{GL}_n(k)$.

Some partial results in this direction are proved in Sections 6 and 7. Finally, in Section 8, we sketch how studying the ring S as a module for kG has been used in calculations in invariant theory.

2. Extended example

In this section we present an extended example to help the reader sort through the details of the proof of [**13**, The main theorem (9.1)]. We take $n = 3$, $q = 3$, and $I = \langle 1, 2\rangle$; i.e. we consider the proof of the theorem for the group U_3 over the field k with three elements. For simplicity of notation we write $U_{1,2}$ instead of $U_{\{1,2\}} = U_3$, and similarly for $U_1 = U_{\{1\}}$ and $U_2 = U_{\{2\}}$ (even though this conflicts with the upper triangular matrices definition of these groups). In an appendix (Section A), we reproduce the relevant definitions and notation from [**13**].

In this case the polynomial ring $S = k[x_1, x_2, x_3]$ contains the invariant ring $S^{U_{1,2}} = k[d_1, d_2, d_3]$, where the degrees of d_1, d_2, and d_3 are 1, 3, and 9 respectively. We study the $kU_{1,2}$-module structure of S by first reducing to the study of the $kU_{1,2}$-module structure of $T = S/(d_3)$, using [**13**, Lemma 5.4] which tells us that, as $kU_{1,2}$-modules, $S = k[d_3] \otimes T$.

Now we examine the statement of [**13**, The main theorem (9.1)] in this case (cf. A.16). Specializing the claims, the result tells us that there is a decomposition

$$T = X_{\langle 1,2\rangle}(\langle 1,2\rangle) \oplus X_{\langle 1\rangle}(\langle 1,2\rangle) \oplus X_{\langle 2\rangle}(\langle 1,2\rangle) \oplus X_{\varnothing}(\langle 1,2\rangle)$$

with certain properties. (In what follows we will eliminate the angle-brackets in the notation.)

Perhaps the simplest property to understand is the statement about Poincaré series. The Poincaré series are given by (see A.16 or [**13**, 9.1]):

$$\mathrm{PS}(X_J(1,2)) = \sum_{(a_1,a_2)\in S_J(I)} \kappa_{a_1}^{(3-1)} \kappa_{a_2}^{(3-2)} t^{a_1+3a_2}.$$

Here the $S_J(1,2)$ are the partition of N_0^2 given by the following inequalities (A.12, or [**13**, 4.6])

$$\begin{aligned} S_{1,2}(1,2) &= \{(a_1,a_2) \mid a_2 < 2 \text{ and } a_1 + 2a_2 < 6\}, \\ S_1(1,2) &= \{(a_1,a_2) \mid a_2 \geq 2 \text{ and } a_1 < 4\}, \\ S_2(1,2) &= \{(a_1,a_2) \mid a_2 < 2 \text{ and } a_1 + 2a_2 \geq 6\}, \quad \text{and} \\ S_\varnothing(1,2) &= \{(a_1,a_2) \mid a_2 \geq 2 \text{ and } a_1 \geq 4\}. \end{aligned}$$

From the definition of the κs (A.14, or [**13**, 7.5]), the sequence $\kappa_i^{(3-1)}$ is 1, 3, 6, 8, 9, 9, 9, ... and the sequence $\kappa_j^{(3-2)}$ is 1, 2, 3, 3, 3, ... We are interested in the matrix of products $\kappa_{a_1}^{(3-1)}\kappa_{a_2}^{(3-2)}$, which looks like this:

1	3	6	8	9	9	9	9	9
2	6	12	16	18	18	18	18	18
3	9	18	24	27	27	27	27	27
3	9	18	24	27	27	27	27	27

By noting the regions corresponding to the $S_J(I)$, we can obtain explicitly the Poincaré Series for the $X_J(I)$ claimed in the result (A.16 or [**13**, 9.1]). For example, the Poincaré Series of $X_{1,2}(1,2)$ is $1+3t+6t^2+(8+2)t^3+(9+6)t^4+(9+12)t^5+16t^6$. This is the only Poincaré series which is a polynomial; the others are

$$\begin{aligned} \mathrm{PS}\big(X_2(1,2)\big) &= \frac{9t^6+18t^7}{1-t}, \\ \mathrm{PS}\big(X_1(1,2)\big) &= \frac{3t^6+9t^7+18t^8+24t^9}{1-t^3}, \quad \text{and} \\ \mathrm{PS}\big(X_\varnothing(1,2)\big) &= \frac{27t^{10}}{(1-t)(1-t^3)}. \end{aligned}$$

It follows from [**13**, Lemma 7.8] that these four Poincaré series add up to the Poincaré series of T; here this can be checked directly by adding the four rational functions.

Properties (1), (2), (3), and (4) of the decomposition of A.16 (or [**13**, 9.1]) will be treated as we examine the proof of this result in the case at hand. (Observe that thus far we have only written down the desired Poincaré series, rather than constructing modules which actually have the corresponding series, so further discussion of (4) is required.) We note that the proof of this decomposition depends through the induction on other decompositions (for smaller groups), and so we describe these now.

We consider S as a module for the group U_1 (of order 9); to understand its structure we study $T(1) = k[x_1, x_2, x_3]/(d_2, e_3)$, where $S^{U_1} = k[x_1, d_2, e_3]$. (In fact e_3 may be written explicitly as $x_3^3 - x_1^2x_3$.) As a kU_1-module, $T(1)$ splits into a projective part and a non-projective part $B_1(1)$ with Poincaré series $1+3t+6t^2+8t^3$. The module $B_1(1)$ is easy to describe: it is just the part of $T(1)$ with degree less than 4.

The $kU_{1,2}$ module $\overline{X}_1(1,2)$ is obtained by induction from $B_1(1)$: as noted in the proof of [**13**, Proposition 8.4], there is a map $\mathrm{Ind}_{U_1}^{U_{1,2}} B_1(1) \to T$ defined by $g \otimes b \mapsto g \cdot \big(G(1,2 \mid 1) \cdot b\big)$. (For the definition of $G(\cdot \mid \cdot)$ see A.15 or [**13**, 8.2].) The proof that this map is injective requires the Trace Lemma ([**13**, 15.1]); in this special case that lemma states that

$$\mathrm{Tr}_{U_2} \frac{d_3}{e_3} \cdot x_1^i = -d_2^2 \cdot d_1^i,$$

for $i = 0, 1, 2, 3$. We omit any discussion of filtration and terms lying in the "error ideal" since in this case, and in the other cases of the trace lemma being treated here, there are no such terms, and we may arrange the modules coming from lower stages of the induction to have monomial socles also. The proof of the injectivity gives an identification (generally in the sense of leading terms, but here exactly) of the socle of $\overline{X}_1(1,2)$; since the module $X_1(1,2)$ is obtained from $\overline{X}_1(1,2)$ by propagating by d_2, it follows that we have identified the socle of $X_1(1,2)$.

This means that $X_1(1,2)$ has the properties (1), (2), (3), and (4) required by the statement of A.16 (or [**13**, 9.1]):

(1) is automatic from the propagation construction,
(2) follows from the construction of $\overline{X}_1(1,2)$ by mapping in an induced module, while
(3) comes from the proof of injectivity in [**13**, Proposition 8.4], which gives an identification of the socle of $\overline{X}_1(1,2)$.

Finally, property (4) follows from our knowledge of the Poincaré series of $B_1(1)$.

The constructions of $X_2(1,2)$ and $X_\varnothing(1,2)$ are similar; we abbreviate the treatments because we have already discussed $X_1(1,2)$ in detail.

To construct $\overline{X}_2(1,2)$ we start with the preliminary module $B_2(2)$, which is the part of $T(2)$ with degree less than 2; this module has Poincaré series $1+2t$. Note that $T(2) = k[x_1,x_2,x_3]/(x_2,f_3)$, where $f_3 = x_3^3 - x_2^2x_3$. $B_2(2)$ is a U_2-module, which we induce to $U_{1,2}$, and map to T by $g\otimes b \mapsto g\cdot\bigl(G(1,2\mid 2)\cdot b\bigr)$. We prove that this map is injective and identify (in the sense of leading terms) the socle using the following special case of the Trace Lemma:

$$\mathrm{Tr}_{U_1}\frac{d_3}{x_3^3 - x_2^2x_3}\cdot x_2^i = (-1)^{i+1}d_1^{6-2i}\cdot d_2^i,$$

where $i = 0,1$. We then propagate by d_1; the resulting module $X_2(1,2)$ has the desired properties.

To construct $\overline{X}_\varnothing(1,2)$ we need only note that the following special case of the Trace Lemma implies the existence of a projective module in the degree 10 part of T:

$$\mathrm{Tr}_{U_{1,2}}G(1,2\mid\varnothing) = \mathrm{Tr}_{U_{1,2}}\frac{d_3}{x_3}\frac{d_2}{x_2} = -d_1^4\cdot d_2^2.$$

We then propagate by d_1 and d_2; the resulting module $X_\varnothing(1,2)$ has the desired properties.

We turn to the construction of $X_{1,2}(1,2)$, the complement to the direct sum of the other three. The existence and properties of this module follow from [**13**, Proposition 8.1]. We review the proof of this proposition to call attention to a subtle point. To show the existence of the complement $X_{1,2}(1,2)$, we note that the other modules we have constructed do not intersect (because it is clear from our bookkeeping that their socles do not), and in large degrees, their Poincaré series add up to the Poincaré series of T. It follows that the modules $X_1(1,2)$, $X_2(1,2)$, and $X_\varnothing(1,2)$ are summands of T in large degrees, and from this it easily follows that they are *individually* summands in small degrees ([**13**, 6.18]). The subtle point is that our control of the socles is sufficient to show that these modules are *simultaneously* summands in small degrees, i.e. that the complement $X_{1,2}(1,2)$ exists. This is handled in Lemma 6.19 of [**13**].

From the decomposition of T we have studied, a similar decomposition for S follows, using [**13**, 5.4]. This completes our study of the extended example.

3. Remarks on the Trace Lemma

The proof of the main result of [**13**] is a very detailed exercise in bookkeeping, and it is natural to wonder if the effort required could somehow be reduced by a slightly different approach. Many of the complications of the trace lemma

TABLE 1. $\mathrm{Tr}_{U_1} G(1,2,3 \mid 2,3) \cdot e_x^i e_y^j$

	$i=0$	1	2	3	4	5
$j=0$	$-d_w^{18}$	$d_w^{16}d_x + d_w^{10}d_x^3$	$-d_w^{14}d_x^2 + d_w^8 d_x^4$	$d_w^{12}d_x^3$	$-d_w^{10}d_x^4$	$d_w^8 d_x^5$
$j=1$	$d_w^{12}d_y - d_w^6 d_x^2 d_y$	$-d_w^{10}d_x d_y$	$d_w^8 d_x^2 d_y$	$-d_w^6 d_x^3 d_y$	0	0

([**13**, 15.1]) are connected with the "higher terms" which must be handled in the accounting.

The reader of Section 2 may wonder if such higher terms are ever nonzero. Below we give an explicit example in which these higher terms are not zero.

EXAMPLE 3.1. Let $n = 4$, $q = 3$. We write U for the upper-triangular group over k, which we take to act on $k[w,x,y,z]$ in such a way as to preserve the flag of subspaces $\langle w \rangle \subset \langle w,x \rangle \subset \langle w,x,y \rangle \subset \langle w,x,y,z \rangle$. We write d_w, d_x, d_y, and d_z for the orbit products of w, x, y, and z. In the notation of [**13**], we will study $\mathrm{Tr}_{U_1} : k[w,x,y,z]^{U_{23}} \to k[w,x,y,z]^U$. We recall for the reader's convenience that U_1 is the "add w" group, isomorphic to $(k^+)^3$, and U_{23} is the "add x,y" group, isomorphic to the group of upper-triangular 3×3 matrices over k. This group acts on $k[x,y,z]$; its invariants are generated by the orbit products (under U_{23}!) of x, y, and z, which we write e_x, e_y, and e_z. In this notation we have $G(1,2,3 \mid 2,3) = d_z/e_z$. Table 1 is a short list of such traces.

In this table of values there are three examples where the trace is *not* a monomial in the chosen generators of the invariant ring; i.e. where the "higher terms" are nonzero. The energetic reader may wish to verify that the leading terms are as specified in the trace lemma ([**13**, 5.1]).

The proof of the trace lemma would be vastly simplified if one could produce a basis with the convenient property that the required traces took basis elements to basis elements (as opposed to linear combinations of these). Such a basis would eliminate all of the complications of "higher terms". Needless to say, the authors have not managed to produce such basis.

The authors noted in [**12**] that a very appealing proof of the main theorem of [**13**], based on a "factorization" of the module structure, was possible for $n = 3$, $q = 2$, but that this proof did not generalize to other values of n and q. (The proof was given in detail in [**2**].) One possible avenue to creating a generalization of the proof in [**2**] would be to take the components of the factorization, which are n submodules, each with Krull dimension 1, then invent a construction which takes these components as input and yields the full polynomial ring. However, it is not clear what sort of construction this might be. For example, when $n = 3$ and $q = 3$, one would expect to construct $k[x,y,z]$ out of three components, one having Poincaré series $(1+3t+6t^2+8t^3+9t^4+9t^5+\cdots)$, and corresponding to the successive radicals of the module $k[U_{1,2}/U_2]$ in the notation of Section 2. Thus, one would expect the Jacobson radical J_8 of $k[U_{1,2}/U_2]$ to figure in the construction of S^3. However, it can be verified with `Magma` that J_8 is neither a submodule nor a quotient of S^3.

4. Examples of possible generalizations

The reader of [**13**] will note that the proof of the main theorem in that paper seems to depend on a remarkable collection of coincidences; it is not easy to see how to generalize the module-structure finiteness result.

In this section we give some examples which suggest that the phenomenon of module-structure finiteness may hold in a more general setting than that of our theorem. There are limits to this generality, as seen in Examples 4.4 and 4.5.

EXAMPLE 4.1 (Hopf Algebras). Let $B = \Lambda(a,b)$, an exterior algebra on two generators over the field of two elements, which we write k in this example. We regard B as an algebra of cohomology operations by taking a to act as Sq^1 and b to act as $Sq^3 + Sq^2\,Sq^1$. We let this algebra act on the cohomology of $\mathbb{Z}_2 \times \mathbb{Z}_2$, which we write $k[x,y]$. We have the following module-structure decomposition:

$$k[x,y] \cong k \oplus (k[x^2,y^2] \otimes F) \oplus (L_1 \oplus L_2 \oplus L_3),$$

where $k = \langle 1 \rangle$ is the trivial B-module, $F = \langle xy, x^2y + xy^2, x^4y + xy^4, x^4y^2 + x^2y^4 \rangle$ is a free B-module of rank 1, $L_1 = \langle x, x^2, x^3, \ldots \rangle$, $L_2 = \langle y, y^2, y^3, \ldots \rangle$, and $L_3 = \langle x^2y, x^2y^2, x^2y^3, \ldots \rangle$. (We have given vector-space bases for each module appearing.)

We note that, as B-modules, $L_1 \cong L_2 \cong L_3$, thus only the three isomorphism classes k, F, and L_1 appear.

We sketch a proof of the above assertions. Note that the Poincaré series of all these modules add up to the Poincaré series of the polynomial ring. So, to get the decomposition result, we need only prove the surjectivity of the map from the decomposition to the polynomial ring.

Using the given basis of the module L_3, plus an induction, we can show that the free module $x^{2i}y^{2j}F$ lets us obtain the monomials $x^{2i+1}y^{2j+1}$, $x^{2i+1}y^{2j+2}$, $x^{2i+4}y^{2j+1}$, and $x^{2i+4}y^{2j+2}$. This gives us every monomial of $x^\alpha y^\beta$, where $\alpha > 0$ and $\beta > 0$. The rest of the monomials are covered by L_1, L_2, and k.

We need to show that the L_i are indecomposable. Perhaps the simplest method of doing this is to use the isomorphism $\Lambda(a,b) \cong \mathbb{F}_2[\mathbb{Z}_2 \times \mathbb{Z}_2]$ and quote the results on infinitely generated modules for this group ring in [**5**].

REMARK 4.2. The reader should be aware that there is some question of whether a module-structure decomposition is well-defined in a context involving modules which are not finite-dimensional. Since we have not chosen hypotheses or made specific conjectures, we will not address this issue.

The finiteness corollary still holds over the algebraic closure of a finite field, because any representation over such a field can be written inside a finite field. It is not clear that this remains the case for other infinite fields.

The next example, which was considered independently by Derksen and Kemper [**10**], shows that the finiteness result can continue to hold even when the field contains transcendentals.

EXAMPLE 4.3 (Infinite fields). Let $K = k(s,t) \supset k$ be the extension of k generated by two algebraically independent elements s, t. Then $G = \{y \mapsto y + (\lambda s + \mu t)x, x \mapsto x \mid \lambda, \mu \in k\}$ is a group of automorphisms of $S = K[x,y]$. We note that $S^G = K[x, d_y]$, where $d_y = \prod_{(\lambda,\mu)\in k^2}(y + (\lambda s + \mu t)x)$.

We will show, by the method of [**13**], that

$$S \cong \{K[x, d_y] \otimes S^{q^2-1}\} \oplus \{K[d_y] \otimes (\oplus_0^{q^2-2} S^i)\},$$

i.e. we will prove that the main theorem of [**13**] holds in this context.

To see this, note that by [**7**, Proposition 9.5], we have that $\mathrm{Tr}_G\, y^{q^2-1} = \alpha x^{q^2-1}$, where α is a nonzero element of K. Thus, the submodule of S^{q^2-1} generated by y^{q^2-1} is free. By dimension-counting, this module is S^{q^2-1} itself. In this setting, projective and injective are the same, so we may propagate S^{q^2-1} by any monomial in x and d_y to obtain new summands, which do not intersect as their socles are different.

The other summand in the decomposition, $K[d_y] \otimes (\oplus_0^{q^2-2} S^i)$, can be mapped to the polynomial ring by multiplication, and does not intersect the first summand. (Again, consider the socles.) Thus, the map we have constructed from the decomposition to the polynomial ring is injective. By counting graded dimensions using Poincaré series, we see that the map is surjective and therefore an isomorphism.

The point of this example is not so much that the finiteness result continues to hold, but that the method of proof of [**13**] still works. As noted in [**12**], cases where there are three or fewer variables in the polynomial ring have special characteristics which may make alternate proofs of the finiteness result possible. In fact, in Section 6 we will prove that the finiteness result holds for the action of a finite group on a polynomial ring in three variables over *any* field. Thus, it is not the result but the success of the method that is interesting in Example 4.3.

Before treating our next two examples, we review some facts about Heller translates. If G is any finite group, and M a finitely-generated kG-module, then M can be embedded in a finitely-generated free kG-module F. The *first Heller translate* $\Omega^1 M$ of M is, by definition, the quotient F/M. The $(n+1)$-st Heller translate $\Omega^{n+1} M$ (of M) is defined by induction to be $\Omega^1(\Omega^n M)$. These modules are defined only up to a finitely-generated projective summand. Often when working concretely with Heller translates, one chooses a convenient representative module, for example, one of minimal dimension over k.

We now show from this definition that for any M, $\Omega^1 k \otimes M \cong \Omega^1 M$, i.e. $\Omega^1 k \otimes M$ fits in an exact sequence as described in the definition (above) of the Heller translate of M. This exact sequence is obtained from the exact sequence $0 \to k \to kG \to \Omega^1 k \to 0$ by tensoring (over k) with M, and noting that $kG \otimes_k M$ is free.

It follows from this that $\Omega^1 k \otimes \Omega^1 k \cong \Omega^2 k$ and that $(\Omega^1 k)^{\otimes n} \cong \Omega^n k$, where the isomorphism symbol is to be interpreted in the sense of the definition of Ω^1, i.e. up to a finitely-generated projective summand. This concludes our review of Heller translates.

The next example shows that the finiteness result fails for free modules over a polynomial ring.

EXAMPLE 4.4 (Free Modules over a polynomial ring). Let k be the field of two elements and let $V = \langle \alpha, \beta \rangle \cong \mathbb{Z}_2 \times \mathbb{Z}_2$ act on $k[x, y]\langle 1, z \rangle$ by $\alpha\colon z \mapsto z + x$ and $\beta\colon z \mapsto z + y$. We can regard $k[x, y]\langle 1, z \rangle$ as a subset of $k[x, y, z]$ or as a free $k[x, y]$-module of rank two.

If $n \in \mathbb{N}_0$, we write $\Omega^n k$ for the nth Heller translate of the trivial kV-module k. These modules are pairwise nonisomorphic, since their ranks over k are all different.

With some work, $\Omega^n k$ can be pictured using the diagrams popularized by Benson and Carlson [**4**]. Writing $a = 1 + \alpha \in kV$, $b = 1 + \beta$ we have:

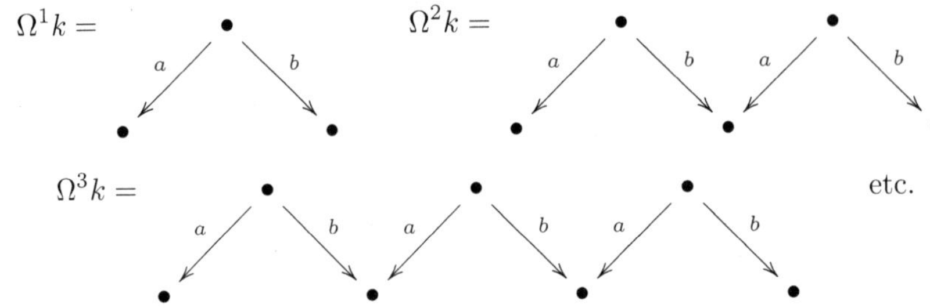

The dots represent basis vectors and the arrows the action of the group ring; see [**4**].

Now we note that if we attach a grading to our module by assigning x, y, and z grading 1, then $M = k[x, y]\langle 1, z\rangle$ is the direct sum $\oplus_{i=0}^{\infty} M_i$. Further, the basis of M_i consisting of the monomials of total degree i has exactly the same diagram as $\Omega^i k$:

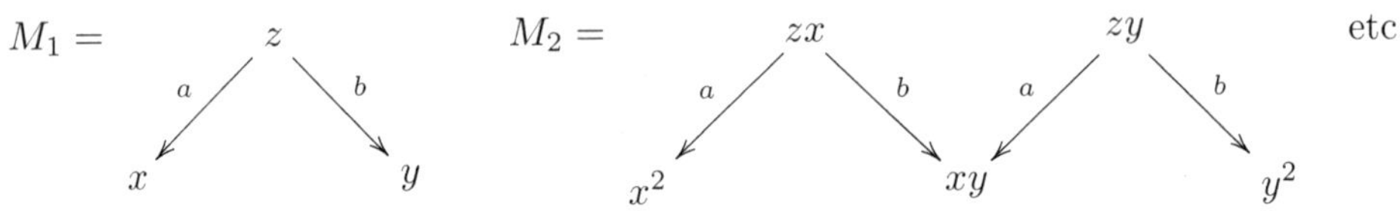

This gives an explicit isomorphism $M_i \cong \Omega^i k$, and therefore

$$M \cong \oplus_{i=0}^{\infty} M_i \cong \oplus_{i=0}^{\infty} \Omega^i k.$$

Thus, the module-structure finiteness property does not hold for M.

EXAMPLE 4.5 (Tensor algebras). In this example once again k is the field of two elements and V the elementary abelian group of order 4. We write T for the tensor algebra of $\Omega^1 k$, so that $T_n = (\Omega^1 k)^{\otimes n}$. Then $T_n \cong \Omega^n k \oplus F$, where F is a free kV-module. As noted above, the modules $\Omega^n k$ are pairwise nonisomorphic, so that the finiteness result does not hold in this context either.

Various results about the modules for which a finiteness result does hold for the tensor algebra are given by Feit in [**11**, II.5]. The obvious generalization of the example above to any prime and the results of Feit show that the only groups for which there is a finiteness result for all modules are those of finite representation type.

5. The general linear group

In this section we indicate the form a module-structure theorem for the general linear group should take by giving two examples. We will study the cases of two and three variables over the field of two elements. Thus, throughout this section, we take $k = \mathbb{F}_2$.

Our first goal is to handle the case of two variables. We want to describe the module structure that results when $G = \mathrm{GL}_2(k)$ acts on $k[y, z]$ by homogeneous linear substitutions. We begin by recalling the following standard result:

LEMMA 5.1. *There are exactly three indecomposable kG-modules: the Steinberg module St, the trivial module k and the free module for the quotient C_2, M.*

Variations on this lemma can be found in most texts on modular representation theory, but at the suggestion of the referee we outline here a proof that the only indecomposable kG-modules are M, k, and St.

SKETCH OF PROOF. Write $G = \langle a, b \mid a^3 = b^2 = 1, bab = a^2 \rangle$. Then $e = 1 + a + a^2$ and $f = 1 - e = a + a^2$ are orthogonal central idempotents, with kGe and kGf isomorphic as rings to kC_2 and $M_2(k)$. Further, any kG module X splits as $eX \oplus fX$, where eX is a kGe-module and fX is a kGf-module.

If X is indecomposable, either eX or fX is zero, so we need only analyse the possible indecomposable kC_2-modules and $M_2(k)$-modules. Since $M_2(k)$ is a matrix ring, every $M_2(k)$-module is a direct sum of copies of the natural module (i.e. column vectors), from which we get St. Since kC_2 is a principal ideal domain, we get the modules k, M from specializing the classification of modules over a PID to the current setting. □

Now we identify instances of these modules in $k[y, z]$. Recall that $k[y, z]^G = k[c_2, c_3]$, where $c_2 = y^2 + yz + z^2$, $c_3 = y^2 z + yz^2$.

LEMMA 5.2. *$\langle y, z\rangle \cong \langle y^2, z^2\rangle \cong St$, $\langle y^2+yz+z^2\rangle \cong k$, and $\langle y^3+y^2z+z^3\rangle \cong M$.*

Writing $St^{(1)}$, $St^{(2)}$ and $M^{(3)}$ for the copies of these modules we have identified in $k[y, z]$, we have

PROPOSITION 5.3. *$k[y, z] \cong k[c_2] \oplus k[c_2, c_3] \otimes (St^{(1)} \oplus St^{(2)}) \oplus k[c_2, c_3] \otimes M^{(3)}$.*

PROOF. We construct a map from the decomposition on the right-hand side to the polynomial ring using the inclusion of the submodules exhibited in Lemma 5.2 and multiplication by the invariants. We note that the Poincaré series of the right-hand side is

$$\frac{2t + 2t^2}{(1 - t^2)(1 - t^3)} + \frac{2t^3}{(1 - t^2)(1 - t^3)} + \frac{1}{1 - t^2} = \frac{1}{(1 - t)^2},$$

i.e. the same as the Poincaré series of the left-hand side. So, to prove that the map is an isomorphism, it suffices to show that it is injective.

To show this, we show that there is no linear relation between the socles of the three parts of our decomposition. Such a relation would have the form

$$f(c_2) + c_3 \cdot g(c_2, c_3) + (ax + by + cx^2 + dy^2) \cdot h(c_2, c_3) = 0,$$

where f, g, and h are polynomials and a, b, c, and d are elements of k. If we rearrange the equation so that all the invariants are on one side, we see that $(ax + by + cx^2 + dy^2)$ must be invariant, and it follows that a, b, c, and d are all zero. Then we have reduced to $f(c_2) + c_3 \cdot g(c_2, c_3) = 0$, and since the Dickson invariants are algebraically independent, it follows that f and g are zero.

This proves that the map is injective and therefore an isomorphism. □

Now we turn to the result for three variables, that is, we study the module structure resulting from the action of $G = \mathrm{GL}_3(k)$ on $k[x, y, z]$ by homogeneous linear transformations. Recall that the Dickson invariants $k[x, y, z]^G$ are a polynomial algebra $k[c_4, c_6, c_7]$ on homogeneous generators of degree 4, 6, and 7.

The result is quite complex, so at the suggestion of the referee, we give a statement which is less detailed than that of Proposition 5.3. The proof, which we omit, involves specifically identifying the modules which below are only described by their graded dimensions, and is handled using `Magma`.

Our decomposition is:

PROPOSITION 5.4. *There is an isomorphism of* $\mathrm{GL}_3(k)$*-modules:*

$$k[x,y,z] \cong [k[c_4] \otimes Z] \oplus [k[c_4, c_6] \otimes X] \oplus [k[c_4, c_6, c_7] \otimes P],$$

where Z *has Poincaré series* $1+3t$, X *has Poincaré series* $6t^2+10t^3+6t^4+10t^5+14z^6+10t^7$, P *has Poincaré series* $8t^4+8t^5+8t^6+16t^7+24t^8+24t^9+24t^{10}+16t^{11}+16t^{12}+16t^{13}+8t^{14}$, *and* P *is projective.*

These two examples suggest that the result for $\mathrm{GL}_n(k)$, where $k = \mathbb{F}_q$, should have n parts, propagated by the subsets of the Dickson invariants

$$\{c_{q^n-q^{n-1}}\}, \{c_{q^n-q^{n-1}}, c_{q^n-q^{n-2}}\}, \ldots, \{c_{q^n-q^{n-1}}, c_{q^n-q^{n-2}}, \ldots, c_{q^n-1}\}.$$

One can make more specific conjectures, but since we do not have a completely precise conjecture on the "best possible" decomposition of $k[x_1, \ldots, x_n]$ as a $\mathrm{GL}_n(k)$-module, we will stop here.

6. A general proof for three variables

Here we present a version of the structure theorem for three variables that applies to any finite group and any field. In particular, this result (6.1) gives a decomposition for the action of $\mathrm{GL}_3(k)$ on $k[x_1, x_2, x_3]$, where k is a finite field (see Proposition 5.4 and the remarks following). Further, Theorem 6.1 applies to a field that contains transcendentals (see also Example 4.3).

THEOREM 6.1. *If a finite group* G *acts on* $S = k[x_1, x_2, x_3]$ *by homogeneous transformations then there are homogeneous invariant elements* $u, v \in S^G$ *and graded* kG*-submodules* P, U, B *of* S *such that* P *is projective (but infinite dimensional) and* U, B *are finite dimensional and there is an isomorphism of* kG*-modules*

$$S \cong P \oplus (U \otimes k[u,v]) \oplus (B \otimes k[u]).$$

The proof depends on several lemmas, which hold for any number of variables. There is no content when k has characteristic 0, so we assume that the characteristic is p.

The next lemma will be applied in the proof of Theorem 6.1 by taking $R = S^G$ and M to be a quotient of S.

LEMMA 6.2. *Let* R *be a commutative graded ring, finitely-generated as an algebra over* $R_0 = k$ *by homogeneous elements of positive degree. Let* M *be a finitely generated graded* RG*-module for which there is a number* n *such that* $\dim_k M_r \leq n$ *for all* $r \in \mathbb{Z}$. *Then there are finite-dimensional graded* kG*-submodules* B, U *of* M *and a homogeneous element* $v \in R$ *such that* $M = B \oplus (U \otimes k[v])$.

PROOF. By Noether normalization there is a homogeneous element $v \in R$ of degree s, say, such that M is finitely-generated over $k[v]$. We may take M to be generated by $\langle a_i \mid i = 1, \ldots, N \rangle$, and we may further take the a_i to be homogeneous. Let n be the maximum of the degrees of the a_i. If $m \in M_{r+s}$, where $r \geq n$, then we have $m = \sum_i f_i a_i$, where we take the f_i to be homogeneous elements of $k[v]$ since m and the a_i are homogeneous. Looking at degrees, we see that $\deg f_i \geq s$, so $f_i = v g_i$ for some $g_i \in k[v]$, so that each $m = v(\sum_i g_i a_i)$. It follows that for n as defined above, if $r \geq n$ then $M_{r+s} = vM_r$. Let $B = \bigoplus_{r<n} M_r$ and $U = \bigoplus_{r=n}^{n+s-1} M_r$. $\square$

LEMMA 6.3. *If G acts on $S = k[x_1, \dots, x_n]$ then there is an invariant homogeneous element $u \in S^G$ such that, as kG-modules, $S \cong (S/uS) \otimes k[u]$.*

PROOF. Let Q be the Sylow p-subgroup of G. We may assume that the matrices have triangular form with respect to the basis $x_1, \dots, x_n$ (such that x_1 is fixed). Let $R = S^G$. Since S is finitely generated over R and S contains x_n^r for all $r \in \mathbb{N}$ there must be a homogeneous element $u \in R$ that contains the monomial x_n^r for some $r \in \mathbb{N}$.

Let T denote the k-subspace of S spanned by all monomials not divisible by x_n^r. Then $S = T \oplus uS$ as kQ-modules. Thus the inclusion of uS in S is split over Q, so it is split over G: let T' be a complement as a kG-module. Then $S = T' \oplus uS = T' \otimes k[u]$. □

REMARK 6.4. If G is a p-group in triangular form then we may take u to be the orbit product of x_n. If k is finite we may take u to be the bottom Dickson invariant for $\mathrm{GL}_n(k)$.

LEMMA 6.5. *Returning to G acting on $S = k[x_1, x_2, x_3]$, with u as in Lemma 6.3, the non-projective part of S/uS, which we denote by $(S/uS)/(\mathrm{proj})$, has dimension bounded by some N in each degree.*

PROOF. Since $S/(\mathrm{proj}) \cong ((S/uS)/(\mathrm{proj})) \otimes k[u]$ and $\dim_k (S/(\mathrm{proj}))_r$ is bounded by a linear function in r by [**20**], the result follows. □

PROOF OF THEOREM 6.1. By Lemma 6.3 we can write $S = (S/uS) \otimes k[u]$ and by Lemma 6.5, we can apply Lemma 6.2 to obtain $(S/uS)/(\mathrm{proj}) = B \oplus (U \otimes k[v])$. □

7. Conlon's induction theorem

One approach to obtaining a structure theorem for general G is to use Conlon's Induction Theorem to reduce it to the case of subgroups that are cyclic modulo p. A group H is cyclic modulo p if there is a p-group P which is a normal subgroup of H for which the quotient H/P is cyclic.

Let $T_n = T_n(k)$ be the subgroup of $\mathrm{GL}_n(k)$ consisting of upper triangular matrices where any non-zero entries are allowed on the diagonal. The generators d_i of the invariants under U_n are eigenvectors for T_n. Observe also that if C is a subgroup of $\mathrm{GL}_n(k)$ which is cyclic modulo p, then C is conjugate in $\mathrm{GL}_n(k)$ to a subgroup of T_n.

PROPOSITION 7.1. *The main structure formula* (1.1) *remains valid over T_n, with T_n acting on the d_i as above.*

PROOF. Examine the proof in [**13**] to verify that the pieces $\overline{X}_I(J)$ can be defined over T_n. □

Rather than a structure theorem we will prove a result that would be one of its main consequences: a formula for the multiplicity as a summand of any indecomposable kG-module.

For any indecomposable kG-module M let $\mathrm{PS}_M(S,t)$ be the power series in which the coefficient of t^r is the multiplicity of M as a summand of S_r. Knowing these power series $\mathrm{PS}_M(S,t)$ for each indecomposable M which is a summand of S would give the multiplicity formulas desired.

By Conlon's Induction Theorem ([**8**]; [**9**, 80.61]), there are cyclic modulo p subgroups C_i of G and rational numbers a_i such that

$$S_r = \sum_i a_i \operatorname{Ind}_{C_i}^G \operatorname{Res}_{C_i}^G S_r$$

in the Green ring for each $r = 0, 1, 2, \ldots$. We will use a graded version of Conlon's result; this trivial consequence of the original combines the above family of equations, yielding

$$S = \sum_{r=0}^{\infty} S_r t^r = \sum_{r=0}^{\infty} \sum_i a_i \operatorname{Ind}_{C_i}^G \operatorname{Res}_{C_i}^G S_r t^r. \tag{7.1}$$

Another way to say this is that we regard S as formal power series whose coefficients lie in the Green ring.

Let ℓ be an extension of k such that all the p' elements of G become diagonalizable in ℓ. We write $\ell S = \ell \otimes_k S$. The image of each C_i in $\mathrm{GL}_n(\ell)$ can be conjugated into $T_n(\ell)$, so we can apply 7.1 to $\operatorname{Res}_{C_i}^G \ell S$ to obtain

$$\operatorname{Res}_{C_i}^G(\ell S) = \sum_j \mathrm{PS}_{N_{ij}}(\ell S, t) N_{ij}, \tag{7.2}$$

where N_{ij} runs through the finite set of ℓC_i-modules which are indecomposable summands of $\operatorname{Res}_{C_i}^G \ell S$. We remind the reader that in Eq. (7.2), $\operatorname{Res}_{C_i}^G(\ell S)$ is a shorthand for $\sum_r \operatorname{Res}_{C_i}^G(\ell S)_r t^r$. Combining this with Eq. (7.1) we obtain

$$\ell S = \sum_{ij} a_i \mathrm{PS}_{N_{ij}}(\ell S, t) \operatorname{Ind}_{C_i}^G N_{ij}.$$

Now we restrict scalars and, to simplify the notation, we let $b_{ij}(M)$ denote the multiplicity of the indecomposable kG-module M as a summand of $\operatorname{Ind}_{C_i}^G \operatorname{Res}_k^\ell N_{ij}$. We obtain:

THEOREM 7.2.

$$\mathrm{PS}_M(S, t) = \frac{1}{[\ell : k]} \sum_{ij} a_i b_{ij}(M) \mathrm{PS}_{N_{ij}}(\ell S, t),$$

where $\mathrm{PS}_{N_{ij}}(\ell S, t)$ *encodes the multiplicity of* N_{ij} *as a summand of* ℓS *as an* ℓ C_i*-module.*

To use this formula involves calculating the C_i and the a_i, which is routine, and also the $\mathrm{PS}_{N_{ij}}(\ell S, t)$. The latter can all be calculated by restricting the formula for the decomposition of ℓS as an $\ell T_n(\ell)$-module. According to [**13**] this involves calculating ℓS up to degree $r^{n-1} - n$, where r is the order of ℓ.

Such a procedure is probably uneconomical, but it does yield an explicit algorithm.

8. Applications to vector invariants

In this section we present an application of module-structure theory to invariant theory. We show how to obtain the Poincaré series of the vector invariants $S^*(V \oplus V \oplus \cdots \oplus V)^G$ using the G-module structure of S^*V. We also show how to take this technique further and in some cases obtain a Poincaré series for what we call *tensor* invariants, that is, $S^*(V \otimes W)^{G \times H}$, where W carries an H-module structure. Most of these techniques were shown to the authors by Milgram [**15**], who states that their use was standard in certain types of homotopy-theoretic calculations about thirty years ago, but offers no specific attribution.

The basic idea for vector invariants is simple: the G-module structure of S^*V determines the Poincaré series of S^*V^G. Since the G-module structure of S^*V determines the G-module structure of $S^*(V^{\oplus n}) \cong [S^*(V)]^{\otimes n}$, it also determines the Poincaré series of $S^*(V^{\oplus n})^G$. Similarly, if W is a monomial representation of H, we will be able to determine the $G \times H$-module structure of $S^*(V \otimes W)$ in terms of the G-module structure of S^*V and the H-module structure of S^*W; the vector invariants are simply the special case where W is the trivial H-module of dimension n.

Because the idea is not deep, and because the appropriate implementation will vary with the particular application, we will confine ourselves to illustrating the method by the study of certain special cases. We begin with the case of vector invariants of the regular representation F of the cyclic group of order 2, C_2, over the field of two elements k. The referee has noted that this ring of invariants is well-known, and offers the reference [**18**, p. 31].

EXAMPLE 8.1 (Vector invariants: the regular representation of $G = C_2$, $n = 2$.). The C_2-module structure of the polynomial ring S^*F is $(k[d_2] \otimes k) \oplus (k[d_1, d_2] \otimes F)$. From this fact we have

$$S^*(F \oplus F) \cong S^*F \otimes S^*F \cong \{(k[d_2] \otimes k) \oplus (k[d_1, d_2] \otimes F)\} \otimes \{(k[e_2] \otimes k) \oplus (k[e_1, e_2] \otimes F)\}.$$

Noting that $F \otimes F \cong F \oplus F$, we may expand this decomposition to yield

$$\begin{aligned} S^*(F \oplus F) \cong {} & (k[d_2, e_2] \otimes k) \oplus (k[d_1, d_2, e_2] \otimes F) \oplus (k[d_2, e_1, e_2] \otimes F) \\ & \oplus (k[d_1, d_2, e_1, e_2] \otimes F \oplus F). \end{aligned} \tag{8.1}$$

Now we note that $\dim_k F^{C_2} = \dim_k k^{C_2} = 1$, so that (8.1) gives us an expression for the Poincaré series of $S^*(F \oplus F)^{C_2}$:

$$\begin{aligned} & \frac{1}{(1-t^2)^2} + \frac{t}{(1-t)(1-t^2)^2} + \frac{t}{(1-t)(1-t^2)^2} + \frac{2t^2}{(1-t)^2(1-t^2)^2} \\ & = \frac{1+t^2}{(1-t)^2(1-t^2)^2}. \end{aligned} \tag{8.2}$$

The series in Eq. (8.2) has been checked using Kemper's software [**14**], which is now incorporated into `Magma` [**6**].

Now that we have illustrated the basic idea, we proceed to some refinements. Our next example studies tensor invariants; in the notation of the opening paragraph of this section we take $G \cong H \cong C_2$ and V and W to be the regular representations of G and H.

EXAMPLE 8.2 (Tensor invariants: $G = C_2$, $H = C_2'$, $V = F$, and $W = F'$.). We regard $F \otimes F' \cong F \tilde{\oplus} F$ as a $C_2 \times C_2'$ module, where the C_2' exchanges the Fs. This is the same as the $C_2 \times C_2'$-module obtained by taking the tensor product of the regular representations F, F' of C_2, C_2'. We study $S^*(F \otimes F')$ as a $C_2 \times C_2'$-module by writing $S^*(F \otimes F') \cong S^*(F \tilde{\oplus} F) \cong S^*(F) \tilde{\otimes} S^*(F)$, where the tilde indicates that the action of C_2' switches the factors in the direct sum or tensor product. Of course underneath all the notation we are just studying the regular representation of $C_2 \times C_2'$; the point is to illustrate the method.

TABLE 2. C_2-module structure of $S^*(F)\widetilde{\otimes}S^*(F)$

	Degree	0	1	2	2
Degree	Module	k	F	F	k
0	k	$k\widetilde{\otimes}k$	$\boxed{k \otimes F}$	$k \otimes F$	$\underline{k \otimes k}$
1	F	$\boxed{F \otimes k}$	$F\widetilde{\otimes}F$	$F \otimes F$	$F \otimes k$
2	F	$F \otimes k$	$F \otimes F$	$F\widetilde{\otimes}F$	$F \otimes k$
2	k	$\underline{k \otimes k}$	$k \otimes F$	$k \otimes F$	$k\widetilde{\otimes}k$

TABLE 3. Poincaré series for the multiplicities of summands in $S^*(F)\widetilde{\otimes}S^*(F)$

Module	Poincaré series
$\mathrm{Ind}_{C_2}^{C_2 \times C_2'}(k \otimes F)$	$\frac{t}{(1-t)(1-t^2)^2}$
$\mathrm{Ind}_{C_2}^{C_2 \times C_2'}(F \otimes F)$	$\frac{t^3}{(1-t)^2(1-t^2)(1-t^4)}$
$F\widetilde{\otimes}F$	$\frac{t^2}{(1-t^2)(1-t^4)}$
$\mathrm{Ind}_{C_2}^{C_2 \times C_2'}(k \otimes k)$	$\frac{t^2}{(1-t^2)(1-t^4)}$
$k\widetilde{\otimes}k$	$\frac{1}{(1-t^4)}$

If we write down the structure of $S^*(F)\widetilde{\otimes}S^*(F)$ as a C_2-module in the first few degrees, we obtain

We note that the action of C_2' swaps the entries in the table across the main diagonal. Thus, as representations of $C_2 \times C_2'$, the boxed modules combine to form $\mathrm{Ind}_{C_2}^{C_2 \times C_2'}(F \otimes k)$, while the underlined modules combine to form $\mathrm{Ind}_{C_2}^{C_2 \times C_2'}(k \otimes k)$. Similar remarks apply to the other off-diagonal entries in the table. The diagonal entries are marked with tildes to indicate that the action of C_2' switches the factors in the tensor product.

These regularities allow us to determine the $C_2 \times C_2'$-module structure of $S^*(F)\widetilde{\otimes}S^*(F)$ in the following way. We first write down for each C_2-isomorphism type $[M]$ of summands of $S^*(F)$ a Poincaré series $P_M(t)$ for its multiplicity. This collection of isomorphism types $[M_i]$ gives rise to three classes of $C_2 \times C_2'$-isomorphism types of summands of $S^*(F)\widetilde{\otimes}S^*(F)$, whose forms are listed below:

1. (boxed type): $\mathrm{Ind}_{C_2}^{C_2 \times C_2'}(M_i \otimes M_j)$.
2. (underlined type): $\mathrm{Ind}_{C_2}^{C_2 \times C_2'}(M_i \otimes M_i)$.
3. (diagonal type): $M_i\widetilde{\otimes}M_i$.

The Poincaré series for the multiplicities of each of these types of modules as summands of $S^*(F)\widetilde{\otimes}S^*(F)$ can be obtained easily in terms of the $P_{M_i}(t)$ as follows:

1. (boxed type): $P_{M_i}(t) \cdot P_{M_j}(t)$.
2. (underlined type): $(1/2)[P_{M_i}(t)^2 - P_{M_i}(t^2)]$.
3. (diagonal type): $P_{M_i}(t^2)$.

In the case at hand we see from our decomposition of $S^*(F)$ that $P_k[t] = (1-t^2)^{-1}$ and that $P_F(t) = t \cdot (1-t)^{-1} \cdot (1-t^2)^{-1}$. We then obtain Poincaré series for the multiplicities of the summands of $S^*(F)\widetilde{\otimes}S^*(F)$:

TABLE 4. Dimensions of invariant subspaces

Module	$\mathrm{Ind}_{C_2}^{C_2\times C_2'}(k\otimes F)$	$\mathrm{Ind}_{C_2}^{C_2\times C_2'}(F\otimes F)$	$F\tilde{\otimes}F$	$\mathrm{Ind}_{C_2}^{C_2\times C_2'}(k\otimes k)$	$k\tilde{\otimes}k$
$\dim_k(M^{C_2\times C_2'})$	1	2	2	1	1

Finally, as in Example 8.1 we need to know the dimension of the invariant subspaces.

This allows us to add rational functions to obtain the Poincaré series of $S^*(F\otimes F')^{C_2\times C_2'}$.

$$\begin{aligned}(8.3)\quad \frac{t}{(1-t)(1-t^2)^2}+\frac{2t^3}{(1-t)^2(1-t^2)(1-t^4)}+\frac{2t^2}{(1-t^2)(1-t^4)}\\ +\frac{t^2}{(1-t^2)(1-t^4)}+\frac{1}{(1-t^4)}=\frac{t^2-t+1}{(1-t)^2(1-t^2)^2}.\end{aligned}$$

This example, like the previous one, has been checked against the result obtained by Kemper's software.

Although the techniques of Example 8.2 may seem specialized to the case of tensor products with the regular representation of C_2, they can be adapted to arbitrary permutation representations by a more sophisticated version of the analysis following Table 2. Next we will see how to handle monomial representations. In the next example we will consider the case of a tensor product with the natural representation St of $\mathrm{GL}_2(k)$, i.e. in the setting of the opening paragraph of this section we take $W = St$ and $H = \mathrm{GL}_2(k)$. Note that $\mathrm{GL}_2(k) \cong S_3 = \langle a, b \mid a^3 = 1, b^2 = 1, bab = a^2\rangle$. We write C_2 for the subgroup $\langle b\rangle \subset S_3$.

Let $K = \{0, 1, \omega, \omega^2\}$ be the field of four elements, and write W for the representation St, defined over K. For the remainder of this section, i.e. the discussion below and the two examples following, we fix this notation ($W = St\otimes_k K$). Notice that although St is defined over k (it is the natural representation of $\mathrm{GL}_2(k)$), if we are only interested in the Poincaré series of the invariant ring, it does not matter whether we work over k or K. Since it is more convenient for our purposes in the rest of this section, we will work over K, so that all tensor products in what follows are over K.

If V is a KG-module, we can describe the $KG\times S_3$-module structure of $S^*(V\otimes W)$ by regarding $V\otimes W$ as $V\tilde{\oplus}V$, where b switches the two copies of V and a acts on the first as multiplication by ω and on the second by multiplication by ω^2.

We can adapt the results of Example 8.2 to the C_2-module structure of $S^*(V\tilde{\oplus}V)$. This gives us three types of submodules of $S^*(V\tilde{\oplus}V)$. The first type is $(V\otimes W)\tilde{\oplus}(W\otimes V)$, the second is $(V\otimes V)\tilde{\oplus}(V\otimes V)$, and the third is $V\tilde{\otimes}V$. Here we have used the tilde to indicate the action of C_2, as in Example 8.2.

We observe that all of these module types are actually S_3-invariant; i.e. working over K, the decomposition of the C_2-module structure is also a decomposition of the S_3-module structure. In the application of this result we used a compact description of the multiplicities of these three types of modules in $S^*(V\tilde{\oplus}V)$ in terms of their multiplicities in $S^*(V)$. We can give a similar description of the multiplicities of the different S_3-module types. To do this, we first refine our description of the decomposition. Modules of the first type are either $W\otimes X\otimes Y$ or $M\otimes X\otimes Y$, regarded as a $S_3 \times G$-module. (Here M is the free module for the quotient C_2 of S_3.) We have the second case if $\deg Y + 2\deg X \equiv 0 \mod 3$, and the first case if $\deg Y + 2\deg X \not\equiv 0 \mod 3$. The same holds for modules of the second type. Modules of the third type always have trivial C_3-action. Note that, for modules of the first and second type, if we are in the first case (i.e. a tensor product with St) there are no invariants, since $St^{S_3} = 0$.)

Observe that if $f(t) = \sum_i a_i t^i$, $g(t) = \sum_j b_j t^j$, then $\sum_{i+2j\equiv 0 \mod 3} a_i b_j t^{i+j}$ can be expressed in terms of f and g as

$$\frac{1}{3}[f(t)g(t) + f(\zeta t)g(\zeta^2 t) + f(\zeta^2 t)g(\zeta t)], \tag{8.4}$$

where $\zeta \in \mathbb{C}$ is a primitive cube root of 1. To see this, write

$$f(t)g(t) = \sum_{i,j} a_i b_j t^{i+j}$$

$$f(\zeta t)g(\zeta^2 t) = \sum_{i,j} a_i b_j t^{i+j} \cdot \zeta^{i+2j}$$

$$f(\zeta^2 t)g(\zeta t) = \sum_{i,j} a_i b_j t^{i+j} \cdot \zeta^{2i+j},$$

and note that $(\zeta^{i+2j})^2 = \zeta^{2i+j}$, so $1 + \zeta^{i+2j} + \zeta^{2i+j}$ is zero if $2i + j \not\equiv 0 \mod 3$ and 3 if $2i + j \equiv 0 \mod 3$. This discussion can be generalized to other roots of 1, and by combining this with arbitrary permutation representations, we can get results for monomial representations. However, the calculations required for such examples quickly become tedious.

We now proceed to an application of this general discussion.

EXAMPLE 8.3 (Tensor invariants: $G = C_2$, $V = F$, $H = S_3$, and $W = St \otimes_k K$, over K). Take G to be the cyclic group C_2 and $V = F$, the regular representation of G. In the decomposition of $S^*(V \otimes W)$ we obtain modules of all five types mentioned above.

We recall from our discussion of the decomposition for $S^*(F)$ that $P_k(t) = 1/(1-t^2)$ and $P_F(t) = t/(1-t)(1-t^2)$. Thus in $S^*(F\tilde{\oplus}F)$, regarded as an $C_2 \times S_3$-module, we have that the modules of type $M \otimes F \otimes k$ have the following multiplicity:

$$\tfrac{1}{3}[P_F(t)P_k(t) + P_F(\zeta t)P_k(\zeta^2 t) + P_F(\zeta^2 t)P_k(\zeta t)].$$

These modules have $C_2 \times S_3$-fixed set of dimension 1, so the function above gives exactly their contribution to the Poincaré series of $S^*(F\tilde{\oplus}F)^{C_2\times S_3}$

Next we have to consider modules of type $M \otimes F \otimes F$ and $M \otimes k \otimes k$. In the first case we get the multiplicity

$$\tfrac{1}{2} \cdot \{\tfrac{1}{3}[P_F(t)^2 + 2P_F(\zeta t)P_F(\zeta^2 t)] - P_F(t^2)\}.$$

These modules have $C_2 \times S_3$-fixed set of dimension 2, so we need to multiply their contribution to the Poincaré series by 2. In the second case we have

$$\tfrac{1}{2} \cdot \{\tfrac{1}{3}[P_k(t)^2 + 2P_k(\zeta t)P_k(\zeta^2 t)] - P_k(t^2)\},$$

and the modules have $C_2 \times S_3$-fixed set of dimension 1.

Finally there are two cases of modules of the third type. The $F\tilde{\otimes}F$-modules have multiplicity function $P_F(t^2)$, and $C_2 \times S_3$-fixed set of dimension 2. The $k\tilde{\otimes}k$-modules have multiplicity function $P_k(t^2)$, and $C_2 \times S_3$-fixed set of dimension 1.

Now we add all of this up and get

$$\frac{(t^8-t^7+t^6+t^4+t^2-t+1)(t+1)(t^2+1)}{(1-t^2)(1-t^3)(1-t^4)(1-t^6)},$$

which agrees with the result obtained from Kemper's program.

Now we apply these ideas to an example which is of interest in its own right and not just as an application of these techniques.

EXAMPLE 8.4 ("Double Dickson" invariants). Take $G = \mathrm{GL}_3(k)$ and let V be the natural G-module. We compute the Poincaré series of the tensor invariants $S^*(V\otimes W)^{\mathrm{GL}_3(k)\times \mathrm{GL}_2(k)}$. Using the techniques described above, the decomposition of $S^*(V)$ as a $\mathrm{GL}_3(k)$-module (5.4), and `Magma` we find that our series is

$$\begin{aligned}
&(t^{46}+t^{43}+t^{42}+2t^{41}+t^{40}+3t^{38}+2t^{36}+3t^{35}+5t^{34}+t^{33}+4t^{32}+3t^{31}+3t^{30}\\
&\quad+t^{29}+5t^{28}+4t^{27}+4t^{26}+4t^{25}+5t^{24}+3t^{23}+2t^{22}+3t^{21}+3t^{20}+t^{19}+5t^{18}\\
&\quad+2t^{17}+3t^{16}+t^{15}+2t^{14}+t^{12}+t^{11}+2t^{10}-t^9+2t^8+t^6-t+1)\\
&\cdot(t^{58}-t^{57}-t^{54}+t^{53}-t^{52}+t^{51}+t^{48}-t^{47}-t^{46}+t^{45}-t^{44}+t^{43}+t^{42}-t^{41}\\
&\quad+2t^{40}-2t^{39}+t^{38}-2t^{37}+t^{35}-t^{34}+2t^{33}-t^{30}-t^{28}+2t^{25}-t^{24}+t^{23}-2t^{21}\\
&\quad+t^{20}-2t^{19}+2t^{18}-t^{17}+t^{16}+t^{15}-t^{14}+t^{13}-t^{12}-t^{11}+t^{10}+t^7-t^6+t^5\\
&\quad-t^4-t+1)^{-1}.
\end{aligned}$$

Admittedly this is not very appealing. One recalls the remark of Neusel and Smith ([**17**, p. 25]) that "[Vector invariants] seem to have fallen out of favor, except as a means of providing nasty examples."

Part of the interest of this example is that it is very close to the limit of what can be done with current invariant-theory software. The authors were unable to verify the result using Kemper's software, though it may be possible to do this with better computing resources or algorithms adapted to specific features of this example. However, the series has been expanded and checked for correctness through degree 12.

This example was originally studied in the context of cohomology of groups. One of the maximal elementary abelian subgroups of M_{24}, E, is of rank 6, and the action of its normalizer realizes this representation. Further, if S is the Sylow 2-subgroup of M_{24}, then $E\subset S\subset M_{24}$ is a weakly closed system [**3**], and so the restriction map from $H^*(M_{24},k)$ to $H^*(E,k)^{N_{M_{24}}(E)}$ is surjective, i.e. the cohomology of M_{24} maps onto our invariant ring.

Appendix A. Notation from [13]

As suggested by the referee and the editors, we reproduce here some definitions and notation from our earlier paper [**13**], with the aim of making Section 2 easier to read.

Let k be the finite field of q elements, where $q = p^r$ for some prime p. Let $S = \oplus S_r = k[x_1,\ldots,x_n]$ denote the polynomial ring over k in n variables, graded by degree.

All the groups we study are subgroups of the upper-triangular $n\times n$ matrices over k with 1s on the diagonal, denoted by U_n. This group is a Sylow p-subgroup of $\mathrm{GL}_n(k)$. The group U_n is taken to act on the polynomial ring S in the natural way: if $a_{ij}(\lambda)$ is the matrix with λ in row i, column j, 1s on the diagonal, and zeroes elsewhere, then $a_{ij}(\lambda) \cdot x_j = x_j + \lambda x_i$, while $a_{ij}(\lambda) \cdot x_l = x_l$ if $l \neq j$.

NOTATION A.1. $[A(x_i, x_j),\ A(x_i)]$ $A(x_i, x_j) = \{a_{ij}(\lambda) \mid \lambda \in k\}$. This is a subgroup of U_n consisting of matrices with just one non-zero off-diagonal entry. $A(x_i) = \prod_{j>i} A(x_i, x_j)$. This is a subgroup of U_n with zero off-diagonal entries except in the ith row.

NOTATION A.2. $[I, J, K]$ I, J, K will always be subsets of $\{1, \ldots, n-1\}$ such that $I = J \amalg K$. We enumerate their elements in ascending order: $I = \langle i_1, \ldots, i_t \rangle$, $K = \langle k_1, \ldots, k_s \rangle$, $J = \langle j_1, \ldots, j_r \rangle$.

Thus, by convention, $K = I - J$, and $t = |I|$ etc.

NOTATION A.3. $[U_I]$ U_I is the group $A(x_{i_1}) \cdot A(x_{i_2}) \cdots A(x_{i_t})$. This is a subgroup of U_n with non-zero off-diagonal entries in the rows corresponding to I. When I consists of a single element i, we write U_i instead of $U_{\{i\}}$. Since $i < n$, it should be possible to avoid confusion with U_n.

NOTATION A.4. $[J^{<i}]$ For any $i \in \mathbb{N}$, let $J^{<i} = \{j \in J \mid j < i\}$, and similarly for $>$, $\leq$, $\geq$.

NOTATION A.5. $[\theta(t; I)]$ If $I = \langle i_1, \ldots, i_s \rangle \subseteq \{1, 2, \ldots n-1\}$, we define

$$\theta(t; I) = \prod_{(\lambda_1, \ldots, \lambda_s) \in k^s} (t + \lambda_1 x_{i_1} + \cdots + \lambda_s x_{i_s}).$$

We interpret this to mean that $\theta(t; \varnothing) = t$.

COMMENT 1. Now we fix notation for the invariants of all the different groups we consider. We start with the invariants of U_J, which are denoted by hs, i.e. $h_i = \theta(x_i; J^{<i})$. As we move to larger groups, we move the letters used to denote the invariants down the alphabet, so that gs denote invariants of $U_{J \cup \{k_1\}}$, fs denote invariants of $U_{J \cup K^{\leq k_{m-1}}}$, es denote invariants of $U_{J \cup K^{k_m}}$, and ds denote invariants of U_I. To summarize,

DEFINITION A.6. $[d_i, e_i, f_i, g_i, h_i]$ For given $k_m \in K$, let

$$h_i = \theta(x_i; J^{<i}), \qquad g_i = \theta(x_i; (J \cup \{k_1\})^{<i}),$$
$$f_i = \theta(x_i; (J \cup K^{\leq k_{m-1}})^{<i}), \qquad e_i = \theta(x_i; (J \cup K^{\leq k_m})^{<i}),$$
$$d_i = \theta(x_i; I^{<i}).$$

PROPOSITION A.7. *The d_i, e_i, f_i, g_i, and h_i are algebraically independent and generate the invariant rings of the groups U_I, $U_{J \cup K^{\leq k_m}}$, $U_{J \cup K^{\leq k_{m-1}}}$, $U_{J \cup \{k_1\}}$, and U_J, i.e.:*

$$S^{U_I} = k[d_1, \ldots, d_n], \qquad S^{U_{J \cup \{k_1\}}} = k[g_1, \ldots, g_n],$$
$$S^{U_{J \cup K^{\leq k_m}}} = k[e_1, \ldots, e_n], \qquad S^{U_J} = k[h_1, \ldots, h_n],$$
$$S^{U_{J \cup K^{\leq k_{m-1}}}} = k[f_1, \ldots, f_n].$$

PROOF. It is easy to give a basis that shows that S is free over S^G of rank $|G|$. See [**19**, Theorem 5.5.6], [**16**]. □

DEFINITION A.8. $[\mu]$ Let $Y = \langle y_1, \dots, y_m \rangle$ be a non-empty increasing sequence of integers. Then

$$\mu(Y) = [(n+1-y_m-1)q^{m-1} + (y_m - y_{m-1} - 1)q^{m-2} + \cdots + (y_2 - y_1 - 1)](q-1).$$

(If $m = 1$, then $\mu(Y) = (n - y_1)(q-1)$.)

DEFINITION A.9. $[\psi]$ If $J = \langle j_1, \dots, j_r \rangle$, let $\langle a_{j_1}, \dots, a_{j_r} \rangle$ be indeterminates and define $\psi(A(J)) = (a_{j_1} + qa_{j_2} + \cdots + q^{r-1}a_{j_r})(q-1)$. (So $\psi(A(\varnothing)) = 0$.)

DEFINITION A.10. $[J_l]$ We write J_l for the sequence whose elements are $\{l\} \cup J^{>l}$ in ascending order.

DEFINITION A.11. $[P(I)]$ For $I = \langle i_1, \dots, i_t \rangle$, the parameter space on I is $P(I) = \mathbb{N}_0^I \cong \mathbb{N}_0^t$.

DEFINITION A.12. $[S_J(I), \mathrm{IE}_{J,i}]$ Let $J \subseteq I = \langle i_1, \dots, i_t \rangle$. We associate to J a set $S_J(I) \subseteq P(I)$ defined by t linear inequalities $\mathrm{IE}_{J,i}$, one for each $i \in I$. Let us write $(a_{i_1}, \dots, a_{i_t})$ for the generic element of $P(I)$. Then

$$\mathrm{IE}_{J,i} \text{ is } \begin{cases} a_i + \psi(A(J^{>i})) < \mu(J_i) & \text{if } i \in J, \\ a_i + \psi(A(J^{>i})) \geq \mu(J_i) & \text{if } i \notin J. \end{cases}$$

DEFINITION A.13. $[\gamma(t)]$ Let $\gamma(t) = 1 + t + \cdots + t^{q-1}$.

DEFINITION A.14. $[\kappa_a^{(n-i)}]$ Let $\kappa_a^{(n-i)}$ be the coefficient of t^a in the formal power series $\gamma(t)^{n-i}/1-t = \sum_{a \geq 0} \kappa_a^{(n-i)} t^a$.

DEFINITION A.15. $[G(I \mid J)]$ Let

$$G(I \mid J) = \prod_{k_1 < i \leq n,\ i \notin J} \frac{\theta(x_i; I^{<i})}{\theta(x_i; J^{<i})}.$$

Note that $G(I \mid J)$ is a polynomial, since $\theta(x_i; J^{<i})$ divides $\theta(x_i; I^{<i})$ from the definition.

THEOREM A.16. *There is a direct sum decomposition as graded kU_I-modules $T(I) \cong \oplus_{J \subseteq I} X_J(I)$, where*

1. $X_J(I) \cong k[d_{k_u} \mid k_u \in K] \otimes \overline{X}_J(I)$,
2. $\overline{X}_J(I) \cong \mathrm{Ind}_{U_J}^{U_I} X_J(J)$ *(but with a degree shift)*,
3. $\mathrm{supp}\overline{X}_J(I) = \bar{S}_J(I)$ *and* $\mathrm{supp}\, X_J(I) = S_J(I)$,
4. $\mathrm{PS}(X_J(I)) = \sum_{(a_{i_1}, \dots, a_{i_t}) \in S_J(I)} \kappa_{a_{i_1}}^{(n-i_1)} \cdots \kappa_{a_{i_t}}^{(n-i_t)} \cdot t^{a_{i_1}} \cdots t^{q^{t-1}a_{i_t}}$.

References

1. J. F. Adams, *Stable homotopy and generalised homology*, Chicago Lectures in Math., University of Chicago Press, Chicago, Ill.-London, 1974.
2. A. Adem, J. F. Carlson, D. Karagueuzian, and R. J. Milgram, *The cohomology of the Sylow 2-subgroup of the Higman–Sims group*, J. Pure Appl. Algebra **164** (2001), no. 3, 275–305.
3. A. Adem and R. J. Milgram, *Cohomology of finite groups*, Fundamental Principles of Mathematical Sciences , vol. 309. Springer-Verlag, Berlin, 1994.
4. D. Benson and J. F. Carlson, *Diagrammatic methods for modular representations and cohomology*, Comm. Algebra **15** (1987), no. 1-2, 53–121.
5. D. Benson, *Infinite dimensional modules for finite groups*, Infinite Length Modules (Bielefeld, 1998), Trends Math., Birkhäuser, Basel, 2000, pp. 251–272.

6. W. Bosma, J. Cannon, and C. Playoust, *The Magma algebra system. I. The user language*, Computational Algebra and Number Theory (London, 1993). J. Symbolic Comput. **24** (1997), no. 3-4, 235–265.
7. H. E. A. Campbell, I. P. Hughes, R. J. Shank, and D. L. Wehlau, *Bases for rings of coinvariants*, Transform. Groups **1** (1996), no. 4, 307–336.
8. S. B. Conlon, *Decompositions induced from the Burnside algebra*, J. Algebra **10** (1968), 102–122.
9. C. W. Curtis and I. Reiner, *Methods of representation theory–with applications to finite groups and orders*, vol. II, Wiley, New York 1987.
10. H. Derksen and G. Kemper, *private communication.*
11. W. Feit, *The representation theory of finite groups,* North-Holland, Amsterdam, 1982.
12. D. B. Karagueuzian and P. Symonds, *The module structure of a group action on a polynomial ring*, J. Algebra **218** (1999), 672–692.
13. D. B. Karagueuzian and P. Symonds, *The module structure of a group action on a polynomial ring: A finiteness theorem*, Preprint 2002.
14. G. Kemper, *Calculating invariant rings of finite groups over arbitrary fields*, J. Symbolic Comput. **21** (1996), no. 3, 351–366.
15. R. J. Milgram, *private communication.*
16. H. Mui, *Modular invariant theory and cohomology algebras of symmetric groups*, J. Fac. Sci. Univ. Tokyo Sect. 1A Math. **22** (1975), 319–369.
17. M. D. Neusel and L. Smith, *Invariant theory of finite groups*, Math. Surveys Monogr., vol. 94. Amer. Math. Soc., Providence, RI, 2002.
18. D. Richman, *On vector invariants over finite fields*, Adv. in Math. **81** (1990), 30–65.
19. L. Smith, *Polynomial invariants of finite groups*, Research Notes in Math., vol. 6, A. K. Peters Ltd., Wellesley, MA, 1995.
20. P. Symonds, *Group actions on polynomial and power series rings*, Pacific J. Math. **195** (2000), 225–230.

MATHEMATICS DEPARTMENT, BINGHAMTON UNIV., P.O. BOX 6000, BINGHAMTON, NY13902-6000, USA.
E-mail address: dikran@math.binghamton.edu

DEPARTMENT OF MATHEMATICS, U.M.I.S.T., P.O. BOX 88, MANCHESTER M60 1QD, UK.
E-mail address: dikran@math.binghamton.edu

Centre de Recherches Mathématiques
CRM Proceedings and Lecture Notes
Volume **35**, 2004

An Invariant Theoretic Description of the Primitive Elements of the mod$-p$ Cohomology of a Finite Loop Space which Are Annihilated by Steenrod Operations

N. E. Kechagias

ABSTRACT. We give an invariant theoretic description of the primitive elements in the mod $-\ p$ cohomology of a finite loop space $\Omega^{2n+1}S^{2n+1}$ for p odd. We also calculate the primitive elements which are annihilated by all Steenrod operations.

1. Introduction

In [**2**], the homology and cohomology of $\Omega^{m+1}S^{m+1}$ is described in terms of Dyer–Lashof operations for m finite or infinite. In [**10**] and [**2**], the dual of the Dyer–Lashof algebra was calculated for p even and odd respectively and its connection with modular invariants, namely the Dickson algebra, was known since then by the experts. That connection was explicitly formulated by Mui in [**5**] for $p = 2$ and later by Kechagias for p odd in [**6**]. The advantage with modular invariants is that computations are easier in most cases because the Adem relations are overcome.

Inspired by work of Campbell, Peterson and Selick [**1**], we give an invariant theoretic description of the primitive elements in $H^*(\Omega_0^{2n+1}S^{2n+1};\mathbb{Z}/p\mathbb{Z})$, $PH^*(\Omega_0^{2n+1}S^{2n+1};\mathbb{Z}/p\mathbb{Z})$, and we use that description to calculate those primitive elements which are annihilated by all Steenrod operations for p odd, Theorem 41. That was a key and lengthy calculation in [**1**]. In our approach the advantage is that computations are easier and the disadvantage that it requires some amount of preliminary work because we do not assume any familiarity with the connection between the Dyer–Lashof and Dickson algebras.

Since components of $\Omega^{m+1}S^{m+1}$ are homotopy equivalent, translation in homology is nontrivial namely, if $x \in H_*(\Omega^{m+1}S^{m+1};\mathbb{Z}/p\mathbb{Z})$ and $x_c \in H_0(\Omega^{m+1}S^{m+1};\mathbb{Z}/p\mathbb{Z})$ denotes the component of x then $x \cdot x_c^{-1} \in H_*(\Omega_0^{m+1}S^{m+1};\mathbb{Z}/p\mathbb{Z})$. Thus we will be considering the base point component only.

For each nonnegative integer we define a subalgebra, $D[k] \otimes S(E_k)^{\mathrm{GL}_k}$, of the ring of invariants $H^*\big((BZ/p\mathbb{Z})^k, \mathbb{Z}/p\mathbb{Z}\big)^{\mathrm{GL}_k}$ which was calculated by Mui [**4**]. That subalgebra turns out to be dual to the Dyer–Lashof coalgebra $R[k]$, [**6**]. We define a

2000 *Mathematics Subject Classification.* Primary: 55P10, 55P35; Secondary: 55S10, 13A50.
This is the final form of the paper.

map j_n from $D[k] \otimes S(E_k)^{\mathrm{GL}_k}$ to the quotient algebra of monomials with monomial degree at most n using the natural decomposition

$$\hat{\imath} : H^*\big((BZ/p\mathbb{Z})^k, \mathbb{Z}/p\mathbb{Z}\big)^{\mathrm{GL}_k} \hookrightarrow H^*\big((BZ/p\mathbb{Z})^k, \mathbb{Z}/p\mathbb{Z}\big)^{B_k}.$$

Here B_k stands for the Borel subgroup. Let $j_n\big((d_{k,0}, M_{k;i}, M_{k;0,i})_{\text{Ideal}}\big)$ be the image of the ideal $(d_{k,0}, M_{k;i}, M_{k;0,i})_{\text{Ideal}}$, then $\bigoplus_k j_n\big((d_{k,0}, M_{k;i}, M_{k;0,i})_{\text{Ideal}}\big) \cong PH^*(\Omega_0^{2n+1}S^{2n+1}; \mathbb{Z}/p\mathbb{Z})$, Theorem 45. Because the exterior generators of $(d_{k,0}, M_{k;i}, M_{k;0,i})_{\text{Ideal}}$ are not annihilated by all Steenrod operations, we compute $\operatorname{Ann} j_n\big((d_{k,0})_{\text{Ideal}}\big)$ using the Steenrod algebra action on Dickson generators and properties of the map j_n in Section 3.

A would like to congratulate Eddy Campbell and David Wehlau for organizing that interesting and fertile conference. Finally, I would like to thank the referee very much for his valuable suggestions.

2. A principal ideal in the Dickson algebra

Let V^ℓ denote a $\mathbb{Z}/p\mathbb{Z}$-vector space with basis $\{e_1, \dots, e_\ell\}$ for $1 \le \ell \le k$. Let Σ_{p^k} be the symmetric group which acts on V^k by permutations. V^k can also be considered as a subgroup of Σ_{p^k} acting by translations and we have a representation

$$\rho : W_{\Sigma_{p^k}}(V^k) \longrightarrow \operatorname{Aut}(V^k) \equiv \mathrm{GL}_k$$

V^k is also an elementary abelian p-subgroup and $H^*(BV^k, \mathbb{Z}/p\mathbb{Z}) \cong E(x_1, \dots, x_k) \otimes P[y_1, \dots, y_k]$ where $\{x_1, \dots, x_k\}$ is a fixed basis for the dual of V^k. Here $|y_i| = 2 = 2|x_i|$ and $\beta(x_i) = y_i$, where $\beta(-)$ is the Bockstein operation. The contragradient representation of ρ induces an action of $\operatorname{Aut}(V^k) \equiv \mathrm{GL}_k$ on the graded algebra $E(x_1, \dots, x_k) \otimes P[y_1, \dots, y_k]$. Let $E_k := E(x_1, \dots, x_k)$ and $S_k = P[y_1, \dots, y_k]$.

The following theorems are well known:

THEOREM 1 ([**3**]). *The Dickson algebra* $S_k^{\mathrm{GL}_k} := D[k] = \mathbb{Z}/p\mathbb{Z}[d_{k,0}, \dots, d_{k,k-1}]$ *is a polynomial algebra and degrees are* $|d_{k,i}| = 2(p^k - p^i)$.

Let B_k be the Borel subgroup of GL_k.

THEOREM 2 ([**4**]). $S_k^{B_k} := B[k] = \mathbb{Z}/p\mathbb{Z}[h_1, \dots, h_k]$ *is a polynomial algebra and degrees are* $|h_i| = 2p^{i-1}(p-1)$.

Note that for convenience we call h_i what in the literature stands for h_i^{p-1}.

Let $\hat{\imath} : D[k] \hookrightarrow B[k]$ be the natural inclusion. This map is described by the following relations.

Let $f_{\ell-1}(x) = \prod_{u \in V^{\ell-1}}(x - u)$, then $f_{\ell-1}(x) = \sum_{i=0}^{\ell-1}(-1)^{k-i}x^{p^i}d_{\ell-1,i}$ and $h_\ell = \prod_{u \in V^{\ell-1}}(y_\ell - u)^{p-1}$. Moreover, (see [**6**]),

$$\hat{\imath}(d_{k,k-i}) = \sum_{1 \le j_1 < \cdots < j_i \le k} \prod_{s=1}^{i} (h_{j_s})^{p^{k-i+s-j_s}}. \tag{2.1}$$

Both $B[k]$ and $D[k]$ are algebras over the Steenrod algebra, A, and its action on generators has been given in [**7**].

Mui gave an invariant theoretic description of the cohomology algebra of a symmetric group and calculated rings of invariants involving the exterior algebra E_k as well in [**4**].

THEOREM 3 ([**4**]). 1. *The algebra* $(E_k \otimes S_k)^{B_k}$ *is a tensor product of the polynomial algebra* $B[k]$ *and the* $\mathbb{Z}/p\mathbb{Z}$*-module spanned by the set of elements consisting of the following monomials*:

$$M_{s;s_1,\dots,s_m} L_s^{p-2}; 1 \le m \le k,\ m \le s \le k, \quad \text{and} \quad 0 \le s_1 < \dots < s_m = s-1.$$

Its algebra structure is determined by the following relations:

(a) $(M_{s;s_1} L_s^{p-2})^2 = 0, \quad$ *for* $1 \le s \le k,\ 0 \le s_1 \le s-1$.

(b)
$$M_{s;s_1,\dots,s_m} L_s^{p-2} (L_s^{p-1})^{m-1} = (-1)^{m(m-1)/2} \prod_{q=1}^{m} \Big(\sum_{r=s_q+1}^{s} M_{r;r-1} L_r^{p-2} h_{r+1} \cdots h_s d_{r-1,s_q} \Big).$$

Here $1 \le m \le k,\ m \le s \le k$, *and* $0 \le s_1 < \dots < s_m = s-1$.

2. *The algebra* $(E_k \otimes S_k)^{\mathrm{GL}_k}$ *is a tensor product of the polynomial algebra* $D[k]$ *and the* $\mathbb{Z}/p\mathbb{Z}$*-module spanned by the set of elements consisting of the following monomials*:

$$M_{k;s_1,\dots,s_m} L_k^{p-2}; 1 \le m \le k, \quad \text{and} \quad 0 \le s_1 < \dots < s_m \le k-1.$$

Its algebra structure is determined by the following relations:

(a) $(M_{k;s_1,\dots,s_m} L_k^{p-2})^2 = 0 \quad$ *for* $1 \le m \le k$, *and* $0 \le s_1 < \dots < s_m \le k-1$.

(b) $M_{k;s_1,\dots,s_m} L_k^{(p-2)} d_{k,k-1}^{m-1} = (-1)^{m(m-1)/2} M_{k;s_1} L_k^{p-2} \cdots M_{k;s_m} L_k^{p-2}$.

Here $1 \le m \le k$, *and* $0 \le s_1 < \dots < s_m \le k-1$.

The elements $M_{k;s_1,\dots,s_m}$ above have been defined by Mui in [**4**] as follows:

$$M_{k;s_1,\dots,s_m} = \frac{1}{m!} \begin{vmatrix} x_1 & \cdots & x_1 \\ \vdots & & \vdots \\ x_1 & \cdots & x_k \\ y_1 & \cdots & y_k \\ \vdots & & \vdots \\ y_1^{p^{k-1}} & \cdots & y_k^{p^{k-1}} \end{vmatrix} \quad \text{and} \quad L_k = \begin{vmatrix} y_1 & \cdots & x_k \\ y_1^p & \cdots & y_k^p \\ \vdots & & \vdots \\ y_1^{p^{k-1}} & \cdots & y_k^{p^{k-1}} \end{vmatrix}.$$

Here there are m rows of x_i's and the s_ith's powers are omitted, where $0 \le s_1 < \dots < s_m \le k-1$ in the first determinant.

The degree of elements above are $|M_{k;s_1,\dots,s_m}| = m + 2\big((1+\dots+p^{k-1}) - (p^{s_1} + \dots + p^{s_m})\big)$ and $|L_k^{p-2}| = 2(p-2)(1+\dots+p^{k-1})$.

The rest of this section is devoted to the connection between the dual of the Dyer–Lashof, Dickson and the cohomology algebra of $\Omega^{2n+1}S^{2n+1}$ for p odd. For details between the Dyer–Lashof and Dickson algebras please see [**8**].

DEFINITION 4. Let $S(E_k)^{B_k}$ be the subspace of $(E_k \otimes S_k)^{B_k}$ generated by:

$$M_{s;s-1}(L_s)^{p-2} \qquad \text{for } 1 \le s \le k,$$

$$\prod_{t=1}^{\ell} M_{s_{2t}+1;s_{2t-1},s_{2t}} (L_{s_{2t}+1})^{p-2} \qquad \text{for } 0 \le s_1 < \dots < s_{2\ell} \le k-1,$$

$$\prod_{t=1}^{\ell} M_{s_{2t}+1;s_{2t-1},s_{2t}} (L_{s_{2t}+1})^{p-2} M_{s;s-1}(L_s)^{p-2} \qquad \text{for } 0 \le s_1 < \dots < s_{2\ell} < s \le k,$$

and $S(E_k)^{\mathrm{GL}_k}$ be the subspace of $(E_k \otimes S_k)^{\mathrm{GL}_k}$ generated by:

$$M_{k;s}(L_k)^{p-2} \qquad \text{for } 0 \le s \le k-1,$$

$$\prod_{t=1}^{\ell M_{k;s_{2t-1},s_{2t}}} (L_k)^{p-2} \qquad \text{for } 0 \le s_1 < \cdots < s_{2\ell} \le k-1,$$

$$M_{k;s-1}(L_k)^{p-2} \prod_{t=1}^{\ell} M_{k;s_{2t-1},s_{2t}}(L_k)^{p-2} \quad \text{for } 0 \le s < s_1 < \cdots < s_{2\ell} < k.$$

The map $\hat{\imath} : D[k] \hookrightarrow B[k]$ defined in 2.1 is extended to $\hat{\imath} : D[k] \otimes S(E_k)^{\mathrm{GL}_k} \hookrightarrow B[k] \otimes S(E_k)^{B_k}$ by the following relations:

LEMMA 5.

$$\begin{aligned} M_{k;s}L_k^{p-2} &= M_{s+1;s}L_{s+1}^{p-2}h_{s+1}\cdots h_k \\ &\quad + \sum_{t=2}^{k-s} M_{s+t;s+t-1}L_{s+t}^{p-2}d_{s+t-1,s}h_{s+t+1}\ldots h_k. \end{aligned} \tag{1}$$

$$\begin{aligned} M_{k;s,m}L_k^{p-2} &= M_{m+1;s,m}L_{m+1}^{p-2}h_{m+2}\cdots h_k \\ &\quad + \sum_{t=2}^{k-m} M_{m+t;s,m+t-1}L_{m+t}^{p-2}d_{m+t-1,m}h_{m+t+1}\cdots h_k \\ &\quad - M_{m+t;m,m+t-1}L_{m+t}^{p-2}d_{m+t-1,s}h_{m+t+1}\ldots h_k. \end{aligned} \tag{2}$$

PROOF. We use induction on Mui's formula:

$$M_{k-1;s_1,\ldots,s_m}L_{k-1}^{p-2}h_k = M_{k;s_1,\ldots,s_m}L_k^{p-2} - \sum_{1}^{m}(-1)^{m+i}M_{k;s_1,\ldots,\widehat{s}_i,\ldots,k-1}d_{k-1,s_i}. \quad \square$$

PROPOSITION 6.

1. *$S(E_k)^{\mathrm{GL}_k}$ is not closed under the Steenrod algebra action.*
2. *$D[k] \otimes S(E_k)^{\mathrm{GL}_k}$ is closed under the Steenrod algebra action.*

PROOF. This follows from [**7**, Theorem 14]. □

Let $D := \prod_{k\ge 0} D[k] \otimes S(E_k)^{\mathrm{GL}_k}$ be the induced graded algebra where its identity element is $\prod_k ({}_k1)$. The height of elements in $D[k] \otimes S(E_k)^{\mathrm{GL}_k}$ is defined to be k. The pair degree and height defines uniquely each element of D. The unit in $D[k] \otimes S(E_k)^{\mathrm{GL}_k}$ is of height k and it is not related with units with different height. D is not of finite type but D_+ is. We shall note that if we let $B := \prod_{k\ge 0} B[k] \otimes S(E_k)^{B_k}$, then B_+ is not of finite type because h_t appears with many different heights.

The augmentation $\varepsilon : D \to \mathbb{Z}/p\mathbb{Z}$ is given by $\varepsilon\big(\prod_k \lambda_k({}_k1)\big) = \lambda_0$.

Next we define a coproduct in D and B as follows (see [**6**, p. 277]):

$$\psi : D[k] \otimes S(E_k)^{\mathrm{GL}_k} \to \sum_{k\ge t\ge 0} \big(D[t] \otimes S(E_t)^{\mathrm{GL}_t}\big) \otimes \big(D[k-t] \otimes S(E_{k-t})^{\mathrm{GL}_{k-t}}\big).$$

DEFINITION 7.

(i) $$\psi({}_k1) = \sum_t 1 \otimes_{k-t} 1;$$

(ii) $$\psi(d_{k,i}) = \sum_{(t,j)} d_{t,0}^{p^{k-t}-p^j} d_{t,i-j}^{p^j} \otimes d_{k-t,j};$$

(iii) $$\psi(\Pi d_{k,i}^{m_i}) = \sum_{(t,j_0,\dots,j_{k-1})} \prod_i (d_{t,0}^{m_i(p^{k-t}-p^{j_i})} d_{t,i-j_i}^{m_i p^{j_i}}) \otimes \prod_i d_{k-t,j_i}^{m_i};$$

(iv) $$\psi(M_{k;s}L_k^{(p-2)}) = \sum_{(t,j)} d_{t,0}^{p^{k-t}-p^j} d_{t,s-j}^{p^j} \otimes M_{k-t;j}L_{k-t}^{(p-2)} + \sum_{(t)} d_{t,0}^{p^{k-t}-1} M_{t;s}L_t^{(p-2)} \otimes d_{k-t,0};$$

(v) $$\begin{aligned}\psi(M_{k;s,i}L_k^{(p-2)}) &= \sum_{(t,j,f)} d_{t,0}^{p^{k-t}-p^f-p^j} (d_{t,t+j-i}^{p^j} d_{t,t+f-s}^{p^f} - d_{t,t+f-i}^{p^f} d_{t,t+j-s}^{p^j}) \otimes M_{k-t;f,j}L_{k-t}^{(p-2)} \\ &\quad - \sum_{(t,j)} d_{t,0}^{p^{k-t}-p^j-1} (d_{t,t+j-i}^{p^j} M_{t;t-s}L_t^{(p-2)} - d_{t,t+j-s}^{p^j} M_{t;t-i}L_t^{(p-2)}) \otimes M_{k-t;j}L_{k-t}^{(p-2)} \\ &\quad + \sum_{(t)} d_{t,0}^{p^{k-t}-1} M_{t;t-s,t-i}L_t^{(p-2)} \otimes d_{k-t,0}.\end{aligned}$$

Moreover, the following rule is applied

$$\Big(\sum_t \sum_{l,l'} a_{t,l} \otimes a_{k-t,l'}\Big)\Big(\sum_t \sum_{q,q'} b_{t,q} \otimes b_{k-t,q'}\Big) := \Big(\sum_t \sum_{l,l',q,q'} a_{t,l}b_{t,q} \otimes a_{k-t,l'}b_{k-t,q'}\Big);$$

(vi) $$\psi({}_kh_t) = \sum_{1\le i\le t-1} ({}_ih_i)^{p^{t-1-i}(p-1)} \otimes ({}_{k-i}h_{t-i}) + \sum_{0\le i\le k-t} ({}_{t+i}h_t) \otimes 1_{k-t-i};$$

(vii) $$\psi({}_mM_{k;k-1}L_k^{(p-2)}) = \sum_{1\le i\le k-1} h_i{}^{p^{k-1-i}(p-1)} \otimes_{m-i} (M_{k-i;k-i-1}L_{k-i}^{(p-2)}) + \sum_{0\le i\le m-k} ({}_{k+i}M_{k;k-1}L_k^{(p-2)}) \otimes 1_{m-k-i}.$$

Here ${}_kh_t$ denotes that it is an element of $B[k]$ i.e. of height k. Now, D_+ is an algebra and a coalgebra but not a Hopf algebra because its coproduct on its unit $\prod_k({}_k1)$ is not well defined although it is well defined on $({}_k1)$. The same is true for B.

Let $d_{k,o}^{m_0}\cdots d_{k,k-1}^{m_{k-1}}$ be a typical element of $D[k]$ and $h_1^{j_1}\cdots h_k^{j_k}$ of $B[k]$. We abbreviate those elements by d_k^m and h_k^j respectively, where $m = (m_0,\dots,m_{k-1})$ and $j = (j_1,\dots,j_k)$ are elements of $\mathbb{N}^k$ the submonoid of $\mathbb{Z}^k$. Here $\mathbb{N}$ stands for the set of nonnegative integers and $\mathbb{Z}$ for the integers. Let us also give a left lexicographical ordering in those sequences and call j **admissible**, if $j_t \leq j_{t+1}$ for $1 \leq t \leq k-1$.

We extend the notion of sequences above to include elements of $S(E_k)^{\mathrm{GL}_k}$ and $S(E_k)^{B_k}$. We consider $(\mathbb{N},+)$ as a submonoid of $(\mathbb{Q},+)$ and let $\langle \mathbb{N}, \frac{1}{2}\rangle$ be the submonoid generated by $\mathbb{N}$ and $\frac{1}{2}$ in $\mathbb{Q}$. Let $\langle \mathbb{N}, \frac{1}{2}\rangle^k$ be the submonoid which is the kth Cartesian product of $\langle \mathbb{Q}, \frac{1}{2}\rangle$.

DEFINITION 8. Let $\langle \mathbb{N}, \frac{1}{2}\rangle^k \times (\mathbb{Z}/2Z)^k$ be the Cartesian product of the submonoid $\langle \mathbb{N}, \frac{1}{2}\rangle^k$ and the group $(\mathbb{Z}/2\mathbb{Z})^k$. Let $I = (i_1, \ldots, i_k)$ and $\varepsilon = (e_1, \ldots, e_k)$, then $(I, \varepsilon) \in \langle \mathbb{N}, \frac{1}{2}\rangle^k \times (\mathbb{Z}/2\mathbb{Z})^k$ will be called admissible, if $0 \leq 2i_t - 2i_{t-1} + e_{t-1}$ for $2 \leq t \leq k-1$.

Note that we will be comparing sequences corresponding to elements of the same degree.

DEFINITION 9. If $\mathring{A}$ is a $\mathbb{Z}/p\mathbb{Z}$-algebra, then we define $\beta(\mathring{A})$ to be the monomial basis of $\mathring{A}$.

Thus $\beta\big(D[k] \otimes S(E_k)^{\mathrm{GL}_k}\big)$ and $\beta\big(B[k] \otimes S(E_k)^{B_k}\big)$ denote the vector space bases of monomials in $D[k] \otimes S(E_k)^{\mathrm{GL}_k}$ and $B[k] \otimes S(E_k)^{B_k}$ respectively.

DEFINITION 10. Let $\chi_{\min}$ and $\chi_{\max}$ be the set functions from

$$\beta\big(D[k] \otimes S(E_k)^{\mathrm{GL}_k}\big)\big(\beta\big(B[k] \otimes S(E_k)^{B_k}\big)\big)$$

to the monoid $\langle \mathbb{N}, \frac{1}{2}\rangle^k \times (\mathbb{Z}/2Z)^k$ given by

(1) $$\chi_{\min}(d_{k,i}) = (\underbrace{0, \ldots, 0}_{i}, \underbrace{1, \ldots, 1}_{k-i}) x (0, \ldots, 0) \quad \text{and}$$

$$\chi_{\max}(d_{k,i}) = (p^{k-1}, \ldots, p^{k-i}, 0, \ldots, 0) x (0, \ldots, 0).$$

(2) $$\chi_{\min}(M_{k;s} L_k^{(p-2)}) = (\underbrace{\tfrac{1}{2}, \ldots, \tfrac{1}{2}}_{s}, \underbrace{1, \ldots, 1}_{k-s}) x (\underbrace{0, \ldots, 0}_{s}, 1, \underbrace{0, \ldots, 0}_{k-s-1}) \quad \text{and}$$

$$\chi_{\max}(M_{k;s} L_k^{(p-2)}) = (\underbrace{\tfrac{1}{2}, \ldots, \tfrac{1}{2}}_{s}, \underbrace{1\tfrac{1}{2}, \ldots, 1\tfrac{1}{2}}_{k-s-1}, 1) x (0, \ldots, 0, 1).$$

(3) $$\chi_{\min}(M_{k;s,m} L_k^{(p-2)}) = (\underbrace{0, \ldots, 0}_{s}, \underbrace{\tfrac{1}{2}, \ldots, \tfrac{1}{2}}_{m-s}, \underbrace{1, \ldots, 1}_{k-m}) x (\underbrace{0, \ldots, 0}_{s}, \underbrace{1, 0, \ldots, 0, 1}_{m-s}, \underbrace{0, \ldots, 0}_{k-m}) \quad \text{and}$$

$$\chi_{\max}(M_{k;s,m} L_k^{(p-2)}) = (\underbrace{0, \ldots, 0}_{m}, \underbrace{1\tfrac{1}{2}, \ldots, 1\tfrac{1}{2}}_{k-m-1}, 1) x (\underbrace{0, \ldots, 0}_{m}, \underbrace{1, 0, \ldots, 0, 1}_{k-m})$$

and the rule $\chi_{\min}(dd'MM') = \chi_{\min}(d) + \chi_{\min}(d') + \chi_{\min}(M) + \chi_{\min}(M')$. Here d, $d' \in \beta(D[k])$ and M, $M' \in \beta\big(S(E_k)^{\mathrm{GL}_k}\big)$. The same holds for $\chi_{\max}$.

Note that the function $\chi_{\min}$ is always admissible and $\hat{\imath}(d_{k,i})$ contains a monomial with a unique admissible sequence, namely $h^{\chi_{\min}(d_{k,i})}$, and a monomial with a unique maximal sequence, namely $h^{\chi_{\max}(d_{k,i})}$. The same is true for elements $M_{k;s-1}L_k^{p-2}$ and $M_{k;s,m}L_k^{p-2}$. Moreover, $\hat{\imath}(d_k^m M)$ might contain a number of monomials with admissible sequences and this is the main point of investigation because of its applications.

The function above define a natural function θ^*:

$$\theta^* : \beta\big(D[k] \otimes S(E_k)^{\mathrm{GL}_k}\big) \rightarrowtail \beta\big(B[k] \otimes S(E_k)^{B_k}\big) \tag{2.2}$$

defined by $\theta^*(d^m) = h^{\chi_{\min}(d_k^m)}$, where $\chi_{\min}(d_k^m) = (i_1, \dots, i_n)$ and $i_1 = m_0$, $i_t = m_0 + \cdots + m_{t-1}$. $\theta^*(M_{k;s}L_k^{(p-2)}) = M_{s+1;s}L_{s+1}^{(p-2)}h_{s+2}\cdots h_k$ and $\theta^*(M_{k;s,m}L_k^{(p-2)}) = M_{m+1;s,m}L_{m+1}^{(p-2)}h_{m+2}\cdots h_k$. Finally, $\theta^*(dM) = \theta^*(d)\theta^*(M)$.

Definition 11. Let $R^*[k]$ be the vector space spanned by elements $(Q_{I,\varepsilon})^*$ where $(I,\varepsilon) = \chi_{\min}(d_k^m M)$ for all $d_k^m M$ in $\beta\big(D[k] \otimes S(E_k)^{\mathrm{GL}_k}\big)$.

We define the degree of $(Q_{I,\varepsilon})^*$ or by abuse of notation of the sequence (I,ε) such that the maps θ^*, $\chi_{\min}$ and $\chi_{\max}$ are degree preserving maps.

Definition 12. Let $I = (i_1, \dots, i_k)$ and $\varepsilon = (e_1, \dots, e_k)$, then the degree of $(Q_{I,\varepsilon})^*$ or (I,ε) is $2(p-1)(\sum_{t=1}^k i_t p^{t-1}) - (\sum_{t=1}^k e_t p^{t-1})$.

Next we define an order on $\langle \mathbb{N}, \frac{1}{2}\rangle^k \times (\mathbb{Z}/2\mathbb{Z})^k$ which is compatible with the order defined in $\langle n\mathbb{N}, \frac{1}{2}\rangle^k \equiv \langle \mathbb{N}, \frac{1}{2}\rangle^k \times (0, \dots, 0)$.

Definition 13. Let (I,ε) be a typical element of $\langle \mathbb{N}, \frac{1}{2}\rangle^k \times (\mathbb{Z}/2\mathbb{Z})^k$ and we call $I_t = (i_t, \dots, i_k)$ and $\varepsilon_t = (e_t, \dots, e_k)$ for $t = 1, \dots, k$. We define $(I,\varepsilon) \geq (I', \varepsilon')$, if $2(p-1)i_t - e_t - |I_{t-1}, \varepsilon_{t-1}| \geq 2(p-1)i'_t - e'_t - |I'_{t-1}, \varepsilon'_{t-1}|$ for the smallest t.

Let the vector space isomorphism $\Phi_k : D[k] \otimes S(E_k)^{\mathrm{GL}_k} \to R^*[k]$ be given by

$$\Phi_k(d_k^m M) = \sum_{(I,\varepsilon) \geq \chi_{\min}(d_k^m M)} a_{m,(I,\varepsilon)}(Q_I)^*.$$

Here $a_{m,(I,\varepsilon)}$ is the coefficient of h^I in $\hat{\imath}(d_k^m M)$ and all sequences (I,ε) are admissible.

Remark 14. The map above is upper triangular with one along the main diagonal and it has been studied along with its inverse in [**8**, Theorem 45].

Example 15. Let us consider $\Phi_k(d_{3,0}d_{3,1}^{12}d_{3,2}^4)$ for $p = 3$.
Using MAPLE we get

$$\begin{aligned}
&\hat{\imath}(d_{3,1}^{12}d_{3,2}^4)\\
&\quad = 2h_1^{27}h_2^{15}h_3^{12} + h_1^{45}h_2^{27}h_3^6 + 2h_1^{45}h_2^{18}h_3^9 + 2h_1^{18}h_2^{18}h_3^{12} + h_1^{36}h_2^{27}h_3^7 + h_1^{27}h_2^{30}h_3^7\\
&\qquad + 2h_1^{36}h_2^{18}h_3^{10} + 2h_1^{36}kh_2^{21}h_3^9 + 2h_1^{36}h_2^{36}h_3^4 + 2h_1^{36}h_2^{12}h_3^{12} + h_1^{27}h_2^{33}h_3^6\\
&\qquad + 2h_1^{36}h_2^{30}h_3^6 + h_1^{18}h_2^{27}h_3^9 + h_1^{45}h_3^{15} + h_2^{15}h_3^{15} + h_1^{72}h_2^{36} + h_1^{72}h_3^{12} + h_2^{39}h_1^{63}\\
&\qquad + h_2^{48}h_1^{36} + h_2^{24}h_3^{12} + h_1^{63}h_3^{13} + h_1^{36}h_3^{16} + h_2^{21}h_3^{13} + h_2^{12}h_3^{16}\\
&\qquad + 2h_1^9h_3^{13}h_2^{18} + 2h_1^{36}h_3^{15}h_2^3 + 2h_2^{12}h_3^{15}h_1^9 + h_1^{27}h_3^{15}h_2^6 + h_1^{18}h_3^{15}h_2^9 + h_1^{45}h_2^{45}\\
&\qquad + 2h_1^{45}h_2^{36}h_3^3 + 2h_2^{12}h_3^{13}h_1^{27} + 2h_1^{63}h_3^3h_2^{30} + h_2^{36}h_1^{63}h_3 + h_2^{30}h_1^9h_3^9 + h_2^3h_3^{13}h_1^{54}\\
&\qquad + h_1^{72}h_2^9h_3^9 + h_2^{42}h_3^3h_1^{27} + h_2^6h_3^{12}h_1^{54} + h_2^{33}h_3^3h_1^{54} + h_2^3h_3^{16}h_1^{27} + h_2^{27}h_1^9h_3^{10}\\
&\qquad + h_2^{30}h_3^4h_1^{54} + h_2^{45}h_1^{36}h_3 + 2h_1^{63}h_3^{12}h_2^3 + h_1^{72}h_3^3h_2^{27} + h_1^{63}h_3^4h_2^{27} + h_2^{39}h_3^4h_1^{27}\\
&\qquad + h_2^{12}h_1^{63}h_3^9 + h_2^9h_1^{63}h_3^{10} + h_1^9h_3^{16}h_2^9.
\end{aligned}$$

The admissible sequences for $\Phi_k(d_{3,0}d_{3,1}^{12}d_{3,2}^{4})$ are: $h^{1,13,17}$, $h^{1,16,16}$, $2h^{10,13,16}$ and $h^{10,10,17}$. Thus

$$\Phi_k(d_{3,0}d_{3,1}^{12}d_{3,2}^{4}) = (Q_{1,13,17})^* + (Q_{1,16,16})^* + 2(Q_{10,13,16})^* + (Q_{10,10,17})^*.$$

Of course, $R^*[k]$ inherits a Steenrod algebra structure from the isomorphism above (see [**6**]).

Let $\kappa = [k + 1/2]$ and $\varepsilon = (e_1, \dots, e_k)$, then $S(E_k)^{\mathrm{GL}_k}$ is spanned by at most κ monomials:

$$M^\varepsilon := \begin{cases} M_{k;s_1,s_2}^{[e_1+e_2/2]} L_k^{p-2} \cdots M_{k;s_{k-1},s_k}^{[e_{k-1}+e_k/2]} L_k^{p-2}, & \text{if } k \text{ is even} \\ M_{k;s_1}^{e_1} L_k^{p-2} M_{k;s_2,s_3}^{[e_2+e_3/2]} L_k^{p-2} \cdots M_{k;s_{k-1},s_k}^{[e_{\kappa-1}+e_\kappa/2]} L_k^{p-2}, & \text{if } k \text{ is odd.} \end{cases}$$

DEFINITION 16. Let

$$\sum(m,\varepsilon) := \begin{cases} m_0 + \cdots + m_{k-1} + \left[\frac{e_1+e_2}{2}\right] + \cdots + \left[\frac{e_{k-1}+e_k}{2}\right], & \text{if } k \text{ is even} \\ m_0 + \cdots + m_{k-1} + e_1 + \left[\frac{e_2+e_3}{2}\right] + \cdots + \left[\frac{e_{k-1}+e_k}{2}\right], & \text{if } k \text{ is odd} \end{cases}$$

and we call the monomial degree of $d_k^m M^\varepsilon$ to be the nonnegative integer $\Sigma(m,\varepsilon)$.

Let $\left(D[k] \otimes S(E_k)^{\mathrm{GL}_k}\right)_n$ be the vector space spanned by all monomials with monomial degree n in $D[k] \otimes S(E_k)^{\mathrm{GL}_k}$.

Let $R_n^*[k]$ be the vector space spanned by elements $(Q_{I,\varepsilon})^*$ where $(I,\varepsilon) = \chi_{\min}(d_k^m M^\varepsilon)$ for all $d_k^m M^\varepsilon$ in $\beta(D[k] \otimes S(E_k)^{\mathrm{GL}_k})_n$.

Since we are considering homogeneous elements the monomial degree can be extended to polynomial degree as well.

LEMMA 17. *$(Q_{I,\varepsilon})^* \in R_n^*[k]$ if and only if (I,ε) is admissible and $i_k \leq n$.*

REMARK 18.

1. We should note that if $m > m'$ and $|d_k^m M^\varepsilon| = |d_k^{m'} M^{\varepsilon'}|$, then the polynomial degree of $d_k^{m'} M^{\varepsilon'}$ is greater of that of $d_k^m M^\varepsilon$.
2. We also note that if $\Phi_k^{-1}(Q_I)^* = \sum_{I \leq \chi_{\min}(d_k^m)} a'_{m,I}(d_k^m M^\varepsilon)$, then the polynomial degree of $d_k^m M^\varepsilon$ is greater or equal than of i_k.

We shall make $(D[k] \otimes S(E_k)^{\mathrm{GL}_k})_n$ a quotient subalgebra of $D[k] \otimes S(E_k)^{\mathrm{GL}_k}$ and write its elements for the natural images for convenience.

Let the epimorphism $\pi_n : R^*[k] \to R_n^{ast}[k]$ be given by

$$\pi_n(Q_I)^* = \begin{cases} (Q_I)^*, & \text{if } (Q_I)^* \in R_n^*[k], \\ 0, & \text{otherwise.} \end{cases}$$

LEMMA 19. *The epimorphism $\pi_n : R^*[k] \to R_n^*[k]$ is a map of algebras.*

PROOF. We must show $\pi_n\big((Q_I)^*(Q_J)^*\big) = \pi_n(Q_I)^*\pi_n(Q_J)^*$. But

$$\pi_n(Q_I)^*\pi_n(Q_J)^* = \begin{cases} (Q_I)^*(Q_J)^*, & \text{if } i_k, j_k \leq n, \\ 0, & \text{otherwise.} \end{cases}$$

On the other hand, $\pi_n\big((Q_I)^*(Q_J)^*\big) = \pi_n\Phi_k(\Phi_k^{-1}(Q_I)^* \cdot \Phi_k^{-1}(Q_J)^*)$. Using Remark 18 2, we get that both sides are equal. □

DEFINITION 20. Let $j_n : \left(D[k] \otimes S(E_k)^{\mathrm{GL}_k}\right) \to (D[k] \otimes S(E_k)^{\mathrm{GL}_k})_n$ be the epimorphism defined by

$$j_n(d_k^m M^\varepsilon) = \Phi_k^{-1}\big(\pi_n\big(\Phi_k(d_k^m M^\varepsilon)\big)\big).$$

The map j_n is well defined because of Remarks 14 and 18 and also an algebra epimorphism.

EXAMPLE 21. Let us continue on Example 15 for $n = 16$. We recall that $\Phi_k(d_{3,0}d_{3,1}^{12}d_{3,2}^4) = (Q_{1,13,17})^* + (Q_{1,16,16})^* + 2(Q_{10,13,16})^* + (Q_{10,10,17})^*$. Since $n = 16$, $\pi_n\big(\Phi_k(d_{3,0}d_{3,1}^{12}d_{3,2}^4)\big) = (Q_{1,16,16})^* + 2(Q_{10,13,16})^*$. Because $2h^{10,13,16}$ is an element of $\hat{\imath}((\theta^*)^{-1}h^{1,16,16})$, $\Phi_k^{-1}\big(\pi_n\big(\Phi_k(d_{3,0}d_{3,1}^{12}d_{3,2}^4)\big)\big) = d_{3,0}d_{3,1}^{15}$.

DEFINITION 22. Let $(d_{k,0}, M_{k;i}, M_{k;0,i})_{\text{Ideal}}$ stand for the ideal of $D[k]\otimes S(E_k)^{\mathrm{GL}_k}$ generated by ,$\{d_{k,0}, M_{k;i}, M_{k;0,i}\}$ and $j_n\big((d_{k,0}, M_{k;i}, M_{k;0,i})_{\text{Ideal}}\big)$ its image in $\big(D[k]\otimes S(E_k)^{\mathrm{GL}_k}\big)_n$. Respectively, $j_n\big((d_{k,0})_{\text{Ideal}}\big)$ stands for the image of the principal ideal $(d_{k,0})_{\text{Ideal}}$ in $D[k]_n$.

REMARK 23.

1. Because D_+ is of finite type,
$$\prod_k (d_{k,0}, M_{k;i}, M_{k;0,i})_{\text{Ideal}} \equiv \bigoplus_k (d_{k,0}, M_{k;i}, M_{k;0,i})_{\text{Ideal}}.$$
2. It is immediate from the definition of the coproduct in D that
$$\bigoplus_k (d_{k,0}, M_{k;i}, M_{k;0,i})_{\text{Ideal}} \quad \text{and} \quad \bigoplus_k (d_{k,0})_{\text{Ideal}}$$
are closed under the same coproduct.

Next we define a vector space which turns out to be isomorphic to the primitive elements of the mod $-p$ cohomology of $\Omega^{2n+1}S^{2n+1}$, i.e. the dual of the generators of $H_*(\Omega^{2n+1}S^{2n+1})$. We recall that the natural product in $H^*(\Omega^{2n+1}S^{2n+1})$ is not related with the product of the Dickson algebra. Moreover, the coproduct in D is not related with the Pontryagin product in $H_*(\Omega^{2n+1}S^{2n+1})$ either.

DEFINITION 24. Let $C(n,p)$ be the $\mathbb{Z}/p\mathbb{Z}$-vector space given by
$$\bigoplus_k j_n\big((d_{k,0}, M_{k;i}, M_{k;0,i})_{\text{Ideal}}\big).$$

We are interested in finding all elements in $C(n,p)$ which are annihilated by all Steenrod operations. Those elements shall be computed in the next section.

DEFINITION 25. Let $(d_{k,0})_{\text{Ideal}}$ be the principal ideal in $D[k]$ and $j_n\big((d_{k,0})_{\text{Ideal}}\big)$ its image in $D[k]_n$. Let $\mathrm{Ann}\big(j_n(d_{k,0})_{\text{Ideal}}\big)$ stand for the space generated by all monomials in $j_n(d_{k,0})_{\text{Ideal}}$ which are annihilated by all Steenrod operations.

We close this section by examining the map j_n, because of its applications in the next section.

LEMMA 26. *Let* $d^m = d_{k,0}^{m_0}\cdots d_{k,i}^{m_i}$ *such that* $m_t \equiv 0 \mod p^{k-i}$ *for* $1 \le t \le i$ *and* $m_i \ge p^{k-i}$. *Let* $n = \Sigma(m)$. *Then* $j_n(d^m d_{k,k-1}) \equiv d^m d_{k,i-1}^{p^{k-i}}/d_{k,i}^{p^{k-i}} +$ *(terms of polynomial degree less than* $n+3-p^{k-i}$*).*

PROOF. Let $I = \chi_{\min}(d^m d_{k,k-1})$, then $i_k = n+1$. We are interested in admissible sequences in $\hat{\imath}(d^m d_{k,k-1}) = \prod_{t=0}^{i}\hat{\imath}(d_{k,t}^{p^{k-i}})^{m_t/p^{k-i}}\hat{\imath}(d_{k,k-1})$ such that they have polynomial degree less than $n+1$. Because of the restriction on the degree, we must consider a summand of at least one element $\hat{\imath}(d_{k,t}^{p^{k-i}})$ in $\hat{\imath}(d^m d_{k,k-1})$ such that it is not divisible by h_k. That summand will be a pth power because of formula 2.1. There are two distinct choices:

1. $\hat{\imath}(d_{k,k-1})$ and
2. $\hat{\imath}(d_{k,t}^{p^{k-i}})$ for $t \leq i$.

1. In order for the corresponding sequence to be admissible, we should consider $h_t^{p^{k-t}}$ in $\hat{\imath}(d_{k,k-1})$ for $t \leq i$. The smallest one will be the one with $t = i$ namely:

$$I' = I - I_{k,k-1} + (0, \ldots, 0, p^{k-i}, \underbrace{0, \ldots, 0}_{k-i}).$$

Moreover, $(\theta^*)^{-1} h^{I'} = d^m d_{k,i-1}^{p^{k-i}} / d_{k,i}^{p^{k-i}}$. For

$$I' = I - I_{k,k-1} + (0, \ldots, 0, p^{k-i-t}, \underbrace{0, \ldots, 0}_{k-i-t}),$$

it is not admissible. One more case where the polynomial degree is n is described as follows: let $\prod_{s=1}^{k} h_s^{a_s p^{l_s}}$ be a summand in $\hat{\imath}(d_{k,t}^{p^{k-i}})$ with $a_{s_0} = 0$ for $k > s_0 > i$ and $h_{s_0}^{p^{k-s_0}}$ the corresponding summand from $d_{k,k-1}$. Let I' be the corresponding sequence, then $h^{I'}$ is divisible by $h_s^{p^{k-i}}$ for all s except for s_0. It follows that I' is not admissible.

2. Let $\prod_{s=1}^{k-1} h_s^{a_s p^{l_s}}$ be a summand in $\hat{\imath}(d_{k,t}^{p^{k-i}})$ and the corresponding sequence I'. Because at least $h_k^{p^{k-i}}$ has been divided from $\theta^*(d^m d_{k,k-1})$, $i'_k \leq n+1-(p^{k-i}-1)$.

Up to this point, we have considered the composition $\pi_n\big(\Phi_k(d^m d_{k,k-1})\big)$ and we should take care of Φ_k^{-1}. It is obvious from case 1. and 2. above that the only possible admissible sequences depending on m are of the form

$$I'(t) = I - I_{k,k-1} + (0, \ldots, 0, p^{k-i+t}, \underbrace{0, \ldots, 0}_{k-i+t}).$$

Here $t \geq 0$ and $I'(0)$ is the smallest one. But $\Phi_k\big(\theta^*(h^{I'(0)})\big)$ and $\Phi_k(d^m d_{k,k-1})$ contains $(Q_{I'(t)})^*$ with the same coefficient for $t > 0$. Finally, $j_n(d^m d_{k,k-1}) = d^m d_{k,i-1}^{p^{k-i}} / d_{k,i}^{p^{k-i}}$ + (other terms of degree less than $n + 3 - p^{k-i}$). ☐

COROLLARY 27. *Let* $d^m = d_{k,0}^{m_0} \cdots d_{k,i}^{m_i}$ *such that* $m_t \equiv 0 \mod p^{k-i+l}$ *for* $1 \leq t \leq i$ *and* $m_i = ap^{k-i+l} + p^{k-i+l+1}b$ *for* $1 \leq a \leq p-1$ *and* $b \geq 0$. *Let* $n = \Sigma(m)$. *Then* $j_n(d^m d_{k,k-1}^{ap^l}) \equiv d^m d_{k,i-1}^{ap^{k-i+l}} / d_{k,i}^{ap^{k-i+l}}$ + (*terms of degree less than* $n + 3 - p^{k-i+l}$).

EXAMPLE 28.

1. Let $p = 3$ and $n = 22$. We recall that $j_n = \Phi_k^{-1}\big(\pi_n(\Phi_k)\big)$. Then $\Phi_k^{-1}\big(\pi_n\big(\Phi_k(d_{3,0} d_{3,1}^{21} d_{3,2}^{2})\big)\big) = 0$.
2. Let $n = 19$, then $\Phi_k^{-1}\big(\pi_n\big(\Phi_k(d_{3,0} d_{3,1}^{18} d_{3,2}^{2})\big)\big) = d_{3,0}^{19}$.

REMARK 29. Campbell, Peterson and Selick provided a convenient method for calculating j_n for relatively small n, [**1**].

3. The Steenrod algebra action and $\mathrm{Ann}(j_n(d_{k,0})_{\mathrm{Ideal}})$

We recall some well known facts about the Steenrod algebra action on generators of $H^*\big((BZ/p\mathbb{Z})^k, \mathbb{Z}/p\mathbb{Z}\big)^{\mathrm{GL}_k}$, please see [**7**] for details.

THEOREM 30 ([7]).

(a) $$P^{p^j}(d_{k,i}) = \begin{cases} d_{k,i-1}, & \text{if } j = i-1, \\ -d_{k,i}d_{k,k-1}, & \text{if } j = k-1, \\ 0, & \text{otherwise.} \end{cases}$$

(b) $$P^n(h_k) = \begin{cases} \frac{1}{d_{k-1,0}}\big(P^n(d_{k,0}) - h_k P^n(d_{k-1,0})\big), \\ 0, \quad \text{if } C(n) = \emptyset. \end{cases}$$

Here $C(n)$ stands for all sequences consisting of nonnegative integers $c = [c_0, \dots, c_{k-1}]$ such that

$$n = \sum_0^{k-1} c_t(p^{k-1} + \cdots + p^{k-1-t}).$$

(c) $$\beta M_{k;0} L_k^{p-2} = d_{k,0}.$$

(d) $$\beta M_{k;0,s} L_k^{p-2} = -M_{k;s} L_k^{p-2}, \quad \text{for } s > 0.$$

(e) $$P^{p^{s-1}} M_{k;s} L_k^{p-2} = M_{k;s-1} L_k^{p-2}, \quad \text{for } s > 0.$$

It is immediate from the theorem above that:

PROPOSITION 31.

1. *No-element of $S(E_k)^{\mathrm{GL}_k}$ is annihilated by all Steenrod operations.*
2. *For any monomial M in $S(E_k)^{\mathrm{GL}_k}$ there exists a Steenrod operation P^l such that M and $P^l M$ have the same polynomial degree.*

In order to discuss the effect of the Steenrod algebra action on polynomial degrees we examine the action of P^q on the Dickson algebra generators.

LEMMA 32. *Let f be the Peterson monomial $f = y_k^{p^k} \cdots y_{i+1}^{p^{i+1}} y_i^{p^{i-1}} \cdots y_1$. The Steenrod algebra action on f, $P^q f$, is not identically zero if and only if $q = \sum_{0,t\neq i}^k a_t p^t$ for $0 \le a_t \le 1$. In that case,*

$$P^q f = \prod_{1+i}^{k} y_t^{p^t(1+a_t(p-1))} \prod_1^i y_t^{p^{t-1}(1+a_{t-1}(p-1))}.$$

PROPOSITION 33. *Let the polynomials $L_k = \sum_{\sigma\in\Sigma_k} \mathrm{sign}(\sigma) y_{\sigma(k)}^{p^{k-1}} \cdots y_{\sigma(1)}$ and $L_{k,i} = \sum_{\sigma\in\Sigma_k} \mathrm{sign}(\sigma) y_{\sigma(k)}^{p^k} \cdots y_{\sigma(i+1)}^{p^{i+1}} y_{\sigma(i)}^{p^{i-1}} \cdots y_{\sigma(1)}$ which are the expand of the following determinants:*

$$L_k := \begin{vmatrix} y_1 & \cdots & y_k \\ \vdots & & \vdots \\ y_1^{p^{k-1}} & \cdots & y_k^{p^{k-1}} \end{vmatrix} \quad \text{and} \quad L_{k,i} = \begin{vmatrix} y_1 & \cdots & y_k \\ \vdots & & \vdots \\ y_1^{p^k} & \cdots & y_k^{p^k} \end{vmatrix}.$$

In the last determinant the row $(y_1^{p^i} \ \cdots \ y_k^{p^i})$ is deleted.

1. *$P^q L_k \neq 0$ if and only if $q = \sum_i^{k-1} p^t$ or 0 and in that case and the action is given as follows:*

$$P^q L_k = \sum_{\sigma\in\Sigma_k} \mathrm{sign}(\sigma) y_{\sigma(k)}^{p^k} \cdots y_{\sigma(i+1)}^{p^{i+1}} y_{\sigma(i)}^{p^{i-1}} \cdots y_{\sigma(1)} = L_{k,i}.$$

2. *$P^q L_{k,i} \neq 0$ if and only if*

$$q = \begin{cases} \sum_s^k p^t & \text{for } s \geq i+1, \\ \sum_l^{i-1} p^t & \text{for } l \geq 0, \\ \sum_s^k p^t + \sum_l^{i-1} p^t & \text{for } s \geq i+1 \quad \text{and} \quad l \geq 0 \end{cases}$$

and in that case

$$P^q L_{k,i}$$
$$= \begin{cases} \sum_{\sigma\in\Sigma_k} \operatorname{sign}(\sigma) y_{\sigma(k)}^{p^{k+1}} \cdots y_{\sigma(s)}^{p^{s+1}} y_{\sigma(s-1)}^{p^{s-1}} \cdots y_{\sigma(i+1)}^{p^{i+1}} y_{\sigma(i)}^{p^{i-1}} \cdots y_{\sigma(1)}, \\ \sum_{\sigma\in\Sigma_k} \operatorname{sign}(\sigma) y_{\sigma(k)}^{p^k} \cdots y_{\sigma(i+1)}^{p^{i+1}} y_{\sigma(i)}^{p^i} \cdots y_{\sigma(l+1)}^{p^{l+1}} y_{\sigma(l)}^{p^{l-1}} \cdots y_{\sigma(1)}, \\ \sum_{\sigma\in\Sigma_k} \operatorname{sign}(\sigma) y_{\sigma(k)}^{p^{k+1}} \cdots y_{\sigma(s)}^{p^{s+1}} y_{\sigma(s-1)}^{p^{s-1}} \cdots y_{\sigma(i+1)}^{p^{i+1}} y_{\sigma(i)}^{p^i} \cdots y_{\sigma(l+1)}^{p^{l+1}} y_{\sigma(l)}^{p^{l-1}} \cdots y_{\sigma(1)}. \end{cases}$$

PROOF. We prove 1. and the proof for 2. is identical. Let q and q' as in Lemma 32 such that $q = q' + p^{i+s} + p^i$ with $s > 1$. Then

$$P^q f = \sum_{\sigma\in\Sigma_k} \operatorname{sign}(\sigma) P^{q'+p^{i+s}} \left(\frac{y_{\sigma(k)}^{p^{k-1}} \cdots y_{\sigma(1)}}{y_{\sigma(i+2)}^{p^{i+1}} y_{\sigma(i+1)}^{p^i}} \right) y_{\sigma(i+2)}^{p^{i+1}} y_{\sigma(i+1)}^{p^{i+1}}.$$

There exists a σ' such that $\sigma' = (i+1, i+2)\sigma$ and $\operatorname{sign}(\sigma') = -\operatorname{sign}(\sigma)$. Thus $P^q f = 0$. □

COROLLARY 34.

1. *Let $g = f^{p^l}$ and $f = L_k$ or $L_{k,i}$, then $P^q g = (P^{q/p^l} f)^{p^l}$.*
2. *Let $g = L_{k,i}$ and $q < p^k$, then $P^q g \neq 0$ if and only if $q = p^{i-1} + \cdots + p^l$ for $l \geq 0$.*

LEMMA 35. $\prod_i^{k-2} \binom{a_{t+1}}{a_t} = \sum_1^{k-i} \binom{a_{k-1}-1}{a_{k-2}-1} \cdots \binom{a_{k-t}-1}{a_{k-t-1}} \binom{a_{k-t-1}}{a_{k-t-2}} \cdots \binom{a_{k-(k-i-1)}}{a_{k-(k-i)}}$.

THEOREM 36.

1. *Let $q = \sum_i^{k-1} a_t p^{t+l}$ such that $p-1 \geq a_t \geq a_{t-1} > a_{i-1} = 0$. Then*

$$P^q d_{k,0}^{p^l} = d_{k,0}^{p^l} (-1)^{a_{k-1}} \prod_i^{k-1} \binom{a_t}{a_{t-1}} d_{k,t}^{p^l(a_t - a_{t-1})}.$$

Otherwise, $P^q d_{k,0}^{p^l} = 0$.

2. *Let $q = \sum_s^{k-1} a_t p^{t+l} > 0$ such that $p-1 \geq a_t \geq a_{t-1} \geq a_i \geq 0$ and $a_i + 1 \geq a_{i-1} \geq a_t \geq a_{t-1} \geq 0$. Then*

$$P^q d_{k,i}^{p^l} = d_{k,i}^{p^l} (-1)^{a_{k-1}} \left(\prod_{i+1}^{k-1} \binom{a_t}{a_{t-1}} \right) \binom{a_i+1}{a_{i-1}} \left(\prod_s^{i-1} \binom{a_t}{a_{t-1}} \right) \prod_s^{k-1} d_{k,t}^{p^l(a_t - a_{t-1})}.$$

Here $a_{s-1} = 0$. Otherwise, $P^q d_{k,0}^{p^l} = 0$.

REMARK 37. Please note that the case $a_i = 0$ and $a_{i-1} = 1$ is allowed.

PROOF. The idea of the proof has been used in [**9**]. We use induction and the Cartan formula on the identity:

$$d_{k,i}^{p^l} L_k^{p^l} = L_{k,i}^{p^l}. \tag{3.1}$$

For convenience we shall use the notation $(a_{k-1},\ldots,a_i)f := P^{a_{k-1}p^{k-1+l}+\cdots+a_ip^{i+l}}f$ in the sequel of the proof.

1. We prove the claimed formula for $(1,\ldots,1)d_{k,0}^{p^l}$ using formula (3.1), Proposition 33 1. and Corollary 34 1. Using induction hypothesis, formula (3.1) and Lemma 35 the formula follows.
2. If $a_{i-1}=0$, then the proof is identical as in 1. If $a_i=0$ and $a_{i-1}>0$, then we use formula (3.1), Proposition 33 1. and 2. and Corollary 34. Let $a_ia_{i-1}>0$. We prove the claimed formula for the following sequences using formula (3.1), Proposition (33) and Corollary 34:

$$\Big(\underbrace{1,\ldots,1}_{k-i},\underbrace{1,\ldots,1}_{i-s}\Big),\ \Big(\underbrace{1,\ldots,1}_{k-i},\underbrace{2,\ldots,2}_{i-s}\Big),\ \Big(\underbrace{1,\ldots,1}_{k-i},\underbrace{2,\ldots,2,1,\ldots,1}_{i-s}\Big).$$

Using induction hypothesis, formula 3.1 and Lemma 35 the formula follows.

□

The following theorem is an application of last theorem and the Cartan formula.

THEOREM 38. *Let* $d=d_{k,0}^{ap^l}d_{k,i}^{ap^{k-i+l}}$ *for* $1\le a\le p-1$. *Then*

$$P^qd=\begin{cases}\sum_{c=0}^{b}(-1)^c\binom{a}{c}\binom{a}{b-c}d_{k,0}^{ap^l}d_{k,k-t}^{cp^l}d_{k,i-t}^{(b-c)p^{k-i+l}}d_{k,i}^{(a-b+c)p^{k-i+l}}\\ \qquad \textit{for } q=b(p^{k-1+l}+\cdots+p^{k-t+l}),\, b\le 2a.\\ (-1)^b\binom{a}{b}d_{k,0}^{ap^l}d_{k,k-1}^{bp^l}d_{k,i}^{ap^{k-i+l}} \qquad \textit{for } q=bp^{2k-1-i+l},\, b\le a.\\ (-1)^c\binom{a}{c}d_{k,k-1}^{cp^{k-i+l}}d_{k,i}^{cp^{k-i+l}}P^{bp^{k-1+l}}(d_{k,0}^{ap^l}d_{k,i}^{(a-c)p^{k-i+l}})\\ \qquad \textit{for } q=bp^{k-1+l}+cp^{2k-1-i+l},\ b,c\le a.\end{cases}$$

PROOF. We use Theorem 30 and Cartan formula. □

COROLLARY 39. *Let* $d=d_{k,0}^{ap^l}d_{k,i}^{ap^{k-i+l}}$ *for* $1\le a\le p-1$ *and* $n=a(p^l+p^{k-i+l})$. *Then* $j_n(P^qd)$ *is nonzero for* $a>1$, $q=b(p^{k-1+l}+\cdots+p^{k-t+l})$ *and* $b=2$. *Moreover,* $j_n(P^qd)=0$ *for all* $q\ge 0$ *if and only if* $a=1$.

EXAMPLE 40. Let $p=3$, $n=2(1+p^{k-i})$ and $d=d_{k,0}^2d_{k,i}^{2p^{k-i}}$. Let $q=2p^{k-1}$, then $P^qd=d_{k,0}^2d_{k,k-1}^2d_{k,i}^{2p^{k-i}}+d_{k,0}^2d_{k,i-1}^{2p^{k-i}}-4d_{k,0}^2d_{k,k-1}d_{k,i-1}^{p^{k-i}}d_{k,i}^{p^{k-i}}$ and $j_n(P^qd)=d_{k,0}^2d_{k,i-1}^{2p^{k-i}}$.

THEOREM 41. *Let* $m=(m_0,\ldots,m_{k-1})$ *and* $d^m=\Pi d_{k,i}^{m_i}$ *such that* $\Sigma m_i=n$.

1. *Let* $m_0=n$, *then* $d_{k,0}^n\in \mathrm{Ann}(j_n(d_{k,0})_{\mathrm{Ideal}})$.
2. *Let* $m_0=\sum_{0\le t\le l(0)}m_{0,s_t}p^{s_t}<n$ *and* $m_i=0$ *for* $1\le i\le k-1$. *then* $d_{k,0}^n\in\mathrm{Ann}(j_n(d_{k,0})_{\mathrm{Ideal}})$ *if and only if* $m_0+p^{s_0-1}\le n<m_0+p^{s_0}$.
3. *Let* $m_0<n$ *and* $\exists i$ *such that* $m_i\ne 0 \mod p$, *then* $d^m\notin\mathrm{Ann}(j_n(d_{k,0})_{\mathrm{Ideal}})$.
4. *Let* $m_0=\sum_{0\le t\le l(0)}m_{0,s_t}p^{s_t}$ *with* $\prod_{0\le t\le l(0)}m_{0,s_t}\ne 0$.

$$d^m\in\mathrm{Ann}(j_n(d_{k,0})_{\mathrm{Ideal}})\quad \textit{if and only if}\quad m_{0,s_t}=1\quad \textit{for all } t$$
$$\textit{and}\quad d^m=\prod_{0\le t\le l(0)}d_{k,0}^{m_{0,s_t}p^{s_t}}d_{k,i_{s_t}}^{m_{0,s_t}p^{k-i_{s_t}+s_t}}.$$

PROOF.

1. It follows from Theorem 30 that $P^a(d^n_{k,0}) \in (d^n_{k,0})_{\text{Ideal}}$. Moreover, if $a > 0$ and $P^a(d^n_{k,0}) \neq 0$, then its polynomial degree is greater than n. Thus $d^n_{k,0} \in \operatorname{Ann}\big(j_n(d_{k,0})_{\text{Ideal}}\big)$.
2. This case is similar to the previous one taking on to account binomial coefficients as well. If n is outside the given range, then $P^{ap^{k-1}} d^{m_0}_{k,0} = m_{0,s_0} d^{m_0}_{k,0} d^a_{k,k-1}$ for $a = p^{s_0}$.
3. Let i be minimal with this property, then $P^{p^{i-1}} d^m = m_{i,0} d^m d_{k,i-1}/d_{k,i}$.
4. Let $\mu(m_i)$ such that $m_i \equiv 0 \mod p^{\mu(m_i)}$ and $m_i \not\equiv 0 \mod p^{\mu(m_i)+1}$. Let $T(m, \min) = \big\{i | i > 0, i-1+\mu(m_i) = \min\{t-1+\mu(m_t) | t > 0\}\big\}$. Let $t_0 \in T(m,\min)$ and $t_0 - 1 + \mu(m_{t_0}) < k-1+\mu(m_0)$, then using Theorem 30 we get $j_n(P^{p^{t_0-1+\mu(m_{t_0})}} d^m) \neq 0$.

Let $t_0 = \max T(m,\min)$ and $t_0 - 1 + \mu(m_{t_0}) > k-1+\mu(m_0)$, then using Theorem 30 and Corollary 27 we get $j_n(P^{p^{k-1+\mu(m_0)}} d^m) = -m_{0,\mu(m_0)} d^m d^{p^{k-t_0+\mu(m_0)}}_{k,t_0-1} / d^{p^{k-t_0+\mu(m_0)}}_{k,t_0}$. Here we also use $k - t_0 + \mu(m_0) < \mu(m_{t_0})$.

Let $t_0 = \max T(m,\min)$ and $t_0 - 1 + \mu(m_{t_0}) = k - 1 + \mu(m_0)$, then using Theorem 30 and Corollary 27 we get

$$
\begin{aligned}
j_n(P^{p^{k-1+\mu(m_0)}} d^m) &= (m_{t_0,\mu(m_{t_0})} - m_{0,\mu(m_0)}) d^m d^{p^{k-t_0+\mu(m_{t_0})}}_{k,t_0-1} / d^{p^{k-t_0+\mu(m_{t_0})}}_{k,t_0} \\
&\quad + \sum_{t_0' \in T(m,\min)-\{t_0\}} m_{t_0',\mu(m_{t_0'})} d^m d^{p^{k-t_0'+\mu(m_{t_0'})}}_{k,t_0'-1} / d^{p^{k-t_0'+\mu(m_{t_0'})}}_{k,t_0'}.
\end{aligned}
$$

The last equation is zero only if $m_{t_0,\mu(m_{t_0})} = m_{0,\mu(m_0)}$ and $T(m,\min) = \{t_0\}$. Thus if $d^m \in \operatorname{Ann}(j_n(d_{k,0})_{\text{Ideal}})$, then d^m is divisible by $d^{m_{0,\mu(m_0)} p^{\mu(m_0)}}_{k,0} d^{m_{0,\mu(m_0)} p^{k-t_0+\mu(m_0)}}_{k,t_0}$.

Following the same method for the next nonzero coefficient, m_{0,s_1}, we deduce that if $d^m \in \operatorname{Ann}(j_n(d_{k,0})_{\text{Ideal}})$, then the following hypothesis holds:

(i) $m_0 = m_0' + m_{0s_1} p^{s_1} + m_{0,\mu(m_0)} p^{\mu(m_0)}$,

(ii) $\exists! t_0$ such that $m_i \equiv 0 \mod p^{\mu(m_{t_0})-t}$ for $1 \leq t \leq \mu(m_{t_0}) - 1$ for all i and $m_{t_0} \equiv m_{t_0,\mu(m_{t_0})} \mod p^{\mu(m_{t_0})}$, $m_i \equiv 0 \mod p^{\mu(m_{t_0})}$, $i \neq t_0$ and $m_{t_0,\mu(m_{t_0})} = m_{0,\mu(m_0)}$,

(iii) $\exists! t_1$ such that $m_i \equiv 0 \mod p^{\mu(m_{t_0})+t}$ for $1 \leq t \leq s_{t_1} - 1$ for all $i \neq t_0$, $m_{t_0} - m_{t_0,\mu(m_{t_0})} p^{\mu(m_{t_0})} \equiv 0 \mod p^{\mu(m_{t_0})+t}$, $m_{t_1} \equiv m_{t_1,s_{t_1}} \mod p^{p^{s_{t_1}}}$, $m_i \equiv 0 \mod p^{s_{t_1}}$, $i \neq t_0, t_1$, $m_{t_0} - m_{t_0,\mu(m_{t_0})} p^{\mu(m_{t_0})} \equiv 0 \mod p^{s_{t_1}}$ and $m_{t_1,s_{t_1}} = m_{0,s_1}$.

Finally, applying the same procedure for all nonzero coefficients of m_0 in its p-adic expansion we prove that: if $d^m \in \operatorname{Ann}\big(j_n(d_{k,0})_{\text{Ideal}}\big)$, then d^m is divisible by $d^{m_{0,s_t} p^{s_t}}_{k,0} d^{m_{0,s_t} p^{k-i_{s_t}+s_t}}_{k,i_{s_t}}$ for all m_{0,s_t} in m_0.

For the other direction we use induction on the number of nonzero coefficients of m_0, Cartan formula, and Corollary 39. □

EXAMPLE 42.

1. Let $n = 1$. Then

$$j_1\big((d_{k,0}, M_{k;i}, M_{k;0,i})_{\text{Ideal}}\big) = \{d_{k,0}, M_{k;i}, M_{k;0,t} \mid 0 \le i \le k-1.1 \le t \le k-1\}$$
$$\text{and} \quad \text{Ann}(j_1(d_{k,0})_{\text{Ideal}}) = \{d_{k,0}\}.$$

2. Let $n = 2$. Then

$$j_2\big((d_{k,0}, M_{k;i}, M_{k;0,i})_{\text{Ideal}}\big)$$
$$= \{d_{k,0}d_{k,s}, d_{k,s}M_{k;i}, d_{k,s}M_{k;0,t}, M_{k;i}M_{k;s,t}, d_{k,0}M_{k;s,t} \mid 0 \le i, s, q \le k-1,$$
$$1 \le t \le k-1\}\text{and} \quad \text{Ann}(j_2(d_{k,0})_{\text{Ideal}}) = \{d_{k,o}^2\}.$$

3. Let $n = p+1$. Then

$$j_n\big((d_{k,0})_{\text{Ideal}}\big) = \{d_{k,0}^{a_0} d_{k,i_1}^{a_{i_1}} \cdots d_{k,i_l}^{a_{i_l}} \mid 0 \le a_t \le p, 0 < a_0, \Sigma a_t \le n\}$$
$$\text{and} \quad \text{Ann}(j_n(d_{k,0})_{\text{Ideal}}) = \{d_{k,o}^n, d_{k,o}^p, d_{k,o}d_{k,k-1}^p\}.$$

4. $H_*(\Omega^{m+1}S^{m+1};\mathbb{Z}/p\mathbb{Z})$ and Dyer–Lashof operations

We recall the description of $H_*(\Omega^{m+1}S^{m+1};\mathbb{Z}/p\mathbb{Z})$ in terms of lower Dyer–Lashof operations from [**2**] and [**1**].

DEFINITION 43. Let $R_n = \big\langle Q_{(I,\varepsilon)} \mid (I,\varepsilon) \in \langle \mathbb{N}, \frac{1}{2}\rangle^k \times (\mathbb{Z}/2Z)^k$ with $i_k \le n$ and admissible as in Definition 8 for $k \ge 0\big\rangle$. Let $Q_0R_n = \langle Q_{(I,\varepsilon)} \in R_n$ with $i_1 = 0\rangle$.

$H_*(\Omega^{m+1}S^{m+1};\mathbb{Z}/p\mathbb{Z})$ is the free commutative graded algebra on R_n/Q_0R_n.

We recall that R_n/Q_0R_n is an opposite Steenrod algebra A_* module through Nishida relations. Let consider all elements in R_n with fixed length k:

$$R_n[k] = \big\langle Q_{(I,\varepsilon)} \in R_n, (I,\varepsilon) \in \langle \mathbb{N}, \tfrac{1}{2}\rangle^k \times (\mathbb{Z}/2Z)^k\big\rangle$$

and $Q_0R_n[k] = \langle Q_{(I,\varepsilon)} \in R_n[k] \cap Q_0R_n, (I,\varepsilon)\rangle$. Those are A_* coalgebras.

Dualizing, we get a Steenrod algebra module isomorphism:

$$PH^*(\Omega^{m+1}S^{m+1};\mathbb{Z}/p\mathbb{Z}\bigoplus_k(R_n[k]/Q_0R_n[k])^*.$$

Next we recall the relation between $R[k]$ and modular invariants. First we note that the hom-dual of $R[k]$, $R[k]^*$, coincides with what we defined in 11.

THEOREM 44 ([**6**]). *$R[k]^* \approxeq D[k] \otimes S(E_k)^{\mathrm{GL}_k}$ as algebras over the Steenrod algebra.*

Using [**1**, Theorem 1.7] we deduce the following theorem.

THEOREM 45.

1. $(R[k]/Q_0R[k])^* \approxeq \big((d_{k,0}, M_{k;i}, M_{k;0,i})_{\text{Ideal}}\big)$.
2. $PH^*(\Omega_0^{2n+1}S^{2n+1};\mathbb{Z}/p\mathbb{Z}) \approxeq j_n\big((d_{k,0}, M_{k;i}, M_{k;0,i})_{\text{Ideal}}\big)$.

We quote Campbell, Peterson and Selick's from [**1**, Theorem 2.5].

THEOREM 46 ([**1**]). *$\text{Ann}\big(PH^*(\Omega_0^{2n+1}S^{2n+1};\mathbb{Z}/p\mathbb{Z})\big)$ is generated by $\{(Q_I)^* \mid I$ admissible and satisfies $i_s + p^t \le n \Rightarrow 2i_s + |I_{s-1}| \equiv 0 \mod p^{t+1}$ for all $t \ge 0$ and $1 \le s \le k\}$.*

Our Theorem 41 describes the isomorphic image of $\mathrm{Ann}\big(PH^*(\Omega_0^{2n+1}S^{2n+1};\mathbb{Z}/p\mathbb{Z})\big)$ in terms of Dickson invariants. It is not obvious how to decompose an element of the hom-dual basis of $R[k]^*$ with respect to the monomial basis of $D[k]\otimes S(E_k)^{\mathrm{GL}_k}$. We describe an algorithm regarding this passage in [**8**].

References

1. H. E. A. Campbell, F. P. Peterson, and P. S. Selick, *Self-maps of loop spaces.* I, Trans. Amer. Math. Soc. **293** (1986), no. 1, 1–39.
2. F. Cohen, T. Lada, and P. J. May, *The homology of iterated loop spaces*, Lecture Notes in Math. **533**, (1975).
3. L. E. Dickson, *A fundamental system of invariants of the general modular linear group with a solution of the form problem*, Trans. Amer. Math. Soc. **12** (1911), 75–98.
4. Mui Huyhn, *Modular invariant theory and the cohomology algebras of the symmetric groups*, J. Fac. Sci. Univ. Tokyo **IA** (1975), 319–369.
5. ———, *Homology operations derived from modular coinvariants*, Lecture Notes in Math. **172** (1985), 85–115.
6. E. Nondas Kechagias, *Extended Dyer–Lashof algebras and modular coinvariants*, Manuscripta Math. **84** (1994), 261–290.
7. ———, *The Steenrod algebra action on generators of rings of invariants of subgroups of* $\mathrm{GL}_n(\mathbb{Z}/p\mathbb{Z})$, Proc. Amer. Math. Soc. **118** (1993), 943–952.
8. ———, *Adem relations in the Dyer–Lashof algebra and modular invariants*, Preprint, 2002.
9. ———, *Homology operations and modular invariant theory*, Ph.D. Thesis, Queen's University, 1990.
10. I. Madsen, *On the action of Dyer–Lashof algebra in $H_*(B)$*, Pacific J. Math. **60** (1975), 235–275.

Department of Mathematics, University of Ioannina, P.O. 1186, Ioannina 45110, Greece.

E-mail address: nkechag@cc.uoi.gr

Centre de Recherches Mathématiques
CRM Proceedings and Lecture Notes
Volume **35**, 2004

On Noether's and Weyl's Bound in Positive Characteristic

F. Knop

ABSTRACT. In this note we generalize several well-known results concerning invariants of finite groups from characteristic zero to positive characteristic not dividing the group order. The first is Schmid's relative version of Noether's theorem. That theorem compares the degrees of generators of a group with those of a subgroup. Then we prove a suitable positive characteristic version of Weyl's theorem on vector invariants: polarization works in small degrees. Using that we show that the regular representation has the "most general" ring of invariants, thereby generalizing theorems of Schmid and Smith.

1. Introduction

Let G be a finite group acting linearly on a k-vector space V. If $k = \mathbb{C}$, Noether showed in [**5**] that the ring $k[V]^G$ of G-invariants is generated by invariants of degree $\leq |G|$. This is called *Noether's bound.* An inspection of her proof reveals that it stays valid for any field k of characteristic 0 or characteristic $p > |G|$. It has been a long standing question whether Noether's bound holds under the weaker assumption that p does not divide $|G|$. This has been proved only recently by Fleischmann [**2**] and Fogarty [**4**].

Meanwhile, Schmid showed in [**7**] that Noether's bound is sharp only for cyclic groups. Again, her proof is valid if either $p = 0$ or $p > |G|$. The only reason why Schmid's proof doesn't carry over to characteristics not dividing $|G|$ is a certain refinement of Noether's theorem which compares degrees of generators of G-invariants with those of H-invariants where H is any subgroup of G. The purpose of this paper is to provide this tool for $p \nmid |G|$ (see Theorem 3.3 for a precise statement).

After a first draft of this note has been completed it came to my attention that this has also been done independently by Fleischmann [**2**], and Sezer [**8**]. Nevertheless, we find it still worthwhile to include a proof. First, our proof is an adaptation of Fogarty's (as opposed to Fleischmann's) proof of Noether's bound. Secondly, we introduce another numerical invariant of independent interest. More precisely, in Section 2 we look at the zero fiber of the quotient map $V \to V/G$ (the null cone) which is the spectrum of a local Artinian ring. We study the degree of

---

2000 *Mathematics Subject Classification.* Primary: 13A50.
Supported by an NSA grant.
The author would like to thank the referee for very carefully reading the manuscript.
This is the final form of the paper.

nilpotence of its maximal ideal in Theorem 2.1 and show that it is closely related to degrees of generating invariants.

In a further section, we study degree bounds for algebras over arbitrary ground rings k. We show that it suffices to consider for k a field and even there the problem depends only on the characteristic of k with $\operatorname{char} k = 0$ giving the lowest bounds.

The second major thread of this paper is Weyl's theorem on vector invariants. It asserts that one can obtain all invariants on any number of copies of a representation V from invariants which live on only $\dim V$ copies. The process allowing this is called polarization and works in that generality only in characteristic zero. Nevertheless, in positive characteristic it is plausible that polarization still works in "low" degrees. Theorem 5.1 makes this statement precise. On the other hand, generating invariants for finite groups have "low" degree by the theorem of Noether–Fleischmann–Fogarty. This yields Weyl type theorems for vector invariants in positive characteristic. In particular, we strengthen theorems of Schmid and Smith according to which the regular representation has the most general ring of invariants.

It should be added that even in characteristic zero our approach is not without interest. It yields a completely elementary proof of Weyl's theorem for the (special but most important) case that G is linearly reductive. In particular, we avoid the use of Capelli's identity and representation theory of $\mathrm{GL}(n)$.

2. The degree of nilpotence of the null cone

Let A be a commutative ring. For a subset S let $\langle S\rangle$ be the ideal generated by S. Now assume that the finite group G acts on A. For a G-invariant ideal $\mathfrak{m} \subseteq A$ we define

$$\eta_G(\mathfrak{m}) := \inf\{d \in \mathbb{N} \mid \mathfrak{m}^d \subseteq \langle \mathfrak{m}^G\rangle\}. \tag{2.1}$$

It is well-known that for finite groups, the fibers of $\operatorname{Spec} A \to \operatorname{Spec} A^G$ are precisely the G-orbits. Thus $\mathfrak{m}$ and $\langle \mathfrak{m}^G\rangle$ define the same vanishing set in $\operatorname{Spec} A$, which implies $\eta_G(\mathfrak{m}) < \infty$ whenever A is Noetherian (see also Corollary 2.2 below). The most important case will be when A is the ring of polynomial functions on a representation of G and $\mathfrak{m}$ the maximal ideal corresponding to the origin. Then $\langle \mathfrak{m}^G\rangle$ defines the null cone and $\eta_G(\mathfrak{m}) - 1$ is the degree of nilpotence of the maximal ideal of $A/\langle \mathfrak{m}^G\rangle$.

THEOREM 2.1. *Let $H \subseteq G$ be a subgroup and assume $[G : H] \in A^\times$. Then*

$$\eta_G(\mathfrak{m}) \le [G : H]\, \eta_H(\mathfrak{m}). \tag{2.2}$$

PROOF. Let $d := [G : H]$. Choose d arbitrary elements $a_u \in \mathfrak{m}^H$ which are indexed by $u \in G/H$. In the following, we will regard u as an element of G representing its coset. Consider the expression

$$\Phi := \sum_{u \in G/H} \prod_{v \in G/H} (va_v - ua_v). \tag{2.3}$$

Since every product contains a zero factor, namely the one corresponding to $v = u$, we have $\Phi = 0$. On the other side, we can expand the product and get

$$\Phi = \sum_S (-1)^{|S|} \Phi_S, \tag{2.4}$$

where S ranges over all subsets of G/H and

$$\Phi_S = \sum_{u\in G/H} \prod_{v\notin S}(va_v)\prod_{v\in S}(ua_v) = \prod_{v\notin S}(va_v)\left[\sum_{u\in G/H} u\left(\prod_{v\in S} a_v\right)\right]. \tag{2.5}$$

Thus, $\Phi_S \in \langle \mathfrak{m}^G\rangle$ unless $S = \emptyset$ when $\Phi_\emptyset = d\prod_{v\in G/H} va_v$. Since the a_v were arbitrary and $d \in A^\times$ we get

$$\prod_{v\in G/H} v\langle \mathfrak{m}^H\rangle \subseteq \langle \mathfrak{m}^G\rangle. \tag{2.6}$$

With $e := \eta_H(\mathfrak{m})$ we have by definition $\mathfrak{m}^e \subseteq \langle \mathfrak{m}^H\rangle$. Thus,

$$\langle \mathfrak{m}^G\rangle \supseteq \prod_{v\in G/H} v\mathfrak{m}^e = \mathfrak{m}^{de}. \tag{2.7}$$

This implies $\eta_G(\mathfrak{m}) \le de = [G:H]\,\eta_H(\mathfrak{m})$. □

For $H = 1$ we get:

COROLLARY 2.2. *Assume $|G| \in A^\times$. Then $\eta_G(\mathfrak{m}) \le |G|$.*

3. The degree of generators

Let $A = \oplus_{d=0}^{\infty} A_d$ be an $\mathbb{N}$-graded commutative ring. We use the notation $A_{\le n} := \oplus_{d=0}^{n} A_d$ and analogously $A_{\ge n} := \oplus_{d=n}^{\infty} A_d$. Of particular interest is the ideal $A_{\ge 1}$ which we denote by $\mathfrak{m}$. We define

$$\beta(A) := \inf\{n \in \mathbb{N} \mid A \text{ is generated as a ring by } A_{\le n}\}, \tag{3.1}$$

which is the smallest degree of a generator of A. Now assume that the finite group G acts on A by degree preserving automorphisms.

LEMMA 3.1. *Assume $\beta(A) \le 1$ and $|G| \in A^\times$. Then $\beta(A^G) \le \eta_G(\mathfrak{m})$.*

PROOF. Let $e := \eta_G(\mathfrak{m})$. By the classical argument of Hilbert, any set of G-invariant generators of the ideal $\langle \mathfrak{m}^G\rangle$ also generates the ring A^G. Thus, we have to show that $\langle \mathfrak{m}^G\rangle$ is generated by invariants of degree $\le e$. To that end, it suffices to show that $\langle \mathfrak{m}^G\rangle$ is generated by elements (invariant or not) of degree $\le e$ since each of them can be expressed in terms of invariants of degree $\le e$.

By definition, $\mathfrak{m}^e \subseteq \langle \mathfrak{m}^G\rangle$. Since $\beta(A) \le 1$ we have $\mathfrak{m}^e = A_{\ge e}$ and that ideal is generated by A_e. Thus, $\langle \mathfrak{m}^G\rangle$ is generated by $A^G_{\le e-1}$ and A_e. This implies that $\langle \mathfrak{m}^G\rangle$ is generated by $A^G_{\le e}$. □

For the next statement let $A(G)$ be the polynomial ring over A in $|G|$ variables $\{x_g \mid g \in G\}$ with the G-action extended by $gx_h := x_{gh}$. It is also $\mathbb{N}$-graded with $\deg x_g = 1$. We can write $A(G) = A \otimes_{\mathbb{Z}} \mathbb{Z}(G)$ where $\mathbb{Z}(G) = S^*(\mathbb{Z}[G])$ is the symmetric algebra over the (integral) regular representation of G. Using this, we see that $A(G)$ is actually bigraded where $a \in A_d$ and x_g have bidegree $(d,0)$ and $(0,1)$, respectively. The following result is a partial converse of Lemma 3.1.

LEMMA 3.2. *For any A we have that $\eta_G(\mathfrak{m}) \le \beta\big(A(G)^G\big)$.*

PROOF. Let $b := \beta(A(G)^G)$. For $a \in A_e$ with $e \geq b$ let

$$\bar{a} := \sum_{u \in G} u(ax_1) = \sum_{u \in G} (ua)x_u \in A(G)^G. \tag{3.2}$$

Its bidegree is $(e, 1)$. Since $A(G)^G$ is generated by elements of degree $\leq b$ we can write $\bar{a}$ as a sum of products $p \cdot q$ where p and q are bihomogeneous G-invariants of bidegree $(c, 0)$ and $(e-c, 1)$ respectively. We may assume that $\deg q = e-c+1 \leq b$. This implies $c \geq e - b + 1 \geq 1$. Thus, $p \in \mathfrak{m}^G$. Comparing the coefficient of x_1 in the expression $\bar{a} = \sum pq$, we obtain $a \in \langle \mathfrak{m}^G \rangle$. Thus $\mathfrak{m}^b \subseteq A_{\geq b} \subseteq \langle \mathfrak{m}^G \rangle$. □

Combining these comparison lemmas with Theorem 2.1, we obtain:

THEOREM 3.3. *Assume $\beta(A) \leq 1$ and $|G| \in A^\times$. Let $H \subseteq G$ be a subgroup. Then*

$$\beta(A^G) \leq [G : H]\beta(A(H)^H). \tag{3.3}$$

PROOF. Indeed,

$$\begin{aligned} \beta(A^G) &\leq \eta_G(\mathfrak{m}) && \text{(by Lemma 3.1)} \\ &\leq [G : H]\eta_H(\mathfrak{m}) && \text{(by Theorem 2.1)} \\ &\leq [G : H]\beta(A(H)^H) && \text{(by Lemma 3.2).} \end{aligned} \tag{3.4}$$

□

REMARK. Since Theorem 2.1 holds under the weaker assumption $[G : H] \in A^\times$ one might wonder whether also Theorem 3.3 holds in that generality.

Taking for H the trivial subgroup, we obtain Fleischmann's and Fogarty's theorem:

COROLLARY 3.4. *Assume $\beta(A) \leq 1$ and $|G| \in A^\times$. Then $\beta(A^G) \leq |G|$.*

This suggests a definition. For a fixed ground ring k let $\beta_k(G) := \sup_A \beta(A^G)$ where A runs through all $\mathbb{N}$-graded k-algebras with G-action and with $\beta(A) \leq 1$. The following generalizes a theorem of Schmid [7] for $k = \mathbb{C}$:

COROLLARY 3.5. *Assume $|G| \in k^\times$ and let $H \subseteq G$ be a subgroup. Then*

$$\beta_k(G) \leq [G : H]\beta_k(H). \tag{3.5}$$

In particular, $\beta_k(G) \leq |G|$.

For arbitrary $\mathbb{N}$-graded algebras we have

THEOREM 3.6. *Assume $|G| \in k^\times$ and let A be a k-algebra. Then $\beta(A^G) \leq \beta_k(G)\beta(A)$.*

PROOF. Let $b := \beta(A)$ and let $\bar{A}$ be the free commutative A_0-algebra over the A_0-module $\oplus_{d=1}^b A_d$. Then $\bar{A}$ has two degree functions, deg and deg′, where an element of A_d has degree d or 1, respectively. Clearly, we have $\deg \leq b \deg'$. If we denote $\bar{A}$ equipped with the grading deg′ by $\tilde{A}$ then $\beta(\tilde{A}) \leq 1$.

There is a natural map of $\bar{A}$ onto A. Hence $\bar{A}^G \to A^G$ is also onto. Therefore,

$$\beta(A^G) \leq \beta(\bar{A}^G) \leq b\,\beta(\tilde{A}^G) \leq b\,\beta_k(G). \quad \square \tag{3.6}$$

The following consequence is due to Schmid [7] for $k = \mathbb{C}$ and to Fleischmann [2] in general:

COROLLARY 3.7. *Let $N \lhd G$ be a normal subgroup with $[G:N] \in k^\times$ and A a k-algebra. Then*

$$\beta(A^G) \leq \beta_k(G/N)\beta(A^N). \tag{3.7}$$

In particular, $\beta_k(G) \leq \beta_k(G/N)\beta_k(N)$.

PROOF. Apply Theorem 3.6 to G/N acting on A^N. □

Now, we can state the following generalization of Schmid's result:

THEOREM 3.8. *Let G be a non-cyclic group with $|G| \in k^\times$. Then $\beta_k(G) < |G|$.*

PROOF. A close inspection shows that Schmid's proof works now (i.e., after having proved Corollary 3.5) for any $\mathbb{N}$-graded k-algebra. □

For completeness we state two more results without proof. In characteristic zero, Domokos and Hegedűs strengthened Schmid's theorem to:

THEOREM 3.9 ([**1**]). *Assume $k = \mathbb{C}$. Let G be a non-cyclic finite group. Then*

$$\beta_k(G) \leq \begin{cases} \frac{3}{4}|G| & \text{if } |G| \text{ is even} \\ \frac{5}{8}|G| & \text{if } |G| \text{ is odd.} \end{cases} \tag{3.8}$$

The only part of the proof which cannot be easily carried over to positive characteristic is Schmid's calculation of β_k for dihedral groups. This has been done by Sezer who thereby proved

THEOREM 3.10 ([**8**]). *Theorem 3.9 holds for any field k whose characteristic does not divide $|G|$.*

In analogy to $\beta_k(G)$ we can define $\eta_k(G)$ as $\sup_{A,\mathfrak{m}} \eta_G(\mathfrak{m})$ where A runs through all (not necessarily graded) k-algebras with G-action and $\mathfrak{m}$ through all G-invariant ideals of A. Then Corollary 2.2 says $\eta_k(G) \leq |G|$ whenever $|G| \in k^\times$. In fact, we have:

THEOREM 3.11. *For all finite groups G with $|G| \in k^\times$ holds $\eta_k(G) = \beta_k(G)$.*

PROOF. Lemma 3.1 shows $\beta_k(G) \leq \eta_k(G)$. For the converse inequality, let $\mathfrak{m}$ be a G-invariant ideal in the G-ring A. Let $\tilde{A}$ be the blow-up algebra of A along $\mathfrak{m}$, i.e., $\tilde{A}$ is the $\mathbb{N}$-graded subring $\oplus_{d=0}^{\infty}\mathfrak{m}^d t^d$ of $A[t]$. Let $\widetilde{\mathfrak{m}} := \tilde{A}_{\geq 1}$. Then Lemma 3.2 implies

$$\eta_G(\widetilde{\mathfrak{m}}) \leq \beta\big(\tilde{A}(G)^G\big) \leq \beta_k(G). \tag{3.9}$$

Let $\pi : \tilde{A} \to A$ be the composition of $\tilde{A} \hookrightarrow A[t] \twoheadrightarrow A$ where t maps to 1. By linear reductivity, we have $\pi(\langle\widetilde{\mathfrak{m}}^G\rangle) = \langle\mathfrak{m}^G\rangle$ and (obviously) $\pi(\widetilde{\mathfrak{m}}^d) = \mathfrak{m}^d$. This implies $\eta_G(\mathfrak{m}) \leq \eta_G(\widetilde{\mathfrak{m}})$, hence $\eta_k(G) \leq \beta_k(G)$. □

REMARK. If $\operatorname{char} k$ divides $|G|$ then Richman has shown $\beta_k(G) = \infty$ (see Theorem 4.3 below). It is feasible that this is also true for $\eta_k(G)$. Then one could remove the assumption $|G| \in k^\times$ in Theorem 3.11. As an example, we compute $\eta_k(G)$ where $G = \{1, \sigma\}$ is the group of order 2 and $\operatorname{char} k = 2$. Let $V = k^{2n}$ with coordinates $x_1, \dots, x_n, y_1, \dots, y_n$ and assume $\sigma(x_i) = y_i$. Let $A = k[V]$. Then A^G is generated by all $x_i y_i$, $1 \leq i \leq n$ and $x^\alpha + y^\alpha$, $\alpha \in \mathbb{N}^n$. In particular, $\mathfrak{m}^G$ contains $x_i + y_i$. That means $A/\langle\mathfrak{m}^G\rangle \cong k[x_1, \dots, x_n]/I$ where I is the ideal generated by all $-x_i^2$ and all $\big(1 + (-1)^{|\alpha|}\big)x^\alpha$ with $|\alpha| > 0$. In characteristic two, the second kind of elements are all zero. Therefore, $A/\langle\mathfrak{m}^G\rangle = k[x_1,\ldots,x_n]/(x_1^2,\ldots,x_n^2)$. Since $x_1\ldots x_n \notin I$ we have $\eta_G(\mathfrak{m}) = n+1$. Thus, $\eta_k(G) = \infty$.

4. Some functoriality properties

In this section, we develop some machinery to compute $\beta_k(A)$ for arbitrary ground rings k.

THEOREM 4.1. *Let $\varphi : k \to K$ be a ring homomorphism. Then $\beta_k(G) \geq \beta_K(G)$ with equality if φ is faithfully flat.*

PROOF. Let A be a graded K-algebra with G-action and $\beta(A) \leq 1$. By means of φ, we can consider A as a k-algebra. Since $\beta(A)$ and $\beta(A^G)$ depend only on the underlying ring structure, we get $\beta_k(G) \geq \beta_K(G)$.

Now assume that φ is faithfully flat and let A be a graded k-algebra with G-action and $\beta(A) \leq 1$. Put $B := A^G$, $\bar{A} := A \otimes_k K$ and $\overline{B} := B \otimes_k K$. Then $\overline{B} = \bar{A}^G$. Indeed, by flatness, the exact sequence

$$0 \to B \to A \xrightarrow{ga-a} \oplus_{g\in G} A \tag{4.1}$$

stays exact upon tensoring with K. Now we claim $\beta(\overline{B}) \geq \beta(B)$. For $b := \beta(\overline{B})$ and any $d \in \mathbb{N}$ let C be the cokernel of

$$\bigoplus_{\substack{\sum i_j=d\\ 0\leq i_j\leq b}} B_{i_1} \otimes \cdots \otimes B_{i_s} \to B_d. \tag{4.2}$$

This map becomes surjective after tensoring with K. Thus, $C \otimes_k K = 0$. Faithful flatness implies $C = 0$ and the claim follows.

We conclude $\beta(A^G) \leq \beta(\bar{A}^G) \leq \beta_K(G)$ for arbitrary A, hence $\beta_k(G) \leq \beta_K(G)$. □

COROLLARY 4.2. *Let k be a field. Then $\beta_k(A)$ depends only on the characteristic of k.*

In view of this result, we write $\beta_p(G) := \beta_k(G)$ where k is any field of characteristic p.

Next, we cite a result of Richman [**6**], which shows that in the modular case there is no universal bound on the degree of generators:

THEOREM 4.3. *Let k be a field whose characteristic divides $|G|$. Then $\beta(S_k^*(U)^G)$ is unbounded, where U runs through all finitely generated $k[G]$-modules. In particular, $\beta_k(G) = \infty$.*

This is being used in the proof of:

PROPOSITION 4.4. *Let k be an arbitrary ring. Then, to compute $\beta_k(G)$ it suffices to take the supremum over all $\beta\big(S_k^*(U)^G\big)$ where U is a finitely generated $k[G]$-module.*

PROOF. First assume that $|G|$ is not invertible in k. Then there is a prime divisor p of $|G|$ with $p \notin k^\times$, i.e., $pk \neq k$. Let $\mathfrak{m} \subset k$ be a maximal ideal containing pk and $K := k/\mathfrak{m}$. Then $\operatorname{char} K = p$ and the assertion follows from Theorem 4.3.

Thus, we may assume $|G| \in k^\times$. Let A be any graded k-algebra with G-action and $\beta(A) = 1$. Let $b := \beta(A^G)$. Then there is $f \in A_b^G$ which is not a polynomial of lower degree invariants. Clearly, we may assume $b \geq 1$. Hence, f is a polynomial of finitely many elements of A_1 with coefficients in k. Let U be the $k[G]$-module generated by these elements. Then we get an algebra homomorphism $S^*(U) \to A$ whose image contains f. Since G is linearly reductive, we can lift f to an invariant $\tilde{f} \in S_k^b(U)^G$. Clearly, we can't obtain $\tilde{f}$ as a polynomial of lower degree invariants, either. Hence $\beta(S^*(U)^G) \geq \beta(A^G)$ which shows the assertion. □

Now we can reduce the computation of $\beta_k(G)$ for an arbitrary ring k to fields. For this we define

$$\operatorname{char}(k) := \begin{cases} \{0\} & \text{if } \mathbb{Q} \subseteq k; \\ \{p \mid p \notin k^\times\} & \text{otherwise.} \end{cases} \tag{4.3}$$

THEOREM 4.5. *Let k be any ring. Then*

$$\beta_k(G) = \sup\{\beta_p(G) \mid p \in \operatorname{char}(k)\}. \tag{4.4}$$

PROOF. If $\mathbb{Q} \subseteq k$ then k is faithfully flat over $\mathbb{Q}$ and the assertion follows from Theorem 4.1. Thus, assume $\mathbb{Q} \not\subseteq k$. If p is not invertible in k then there is a maximal ideal $\mathfrak{m}$ such that $k/\mathfrak{m}$ has characteristic p. Thus, $\beta_k(G)$ is at least as big as the left hand side of (4.4). In particular, we are done if one of the $\beta_p(G)$ is ∞. Thus we may assume that $|G|$ is invertible in k.

Let A be a $k[G]$-algebra which, by Proposition 4.4, we may assume to be of the form $A = S_k^*(U)$ where U is a finitely generated $k[G]$-module. This implies that each homogeneous component A_d is a finitely generated k-module. Being a direct summand the same holds for A_d^G.

Let $\mathfrak{m} \subset k$ be a maximal ideal. Then we claim that

$$A^G \otimes_k k/\mathfrak{m} \to (A \otimes_k k/\mathfrak{m})^G \tag{4.5}$$

is an isomorphism. Indeed, injectivity follows from the fact that A^G is a direct summand of A. Moreover, $A^G \to (A/\mathfrak{m}A)^G = (A \otimes_k k/\mathfrak{m})^G$ is surjective since $|G|$ is invertible in k.

Let b be the right hand side of (4.4) and let $C(b,d)$ be the cokernel of (4.2). Then $C(b,d) \otimes_k k/\mathfrak{m} = 0$ for every maximal ideal. Since $C(b,d)$ is finitely generated this implies $C(b,d) = 0$ (Nakayama Lemma). Thus $\beta(A^G) \leq b$ and therefore $\beta_k(G) \leq b$. □

COROLLARY 4.6. *For any ring k holds $\beta_k(G) < \infty$ if and only if $|G| \in k^\times$.*

A similar argument shows that $k = \mathbb{Q}$ is the best possible case:

THEOREM 4.7. *For all primes p holds $\beta_p(G) \geq \beta_0(G)$. For almost all primes holds equality.*

PROOF. If p divides $|G|$ there is nothing to prove since then $\beta_p(G) = \infty$ (Theorem 4.3). Thus assume $p \nmid |G|$. Let $k = \mathbb{Z}_p$ be the ring of p-adic integers with field of fractions $K = \mathbb{Q}_p$ and residue field $\mathbb{F}_p$. By Proposition 4.4, there is a finite dimensional $K[G]$-module U_K with $\beta(S_K^*(U_K)^G) = \beta_0(G) =: b_0$. Let $U_k \subseteq U_K$ be a G-stable lattice and $A := S_k^*(U_k)$. Let $\bar{A} := A \otimes_k \mathbb{F}_p$. Since $\bar{A}$ is an $\mathbb{F}_p$-algebra, it suffices to prove $\beta(\bar{A}^G) \geq b_0$.

The same argument as for (4.5) shows that

$$A^G \otimes_k \mathbb{F}_p \to (A \otimes_k \mathbb{F}_p)^G \tag{4.6}$$

is an isomorphism. Now let $B := A^G$ and consider for any b again the cokernel $C(b,d)$ of the map (4.2). For any $b < b_0$ there is a $d > 0$ such that $C(b,d) \otimes_k K \neq 0$.

But $C(b,d)$ is a finitely generated k-module which implies $C(b,d) \otimes_k \mathbb{F}_p \neq 0$. This implies that $\beta(\bar{A}^G)$ can not be smaller than b.

Now we show that conversely $\beta_p(G) \leq \beta_0(G)$ for almost all primes. Let $k(G)$ be the symmetric algebra over the regular k-representation. Let $f_1, \dots, f_r$ be a minimal set of generators of $\mathbb{Q}(G)^G$. Let $h_1, \dots, h_s$ be generators of $\mathbb{Z}[1/g](G)^G$ where $g := |G|$. Then there is a multiple N of g such that the f_i are defined over $k := \mathbb{Z}[1/N]$ and such that all h_j are in the k-algebra generated by the f_i. This implies that $k(G)^G$ is also generated by the f_i. The homomorphism $k(G)^G \to \mathbb{F}_p(G)^G$ is surjective where p is any prime not dividing N. This implies $\beta(\mathbb{F}_p(G)^G) \leq \beta(\mathbb{Q}(G)^G)$. We conclude with theorems of Schmid [**7**] and Smith [**9**] (see also its generalization, Corollary 7.5, further down) which assert $\beta(\mathbb{Q}(G)^G) = \beta_0(G)$ and $\beta(\mathbb{F}_p(G)^G) = \beta_p(G)$ for $p > |G|$. □

REMARK. Presently, no group G and prime p not dividing $|G|$ with $\beta_p(G) > \beta_0(G)$ seems to be known.

5. Polarization in positive characteristic

From now on, k will be a field which, for convenience, we assume to be infinite. We set $p = \operatorname{char} k$ if positive and $p = \infty$ otherwise. Let V be a k-vector space of dimension $\ell < \infty$. Let $k[V]_d = S^d(V^*)$ be the space of polynomial functions which are homogeneous of degree d. We are interested in the space $V^{\oplus n}$ of n-tuples of vectors in V.

There is a natural action of the algebraic group $\mathrm{GL}(n)$ on $V^{\oplus n} = V \otimes_k k^n$. Let S be a subset of $k[V^{\oplus n}]$. Then the GL_n-submodule $\langle S \rangle_{\mathrm{GL}(n)}$ generated by S is called the *polarization* of S. The action of the Lie algebra is given by the *polarization operators*

$$P_{jj'} := \sum_{i=1}^{\ell} x_{ij} \frac{\partial}{\partial x_{ij'}}, \quad 1 \leq j,\ j' \leq n \tag{5.1}$$

where $x_{1j}, \dots, x_{\ell j}$ are coordinates of the jth copy of V in $V^{\oplus n}$. If $m \leq n$ then we consider $V^{\oplus m}$ as a quotient of $V^{\oplus n}$ by forgetting the last $n-m$ components. In characteristic zero, it follows from work by Weyl ([**10**] II.5) that every function on V^n with $n \geq \ell := \dim V$ can be obtained by polarization from functions on V^ℓ. In positive characteristic this is true in sufficiently small degrees.

THEOREM 5.1. *Let $n \geq m \geq \ell := \dim V$ and $d \leq (p-1)m$. Then*

$$k[V^{\oplus n}]_d = \langle k[V^{\oplus m}]_d \rangle_{\mathrm{GL}(n)}. \tag{5.2}$$

PROOF. First, by replacing V with $V \oplus k^{m-\ell}$ we may assume that $m = \ell$. Moreover, it clearly suffices to treat the case $n = m+1 = \ell+1$. The proof will proceed by induction on ℓ starting at the trivial case $\ell = 0$.

We label the copies of $V^{\oplus \ell+1}$ by $j = 0, \dots, \ell$ such that $V^{\oplus \ell}$ corresponds to $j = 1, \dots, \ell$. The coordinates of the jth copy are denoted by $x_{1j}, \dots, x_{\ell j}$. A monomial in the x_{ij} can be represented by an $\mathbb{N}$-valued matrix

$$A = \begin{pmatrix} a_{10} & a_{11} & \dots & a_{1\ell} \\ \vdots & \vdots & \iddots & \vdots \\ a_{\ell 0} & a_{\ell 1} & \dots & a_{\ell\ell} \end{pmatrix} \tag{5.3}$$

where $x^A := \prod_{ij} x_{ij}^{a_{ij}}$. Its degree is $\deg A := \sum_{ij} a_{ij}$.

We introduce a total order on the set of matrices A as follows. Let $r_i = r_i(A) = \sum_j a_{ij}$ and $c_j = c_j(A) = \sum_i a_{ij}$ be the row and column sums, respectively. Then we define the *index* of A as the vector

$$\operatorname{ind} A = (c_0, \ldots, c_\ell, r_1, \ldots, r_\ell, a_{10}, \ldots, a_{1\ell}, a_{20}, \ldots, a_{2\ell}, \ldots, a_{\ell 0}, \ldots, a_{\ell\ell}). \tag{5.4}$$

In words: $\operatorname{ind} A$ starts with the column sums followed by the row sums followed by the coefficients read from left to right, top to bottom. We say $A < B$ if $\operatorname{ind} A$ is lexicographically smaller that $\operatorname{ind} B$. We have to show that $\deg A = d \le (p-1)\ell$ implies $x^A \in X(\ell, d) := \langle k[V^{\oplus \ell}]_d \rangle_{\mathrm{GL}(\ell+1)}$. Let A be the smallest counterexample.

First, observe that $X(\ell, d)$ is invariant under both $\mathrm{GL}(\ell)$ (acting on V) and $\mathrm{GL}(\ell+1)$ (by definition). Thus, the property $x^A \in X(\ell, d)$ is invariant under permutations of the rows or the columns of A. The minimality of A implies

$$c_0(A) \le \cdots \le c_\ell(A) \quad \text{and} \tag{5.5}$$

$$r_1(A) \le \cdots \le r_\ell(A). \tag{5.6}$$

From $d = \sum_i r_i(A) \le (p-1)\ell$ we obtain

$$r_1(A) \le p - 1. \tag{5.7}$$

The matrix A must have a non-zero entry in the first column since otherwise $x^A \in k[V^{\oplus \ell}]_d \subseteq X(\ell, d)$. A fortiori, there is non-zero entry a_{ij} with $i + j \le \ell$ (i.e., which is strictly above the dotted diagonal in (5.3)). Let a_{ij} be the first one, where we read from left to right and top to bottom. Then a_{ij} is not in the last column and the entries above a_{ij+1} are all zero.

Let $\bar{A}$ be the matrix obtained from A by replacing the entries a_{ij} and a_{ij+1} by $a_{ij} - 1$ and $a_{ij+1} + 1$, respectively. Since $c_j(\bar{A}) < c_j(A)$, the minimality of A implies $x^{\bar{A}} \in X(\ell, d)$. Now apply the polarization operator P_{jj+1} to $x^{\bar{A}}$. Then we get

$$P_{jj+1}(x^{\bar{A}}) = (a_{ij+1} + 1)x^A + a_{i+1j+1}x^{A_{i+1}} + \cdots + a_{\ell j+1}x^{A_\ell} \in X(\ell, d). \tag{5.8}$$

Here, A_ν is the matrix

$$\begin{pmatrix} & \vdots & \vdots & \\ \cdots & a_{ij} - 1 & a_{ij+1} + 1 & \cdots \\ & \vdots & \vdots & \\ \cdots & a_{\nu j} + 1 & a_{\nu j+1} - 1 & \cdots \\ & \vdots & \vdots & \end{pmatrix}. \tag{5.9}$$

Since all A_ν are smaller than A we obtain[1] $a_{ij+1} + 1 = 0$ in k. In particular, we get

$$a_{ij} \ge 1 \quad \text{and} \quad a_{ij+1} \ge p - 1. \tag{5.10}$$

[1]In characteristic 0, we are done at this point: A can't exist.

Let $\ell' := \ell + 1 - i \geq j + 1$. Then $a_{i\ell'}$ is on the diagonal starting from $a_{1\ell}$:

$$A = \begin{pmatrix} 0 & \cdots & \cdots & \cdots & \cdots & 0 & \cdots & a_{1\ell} \\ \vdots & & & & & \vdots & \iddots & \vdots \\ 0 & \cdots & a_{ij} & a_{ij+1} & \cdots & a_{i\ell'} & & \vdots \\ \vdots & & & & \iddots & \vdots & & \vdots \\ \vdots & & & \iddots & & \vdots & & \vdots \\ \vdots & & \iddots & & & \vdots & & \vdots \\ a_{\ell 0} & a_{\ell 1} & \cdots & \cdots & \cdots & a_{\ell\ell'} & \cdots & a_{\ell\ell} \end{pmatrix}. \tag{5.11}$$

Let A' be the submatrix of A consisting of columns $0, \ldots, \ell'$. From (5.10) we get $r_i \geq p$ which implies $\ell' < \ell$ by (5.7). Moreover we have $p - 1 \leq c_{j+1} \leq c_{\ell'+1} \leq \ldots \leq c_\ell$. Hence

$$\deg A' = c_0 + \cdots + c_{\ell'} \leq (p-1)\ell - (p-1)(\ell - \ell') = (p-1)\ell'. \tag{5.12}$$

Now observe that the first $i - 1 = \ell - \ell'$ rows of A' are zero. Hence, the monomial $x^{A'}$ "lives" on $(V')^{\oplus \ell'+1}$ where $\dim V' = \ell'$. Thus we can use induction and conclude that $x^{A'}$ can be obtained by polarization from $(V')^{\oplus \ell'}$. But then the same polarization process produces x^A from $V^{\oplus \ell'}$ in contradiction to the choice of A. $\square$

REMARKS.

1. The proof shows that under the given conditions it suffices to apply just polarization operators and column permutations.
2. One can show that the monomial $x_{10}(x_{11} \ldots x_{1\ell})^{p-1}$ cannot be obtained from polarization of polynomials on fewer than $\ell + 1$ copies of V. This shows that the given degree bound is optimal.

6. Weyl's theorem

Let k and p be as in the preceding section. We are going to apply polarizations to invariant theory. Let U and V be two finite dimensional representations of a finite group G. Then the $\mathrm{GL}(n)$-action on $U \oplus V^{\oplus n}$ commutes with G, hence we get an $\mathrm{GL}(n)$-action and a notion of polarization on $k[U \oplus V^{\oplus n}]^G$. If $m \leq n$ we can restrict invariants from $U \oplus V^{\oplus n}$ to $U \oplus V^{\oplus m}$. This process is kind of inverse to polarization and is called *restitution*.

THEOREM 6.1. *Let S be a generating set of G-invariants on $U \oplus V^{\oplus n}$ where $n \geq \max\big(\dim V, \beta_k(G)/(p-1)\big)$. Then the polarization of S generates the ring of invariants on $U \oplus V^{\oplus m}$ for any $m \geq n$.*

PROOF. First, let us remark that G is linearly reductive. Indeed, otherwise $\beta_k(G) = \infty$ and m would not exist.

Let $\mathcal{P}_m \subseteq \operatorname{End} k[U \oplus V^{\oplus m}]$ be the subalgebra generated by all polarization and permutation operators. Let $A(m)$ be the algebra generated by $\mathcal{P}_m \cdot S \subseteq k[U \oplus V^{\oplus m}]$. Since $\mathcal{P}_m$ commutes with G we have $A(m) \subseteq k[U \oplus V^{\oplus m}]^G$. We have to show equality.

The formulas $P(fg) = P(f)g + fP(g)$ for any polarization operator and $\pi(fg) = \pi(f)\pi(g)$ for any permutation imply that

$$\mathcal{P}_m(fg) \subseteq \mathcal{P}_m(f)\mathcal{P}_m(g) \tag{6.1}$$

for all f, g. This shows that $A(m)$ is a $\mathcal{P}_m$-module. Let $d \leq \beta_k(G)$. Then the assumption on n and Theorem 5.1 imply that the map

$$\mathcal{P}_m \otimes k[U \oplus V^{\oplus n}]_d \to k[U \oplus V^{\oplus m}]_d \tag{6.2}$$

is surjective. By linear reductivity of G we obtain

$$\mathcal{P}_m \cdot k[U \oplus V^{\oplus n}]^G_d = k[U \oplus V^{\oplus m}]^G_d. \tag{6.3}$$

On the other hand,

$$\mathcal{P}_m \cdot k[U \oplus V^{\oplus n}]^G_d = \mathcal{P}_m \cdot k[S]_d \subseteq \mathcal{P}_m A(m)_d = A(m)_d. \tag{6.4}$$

Thus, we have proved that $A(m)$ contains all G-invariants of degree $\leq \beta_k(G)$ which implies that it consists of all invariants. □

This theorem provides a means to construct a representation of G which has the "most general" ring of invariants. To make this more precise, we extend our notion of polarization. Let U be a G-module. For any subset S of $k[U]$ let $\mathrm{Pol}^U(S)$ be the $\mathrm{Aut}^G(U)$-module generated by it. Observe, that if $|G| \in k^\times$ then $\mathrm{Aut}^G(U)$ is just a product of general linear groups corresponding to the isotypic components of U. There are two more operations: If $V \subseteq U$ then $\mathrm{Res}^U_V(S)$ be the set of restrictions of elements of S to V. Conversely, if U is a quotient of V let $\mathrm{Ext}^V_U(S)$ be the set of pullbacks. Now, we define

DEFINITION. A G-module U has *universal invariants* if for every finite dimensional G-module V the set

$$\mathrm{Res}_V \mathrm{Pol}^{U\oplus V} \mathrm{Ext}^{U\oplus V}_U(S) \tag{6.5}$$

generates $k[V]^G$ where S is any generating set of $k[U]^G$.

In other words, if U has universal invariants then one obtains a generating set of invariants for any other G-module by the following process:

1. Start with generators for $k[U]^G$.
2. Think of them as invariants on $U \oplus V$.
3. Apply $\mathrm{Aut}^G(U \oplus V)$ to S.
4. Restrict the ensuing invariants to V.

One of the main properties of a module with universal invariants is:

THEOREM 6.2. *Assume U has universal invariants. Then $\beta_k(G) = \beta\big(k[U]^G\big)$.*

PROOF. First, $\beta_k(G)$ is the supremum of the $\beta(k[V]^G$ where V runs through all finite dimensional G-modules (Proposition 4.4). Then the assertion follows from the fact that the process of extension, polarization, and restriction does not change degrees. □

Our main criterion for having universal invariants is:

THEOREM 6.3. *Assume $|G| \in k^\times$. Let U be a representation of G in which every irreducible module M appears with multiplicity at least* $\max\big(\dim M, \beta_k(G)/(p-1)\big)$. *Then U has universal invariants.*

PROOF. Let S be a generating set of $k[U]^G$ and let V be a finite dimensional G-module. We show by induction on the number of isotypic components of V the more general statement that $\mathrm{Aut}^G(U \oplus V) \cdot S$ generates $k[U \oplus V]^G$. We start with $V = 0$ where the assertion is trivial.

Assume that the simple module M appears in V. Write $U = U' \oplus M^{\oplus m}$ and $V = V' \oplus M^{\oplus n}$ such that M is not contained in U' and V'. The set $S' := \mathrm{Aut}^G(U \oplus V') \cdot S$ generates $k[U \oplus V']^G$ by the induction hypothesis. We have $U \oplus V' = (U' \oplus V') \oplus M^{\oplus m}$ while $U \oplus V = (U' \oplus V') \oplus M^{\oplus m+n}$. Theorem 6.1 implies that $k[U \oplus V]^G$ is generated by $\mathrm{Gl}(m+n) \cdot S' = \mathrm{Gl}(m+n) \cdot \mathrm{Aut}^G(U \oplus V') \cdot S = \mathrm{Aut}^G(U \oplus V) \cdot S$. □

COROLLARY 6.4. *Assume* $|G| \in k^\times$. *Then* G *has a finite dimensional representation with universal invariants. More precisely, assume that* U *contains every irreducible module with multiplicity at least* $|G|$. *Then* U *has universal invariants.*

PROOF. Let M be an irreducible representation of G. Since M is a quotient of the regular representation, we have $\dim M \leq |G|$. On the other hand, $\beta_k(G)/(p-1) \leq |G|$ by Corollary 3.5. □

COROLLARY 6.5. *Assume* $p > \beta_k(G)$. *Then the regular representation of* G *has universal invariants.*

PROOF. We may assume that k is algebraically closed. Then the regular representation contains each simple module W with multiplicity $\dim W$. The assertion follows from $\beta_k(G)/(p-1) \leq 1 \leq \dim W$. □

7. A further improvement

One can try to obtain better results by generalizing the polarization process. The point is that in positive characteristic, the commutator algebra of $\mathrm{GL}(V)$ in $\mathrm{End}(k[V^{\oplus m}])$ is not generated by the image of $\mathrm{GL}(m)$ (not even in a topological sense).

In general, one can replace $\mathcal{P}_m$ by any subalgebra $\overline{\mathcal{P}}_m$ of endomorphisms of $k[V^{\oplus m}]$ which commutes with the $\mathrm{GL}(V)$-action and which satisfies the multiplicativity property (6.1). We will not pursue this direction in any detail but illustrate it by the following easy fact:

THEOREM 7.1. *Assume* $\dim V = 1$ *and* $\overline{\mathcal{P}}_m := \mathrm{End}^{\mathrm{GL}(V)}(k[V^{\oplus m}])$. *Then for any* $d \geq 0$ *holds*

$$k[V^{\oplus m}]_d = \langle k[V]_d \rangle_{\overline{\mathcal{P}}_m}. \tag{7.1}$$

PROOF. The group $\mathrm{GL}(V) \cong \mathrm{GL}(1)$ acts on $W_d := k[V^{\oplus m}]_d$ by the character $t \mapsto t^d$. Hence, $\overline{\mathcal{P}}_m = \prod_{d=0}^{\infty} \mathrm{End}_k W_d$. This implies that any non-zero element generates W_d as a $\overline{\mathcal{P}}$-module. □

REMARK. This theorem also serves as an example that $\overline{\mathcal{P}}_m$ may be bigger than $\mathcal{P}_m$. In fact, $k[V^{\oplus m}]_d$ is a simple $\overline{\mathcal{P}}_m$-module but, in general, it is not a simple $\mathrm{GL}(m)$-module.

LEMMA 7.2. *Let* $\overline{\mathcal{P}}_m$ *be as in Theorem* 7.1. *Then for any* $f, g \in k[V^{\oplus m}]$ *holds* $\overline{\mathcal{P}}_m(fg) \subseteq \overline{\mathcal{P}}_m(f)\overline{\mathcal{P}}_m(g)$.

PROOF. Let $f = \sum_d f_d$ be the decomposition into homogeneous components. Since $\overline{\mathcal{P}}_m$ contains the projection onto $k[V^{\oplus m}]_d$ we have $f_d \in \overline{\mathcal{P}}_m(f)$. Hence, we may assume that f and g are homogeneous of degree, say, d and e, respectively. But then $\overline{\mathcal{P}}_m(f) = k[V^{\oplus m}]_d$ and $\overline{\mathcal{P}}_m(g) = k[V^{\oplus m}]_e$, hence $\overline{\mathcal{P}}_m(f)\overline{\mathcal{P}}_m(g) = k[V^{\oplus m}]_{d+e} \supseteq \overline{\mathcal{P}}_m(fg)$. □

Now assume $|G| \in k^\times$. For an arbitrary G-module V we generalize the polarization process as follows. Let $V = \oplus_i M_i^{m_i}$ be the isotypic decomposition of V. Let $\overline{\mathcal{P}}_V \subseteq \operatorname{End} k[V]$ be the subalgebra which is generated by $\mathrm{GL}(m_i)$ if $\dim M_i > 1$ and by $\overline{\mathcal{P}}_{m_i}$ if $\dim M_i = 1$. For a subset $S \subseteq k[V]$ let $\overline{\mathrm{Pol}}^V(S) := \overline{\mathcal{P}}_V(S)$.

We say a module U has *weakly universal invariants* if $k[V]^G$ is generated by

$$\operatorname{Res}_V \overline{\mathrm{Pol}}^{U\oplus V} \operatorname{Ext}_U^{U\oplus V}(S), \tag{7.2}$$

where V is any finite dimensional G-module and S is any generating set of $k[U]^G$. Then we get the following analogue of Theorem 6.3.

THEOREM 7.3. *Assume $|G| \in k^\times$. Let U be a representation of G in which every one-dimensional module appears at least once and every other irreducible module M appears with multiplicity at least* $\max(\dim M, (\beta_k(G))/(p-1))$. *Then U has weakly universal invariants.*

The analogue of Corollary 6.5 is

COROLLARY 7.4. *Assume k is algebraically closed. Let*

$$\ell := \inf\{\dim W \mid W \textit{ simple } G\textit{-module}, \dim W > 1\} \in [2, \infty]. \tag{7.3}$$

Assume $p \geq \beta_k(G)/\ell + 1$. Then the regular representation R of G has weakly universal invariants. In particular, $\beta_k(G) = \beta(k[R]^G)$.

If we combine this result with the bound of Domokos–Hegedűs–Sezer we obtain the following strengthening of theorems of Schmid [**7**], and Smith [**9**].

COROLLARY 7.5. *Assume $|G| \in k^\times$ and $p \geq 3/8\ |G| + 1$. Then $\beta_k(G) = \beta(k[R]^G)$.*

PROOF. If G is cyclic then $\ell = \infty$. Otherwise $\ell \geq 2$ and $\beta_k(G) \leq 3/4\ |G|$ by Theorem 3.10. □

REMARK. The condition in [**7**] is $\operatorname{char} k = 0$. In [**9**] it is $p > |G|$.

References

1. M. Domokos and P. Hegedűs, *Noether's bound for polynomial invariants of finite groups*, Arch. Math. **74** (2000), 161–167.
2. P. Fleischmann, *The Noether bound in invariant theory of finite groups*, Adv. Math. **156** (2000), 23–32.
3. ———, *On invariant theory of finite groups*, this volume.
4. J. Fogarty, *On Noether's bound for polynomial invariants of a finite group*, Electron. Res. Announc. Amer. Math. Soc. **7** (2001), 5–7 (electronic).
5. E. Noether, *Der Endlichkeitssatz der Invarianten endlicher Gruppen*, Math. Ann. **77** (1916), 89–92.
6. D. Richman, *Invariants of finite groups over fields of characteristic p*, Adv. Math. **124** (1996), 25–48.
7. B. Schmid, *Finite groups and invariant theory*, Topics in Invariant Theory (Paris, 1989–1990) (M.-P. Malliavin, ed.), Lecture Notes in Mathematics, vol. 1478, Springer-Verlag, Berlin, 1991, pp. 35–66.
8. M. Sezer, *Sharpening the generalized Noether bound in the invariant theory of finite groups*, J. Algebra **254** (2002), 252–263.
9. L. Smith, *On a theorem of Barbara Schmid*, Proc. Amer. Math. Soc. **128** (2000), no. 8, 2199–2201.
10. H. Weyl, *The classical groups, their invariants and representations*, Princeton Mathematical Series, vol. 1, Princeton University Press, Princeton, 1939.

Department of Mathematics, Rutgers University, 110 Frelinghuysen Road, Piscataway NJ 08854-8019, USA.
E-mail address: knop@math.rutgers.edu

Centre de Recherches Mathématiques
CRM Proceedings and Lecture Notes
Volume **35**, 2004

Comparing the Depths of Rings of Invariants

M. D. Neusel

ABSTRACT. We consider linear representations $G \hookrightarrow \mathrm{GL}(n, \mathbb{F})$ of finite groups G such that the vector space $\mathbb{F}^n = V = W \oplus W'$ has a direct sum decomposition as $\mathbb{F}G$-module. We compare the Cohen–Macaulay defect of the rings of polynomial invariants $\mathbb{F}[V]^G$ and $\mathbb{F}[W]^G$.

1. Introduction

Let $\mathbb{F}$ be a field, $V = \mathbb{F}^n$ the n-dimensional vector space over $\mathbb{F}$. Let

$$\rho : G \hookrightarrow \mathrm{GL}(n, \mathbb{F})$$

be a faithful representation of a finite group G. Assume that V decomposes into a direct sum

$$V = W \oplus W' \tag{1.1}$$

as $\mathbb{F}G$-modules. Set $\dim(W) = k$ and $\dim(W') = k'$ with $k + k' = n$.

Let $\mathbb{F}[V]$ be the symmetric algebra on the dual space $V^{\#}$. Denote by

$$\mathrm{pr} : \mathbb{F}[V] \to \mathbb{F}[W]$$

the canonical projection mapping the last k' variables to zero. Since V has a direct sum decomposition (1.1), the canonical inclusion $\mathbb{F}[W] \hookrightarrow \mathbb{F}[V]$ hands us the inclusion

$$\mathbb{F}[W]^G \hookrightarrow \mathbb{F}[V]^G.$$

We want to compare the depths of these rings of invariants. For that we choose polynomials

$$f_1, \dots, f_s \in \mathbb{F}[W]^G$$

such that the ring of invariants $\mathbb{F}[W]^G$ is finite over the subring generated by the f_i's, $i = 1, \dots, s$. We extend this set of polynomials to the set

$$f_1, \dots, f_s, \quad f_{s+1}, \dots, f_r \in \mathbb{F}[V]^G$$

such that $\mathbb{F}[V]^G$ is finite over the subring generated by $f_1, \dots, f_r$. Denote by

$$S := \mathbb{F}[F_1, \dots, F_s] \hookrightarrow R := \mathbb{F}[F_1, \dots, F_r]$$

2000 *Mathematics Subject Classification.* 13A50.
This is the final version of the paper.

polynomial rings over $\mathbb{F}$. Then S, resp. R, acts on $\mathbb{F}[W]^G$, resp. $\mathbb{F}[V]^G$, by the remembering map.

PROPOSITION 1.1. *With the above notation*

$$\operatorname{projdim}_R(\mathbb{F}[V]^G) \geq \operatorname{projdim}_S(\mathbb{F}[W]^G),$$

where $\operatorname{projdim}_R(--)$ *denotes the projective dimension of the R-module* $--$.

PROOF. Since R is free over S (see, e.g. [**5**, Lemma 4.3.13]), the projective dimension $\operatorname{projdim}_S(R) = 0$. Thus we find

$$\operatorname{projdim}_R(\mathbb{F}[V]^G) \overset{(1)}{\geq} \operatorname{projdim}_S(\mathbb{F}[V]^G) - \operatorname{projdim}_S(R) = \operatorname{projdim}_S(\mathbb{F}[V]^G),$$

where (1) is true for any change of rings $S \hookrightarrow R$ (see, e.g. [**5**, Theorem 4.3.1]).

The canonical projection pr gives us a projection $\mathbb{F}[V]^G \to \mathbb{F}[W]^G$ of S-modules with kernel K. We obtain a short exact sequence of S-modules

$$0 \to K \to \mathbb{F}[V]^G \to \mathbb{F}[W]^G \to 0$$

with a splitting provided by the canonical inclusion $\mathbb{F}[W]^G \hookrightarrow \mathbb{F}[V]^G$. Thus (see, e.g.,[**5**, Exercise 4.1.2])

$$\begin{aligned}\operatorname{projdim}_R(\mathbb{F}[V]^G) &\geq \operatorname{projdim}_S(\mathbb{F}[V]^G)\\ &= \max\big(\operatorname{projdim}_S(\mathbb{F}[W]^G), \operatorname{projdim}_S(K)\big) \geq \operatorname{projdim}_S(\mathbb{F}[W]^G)\end{aligned}$$

as desired. □

Since S, R are polynomial rings over the field $\mathbb{F}$ their global dimensions are finite by Hilbert's syzygy theorem (see, e.g. [**5**, Corollary 4.3.8]). Thus we can apply the Auslander–Buchsbaum formula (see, e.g. [**5**, Theorem 4.4.15]) and translate the result of the preceding proposition into a statement about the depths involved:

$$\operatorname{depth}(R) - \operatorname{depth}_R(\mathbb{F}[V]^G) \geq \operatorname{depth}(S) - \operatorname{depth}_S(\mathbb{F}[W]^G),$$

where $\operatorname{depth}_R(--)$ denotes the depth of the R-module $--$ and $\operatorname{depth}(--)$ denotes the depth of the ring $--$.

Since $\mathbb{F}[V]^G$ is finite over R and $\mathbb{F}[W]^G$ is finite over S, we find (see, e.g. [**1**, Exercise 1.2.26])

$$r - \operatorname{depth}(\mathbb{F}[V]^G) \geq s - \operatorname{depth}(\mathbb{F}[W]^G). \tag{1.2}$$

Indeed we have the following result.

THEOREM 1.2. *With the above notation*

$$\dim(W') \geq \operatorname{depth}(\mathbb{F}[V]^G) - \operatorname{depth}(\mathbb{F}[W]^G).$$

In particular, $\mathbb{F}[W]^G$ *is Cohen–Macaulay whenever* $\mathbb{F}[V]^G$ *is.*

PROOF. Choose the polynomials $f_1, \ldots, f_s$ to be a system of parameters for $\mathbb{F}[W]^G$, i.e., $s = k$. Extend this to a system of parameters $f_1, \ldots, f_r$ for $\mathbb{F}[V]^G$, i.e., $r = n$. This is possible, because V has a direct sum decomposition (1.1). Then inequality (1.2) gives

$$n - \operatorname{depth}(\mathbb{F}[V]^G) \geq k - \operatorname{depth}(\mathbb{F}[W]^G). \tag{1.3}$$

Since $\dim(W') = k' = n - k$ we are done. □

REMARK. Denote by trzdeg the transcendence degree. Then

$$r - s \geq \operatorname{trzdeg}(\mathbb{F}[V]^G || \mathbb{F}[W]^G) = n - k = \dim(W').$$

Thus Theorem 1.2 gives the strongest version of inequality (1.2).

REMARK. The result of Theorem 1.2 is sharp: if the characteristic of $\mathbb{F}$ is prime to the group order $|G|$, then we get from Theorem 1.2

$$k' \geq k',$$

since rings of nonmodular invariants are always Cohen–Macaulay, see [**4**, Theorem 5.5.2].

Theorem 1.2 is particularly useful when dealing with vector invariants as we illustrate next.

EXAMPLE 1.3. Let $\rho : G \hookrightarrow \mathrm{GL}(n, \mathbb{F})$ be a modular faithful representation of some finite group G. We denote by ρ_i the i-fold vector representation afforded by block matrices of the form

$$\begin{pmatrix} \rho(g) & 0 & \dots & 0 \\ 0 & \rho(g) & \dots & 0 \\ \vdots & & \ddots & \\ 0 & \dots & 0 & \rho(g) \end{pmatrix} \quad \text{for } g \in G.$$

It is known that the ring of invariants of ρ_i is not Cohen–Macaulay for some large enough i no matter how "good" the ring of invariants of $\rho = \rho_1$ is, see [**2**] for the case of p-groups and [**3**] for the general case. "Good" means in this context how close the ring is to being Cohen–Macaulay, i.e., how small the Cohen–Macaulay defect (i.e., dim $-$depth) is.

The inequality (1.3) shows that the transition from ρ to any ρ_i, or more generally from ρ_i to ρ_j for $j > i$, cannot improve the situation:

Take $V = (\mathbb{F}^n)^j$ and $W = (\mathbb{F}^n)^i$

$$jn - \operatorname{depth}(\mathbb{F}[V]^G) \geq in - \operatorname{depth}(\mathbb{F}[W]^G),$$

i.e., the Cohen–Macaulay defect can never decrease.

Acknowledgement

This work was done during my stay at the École Polytechnique Fédérale de Lausanne during June 2002. I thank Eva Bayer for the invitation, Ragnar Buchweitz, Elisa Gorla, and Gérard Maze for good discussions and Irena Swanson for an interesting question that led to this paper. Thanks also to the referee for useful comments.

References

1. W. Bruns and J. Herzog, *Cohen–Macaulay rings*, Cambridge Stud. Adv. Math., vol. 39, Cambridge University Press, Cambridge, 1993.
2. H. E. A. Campbell, A. V. Geramita, I. P. Hughes, R. J. Shank, and D. L. Wehlau, *Non-Cohen–Macaulay vector invariants and a Noether bound for a Gorenstein rings of invariants*, Canad. Math. Bull. **42** (1999), 155–161.
3. G. Kemper, *On the Cohen–Macaulay property of modularinvariant rings*, J. Algebra **215** (1999), 330–351.
4. M. D. Neusel and L. Smith, *Invariant theory of finite groups*, Math. Surveys Monogr., vol. 94, Amer. Math. Soc., Providence, RI, 2002.

5. C. Weibel, *An introduction to homological algebra*, Cambridge Stud. Adv. Math., vol. 38, Cambridge University Press, Cambridge, 1994.

Department of Mathematics and Statistics, Texas Tech. University, Mail Stop 1042, Lubbock, TX 79409-1042, USA.

E-mail address: mneusel@math.ttu.edu

Centre de Recherches Mathématiques
CRM Proceedings and Lecture Notes
Volume **35**, 2004

Moment Polytopes of Nilpotent Orbit Closures; Dimension and Isomorphism of Simple Modules; and Variations on the Theme of J. Chipalkatti

V. L. Popov

Notation

- k an algebraically closed field of characteristic zero.
- G a connected reductive algebraic group over k.
- T a maximal torus of G.
- B a Borel subgroup of G containing T.
- U the unipotent radical of B.
- X the lattice of characters of T in additive notation.
- $R_+ \subset \mathrm{X}$ the system of positive roots of G defined by B.
- $E := \mathrm{X} \otimes_{\mathbb{Z}} \mathbb{Q}$.
- E_+ the Weyl chamber in E defined by B.
- X_{++} the monoid of dominant weights of G defined by T and B.
- $V(\lambda)$ a simple G-module over k with the highest weight $\lambda \in \mathrm{X} \cap \mathrm{P}_{++}$.
- $G \times V(\lambda) \to V(\lambda)$, $(g, v) \mapsto g \cdot v$, the G-module structure on $V(\lambda)$.
- λ^* the highest weight of the G-module $V(\lambda)^*$ dual to $V(\lambda)$.
- $m(\lambda, M)$ the multiplicity of $V(\lambda)$ in an algebraic G-module M.

1. Describing moment polytopes of nilpotent orbit closures

Let $G = \mathrm{SL}_n$. The Lie algebra $\operatorname{Lie} G$ of G is $\{X \in \mathrm{Mat}_{n\times n} \mid \operatorname{tr} X = 0\}$, and nilpotent $\operatorname{Ad} G$-orbits in $\operatorname{Lie} G$ are classified by partitions of n (cf., e.g., [**14**]): if $\sigma = (\sigma_1 \geqslant \cdots \geqslant \sigma_t)$ is such a partition, then the orbit $\mathcal{O}_\sigma$ corresponding to σ is the variety of all $n \times n$-matrices whose Jordan normal form is

$$\operatorname{diag}(J_{\sigma_1}, \dots, J_{\sigma_t}), \quad \text{where } J_d := \begin{bmatrix} 0 & 1 & 0 & \dots & 0 \\ 0 & 0 & 1 & \dots & 0 \\ 0 & 0 & 0 & \dots & 0 \\ \vdots & \vdots & \vdots & \ddots & \vdots \\ 0 & 0 & 0 & \dots & 0 \end{bmatrix} \in \mathrm{Mat}_{d\times d}\,.$$

2000 *Mathematics Subject Classification.* Primary: 20G05; Secondary: 14L30, 14N99.
This is the final form of the paper.

As the closure $\overline{\mathcal{O}_\sigma}$ of $\mathcal{O}_\sigma$ in Lie G is stable under scalar multiplications, the algebra $k[\overline{\mathcal{O}_\sigma}]$ of regular functions on $\overline{\mathcal{O}_\sigma}$ is $\mathbb{N}$-graded:

$$k[\overline{\mathcal{O}_\sigma}] = \bigoplus_d k[\overline{\mathcal{O}\mathcal{O}_\sigma}]_d \quad \text{where}$$

$$k[\overline{\mathcal{O}_\sigma}]_d := \{f \in k[\overline{\mathcal{O}_\sigma}] \mid f(\mu x) = \mu^d f(x) \ \forall\ \mu \in k^\times, x \in \overline{\mathcal{O}_\sigma}\}.$$

Put

$$\mathcal{C}_\sigma := \left\{\frac{\lambda}{d} \in E_+ \mid \lambda \in \mathrm{X}_{++}, d \in \mathbb{N}, d \neq 0, m(\lambda^*, k[\overline{\mathcal{O}_\sigma}]_d) \neq 0\right\}.$$

It is known, [**1**], that $\mathcal{C}_\sigma$ is the convex hull of a finite set of points in E, and if $k = \mathbb{C}$, then $\mathcal{C}_\sigma = E_+ \cap \mu(\overline{\mathcal{O}_\sigma})$ where μ is the moment map.

PROBLEMS.

1. Let σ be a partition of n.
   - (i) Find all vertices of the polytope $\mathcal{C}_\sigma$.
   - (ii) Is there a representation theoretic characterization of them?
   - (iii) Find a system of linear inequalities defining $\mathcal{C}_\sigma$ in E.
2. Does $\overline{\mathcal{O}_\sigma} \supset \mathcal{O}_\tau$ imply a geometric interrelation between $\mathcal{C}_\sigma$ and $\mathcal{C}_\tau$ other than the evident inclusion $\mathcal{C}_\sigma \supseteq \mathcal{C}_\tau$?

There is a distinguished polytope, $\mathcal{C}_{(n)}$, among all $\mathcal{C}_\sigma$'s. It corresponds to the cone $\overline{\mathcal{O}_{(n)}} = \mathfrak{N} := \{A \in \mathrm{Mat}_{n\times n} \mid A^n = 0\}$ of all nilpotent elements in Lie G. It is natural first to study the problems (i)–(iii) for $\mathcal{C}_{(n)}$. In this case, the $\mathbb{N}$-graded G-module $k[\mathfrak{N}]$ is isomorphic to the module of harmonic polynomials on Lie G, and the multiplicities $m(\lambda, (k[\mathfrak{N}])_d)$ can be determined via the eigenvalues of $\mathrm{diag}(n-1, n-3, \dots, 3-n, -n+1)$ on $V(\lambda)$, see [**7**].

The formulated problems can be generalized to the case of arbitrary connected semisimple group G. Namely, there are only finitely many nilpotent Ad G-orbits in Lie G, and each nonzero orbit $\mathcal{O}$ is uniquely determined by its *characteristic* $h \in E^*$, cf., e.g., [**14**]; we write $\mathcal{O} = \mathcal{O}_h$. As the closure $\overline{\mathcal{O}_h}$ is stable under scalar multiplications, the algebra $k[\overline{\mathcal{O}_h}]$ is $\mathbb{N}$-graded. The set $\mathcal{C}_h := \{\lambda/d \in E_+ \mid \lambda \in \mathrm{X}_{++}, d \in \mathbb{N}, d \neq 0, m(\lambda^*, k[\overline{\mathcal{O}_h}]_d) \neq 0\}$ is a polytope, [**1**]. Replacing partitions by characteristics one obtains a reformulation of the above problems for G. The cone $\mathfrak{N}$ of all nilpotent elements in Lie G is the closure of the *principal nilpotent* orbit $\mathcal{O}_{h_{\mathrm{pr}}}$. The multiplicities $m\big(\lambda, (k[\mathfrak{N}])_d\big)$ can be found via the eigenvalues of h_{pr} on $V(\lambda)$, [**7**]. In [**2**] there is a discussion of $m(\lambda, k[\overline{\mathcal{O}_h}])$ for some other h. If $\mathcal{O}_h$ is the minimal nonzero orbit, i.e., $\mathcal{O}_h = \mathrm{Ad}\, G \cdot e_\alpha$ where α is the highest root, then [**15**] implies that $\mathcal{C}_h = \alpha$. See also [**13**].[1]

2. Dimension and isomorphism of simple modules

Let $X(\lambda)$ be the cone of primitive vectors in $V(\lambda)$, i.e., the closure of the G-orbit of a highest vector. By [**15**], the varieties $X(\lambda)$ and $\mathrm{X}(\mu)$ are G-isomorphic

[1] *Note added in proof*: In the preprint of November 28, 2002, A. Smirnov, *Decomposition of symmetric powers of irreducible representations of simply connected semisimple complex Lie algebras and Brion's polytope* (in Russian), the polytope $\mathcal{C}_{h_{\mathrm{pr}}}$ is described for all simple G of the types A_l, $l = 1, 2, 3, 4$; B_l; C_l; D_l; G_2; F_4; E_7 and E_8. This paper will appear in Trans. Moscow Math. Soc.

if and only if $\lambda = \mu$. Here we are interested in their isomorphism as abstract varieties.

If $\sigma \in \operatorname{Aut} G$, then the simple G-module obtained from $V(\lambda)$ by "σ-twisting" of the G-module structure is defined by the formula $g * v := \sigma(g) \cdot v,\ g \in G,\ v \in V(\lambda)$. Denoting its dominant weight by $\sigma(\lambda)$, one obtains an action of $\operatorname{Aut} G$ on X_{++}. Every inner automorphism lies in its kernel, so actually it yields an action of the group of outer automorphisms of G on X_{++}. If the weights λ and $\mu \in \mathrm{X}_{++}$ lie in the same orbit of this action, we write $\lambda \sim \mu$. Clearly $\lambda \sim \mu$ implies that $X(\lambda)$ and $X(\lambda^*)$ are isomorphic as abstract varieties.

CONJECTURE 1. Assume that G is simple. Then $X(\lambda)$ and $X(\mu)$ are isomorphic as abstract varieties if and only if $\lambda \sim \mu$.

This conjecture is equivalent to either of two conjectures on dimensions of simple modules formulated below.

By [**9**], the variety $X(\lambda)$ is nonsingular if and only if $X(\lambda) = V(\lambda)$, and if it is singular, its singular locus is the single point 0. Let $\mathfrak{m}$ be the maximal ideal of the local ring $\mathcal{O}_{0,X(\lambda)}$ of this point. By [**9**], the G-module $\mathfrak{m}^d/\mathfrak{m}^{d+1}$ is isomorphic to $V(d\lambda^*)$, whence the Poincaré series of $\mathcal{O}_{0,X(\lambda)}$ is of the form

$$P(\mathcal{O}_{0,X(\lambda)}, t) := \sum_{d \geqslant 0} \dim(\mathfrak{m}^d/\mathfrak{m}^{d+1})\, t^d = \sum_{d \geqslant 0} \dim V(d\lambda^*)\, t^d, \tag{2.1}$$

where $\mathfrak{m}^0 := \mathcal{O}_{0,X(\lambda)}$.

From the aforesaid it follows that if $X(\lambda)$ and $X(\mu)$ are isomorphic as abstract varieties, then $P(\mathcal{O}_{0,X(\lambda)}) = P(\mathcal{O}_{0,X(\mu)})$. Given (2.1), this implies that Conjecture 1 is equivalent to

CONJECTURE 2. Assume that G is simple. Let λ and $\mu \in \mathrm{X}_{++}$. Then

$$\dim V(d\lambda) = \dim V(d\mu)\ \forall\, d \in \mathbb{N} \iff \lambda \sim \mu.$$

For some G, Conjecture 2 is proved in [**9**]. Note that Weyl's formula for dimension of simple module yields the following presentation of $\dim V(d\lambda)$ as a polynomial in d:

$$\dim V(d\lambda) = \prod_{\alpha \in R_+} \left(1 + d\frac{(\lambda, \alpha)}{(\rho, \alpha)}\right), \tag{2.2}$$

where ρ is the half-sum of all positive roots and $(\ ,\ )$ a positive definite inner product on E invariant with respect to the Weyl group. It follows from (2.2) that Conjecture 2 is equivalent to

CONJECTURE 3. Assume that G is simple. Let λ and $\mu \in \mathrm{X}_{++}$. Then

$$\left\{ \frac{(\lambda, \alpha)}{(\rho, \alpha)} \;\middle|\; \alpha \in R_+ \right\} = \left\{ \frac{(\mu, \alpha)}{(\rho, \alpha)} \;\middle|\; \alpha \in R_+ \right\} \iff \lambda \sim \mu.$$

Let $\omega_1, \dots, \omega_r$ be the system of fundamental dominant weights. Then $\lambda = \sum_{i=1}^r n_i \omega_i$ for some $n_i \in \mathbb{N}$. Take some $\alpha \in R_+$. It is not hard to see that $(\lambda, \alpha)/(\rho, \alpha)$ for any $\alpha \in R_+$ is contained in the convex hull of $n_1, \dots, n_r$, whence

$$\min\{n_i \mid 1 \leqslant i \leqslant r\} \leqslant \frac{(\lambda, \alpha)}{(\rho, \alpha)} \leqslant \max\{n_i \mid 1 \leqslant i \leqslant r\},$$

and that for some α's these minimum and maximum are attained. This immediately implies that Conjecture 3 holds for $r = 2$ (this proof is due to T. Springer).

REMARK 2.1. The assumption that G is simple is essential. For instance, let G be of the type $\mathrm{A}_1 + \mathrm{C}_2$. If $\lambda = 4\omega_1 + 2\omega_2 + 8\omega_3$ and $\mu = 6\omega_1 + 8\omega_2 + 2\omega_3$ where ω_1 is the fundamental weight of A_1 and ω_2, ω_3 are the fundamental weights of C_2, then $\dim V(d\lambda) = \dim V(d\mu)\ \forall\ d \in \mathbb{N}$ but $\lambda \not\sim \mu$.

I am grateful to M. Larsen whose remarks led to fixing some inaccuracies in this section. The idea of the example in Remark 2.1 is due to him.

3. Variations on the theme of J. Chipalkatti

In J. Chipalkatti's talk at this Workshop and in [**3**] some nice varieties associated with the spaces of forms in several variables (in particular, with binary forms) were introduced and studied. I refer to Ciliberto's invited address at the ECM, [**4**], on the Waring and related problems. In this section we generalize this topic to the case of arbitrary G and $\lambda \in \mathrm{X}_{++}$ (J. Chipalkatti's case corresponds to $G = \mathrm{SL}_n$ and $\lambda = \omega_1$). Our approach stems from [**10**].

Consider the homogeneous variety G/U (some experts in representation theory awkwardly call it "principal affine space" of G, see, e.g., [**5**], [**17**], despite the fact that it is not affine variety). The algebra A of regular functions on G/U plays an important role in many theories. It is well known, cf. e.g., [**11**], that $m(\lambda, A) = 1$ for each $\lambda \in \mathrm{X}_{++}$, and that the decomposition

$$A = \bigoplus_{\lambda \in \mathrm{X}_{++}} A(\lambda),$$

where $A(\lambda)$ is the isotypical component of type $V(\lambda)$ of the G-module A, is the grading of algebra such that

$$A(\lambda)A(\mu) = A(\lambda + \mu) \quad \forall \lambda, \mu \in \mathrm{X}_{++}.$$

(a) Fix $\lambda \in \mathrm{X}_{++}$ and positive integers d, s, t. Denote by $G\big(t, A(d, \lambda)\big)$ the Grassmannian of t-dimensional linear subspaces of $A(d\lambda)$ and put

$$W(s, t, d; \lambda) := \overline{\{L \in G\big(t, A(d\lambda)\big) \mid \exists\, l_1, \ldots, l_s \in A(\lambda) \text{ s.t. } L \subseteq \operatorname{span}(l_1^d, \ldots, l_s^d)\}}$$

where bar denotes closure.

PROBLEMS.

1. Describe the ideal of $W(s, t, d; \lambda)$ in the homogeneous coordinate ring of the Grassmannian $G\big(t, A(d, \lambda)\big)$.
2. Study the properties of $W(s, t, d; \lambda)$ (dimension, normality, Cohen–Macaulayness, rationality of singularities, Gorensteinness, etc.)

An interesting particular case is obtained when $s = t = 1$. Then $G\big(t, A(d, \lambda)\big)$ is the projective space $\mathbb{P}\big(A(d\lambda)\big)$, and

$$W(1, 1, d; \lambda) = \mathbb{P}\big(\overline{\{l^d \mid l \in A(\lambda)\}}\big) \subseteq \mathbb{P}\big(A(d\lambda)\big).$$

(b) Another class of interesting varieties is defined by

$$X(s, d; \lambda) := \overline{\{l_1^d + \cdots + l_s^d \mid l_i \in A(\lambda)\ \forall\ i\}} \subseteq A(d\lambda)$$

(actually $\{l_1^d + \cdots + l_s^d \mid l_i \in A(\lambda)\}$ is closed, so the bar may be dropped). For $s = 1$ this gives

$$X(1, d; \lambda) := \overline{\{l^d \mid l \in A(\lambda)\}} \subseteq A(d\lambda),$$

the affine cone over $W(1, 1, d; \lambda)$.

PROBLEMS.

1. Describe the ideal of $X(s, d; \lambda)$ in the ring of polynomials on $A(d\lambda)$.
2. Study the properties of $X(s, d; \lambda)$ (dimension, normality, Cohen–Macaulayness, rationality of singularities, Gorensteinness, etc.)
3. Classify all cases such that $X(s, d; \lambda) = A(d\lambda)$.
4. Classify all cases such that $X(s, d; \lambda)$ is a hypersurface in $A(d\lambda)$.
5. In the latter case find $\deg\bigl(X(s, d; \lambda)\bigr)$ and describe a generator of the ideal of $X(s, d; \lambda)$ from the viewpoint of Invariant theory (this generator is automatically G-invariant).

EXAMPLE. Let V be an $(n+1)$-dimensional vector space over k where $n > 0$. Let $G = \mathrm{SL}(V)$, and $\lambda = \omega_1$. Then $\lambda^* = \omega_n$, and the subalgebra $\bigoplus_{d\in\mathbb{N}} A(d\omega_n)$ of A is isomorphic, as the graded G-algebra, to the symmetric algebra $\mathrm{Sym}\, V^* = \bigoplus_{d\in\mathbb{N}} \mathrm{Sym}^d\, V^*$. In this case, $W(1, 1, d; \omega_n)$ is the minimal G-orbit in $\mathbb{P}(\mathrm{Sym}^d V^*)$, that is the projectivization of the G-orbit of a highest vector in $\mathrm{Sym}^d\, V^*$. Accordingly, $X(1, d; \omega_n)$ is the closure of the latter orbit, and $\mathbb{P}\bigl(X(s, d; \lambda)\bigr)$ is a t-secant variety of the minimal G-orbit in $\mathbb{P}(\mathrm{Sym}^d V^*)$ for some t.

Note that for arbitrary G and λ people studied the 1-secant varieties of the minimal G-orbit in $\mathbb{P}\bigl(V(\lambda)\bigr)$ (e.g., see [**6**] and Addendum by F. Zak to [**8**]). It would be interesting to study its higher secant varieties (dimension, normality, etc., and the orbital structure) as well (see [**16**] related to this topic).

(c) There is a more general class of varieties $Y(\boldsymbol{n}_1, \ldots, \boldsymbol{n}_s; \boldsymbol{\lambda})$, to which the afore-formulated problems for $X(s, d; \lambda)$ are equally related. The definition of $Y(\boldsymbol{n}_1, \ldots, \boldsymbol{n}_s; \boldsymbol{\lambda})$ is the following. Fix $\lambda \in \mathrm{X}_{++}$ and the decompositions (some of them may coincide with each other)

$$\lambda = d_{11}\lambda_1 + \cdots + d_{1r}\lambda_r = \cdots = d_{s1}\lambda_1 + \cdots + d_{sr}\lambda_r \quad d_{ij} \in \mathbb{N},\ \lambda_i \in \mathrm{X}_{++},$$

and put $\boldsymbol{d}_i = (d_{i1}, \ldots, d_{ir})$, $\boldsymbol{\lambda} = (\lambda_1, \ldots, \lambda_r)$. Then

$$Y(\boldsymbol{d}_1, \ldots, \boldsymbol{d}_s; \boldsymbol{\lambda}) := \overline{\{l_1^{d_{11}}, \ldots, l_r^{d_{1r}} + \cdots + l_1^{d_{s1}}, \ldots, l_r^{d_{sr}} \mid l_i \in A(\lambda_i)\ \forall i\}} \subseteq A(\lambda).$$

An interesting particular case is obtained for $s = 1$:

$$Y(\boldsymbol{d}; \boldsymbol{\lambda}) := \overline{\{l_1^{d_1}, \ldots, l_r^{d_r} \mid l_i \in A(\lambda_i)\ \forall i\}} \subseteq A(\lambda),$$

where $\boldsymbol{d} = (d_1, \ldots, d_r)$ (again the bar may be dropped).

Note that if G is simply connected, then the algebra A is factorial, [**12**]. In this case the complement in $A(\lambda)$ to the union of all subvarieties $Y(\boldsymbol{d}; \boldsymbol{\lambda})$ with $\lambda = d_1\lambda_1 + \cdots + d_r\lambda_r$ is the set of all irreducible elements of the factorial algebra A contained in $A(\lambda)$. This topic is considered in [**10**].

References

1. M. Brion, *Sur l'image de l'application moment*, Séminaire d'algèbre Paul Dubreil et Marie–Paule Malliavin, vol. 1296, Springer-Verlag, 1987, pp. 177–192.
2. B. Broer, *Hilbert series in invariant theory*, Thesis, Rejksuniversiteit Utrecht, 1990.
3. E. Carlini and J. Chipalkatti, *On Waring's problem for several algebraic forms*, Preprint, 2002.
4. C. Ciliberto, *Geometric aspects of polynomial interpolation in more variables and of Waring's problem*, European Congress of Mathematics, vol. I (Barcelona, 2000), Progr. Math., vol. 201, Birkhäuser, Basel, 2001, pp. 289–316.
5. I. M. Gel'fand and A. A. Kirillov, *The structure of the Lie field connected with a split semisimple Lie algebra*, Funct. Anal. Appl. **3** (1969), 6–21.

6. H. Kaji, M. Ohto, and O. Yasukura, *Adjoint varieties and their secant varieties*, Indag. Math. (N.S.) **10** (1999), no. 1, 45–57.
7. B. Kostant, *Lie group representations on polynomial rings*, Amer. J. Math. **85** (1963), 327–404.
8. R. Lazarsfeld and A. Van de Ven, *Topics in the geometry of projective space*, with an addendum by Zak. DMV Seminar, vol. 4, Birkhäuser Verlag, 1984.
9. V. L. Popov, *Groups, generators, syzygies, and orbits in invariant theory*, Transl. Math. Mono., vol. 100, Amer. Math. Soc., Providence, RI, 1992.
10. ______, *Some applications of algebra of functions on G/U*, Group Actions and Invariant Theory (Montreal, 1988), CMS Conf. Proc., vol. 10, Amer. Math. Soc., Providence, RI, 1989, pp. 157–166.
11. ______, *Contractions of the actions of reductive algebraic groups*, Math. USSR Sbornik, **58** (1987), no. 2, 311–335.
12. ______, *Picard group of homogeneous spaces of linear algebraic groups and one dimensional homogeneous vector bundles*, Math. USSR-Izv. **8** (1974), 301–327.
13. R. Sjamaar, *Convexity properties of the moment mapping re-examined*, Adv. in Math. **138**, (1998), no. 1, 46–91.
14. T. A. Springer and R. Steinberg, *Conjugacy classes*, Seminar on Algebraic Groups and Related Finite Groups (Princeton, 1968/69), Lecture Notes in Math., vol. 131, 1970, Springer-Verlag, pp. 167–266.
15. E. B. Vinberg and V. L. Popov, *On a class of quasihomogeneous affine varieties*, Math. USSR-Izv **6**, 1972, 743–758.
16. F. L. Zak, *Determinants of projective varieties and their degrees*, Interesting Algebraic Varieties Arising in Algebraic Transformation Groups Theory (Vienna, 2001), Proc. of the ESI Conference (to appear).
17. D. P. Zhelobenko, *Representations of reductive Lie algebras*, Nauka, Moscow, 1994 (in Russian).

Steklov Mathematical Institute, Russian Academy of Sciences, Gubkina 8, Moscow, 117966, GSP-1, Russia.

E-mail address: popovvl@orc.ru

Centre de Recherches Mathématiques
CRM Proceedings and Lecture Notes
Volume **35**, 2004

Compressions of Group Actions

Z. Reichstein

Let G be a finite group. A *G-variety* X is an algebraic variety with a regular G-action; X is *faithful* if every $1 \neq g \in G$ acts non-trivially. I will refer to a dominant G-equivariant rational (respectively, regular) map of faithful G-varieties as a rational (respectively, *regular*) compression. All varieties, actions, vector spaces, maps, etc., are assumed to be defined over a fixed algebraically closed base field k of characteristic zero; all varieties are assumed to be irreducible.

1. Essential dimension

Let V be a faithful linear representation of G and let d be the minimal value of $\dim(X)$, where the minimum is taken over all rational compressions $f\colon V \dashrightarrow X$. Note that

(a) (see [**1**, Theorem 3.1] or [**6**, Theorem 3.4(b)]) d depends only on the group G and not on the choice of V, and
(b) (cf. [**6**, Proposition 7.1]) in the definition of d we may assume that X is a G-invariant subvariety of V, i.e., X is the closure of the image of a rational covariant $f\colon V \dashrightarrow V$.

The number d is called the *essential dimension* of G and is usually denoted by $\mathrm{ed}(G)$. This number has interesting connections with the algebraic form of Hilbert's 13th problem, cohomological invariants, generic polynomials and other topics; these connections are described in [**1**] and [**2**]. The case where $G = \mathrm{S}_n$ is of particular interest. (The notion of essential dimension is also of interest in the context of algebraic groups; see [**6**] and [**7**].)

PROBLEM 1. *Find* $\mathrm{ed}(G)$ *and, in particular,* $\mathrm{ed}(\mathrm{S}_n)$.

The value of $\mathrm{ed}(G)$ is known if G is an abelian group; see [**1**, Theorem 6.1]. For symmetric groups, $\mathrm{ed}(\mathrm{S}_n) \geq [n/2]$; this is proved, in different ways, in [**1**, Theorem 6.5(c)], [**2**, Section 8], [**6**, Example 12.8] and [**7**, Section 9]. (For $n = 5$ this inequality goes back to Felix Klein [**3**].) The best known upper bound on

---

2000 *Mathematics Subject Classification.* Primary: 14L30; Secondary: 14H37.

Z. Reichstein was supported in part by an NSERC research grant.
I would like to thank V. L. Popov for stimulating discussions and for helpful comments on an earlier draft of this note.
This is the final form of the paper.

$\mathrm{ed}(\mathrm{S}_n)$ is $\mathrm{ed}(\mathrm{S}_n) \leq n-3$; see [**1**, Theorem 6.5(c)], and exact values of $\mathrm{ed}(\mathrm{S}_n)$ are only known for $n \leq 6$; see [**1**, Theorems 6.2 and 6.5(d)].

We now consider an analogous notion in the context of regular, rather than rational compressions. Let V be a faithful linear representation of G and let D be the minimal value of $\dim(X)$, where the minimum is taken over all regular compressions $F\colon V \longrightarrow X$ from V to a faithful affine G-variety X. Properties (a) and (b) remain valid in this setting:

(A) D depends only on the group G and not on the choice of V and
(B) in the definition of D we may assume that X is a G-invariant subvariety of V, i.e., X is the closure of the image of a (regular) covariant $F\colon V \longrightarrow V$.

Part (B) is an immediate consequence of [**5**, Theorem 1.7.12]. To prove (A), assume that V_1 and V_2 are faithful representations of G, and $F_i\colon V_i \longrightarrow X_i$ is a regular compression with minimal possible values of $\dim(X_i) = D_i$, $i = 1, 2$. We want to show that $D_1 = D_2$. Choose a point $v_2 \in V_2$ such that $F_2(v_2)$ has a trivial stabilizer in X_2. By [**5**, Theorem 1.7.12] there exists a covariant $\phi\colon V_1 \longrightarrow V_2$ such that v_2 lies in the image of ϕ. Let X_1' be the closure of the image of $F_2 \circ \phi\colon V_1 \longrightarrow X_2$. Then X_1' is a faithful G-variety (because it contains $F(v_2)$) and thus, by the definition of D_1,

$$D_1 \leq \dim(X_1') \leq \dim(X_2) = D_2.$$

By symmetry, we also have $D_2 \leq D_1$, so that $D_1 = D_2$, as claimed.

In view of (A), it is natural to write $\mathrm{eD}(G)$ in place of D.

PROBLEM 2. *Find* $\mathrm{eD}(G)$ *and, in particular,* $\mathrm{eD}(\mathrm{S}_n)$.

Problems 1 and 2 are closely related, because

$$\mathrm{ed}(G) \leq \mathrm{eD}(G) \leq \mathrm{ed}(G) + 1.$$

The inequality $\mathrm{ed}(G) \leq \mathrm{eD}(G)$ is obvious from the definition. To prove that $\mathrm{eD}(G) \leq \mathrm{ed}(G) + 1$, note that if $f\colon V \dashrightarrow V$ is a rational covariant then $F = \alpha f\colon V \longrightarrow V$ is a regular covariant for some $0 \neq \alpha \in k[V]^G$. (That is, $F(v) = \alpha(v)f(v)$ for every $v \in V$.) Then

$$\dim f(V) \leq \dim F(V) \leq \dim f(V) + 1,$$

and the proof is complete.

2. Incompressible varieties

I will call a G-variety X *incompressible* if every rational compression $X \dashrightarrow Y$ is a birational isomorphism. (Note that this definition bears some resemblance to the definition of a relatively minimal model in algebraic geometry; cf., e.g., [**4**, p. 418].)

PROBLEM 3. *Classify incompressible* G*-varieties.*

To motivate this problem, I will discuss three simple examples.

EXAMPLE 4. If $G = \{1\}$ then the only incompressible G-variety is a point.

EXAMPLE 5. If $G = \langle c \rangle$ is a cyclic group of order $n \geq 2$ then there are no incompressible G-varieties.

Indeed, assume, to the contrary, that X is an incompressible G-variety. Diagonalizing the G-action on $k(X)$ (viewed as a finite-dimensional $k(X)^G$-vector space), we can find a non-zero rational function $f \in k(X)$ such that $f(c\cdot x) = \zeta f(x)$, where ζ is a primitive nth root of unity. Then $x \mapsto f(x)$ is a G-compression $X \dashrightarrow \mathbb{A}^1$, where c acts on $\mathbb{A}^1$ by $y \longrightarrow \zeta y''$. Thus X is birationally isomorphic to $\mathbb{A}^1$ (as a G-variety). On the other hand, $\mathbb{A}^1$ is not incompressible; indeed, $y \longrightarrow y^{n+1}$ is a nontrivial compression $\mathbb{A}^1 \longrightarrow \mathbb{A}^1$, a contradiction.

EXAMPLE 6. Let G be a finite group that does not embed in $\mathrm{Aut}(X)$ for any curve X of genus ≤ 1. There are many such groups; see below. Let X be a faithful G-curve of the smallest possible genus $g(X)$. (It is easy to see that faithful G-curves exist for any finite group G; cf., e.g., [**1**, Remark 4.5].) I claim that X is incompressible. Indeed, suppose that $X \dashrightarrow Y$ is a G-compression. By our assumption, $g(Y) \geq 2$. Then by the Hurwitz formula (see, e.g., [**4**, Corollary 4.2.4] or [**9**, Theorem II.5.9]), $g(Y) \leq g(X)$, a contradiction.

To construct a group G that does not embed in $\mathrm{Aut}(X)$ for any curve X of genus ≤ 1, recall the following classical results on automorphisms of rational and elliptic curves.

- The only finite subgroups of $\mathrm{Aut}(\mathbb{P}^1) = \mathrm{PGL}_2$, that are not cyclic or dihedral, are A_4, S_4 and A_5; see [**3**, Chapter 1] or [**8**, Theorem 2.6.1].
- Let E be an elliptic curve. Then there exists an exact sequence
$$\{1\} \longrightarrow E \longrightarrow \mathrm{Aut}(E) \longrightarrow H \longrightarrow \{1\},$$
where E acts on itself by translations (and thus is embedded in $\mathrm{Aut}(E)$) and $|H| = 2$, 4 or 6; see, e.g., [**9**, Theorems III.10.1 and Proposition X.5.1]. (Note that [**9**] uses the symbols $\mathrm{Isom}(E)$ and $\mathrm{Aut}(E)$ in place of my $\mathrm{Aut}(E)$ and H, respectively.)

It is now easy to see that many finite groups cannot be embedded in $\mathrm{Aut}(\mathbb{P}^1)$ or $\mathrm{Aut}(E)$ for any elliptic curve E. Examples of such groups include (i) the symmetric group S_5, (ii) the alternating group $G = \mathrm{A}_6$, and perhaps more surprisingly (in view of Example 5), the abelian groups (iii) $G = (\mathbb{Z}/p)^3$ for $p \geq 5$, (iv) $G = (\mathbb{Z}/3)^4$, and (v) $G = (\mathbb{Z}/2)^5$. Of course, we can also take G to be any group containing a subgroup isomorphic to (i), (ii), (iii), (iv) or (v).

In conclusion, note that Problem 3 can be posed for any algebraic (possibly infinite) group G; here a compression should be defined as a dominant map of *generically free*, rather than faithful, G-varieties (cf. [**6**, Definition 2.15]).

References

1. J. Buhler and Z. Reichstein, *On the essential dimension of a finite group*, Compositio Math. **106** (1997), no. 2, 159–179.
2. J. Buhler and Z. Reichstein, *On Tschirnhaus transformations*, Topics In Number Theory (University Park, 1997), Kluwer Acad. Publ., Dordrecht, 1999, pp. 127–142.
3. F. Klein, *Lectures on the icosahedron and the solution of equations of the fifth degree*, revised ed., Dover Publications Inc., New York, N.Y., 1956; English Transl. by G. G. Morrice.
4. R. Hartshorne, *Algebraic geometry*, Grad. Texts in Math., no. 52. Springer-Verlag, New York, 1977.
5. V. L. Popov, *Sections in invariant theory*, The Sophus Lie Memorial Conference (Oslo, 1992), Scand. Univ. Press, Oslo, 1994, pp. 315–361.
6. Z. Reichstein, *On the notion of essential dimension for algebraic groups*, Transform. Groups **5** (2000), no. 3, 265–304.
7. Z. Reichstein and B. Youssin, *Essential dimensions of algebraic groups and a resolution theorem for G-varieties*, Canad. J. Math. **52** (2000), no. 5, 1018–1056.
8. J. Shurman, *Geometry of the quintic*, John Wiley & Sons Inc., New York, 1997.

9. J. H. Silverman, *The arithmetic of elliptic curves*, Springer-Verlag, New York, 1986.

Department of Mathematics, University of British Columbia, Vancouver, BC V6T 1Z2, Canada.
E-mail address: reichst@math.ubc.ca

Centre de Recherches Mathématiques
CRM Proceedings and Lecture Notes
Volume **35**, 2004

Commutativity of Weakly Commutative Riemannian Homogeneous Spaces

L. G. Rybnikov

ABSTRACT. A Riemannian homogeneous space $X = G/H$ is said to be *commutative* if the algebra of G-invariant differential operators on X is commutative, and *weakly commutative* if the associated Poisson algebra is commutative. Clearly, the commutativity of X implies its weak commutativity. Some ideas of the proof of the backward implication are discussed in this talk.

Suppose that G is a real Lie group, $\mathfrak{g} = \operatorname{Lie} G$, H is a compact subgroup in G, $\mathfrak{h} = \operatorname{Lie} H$. Consider the Riemannian homogeneous space $X = G/H$. Let $D(X)$ be the algebra of differential operators on X.

The algebra $D(X)$ possesses the filtration by order of differential operators: $D(X) = \bigcup_{n=0}^{\infty} D(X)^{(n)}$. The associated graded algebra $\operatorname{gr} D(X)$ is isomorphic to the algebra $P(X)$ of fiberwise polynomial functions on the cotangent bundle T^*X of X; the isomorphism is performed by taking the symbol of a differential operator.

The algebra $P(X)$ thus obtains a natural Poisson bracket; namely, for $r \in D(X)^{(n)}$, $s \in D(X)^{(m)}$, we have $\{\operatorname{gr}_n r, \operatorname{gr}_m s\} = \operatorname{gr}_{n+m-1}[r,s]$. This bracket coincides with the Poisson bracket defined by the standard symplectic structure on T^*X. The main objects of the further considerations are the algebras $D(X)^G$ and $P(X)^G$, i.e., the subalgebras of G-invariants in $D(X)$ and in $P(X)$, respectively. It is clear that the Poisson algebra corresponding to $D(X)^G$ is embedded in $P(X)^G$. Let us give a more explicit description of $D(X)^G$ in terms of the universal enveloping algebra $U(\mathfrak{g})$.

To each element ξ of the Lie algebra $\mathfrak{g} = T_eG$, we put in correspondence the left-invariant first order differential operator on G defined as the differentiation along the left-invariant vector field equal to ξ at the identity element of G. This correspondence extends to an isomorphism of $U(\mathfrak{g})$ onto the algebra of left-invariant differential operators on G. Besides, the order of the differential operator corresponding to an element $u \in U(\mathfrak{g})$ is equal to the minimal degree of a polynomial expressing u in terms of elements of $\mathfrak{g}$, and taking the symbol of a left-invariant differential operator at e is the isomorphism $\operatorname{gr} U(\mathfrak{g}) \to S(\mathfrak{g})$ given by the Poincaré–Birkhoff–Witt theorem.

2000 *Mathematics Subject Classification.* Primary: 22F30; Secondary: 53C30, 17B63.
This is the final version of the paper.

Let $U(\mathfrak{g})^H$ be the subalgebra of $U(\mathfrak{g})$ consisting of all invariant elements with respect to the adjoint action of H. The subalgebra $S(\mathfrak{g})^H$ in $S(\mathfrak{g})$ is defined similarly. The algebras $U(\mathfrak{g})$ and $S(\mathfrak{g})$ are isomorphic as H-modules (moreover, they are isomorphic as G-modules); therefore, the embedding $\operatorname{gr} U(\mathfrak{g})^H \to S(\mathfrak{g})^H$ is an isomorphism. Every element $D \in U(\mathfrak{g})^H$ commutes with the right shifts by the elements of H and hence maps the space of H-right-invariant functions on G to itself. Thus it defines an invariant differential operator D_H on $X = G/H$. Note that the symbol of D_H is the image of the symbol of D under the canonical homomorphism $P(x)^G \to P(x)^G$. In particular, the symbol of D_H at $o = eH \in X$ is the image of the symbol of D at $e \in G$ under the canonical homomorphism $\mathbb{R}[T_e^*G] = S(\mathfrak{g}) \to S(\mathfrak{g}/\mathfrak{h}) = \mathbb{R}[T_o^*X]$.

All differential operators belonging to the left ideal $U(\mathfrak{g})\mathfrak{h}$ in $U(\mathfrak{g})$ vanish on H-right-invariant functions. Since the elements of $U(\mathfrak{g})^H$ commute with the elements of $\mathfrak{h}$, it follows that $(U(\mathfrak{g})\mathfrak{h})^H$ is a two-sided ideal in $U(\mathfrak{g})^H$. Thus we have the homomorphism of associative algebras

$$\mu : U(\mathfrak{g})^H / \big(U(\mathfrak{g})\mathfrak{h}\big)^H \to D(X)^G.$$

Since the group H is compact, $U(\mathfrak{g})\mathfrak{h}$ and $S(\mathfrak{g})\mathfrak{h}$ are isomorphic H-modules. Therefore, $\operatorname{gr}\big(U(\mathfrak{g})\mathfrak{h}\big)^H = \big(S(\mathfrak{g})\mathfrak{h}\big)^H$. Hence, in particular, it follows that $\big(S(\mathfrak{g})\mathfrak{h}\big)^H$ is a Poisson ideal in $S(\mathfrak{g})^H$. Passing from associative algebras to the associated Poisson algebras, we obtain the homomorphism of Poisson algebras

$$\operatorname{gr}\mu : S(\mathfrak{g})^H / \big(S(\mathfrak{g})\mathfrak{h}\big)^H \to \operatorname{gr} D(X)^G \subset P(X)^G.$$

It is known [**1**] that

1. The homomorphism μ is an isomorphism of associative algebras
$$U(\mathfrak{g})^H / \big(U(\mathfrak{g})\mathfrak{h}\big)^H \simeq D(X)^G$$
2. The homomorphism $\operatorname{gr}\mu$ is an isomorphism of Poisson algebras
$$S(\mathfrak{g})^H / \big(S(\mathfrak{g})\mathfrak{h}\big)^H \simeq P(X)^G,$$

Vorobjovy Gory, Moscow, 119899.

Therefore, the Poisson algebra corresponding to $D(X)^G$ coincides with $P(X)^G$.

The Poisson algebra isomorphism

$$(\operatorname{gr}\mu)^{-1} : P(X)^G \to S(\mathfrak{g})^H / \big(S(\mathfrak{g})\mathfrak{h}\big)^H = \mathbb{R}[T_o^*X]^H$$

has an obvious geometric interpretation, namely, this is the homomorphism of restriction to the fiber $T_o^*X = (\mathfrak{g}/\mathfrak{h})^*$ of the cotangent bundle at $o = eH$.

The homogeneous space X is said to be *commutative* if the algebra $D(X)^G$ is commutative and *weakly commutative* if the algebra $P(X)^G$ is commutative with respect to the Poisson bracket. Since $\operatorname{gr} D(X)^G = P(X)^G$, the commutativity of a Riemannian homogeneous space implies its weak commutativity.

It is known that the converse statement (i.e., that the weak commutativity of a Riemannian homogeneous space implies its commutativity) is true in the following two opposite cases: when the group G is reductive [**2, 3**] (in this case the space X is spherical), and when $G = H \ltimes N$, where N is a nilpotent Lie group [**1**].

In my paper [**4**] this statement is proved for the general case. In particular, a new proof for a reductive group G is obtained. The proof is based on the following lemma:

LEMMA 1. *Let* $\mathcal{A} = \bigcup_{n=0}^{\infty} \mathcal{A}_n$ *be a filtered associative algebra, and let* $A = \operatorname{gr} \mathcal{A}$. *Suppose that* $\mathcal{B} \subset \mathcal{A}$ *is a commutative subalgebra,* $B = \operatorname{gr} \mathcal{B}$, *and the following assertions hold*:

1. A *is an integral commutative algebra*;
2. $A \supseteq B$ *is an algebraic extension.*

Then the algebra $\mathcal{A}$ *is commutative.*

PROOF. Consider the sequence of operations $\{\cdot,\cdot\}_k$ on A defined as follows:

1. For $r \in \mathcal{A}_n$, $s \in \mathcal{A}_m$, we set $\{\operatorname{gr}_n r, \operatorname{gr}_m s\}_1 = \operatorname{gr}_{n+m-1}[r,s]$.
2. If the operation $\{\cdot,\cdot\}_{k-1}$ is zero, then, for $r \in \mathcal{A}_n$, $s \in \mathcal{A}_m$, we set $\{\operatorname{gr}_n r, \operatorname{gr}_m s\}_k = \operatorname{gr}_{n+m-k}[r,s]$.

Assume that the algebra $\mathcal{A}$ is not commutative. Then the above sequence terminates, say, on lth step. It is clear that $\{\cdot,\cdot\}_l$ is a nontrivial biderivation of A, and $\{B,B\}_l = 0$.

Each element $s \in B$ determines a derivation of A taking $r \in A$ to $\{r,s\}_l$. It is clear that any derivation of this kind vanishes on B and hence on any algebraic extension of B. Therefore, $\{B,A\}_l = 0$. Hence we conclude in a similar way that $\{A,A\}_l = 0$. Thus the algebra $\mathcal{A}$ is commutative. $\square$

In the case of a reductive group G, the image of the center of the universal enveloping algebra under the map μ is a "sufficiently large" commutative subalgebra in $D(X)$, i.e., satisfies the assumptions of the lemma. The general case can be reduced to the case of a reductive group G with the help of the following structural theorem:

THEOREM 1. [**1**] *Suppose that* $X = G/H$ *is a weakly commutative Riemannian homogeneous space. Then* $G = L \leftthreetimes N$, *where*

1. N *is at most* 2*-step nilpotent Lie group*;
2. L *is a reductive Lie subgroup containing* H;
3. $\mathbb{R}[\mathfrak{n}]^L = \mathbb{R}[\mathfrak{n}]^H$.

Namely, let $\mathfrak{z}$ be the commutator subalgebra of $\mathfrak{n}$ (which is contained in its center), and let $\mathfrak{v}$ be an L-invariant complement subspace. For any linear function $\alpha \in \mathfrak{z}^*$, by $\hat{\alpha}$ denote the bilinear function on $\mathfrak{v}$ defined by

$$\hat{\alpha}(\xi,\eta) = \alpha([\xi,\eta]).$$

Let L_α and H_α be the stabilizers of α in L and H, respectively. Property (3) implies that the quotient of L by the kernel of the action $L : \mathfrak{n}^*$ is compact. Therefore, the group L_α is reductive. The subspace $\operatorname{Ker}\hat{\alpha} \subset \mathfrak{v}$ is clearly L_α-invariant. Let $\mathfrak{v}_\alpha$ be an L_α-invariant complement subspace. The natural linear representation of L_α (and H_α) in the symplectic vector space $\mathfrak{v}_\alpha \simeq \mathfrak{v}/\operatorname{Ker}\hat{\alpha}$ is symplectic.

For any linear function $\alpha \in \mathfrak{z}^*$, we have the natural homomorphism

$$\varphi_\alpha : U(\mathfrak{n}) \to W_\alpha(\mathfrak{v}) = U(\mathfrak{n})/\big(\xi - \alpha(\xi) : \xi \in \mathfrak{z}\big),$$

whose image is the Weil algebra (possibly degenerate) defined by the form $\hat{\alpha}$ on $\mathfrak{v}$.

For any linear function $\beta \in (\operatorname{Ker}\hat{\alpha})^*$, we have the homomorphism

$$\psi_{\alpha,\beta} : W_\alpha(\mathfrak{v}) \to W_{\alpha,\beta}(\mathfrak{v}_\alpha) = W_\alpha(\mathfrak{v})/\big(\xi - \beta(\xi) : \xi \in \operatorname{Ker}\hat{\alpha}\big),$$

whose image is already a nondegenerate Weil algebra.

Note that there is a natural L-equivariant bijection between the set of pairs $\big(\alpha \in \mathfrak{z}^*, \beta \in (\operatorname{Ker}\hat{\alpha})^*\big)$ and the set of N-orbits in $\mathfrak{n}^*$. Namely, let $p : \mathfrak{n}^* \to \mathfrak{z}^*$ be the natural projection dual to the embedding $\mathfrak{z} \to \mathfrak{n}$, and let $p_\alpha : \mathfrak{n}^* \to (\operatorname{Ker}\hat{\alpha})^*$ be the natural projection dual to the embedding $\operatorname{Ker}\hat{\alpha} \to \mathfrak{n}$. The affine space $O_{\alpha,\beta} = p^{-1}(\alpha) \cap p_\alpha^{-1}(\beta)$ is clearly N-invariant, and the Poisson–Lie bracket on it is nondegenerate. Since the symplectic leaves of the Poisson structure on $\mathfrak{n}^*$ are the N-orbits in $\mathfrak{n}^*$, the space $O_{\alpha,\beta}$ contains an open N-orbit. But the group N is unipotent, whence all N-orbits in $\mathfrak{n}^*$ are closed, and hence the space $O_{\alpha,\beta}$ is an N-orbit. Thus an orbit corresponds to each pair $\big(\alpha \in \mathfrak{z}^*, \beta \in (\operatorname{Ker}\hat{\alpha})^*\big)$. On the other hand, if $O = Nx$ is an orbit in $\mathfrak{n}^*$, then $O = O_{p(x),p_{p(x)}(x)}$, whence the map $(\alpha, \beta) \mapsto O_{\alpha,\beta}$ defined above is bijective.

This construction shows that the homomorphism of Poisson algebras

$$\operatorname{gr}(\psi_{\alpha,\beta} \circ \varphi_\alpha) : \mathbb{R}[\mathfrak{n}^*] = S(\mathfrak{n}) \to S(\mathfrak{v}_\alpha) = \operatorname{gr} W_{\alpha,\beta}(\mathfrak{v}_\alpha)$$

has an obvious geometric interpretation: it is the homomorphism of restriction to the orbit $O_{\alpha,\beta} \subset \mathfrak{n}^*$, which is an affine subspace in $\mathfrak{n}^*$.

Let $H_{\alpha,\beta}$ and $L_{\alpha,\beta}$ be the stabilizers of the orbit $O_{\alpha,\beta}$ in H_α and in L_α, respectively (or, which is the same, the stabilizers of the functional β in H_α and in L_α). Since the quotient of L by the kernel of the action $L : \mathfrak{n}^*$ is compact, the group $L_{\alpha,\beta}$ is reductive.

In [**4**] it is shown that, for a generic pair (α, β), the restriction of the homomorphism of Poisson algebras

$$\operatorname{gr} \psi_{\alpha,\beta} \circ \varphi_\alpha : S(\mathfrak{n}) \to S(\mathfrak{v}_\alpha)$$

to $S(\mathfrak{n})^H$ extends to a homomorphism of Poisson algebras

$$S(\mathfrak{g})^H / \big(S(\mathfrak{g})\mathfrak{h}\big)^H \to S(\mathfrak{l}_{\alpha,\beta}/\mathfrak{h}_{\alpha,\beta} \oplus \mathfrak{v}_\alpha)^H.$$

Moreover, the space X is weakly commutative iff, for generic pairs (α, β), the Poisson algebras $S(\mathfrak{l}_{\alpha,\beta}/\mathfrak{h}_{\alpha,\beta} \oplus \mathfrak{v}_\alpha)^H$ are commutative.

This homomorphism of Poisson algebras can be lifted to a homomorphism of filtered associative algebras

$$U(\mathfrak{g})^H / \big(U(\mathfrak{g})\mathfrak{h}\big)^H \to B(\mathfrak{l}_{\alpha,\beta}, \mathfrak{h}_{\alpha,\beta}, \mathfrak{v}_\alpha).$$

The algebra $B(\mathfrak{l}_{\alpha,\beta}, \mathfrak{h}_{\alpha,\beta}, \mathfrak{v}_\alpha)$, which is the image of this map, can be described in terms of universal enveloping algebras and Weil algebras. Besides, the space X is commutative iff, for generic pairs (α, β), the image of this homomorphism is commutative.

The elements of degree 2 in $S(\mathfrak{v}_\alpha)$, with respect to the Poisson bracket, form a Lie algebra isomorphic to $\mathfrak{sp}(\mathfrak{v}_\alpha)$. Thus there is a homomorphism of Poisson algebras $S\big(\mathfrak{sp}(\mathfrak{v}_\alpha)\big) \to S(\mathfrak{v}_\alpha)$. This homomorphism induces a homomorphism of Poisson algebras

$$S\big(\mathfrak{l}_{\alpha,\beta} \oplus \mathfrak{sp}(\mathfrak{v}_\alpha)\big)^{L_{\alpha,\beta} \times Sp(\mathfrak{v}_\alpha)} \to S(\mathfrak{l}_{\alpha,\beta}/\mathfrak{h}_{\alpha,\beta} \oplus \mathfrak{v}_\alpha)^H.$$

This homomorphism of Poisson algebras can be lifted to a homomorphism of filtered associative algebras

$$ZU\big(\mathfrak{l}_{\alpha,\beta} \oplus \mathfrak{sp}(\mathfrak{v}_\alpha)\big) \to B(\mathfrak{l}_{\alpha,\beta}, \mathfrak{h}_{\alpha,\beta}, \mathfrak{v}_\alpha).$$

In [**4**] it is proved that the image of this homomorphism is a "sufficiently large" commutative subalgebra in $B(\mathfrak{l}_{\alpha,\beta}, \mathfrak{h}_{\alpha,\beta}, \mathfrak{v}_\alpha)$, i.e., satisfies the assumptions of the lemma. Hence follows the commutativity of weakly commutative Riemannian homogeneous spaces.

References

1. E. B. Vinberg, *Commutative homogeneous spaces and coisotropic symplectic actions*, Uspekhi Mat. Nauk **56** (2001), no. 1, 3–62 (Russian).
2. I. V. Mikityuk, *Integrability of invariant Hamiltonian systems with homogeneous configuration spaces*, Mat. Sb. (N.S.) **129** (1986), no. 4, 514–534 (Russian).
3. V. Guillemin and S. Sternberg, *Multiplicity-free spaces*, J. Differential Geom. **19** (1984), 31–56.
4. L. G. Rybnikov, *On commutativity of weakly commutative Riemannian homogeneous spaces*, Func. Anal. Appl. **37** (2003), no. 2, pp. 41–51 (Russian).

Chair of High Algebra, Department of Mathematics, Moscow State University, Vorobjovy Gory, Moscow, 119899, Russia.
E-mail address: leo_rybnikov@mtu-net.ru

Centre de Recherches Mathématiques
CRM Proceedings and Lecture Notes
Volume **35**, 2004

Group Actions and Quotients for Compact Lie Groups and Algebraic Groups

G. W. Schwarz

1. Introduction

This article consists of revised notes of my four lectures given at the Workshop on Invariant Theory. The main topics were

1. Actions of Lie groups, especially compact Lie groups.
2. Quotients of finite dimensional representations of compact Lie groups.
3. The differentiable slice theorem and its corollaries. Stratifications of K-manifolds and their quotients, K compact Lie.
4. The Peter–Weyl theorem and consequences.
5. Complex reductive groups viewed as complexifications of compact Lie groups.
6. Quotients of complex reductive group actions, Kempf–Ness theory and applications.
7. Luna's slice theorem and its corollaries. Stratifications of affine G-varieties and their quotients, G complex reductive.
8. Applications of Luna's slice theorem to classification problems.

Because of time constraints, the topics above were presented mostly without proofs, and that is reflected in these notes. For a fuller treatment of most of the topics above with proofs see [**19**] and [**20**].

2. Lie groups

A reference for this section is [**22**].

Definition 2.1. A *Lie group* is a C^∞-manifold G which has a group structure such that $G \times G \to G$, $(g, h) \to gh^{-1}$ is C^∞.

Examples 2.2.

1. $\mathbb{R}^n$ under addition is a Lie group.

2. The matrix group $\mathrm{GL}_n(\mathbb{R})$ is a Lie group. Here $\mathrm{GL}_n(\mathbb{R})$ is an open subset of $\mathrm{M}_n(\mathbb{R})$ and has the induced manifold structure. Matrix multiplication is clearly

---

2000 *Mathematics Subject Classification*. Primary: 20G20, 57S15; Secondary: 22C05.
I thank David Wehlau for taking notes and Jianjun Chuai for typing them up.
The author was partially supported by NSF grant DMS 0070472.
This is the final form of the paper.

C^∞, and for $g \in \mathrm{GL}_n(\mathbb{R})$, $g^{-1} = \big(1/\det(g)\big)\widetilde{g}$, where $\widetilde{g}$ is the adjoint matrix of g. The entries of $\widetilde{g}$ are, up to sign, determinants of minors of g. Hence $g \mapsto g^{-1}$ is C^∞, and $\mathrm{GL}_n(\mathbb{R})$ is a Lie group. Similarly, $\mathrm{GL}_n(\mathbb{C})$ is a Lie group.

3. $\mathrm{O}_n(\mathbb{R}) = \{g \in \mathrm{GL}_n(\mathbb{R}) \mid gg^t = I\}$ is a subgroup of $\mathrm{GL}_n(\mathbb{R})$ where g^t denotes the transpose of $g \in \mathrm{GL}_n(\mathbb{R})$ and I denotes the $n \times n$ identity matrix. Then $\mathrm{O}_n(\mathbb{R})$ is closed in $\mathrm{GL}_n(\mathbb{R})$ and is a Lie group (see 2.6). Similarly, $\mathrm{U}_n(\mathbb{C}) = \{g \in \mathrm{GL}_n(\mathbb{C}) \mid g\bar{g}^t = I\}$ is a Lie group, where $\bar{g}$ denotes the complex conjugate of $g \in \mathrm{GL}_n(\mathbb{C})$. Both $\mathrm{O}_n(\mathbb{R})$ and $\mathrm{U}_n(\mathbb{C})$ are compact since they are closed bounded subsets of matrices.

4. If G, H are Lie groups, then $G \times H$ is a Lie group.

5. Products of $S^1 = \mathrm{U}_1(\mathbb{C})$ are (compact) tori.

A *homomorphism of Lie groups from G to H* is a homomorphism $\sigma : G \to H$ of groups which is C^∞.

REMARK 2.3. A continuous homomorphism of Lie groups is automatically C^∞.

A (*left*) *action of G* on a (Hausdorff) topological space U is a continuous map

$$G \times U \xrightarrow{\tau} U, \quad (g, x) \mapsto gx := \tau(g, x)$$

where we assume that

1. The identity $e \in G$ acts as the identity on U, i.e., $ex = x$ for all $x \in U$.
2. $g(hx) = (gh)x$ for all $x \in U$, g, $h \in G$.

Thus G acts on U as homeomorphisms. We call U a *G-space.* If U is a C^∞-manifold and τ is C^∞, then we call U a *G-manifold.*

EXAMPLES 2.4. There are several natural smooth actions of G on itself. Define actions λ, ρ, and conj by $\lambda(g)x = gx$, $\rho(g)x = xg^{-1}$ and $\mathrm{conj}(g)x = gxg^{-1}$ for g, $x \in G$.

Let $f : U \to U'$ be a map between two G-spaces. We say that f is *equivariant* if $f(gx) = gf(x)$ for all $g \in G$ and $x \in U$.

2.5. Closed subgroups. Let M be a G-manifold and let $x \in M$. Then

$$G_x := \{g \in G : gx = x\}$$

is the *isotropy group at x*. Is G_x a Lie group in some way? It is a closed subgroup of G and we may apply

THEOREM 2.6. *Let H be a closed subgroup of a Lie group G. Then there exists a unique Lie group structure on H such that the inclusion $i : H \hookrightarrow G$ is a homomorphism of Lie groups.*

2.7. Homogeneous spaces. A G-manifold M is *homogeneous* if it consists of a single G-orbit. Choosing $x \in M$ we have a bijection:

$$G/G_x \to M, \quad gG_x \to gx.$$

We want to give G/G_x the structure of a G-manifold so that it is equivariantly diffeomorphic to M: Let H be a closed subgroup of G and let $\pi : G \to G/H$ be the natural map. We give G/H the quotient topology.

THEOREM 2.8.

1. *There exists a unique C^∞-manifold structure on G/H such that the map*
$$G \times G/H \to G/H, \quad (g, g'H) \to gg'H$$
is C^∞, i.e., G/H is a G-manifold.
2. *Every $gH \in G/H$ has a neighborhood U together with a smooth map $s\colon U \to G$ such that $\pi \circ s = \mathrm{Id}_U$.*
3. *Let M be a smooth G-manifold, and let $x \in M$. If Gx is closed in M (e.g., G is compact) then the orbit mapping $g \mapsto gx$ induces a diffeomorphism $\varphi\colon G/G_x \simeq Gx \hookrightarrow M$. If M is homogeneous, then $G/G_x \simeq M$.*

REMARK 2.9. Let s and U be as in 2 above. Then we have an isomorphism of H-manifolds
$$U \times H \simeq \pi^{-1}(U), \quad (x, h) \mapsto \sigma(x)h; \qquad x \in U,\ h \in H,$$
where H acts via ρ on $\pi^{-1}(U)$ and on the factor H of $U \times H$ (see 2.4). In particular, $G \to G/H$ is locally just a projection

EXAMPLES 2.10.

1. Let $n \in \mathbb{N}$, $n \geq 1$. Since $\mathrm{O}_{n+1}(\mathbb{R})$ acts transitively on S^n and since the isotropy group of the point $(1, 0, \dots, 0)$ is isomorphic to a copy of $\mathrm{O}_n(\mathbb{R})$, we have an isomorphism $S^n \simeq \mathrm{O}_{n+1}(\mathbb{R})/\mathrm{O}_n(\mathbb{R})$.
2. Similarly, complex projective space $\mathbb{CP}^n$ is isomorphic to $\mathrm{U}_{n+1}(\mathbb{C})/\big(\mathrm{U}_n(\mathbb{C}) \times \mathrm{U}_1(\mathbb{C})\big)$.

2.11. Algebraic groups. A Lie group G is called a *complex linear algebraic group* if it is an algebraic subgroup (defined by polynomial equations) of some $\mathrm{GL}_n(\mathbb{C})$. *Real linear algebraic groups* are defined similarly. For example, $\mathrm{O}_n(\mathbb{R})$ is a (real) linear algebraic subgroup of $\mathrm{GL}_n(\mathbb{R})$ since it is $\{g \in \mathrm{GL}_n(\mathbb{R}) \mid gg^t = I\}$, while $\mathrm{O}_n(\mathbb{C})$ is a (complex) linear algebraic subgroup of $\mathrm{GL}_n(\mathbb{C})$ defined by the same equations.

If G is a complex linear algebraic group and M is a complex affine variety with algebraic action $\tau : G \times M \to M$, then we say that M *is a complex affine G-variety.* Similarly, we can define *real affine G-varieties* for G a real linear algebraic group. If M is a complex (resp. real) affine G-variety, then we denote the algebra of complex (resp. real) polynomial functions on M by $\mathcal{O}(M)$ (resp. $\mathcal{O}_\mathbb{R}(M)$).

2.12. Representations. Let W be a real vector space and $\sigma : G \to \mathrm{GL}(W)$ a homomorphism of groups. Then we say that σ *is a representation of G*, or that W *is a G-module.* If $\dim W < \infty$, then we always assume that the action is rational, i.e., that σ is C^∞. This is the same as a G-manifold structure on W where each $g \in G$ acts linearly. We often use the notation (W, G) to indicate that W is a G-module. We will always assume our modules are finite dimensional unless specified otherwise.

We say that a representation W is *locally finite* if it is a sum of finite dimensional G-modules. Most of the infinite dimensional G-modules that we consider will be locally finite. For example, if W is a finite dimensional K-module, then the action of K on $\mathcal{O}_\mathbb{R}(W)$; k, $f \mapsto f \circ k^{-1}$, is locally finite (since K preserves the degree of elements of $\mathcal{O}_\mathbb{R}(W)$.)

The definitions above make sense if we consider complex vector spaces instead of real ones. We use the terminology "real G-module" and "complex G-module" to differentiate between the two cases. We leave the definition of modules for real and complex algebraic groups to the reader.

2.13. Haar measure. A very useful tool in the theory of Lie groups is Haar measure, especially in the case of compact Lie groups. Let G be a Lie group. Denote the continuous compactly supported complex valued functions on G by $C_c(G)$. A measure μ on G is *left-invariant* if for all $f \in C_c(G)$ and for all $g \in G$,

$$\int_G f\,d\mu = \int_G f \circ \lambda(g^{-1})\,d\mu.$$

Up to a positive constant, there exists a unique nonzero left invariant measure on G, called the *Haar measure*. If G is compact, we can and will assume that $\int_G 1 d\mu = 1$.

EXAMPLES 2.14.

1. If G is discrete, then Haar measure can be taken to be counting measure.
2. If G is finite, one uses counting measure divided by the order $|G|$ of G. Thus for f a function on G, we have $\int_G f\,d\mu = 1/|G| \sum_{g \in G} f(g)$.
3. If $G = S^1 \cong \mathbb{R}/2\pi\mathbb{Z}$ and $f \in C(S^1)$, then
$$\int_G f d\mu = \frac{1}{2\pi} \int_0^{2\pi} f(x)\,dx.$$
4. If $G = \mathbb{R}^n$ under addition, then μ is any positive multiple of Lebesgue measure.

3. Elementary properties of compact group actions

A general reference for this section is [**2**]. From now on K denotes a compact Lie group.

3.1. Complete reducibility of representations of compact groups. We say that an inner product $(\cdot\,,\cdot)$ on a real K-module W is *K-invariant* if $(v, w) = (kv, kw)$ for all v, $w \in W$ and $k \in K$.

PROPOSITION 3.2. *Let W be a real K-module. Then W admits a K-invariant inner product $(\cdot\,,\cdot)$.*

PROOF. Let $\{\cdot\,,\cdot\}$ be any inner product on W. Set $f_{v,w}(k) = \{k^{-1}v, k^{-1}w\}$; v, $w \in W$, $k \in K$. Define (v, w) to be $\int_K f_{v,w}(k)\,d\mu$. Then it is easy to check that $(\cdot\,,\cdot)$ is indeed an inner product. To check that it is invariant, we need to show that $(lv, lw) = (v, w)$ for any $l \in K$, v, $w \in W$. But $(lv, lw) = \int_K f_{v,w} \circ \lambda(l^{-1})\,d\mu = \int_K f_{v,w}\,d\mu = (v, w)$. □

COROLLARY 3.3. *Let W be a real K-module and W_1 a K-stable submodule. Then there exists a K-stable submodule W_2 such that $W = W_1 \oplus W_2$.*

PROOF. Choose $(\cdot\,,\cdot)$ K-invariant and take $W_2 = W_1^{\perp}$. □

Of course, we have the analog of 3.2 for hermitian inner products on complex K-modules and the analog of 3.3 for complex K-modules.

COROLLARY 3.4. *Every real (resp. complex) K-module is completely reducible, i.e., can be written as a direct sum of irreducible real (resp. complex) K-modules.*

For any locally finite K-module W we have the fixed points W^H and the (equivariant) projection $\rho\colon W \to W^K$ with kernel the sum of the non-trivial K-submodules of W. Then $\rho(w) = \int_K kw\,d\mu$, $w \in W$. The projection ρ is called the *Reynolds operator.*

3.5. Quotients by compact groups. Since K is compact we have

LEMMA 3.6. *Let U be a K-space. Then the action $K \times U \to U$ is a closed mapping.*

LEMMA 3.7. *Let U be a K-space and F a closed invariant subspace. Then F has a neighborhood basis consisting of K-invariant open sets.*

PROOF. If V is a neighborhood of F, then V^c (the complement of V) is closed, hence the image $K \times V^c \to U$, $(k,x) \mapsto kx$, is closed. Thus $(KV^c)^c$ is a K-invariant neighborhood of F contained in V. □

Let U be a K-space. We give U/K (the set of orbits) the quotient topology, i.e., $S \subseteq U/K$ is open if and only if $\pi_U^{-1}(S)$ is open, where $\pi_U : U \to U/K$ is the quotient map. Note that π_U is automatically continuous.

LEMMA 3.8. *Let U be a K-space. Then*

1. *$\pi_U : U \to U/K$ is open.*
2. *π_U is closed.*
3. *U/K is Hausdorff.*

PROOF. If $B \subset U$ is open, then $\pi_U^{-1}(\pi_U(B)) = K \cdot B$, which is open, and we have 1. If $C \subset U$ is closed, then $\pi_U^{-1}(\pi_U(C)) = K \cdot C$, which is closed, giving 2. Finally, if $Kx \neq Ky$ for x, $y \in U$, then Kx and Ky, being compact, can be separated by disjoint open subsets in U. By Lemma 3.7 above, these open sets can be taken to be K-invariant, in which case their images separate Kx and Ky in U/K. □

3.9. Hilbert's finiteness theorem. Let W be a real K-module. We have

THEOREM 3.10 (Hilbert). *The algebra $\mathcal{O}_{\mathbb{R}}(W)^K = \mathbb{R}[W]^K$ has a finite set of homogeneous generators $p_1, \ldots, p_d$.*

PROOF. Let $\mathcal{O}_{\mathbb{R}}(W)_n$ denote the homogeneous elements of degree n in $\mathcal{O}_{\mathbb{R}}(W)$, and similarly for $\mathcal{O}_{\mathbb{R}}(W)_n^K$. Let J denote the ideal in $\mathcal{O}_{\mathbb{R}}(W)$ which is generated by $\oplus_{n>0}\mathcal{O}_{\mathbb{R}}(W)_n^K$. The Hilbert basis theorem gives us finitely many homogeneous elements $h_1, \ldots, h_r \in \mathcal{O}_{\mathbb{R}}(W)^K$ which generate J.

Let $h \in \mathcal{O}_{\mathbb{R}}(W)^K$ be homogeneous of positive degree. Then $h = \sum_{i=1}^r h_i f_i$ where the h_i and f_i are homogeneous (and $\deg f_i + \deg h_i = \deg f$ for all i). Applying the Reynolds operator ρ allows us to replace the f_i by their projections to $\mathcal{O}_{\mathbb{R}}(W)^K$. Note that each $\mathcal{O}_{\mathbb{R}}(W)_n^K$ is K-stable, so ρ preserves degree. Now by induction on degree, each f_i is a polynomial in the h_i, hence so is h. □

3.11. Semialgebraic sets. An *elementary semialgebraic set* is a subset of $\mathbb{R}^n$ defined by simultaneous polynomial equalities and inequalities (e.g., $f(x) > 0$, $g(x) = 0$). A *semialgebraic set* is defined to be a finite union of elementary semialgebraic sets. Tarski and Seidenberg proved that if $U \subset \mathbb{R}^n$ is semialgebraic and $p : \mathbb{R}^n \to \mathbb{R}^d$ is a polynomial map, then $p(U)$ is also semialgebraic.

Let W be a finite dimensional K-module and let $p = (p_1, \dots, p_d) : W \to \mathbb{R}^d$ be a polynomial mapping whose entries are homogeneous generators of $\mathcal{O}_{\mathbb{R}}(W)^K$. Let J denote the ideal of relations of the p_i in $\mathbb{R}[y_1, \dots, y_d]$, that is, $h \in J$ if and only if $h(p_1, \dots, p_d) \equiv 0$. There is a corresponding closed algebraic set Z in $\mathbb{R}^d$, the zeros of J. Let X denote $\operatorname{Im} p \subset Z$. Then, by construction, Z is the Zariski closure of the semialgebraic set X.

Since $p : W \to X$ is constant on K-orbits, there is an induced map $\bar{p} : W/K \to X$ such that $p = \bar{p} \circ \pi_W$.

• THEOREM 3.12. *Let p, X, etc. be as above,*

1. *The maps p and $\bar{p}$ are proper. Hence X is closed in $\mathbb{R}^d$.*
2. *The map p separates disjoint closed K-invariant sets.*
3. *The map $\bar{p}$ is injective and $\bar{p} : W/K \to X$ is a homeomorphism.*

PROOF. We prove 2. If $Kx \neq Ky$, x, $y \in W$, then we can find a continuous function f which is 0 on Kx and 1 on Ky. By Weierstrass we can find a polynomial g such that $g < \frac{1}{2}$ on Kx and $g > \frac{1}{2}$ on Ky. By averaging over K we may assume that g is K-invariant. Since g takes different values on Kx and Ky, so does p, and hence p separates K-orbits. We leave 1 and 3 to the reader. □

EXAMPLES 3.13.

1. $W = \mathbb{R}$, $K = \{\pm 1\}$ acting by multiplication. Then $p = x^2 : W \to X = \{y \in \mathbb{R} : y \geq 0\} \subset Z = \mathbb{R}$.
2. $W = \mathbb{R}^2$, $K = \{\pm 1\}$ acting by multiplication. Then
$$\mathcal{O}_{\mathbb{R}}(W)^K = \mathbb{R}[x, y]^{\pm 1} = \mathbb{R}[x^2, xy, y^2].$$
Let $\alpha = \frac{1}{2}(x^2 + y^2)$, $\beta = \frac{1}{2}(x^2 - y^2)$, $\gamma = xy$. Then there is the relation $\alpha^2 - \beta^2 - \gamma^2 = 0$. Set $p = (\alpha, \beta, \gamma) \colon W \to \mathbb{R}^3$. We have
$$Z = \{(y_1, y_2, y_3) : y_1^2 - y_2^2 - y_3^2 = 0\}$$
and
$$X \subset \{(y_1, y_2, y_3) \in Z : y_1 \geq 0\}.$$
(One can show that the inclusion above is actually an equality.) Again, $X \neq Z$.

We may ask the following questions: How can one find the inequalities defining X? How nice is X as a semialgebraic set? Is it elementary?

Assume $W \cong \mathbb{R}^n$ such that K preserves the standard inner product $(\cdot\,,\cdot)$. Let $A = (\partial p_i/\partial x_k)_{1\leq i\leq d, 1\leq k\leq n}$. Then one computes that $(AA^t)_{ij} = (\operatorname{grad}(p_i), \operatorname{grad}(p_j)) = \sum_k \partial p_i/\partial x_k \partial p_j/\partial x_k \in \mathcal{O}_{\mathbb{R}}(W)^K \simeq \mathcal{O}_{\mathbb{R}}(Z)$; $1 \leq i, j \leq d$. Thus AA^t can be considered as a symmetric matrix of functions on Z, call it Grad.

THEOREM 3.14 (Procesi–Schwarz [**18**]). $X = \{z \in Z : \operatorname{Grad}(z) \geq 0\}$.

By $\operatorname{Grad}(z) \geq 0$ we mean that $\operatorname{Grad}(z)$ is positive semidefinite. If $z = p(w) \in X$, then $\operatorname{Grad}(z) = A(w)A(w)^t$ which is automatically positive semidefinite. Note that the condition $B \geq 0$ for a symmetric matrix B is equivalent to a simultaneous set of inequalities $\{B_\alpha \geq 0\}$ where B_α runs over the determinants of symmetric minors of B. Hence we get:

COROLLARY 3.15. *X is an elementary semialgebraic set.*

EXAMPLE 3.16. Let S_2 (the symmetric group on two letters) act on $\mathbb{R}^2$ by permutations. Then $\mathbb{R}[\mathbb{R}^2]^{S_2}$ is generated by the polynomials $p_1(s,t) = s+t$ and $p_2(s,t) = st$, $(s,t) \in \mathbb{R}^2$. Let $p = (p_1, p_2)\colon \mathbb{R}^2 \to \mathbb{R}^2$. Then $\operatorname{Im} p = \{(b_1, b_2) \in \mathbb{R}^2 \mid x^2 - b_1 x + b_2$ has real roots$\}$. One can compute that Grad is the matrix $\begin{pmatrix} 2 & p_1 \\ p_1 & p_1^2 - 2p_2 \end{pmatrix}$. Then Grad ≥ 0 if and only if $2 \geq 0$, $p_1^2 - 2p_1 \geq 0$ and $p_1^2 - 4p_2 \geq 0$. Thus $x^2 - b_1 x + b_2$ has real roots if and only if $b_1^2 - 4b_2 \geq 0$, not exactly a new result. However, the method of this example generalizes to polynomials of arbitrary degree, see [**18**].

4. Differentiable slice theorem

A general reference for this section is [**2**].

Let M be a K-manifold, $x \in M$, and suppose that $Kx = \{x\}$. Then $T_x M$ is a K-module, where $k \in K$ acts by its differential $(dk)_x$. We show that we can linearize the action of K near x:

LEMMA 4.1. *There are K-invariant neighborhoods U of $x \in M$ and B of $0 \in T_x M$ and a K-equivariant diffeomorphism $\phi : B \to U$.*

PROOF. Choose a smooth function $f : U \to T_x M$, where U is a neighborhood of x, such that $x \mapsto 0$ and such that $df_x : T_x M \to T_0 T_x M \cong T_x M$ is the identity. Set $\hat{f} = \int_K (k \circ f \circ k^{-1})\, d\mu$. Then $\hat{f}$ is K- equivariant and $d\hat{f}_x$ is still the identity. Now apply the inverse function theorem. □

4.2. Twisted product. Let H be a closed subgroup of K and Y an H-manifold. The *twisted product* $K *^H Y$ is the orbit space of $K \times Y$ with respect to the action: $h(k,y) = (kh^{-1}, hy)$; $h \in H$, $k \in K$, $y \in Y$. We denote the orbit of (k,y) by $[k,y]$. We have the canonical map $\gamma : K *^H Y \to K/H$, $[k,y] \to kH$. Then γ is K-equivariant, where $k'[k,y] = [k'k, y]$; k', $k \in K$, $y \in Y$. It follows from 2.9 that $K *^H Y$ is smooth and locally isomorphic to products $U \times Y$, where $U \subset K/H$ is open. Moreover, γ is locally just the projection $U \times Y \to U$. Because of the local product structure, γ is called a *fiber bundle with base space K/H and fiber Y*. (There is actually more structure to a fiber bundle than we have indicated. See [**2**].)

Special case: Suppose that $Y = W$ is an H-module. Then $K *^H W$ is a *homogeneous vector bundle*. Every fiber of $\gamma : K *^H W \to K/H$ is naturally a vector space, and if $k \in K$, then $k\colon \gamma^{-1}(x) \to \gamma^{-1}(kx)$ is linear. (Again, there is more structure than we have indicated.)

Let M be a K-manifold and $x \in M$. Since K_x is compact, $T_x M = T_x(Kx) \oplus N_x$, where N_x is a K_x-module and is a normal space to the orbit at x. If $k \in K$, then $(dk)_x(N_x)$ is a normal space at kx. So $\bigcup_k (dk)_x(N_x) \simeq K *^{K_x} N_x$ is the normal bundle to the orbit Kx. The differentiable slice theorem (DST) just says that a neighborhood of an orbit Kx is equivariantly isomorphic to the normal bundle of Kx. Lemma 4.1 is the special case where $Kx = \{x\}$.

THEOREM 4.3 (DST). *Let K, M, x, etc. be as above. Then there exists a K_x-invariant neighborhood D of 0 in N_x and a K-isomorphism $K *^{K_x} D \to U$, where U is a K-invariant neighborhood of Kx.*

We define the *slice* S to be the image of D in M.

COROLLARY 4.4.

1. *If $k \in K$ and $kS \cap S \neq \emptyset$, then $k \in K_x$.*

2. *Every K-orbit in U intersects S in a K_x-orbit.*
3. $U/K \cong S/K_x \cong D/K_x$.

4.5. Isotropy type. If H is a subgroup of K, let (H) denote the conjugacy class of H in K and define

$$M^{(H)} = \{x \in M : K_x \in (H)\}.$$

Points of $M^{(H)}$ are said to have *isotropy type* (H). We define $(M/K)^{(H)}$ to be the image of $M^{(H)}$ in M/K. There is a partial order on conjugacy classes: $(H_1) \leq (H_2)$ if and only if H_1 is contained in a conjugate of H_2. From Corollary 4.4 2 we have:

COROLLARY 4.6. *Let $x \in M$. Then there is a K-invariant neighborhood U of x such that for all $y \in U$, $(K_y) \leq (K_x)$.*

Our K-manifold M and its quotient space M/K are stratified by isotropy type. We will see in Theorem 4.11 that the strata are smooth manifolds.

4.7. Slice type. We define a stratification which is slightly finer than the one by isotropy type. If $x \in M$, then the action of K_x on $N_x \cong T_xM/T_x(Kx)$ is the *slice representation of K_x at x*. If $H \subseteq K$ and W is a K-module, we say that x has *slice type* (W, H) if

1. $H = K_{kx}$ for some $k \in K$ and
2. the action of H on W is isomorphic to the action of $H = K_{kx}$ on N_{kx}.

We define $M^{(W,H)} = \{x \in M : x \text{ has slice type } (W, H)\}$.

PROPOSITION 4.8. *Let $x \in M$. Then there exists a K-invariant neighborhood U of x such that*

$$M^{(K_x)} \cap U = M^{(N_x,K_x)} \cap U.$$

So, isotropy type is locally the same as slice type. For M a K-module, isotropy type determines slice type and the stratifications by isotropy type and slice type agree.

PROOF. By the DST, we can reduce to the case of a representation of a subgroup of K. So, without loss of generality, assume that $M = W$ is a K-module. Let $x \in W$. Then we have isomorphisms of K_x-modules

$$\begin{aligned} W &\cong T_x(Kx) \oplus N_x, \\ &\cong T_{eK_x}(K/K_x) \oplus N_x \end{aligned}$$

so that isotropy type determines slice type. □

EXAMPLES 4.9.

1. Let $M = \mathbb{RP}^2$, $K = \{\pm 1\}$. For a point $y \in \mathbb{RP}^2$ with homogeneous coordinates (x_0, x_1, x_2) we have $ky = (kx_0, x_1, x_2)$; $k \in K$. Then,
$$M^K = \{(1,0,0)\} \cup \{(0, x_1, x_2)\} \simeq \text{pt} \cup S^1.$$
Clearly the slice representations at $(1,0,0)$ and at $(0,0,1)$ are different.
2. $(W, K) = (\mathbb{R}^n \oplus \mathbb{R}^n, \mathrm{O}_n(\mathbb{R}))$, $n \geq 2$. Take $w = (w_1, w_2) \in W$. There are three cases:
   (a) $w = 0$. The slice representation is (W, K) again.
   (b) $w \neq 0$ and $w_1 \wedge w_2 = 0$. Then the slice representation is $(\mathbb{R}^{n-1} \oplus \theta_2, \mathrm{O}_{n-1}(\mathbb{R}))$. Here θ_t denotes the trivial representation of dimension t, $t \in \mathbb{N}$.
   (c) $w_1 \wedge w_2 \neq 0$. Then the slice representation is $(\theta_3, \mathrm{O}_{n-2}(\mathbb{R}))$.

In 4.9 2 define $\alpha(w_1, w_2) = \frac{1}{2}(|w_1|^2 + |w_2|^2)$, $\beta(w_1, w_2) = \frac{1}{2}(|w_1|^2 - |w_2|^2)$ and $\gamma(w_1, w_2) = w_1 \cdot w_2$. Then the Schwarz inequality says that $\alpha^2 \geq \beta^2 + \gamma^2$. One can show that the orbit space is isomorphic to the closed cone $\{y_1^2 \geq y_2^2 + y_3^2,\ y_1 \geq 0\}$ in $\mathbb{R}^3$. The isotropy type strata correspond to the origin, the boundary of the cone (minus the origin) and the interior of the cone. These strata are also the natural ones from singularity theory. This is not an accident. The two stratifications always coincide by a theorem of Bierstone [**1**].

4.10. The Strata are smooth.

THEOREM 4.11. *Let H be an isotropy group of the K-manifold M. Then*

1. *$M^{(H)}$ and $(M/K)^{(H)}$ have natural manifold structures.*
2. *$M^{(H)} \to (M/K)^{(H)}$ is a fiber bundle with fiber K/H.*

PROOF. Using the DST we may reduce to the case that $M = K *^H W \to K/H$ is a homogeneous vector bundle. Then $M^{(H)} = K *^H W^H \cong K/H \times W^H$ and $(M/K)^{(H)} \cong W^H$, giving 1 and 2. □

For the number of strata we have the following result.

THEOREM 4.12 ([**13**]). *Let M be a K-manifold such that*

1. *The $\mathbb{Z}$-homology of M is finitely generated, and*
2. *M is orientable, or its orientable double cover has finitely generated $\mathbb{Z}$-homology.*

Then there are finitely many isotropy types.

COROLLARY 4.13. *Let M be a K-module or a compact K-manifold. Then the number of isotropy types is finite.*

REMARK 4.14. It is easy to establish the corollary above directly; see [**20**, Ch. III 2.4].

Here is a nice characterization of finiteness of isotropy types.

THEOREM 4.15 ([**14, 17**]). *A K-manifold M embeds equivariantly in a K-module if and only if the number of isotropy types of M is finite.*

COROLLARY 4.16.

1. *A K-manifold M has finitely many isotropy types if and only if it has finitely many slice types.*
2. *If M is compact, then M embeds equivariantly in a K-module W.*

4.17. Principal orbits. Assume that M is connected (or, more generally, that M/K is connected). We find that there is a biggest stratum, hence a smallest isotropy group.

COROLLARY 4.18 (of the DST). *A stratum $M^{(W,H)}$ is open in M if and only if W is a trivial representation of H.*

THEOREM 4.19. *Let M be a K-manifold where M/K is connected. Let (H) be an isotropy class. Then the following are equivalent:*

1. *$M^{(H)}$ is open in M.*
2. *$M^{(H)}$ is dense in M.*
3. *$(M/K)^{(H)}$ is open in M/K.*

4. *$(M/K)^{(H)}$ is dense in M/K.*
5. *The slice representation at points of $M^{(H)}$ is trivial.*
6. *(H) is minimal in the partial order of isotropy classes.*
7. *(H) is the minimum isotropy class.*

REMARKS 4.20. Clearly, from the above, there is a unique (H) satisfying 1–7. Moreover, one can show that the corresponding stratum $(M/K)^{(H)}$ is connected.

DEFINITIONS 4.21.

1. An H as above is called a *principal isotropy group.*
2. Orbits of type (H) are called *principal orbits.*
3. (H) is called the *principal isotropy class.*

EXAMPLES 4.22.

1. For any effective action of a finite group on a connected manifold the (unique) principal isotropy group is the trivial group.
2. For the action of $\mathrm{O}_n(\mathbb{R})$ in 4.9 a principal isotropy group is $\mathrm{O}_{n-2}(\mathbb{R})$.

5. The Peter–Weyl theorem

The book [**3**] is a general reference for this section.

Let M be a K-space. From the action of K on M we obtain an action (not generally locally finite!) on $C(M)$, the continuous complex valued functions on M, by:

$$(k \cdot f)(x) = f(k^{-1}x); \quad f \in C(M),\ k \in K,\ x \in M.$$

We say that $f \in C(M)$ is *K-finite* if f lies in a finite dimensional K-stable subspace of $C(M)$, i.e., if $\mathrm{Span}\{f \circ k^{-1} : k \in K\}$ is finite dimensional. Let $C(M)_F$ denote the set of K-finite functions. Then K acts on $C(M)_F$ and $C(M)_F$ is an algebra. In case $M = K$, we are considering K-finite functions relative to the λ-action of K on K(see 2.4).

Let $\sigma : K \to \mathrm{GL}(W)$ be a finite dimensional complex representation of K. Relative to a basis of W, $\sigma(k)$ has matrix $\big(\sigma_{ij}(k)\big)$; $k \in K$, $1 \le i, j \le \dim W$. The span of the matrix entries σ_{ij} is K-invariant, hence consists of K-finite functions.

LEMMA 5.1. $\mathrm{Span}\{\sigma_{ij}\} \subset C(K)$ *is independent of the basis of W.*

PROOF. Let

$$\begin{aligned} \phi_W : W \otimes W^* &\to C(K), \\ w \otimes w^* &\mapsto \big(k \mapsto w^*(k^{-1}w)\big). \end{aligned}$$

Then $\mathrm{Im}\,\phi_W$ is the span of the matrix entries of the dual of σ. □

We have a natural action of $K \times K$ on $W \otimes W^*$:

$$(k_1, k_2)(w \otimes w^*) = k_1 w \otimes w^* \circ k_2^{-1}; \quad (k_1,\ k_2) \in K \times K,\ w \in W,\ w^* \in W^*.$$

We also have a natural action of $K \times K$ on $C(K)$ coming from the product of the actions λ and ρ of Example 2.4:

$$\big((k_1, k_2)f\big)(x) = f(k_1^{-1} x k_2); \quad (k_1,\ k_2) \in K \times K,\ x \in K,\ f \in C(K).$$

With respect to these actions, the map ϕ_W above is $(K \times K)$-equivariant.

LEMMA 5.2. *Suppose that $W \subseteq C(K)$ is a finite dimensional representation of K. Then $W \subseteq \mathrm{Im}\,\phi_W$.*

PROOF. Let $\tau : W \to \mathbb{C}$ be evaluation at e. Then for $k \in K$ and $w \in W$ we have

$$(w \otimes \tau)(k) = \tau(k^{-1}w) = \tau(w \circ k) = w(ke) = w(k).$$

□

Let Ω denote the set of isomorphism classes of irreducible finite dimensional complex representations of K. For $\omega \in \Omega$, let $V(\omega)$ denote a representative. The span of the collection of finite dimensional submodules of $C(K)$ isomorphic to $V(\omega)$ is denoted $C(K)_\omega$. It is clearly stable for the action of $K \times K$ on $C(K)$. Then we have that, via $\phi_{V(\omega)}$, an isomorphism of $(K \times K)$-modules, $C(K)_\omega \simeq V(\omega) \otimes V(\omega)^*$. Hence

$$C(K)_F = \oplus_{\omega\in\Omega} C(K)_\omega \cong \oplus_{\omega\in\Omega} V(\omega) \otimes V(\omega)^*.$$

Note that $C(K)$ is a Banach space under the sup norm.

THEOREM 5.3 (Peter–Weyl). *$C(K)_F$ is dense in $C(K)$.*

This theorem has several important consequences.

COROLLARY 5.4.

1. *K has a faithful finite dimensional representation.*
2. *$C(K)_F$ is a finitely generated algebra.*
3. *If $K \subseteq \mathrm{O}_n(\mathbb{R})$, then K is a real algebraic subgroup of $\mathrm{O}_n(\mathbb{R})$ and the complex valued polynomial functions on K are $C(K)_F$.*

PROOF.

1. Let W be any representation of K. Let H be the kernel of this representation and assume $H \neq \{e\}$. Since $C(K)_F$ is dense in $C(K)$, there exists a K-module $W' \subset C(K)_F$ which is nontrivial when restricted to H. So $H' := \mathrm{Ker}(W \oplus W')$ is a proper subgroup of H. Note that H' has lower dimension or fewer connected components than H. We may continue the process with H replaced by H', and eventually we arrive at the case $H = \{e\}$, proving 1. We omit the proofs of 2 and 3.

□

REMARK 5.5. The Peter–Weyl Theorem is an easy consequence of 5.4 1.

6. Reductive groups from a compact viewpoint

The books [**3**] and [**8**] are references for this section. We show how to use compact groups to obtain information about actions of associated complex groups.

Our compact Lie group K is an algebraic subgroup of $\mathrm{O}_n(\mathbb{R})$ for some n. Let I_K be the ideal of K, and let $K_\mathbb{C}$ be the complex zeroes of I_K in $\mathrm{O}_n(\mathbb{C})$.

PROPOSITION 6.1.

1. *$K_\mathbb{C}$ is a complex linear algebraic group. Up to isomorphism, it is independent of the choice of the embedding $K \subseteq \mathrm{O}_n(\mathbb{R})$.*
2. $\dim_\mathbb{C} K_\mathbb{C} = \dim_\mathbb{R} K$.
3. *K is dense in $K_\mathbb{C}$ (in the Zariski topology).*
4. *$K_\mathbb{C} = KP \cong K \times P$, where P is diffeomorphic to a Euclidean space.*
5. *K is a maximal compact subgroup of $K_\mathbb{C}$.*

REMARKS 6.2.

1. The decomposition $K_{\mathbb{C}} = KP$ corresponds to the polar decomposition of $\mathrm{GL}_n(\mathbb{C})$. Any $g \in \mathrm{GL}_n(\mathbb{C})$ can be uniquely written as $g = kp$, where $k \in \mathrm{U}_n(\mathbb{C})$ and p is a positive definite Hermitian symmetric matrix. The set of positive definite Hermitian symmetric matrices is diffeomorphic, via the exponential map, to the set of Hermitian symmetric matrices and so is diffeomorphic to a Euclidean space.
2. (See [**5**]). Let G be any Lie group with finitely many components. Then G has maximal compact subgroups and they are all conjugate. Moreover, if K is maximal compact, then G/K is contractible. In particular, K meets every component of G.

DEFINITION 6.3. A complex linear algebraic group G is called *reductive* if $G \simeq K_{\mathbb{C}}$, where K is a compact Lie group. This is not the standard definition of complex reductive group (see [**6**]).

EXAMPLE 6.4. We note that the additive group $(\mathbb{C}, +)$ is not reductive since the only compact subgroup is $\{0\}$. On the other hand, $\mathrm{GL}_n(\mathbb{C})$ is isomorphic to the complexification of $\mathrm{U}_n(\mathbb{C})$, hence $\mathrm{GL}_n(\mathbb{C})$ is reductive.

When we speak of a module V for a complex algebraic group G, unless otherwise stated, we will assume that V is a complex vector space and that the corresponding homomorphism $G \to \mathrm{GL}(V)$ is a homomorphism of complex algebraic groups. That is, the action $G \times V \to V$ is a morphism of complex algebraic varieties. Similarly, by a G-variety we mean a complex algebraic variety with G action such that the morphism $G \times Y \to Y$ is algebraic.

6.5. Weyl's unitary trick. From 6.1 3 we obtain the complex analogues of 3.10 and 3.4.

THEOREM 6.6. *Let $G = K_{\mathbb{C}}$ be a reductive group and V a G-module. Then*

1. *V is completely reducible.*
2. *$\mathcal{O}(V)^G = \mathcal{O}(V)^K$ is finitely generated.*

6.7. Quotients by reductive groups. Let G be reductive, and let V be a G-module. We define $V/\!\!/G$ to be the affine variety with coordinate ring $\mathcal{O}(V)^G$. Let $\pi_V : V \to V/\!\!/G$ be the morphism dual to the inclusion $\mathcal{O}(V)^G \hookrightarrow \mathcal{O}(V)$. We can realize $V/\!\!/G$ as follows: Choose generators $q_1, \dots, q_e$ of $\mathcal{O}(V)^G$ and let

$$q = (q_1, \dots, q_e) : V \to \mathbb{C}^e.$$

Let $Y \subseteq \mathbb{C}^e$ be the variety of relations of the q_i. Then $Y \cong V/\!\!/G$.

PROPOSITION 6.8. *Let q, Y, etc. be as above. Then*

1. $\operatorname{Im} q = Y$.
2. *If $v \in V$, then $\overline{Gv} \setminus Gv$ is a union of orbits of lower dimension.*
3. *q separates disjoint closed G-stable subsets of V.*
4. *Every orbit contains a unique closed orbit in its closure.*
5. *$q : V \to Y$ sets up a bijection between the set of closed orbits in V and the set of points of Y.*

EXAMPLE 6.9. Let $G = \mathbb{C}^*$ act on $\mathbb{C}^2$ via $t(x, y) = (tx, t^{-1}y)$; $t \in \mathbb{C}^*$, $(x, y) \in \mathbb{C}^2$. Then one can take $q(x, y) = xy$, $(x, y) \in \mathbb{C}^2$. The fibers of q over $\mathbb{C} \setminus \{0\}$

are closed orbits, while $q^{-1}(0)$ contains the nonclosed orbits $\{(a,0) : a \neq 0\}$ and $\{(0,b) : b \neq 0\}$. They both contain the closed orbit $\{(0,0)\}$ in their closure.

REMARK 6.10. Let S be an affine G-variety. Then one can show that $\mathcal{O}(S)$ is a locally finite G-module, and it follows that there exists an equivariant embedding $i : S \hookrightarrow V$ for some G-module V. From Theorem 6.6 and local finiteness we obtain that $\mathcal{O}(V)^G|_S = \mathcal{O}(S)^G$ is finitely generated. Let $S/\!\!/G$ denote the affine variety corresponding to $\mathcal{O}(S)^G$ and let $\pi_S\colon S \to S/\!\!/G$ be dual to the inclusion $\mathcal{O}(S)^G \subset \mathcal{O}(S)$. Then the analogue of Proposition 6.8 holds and we have the following commutative diagram, where $\bar{i}$ is induced by i:

$$\begin{array}{ccc} S & \overset{i}{\hookrightarrow} & V \\ \downarrow \pi_S & & \downarrow \pi_V \\ S/\!\!/G & \overset{\bar{i}}{\hookrightarrow} & V/\!\!/G \end{array}$$

6.11. Kempf–Ness theory. Here we see how to view the quotient of a reductive group $K_{\mathbb{C}}$ in terms of quotients of K. References are [**4**] and [**7**].

Let V be a G-module, where $G = K_{\mathbb{C}}$. Let $\langle \cdot\,,\cdot \rangle$ be a K-invariant Hermitian form on V with norm $\|\cdot\|$. Let $v \in V$ and consider $F_v : G \to \mathbb{R}$, $g \to \|gv\|^2$.

THEOREM 6.12 (Kempf–Ness).

1. *F_v has a critical point if and only if Gv is closed.*
2. *All critical points of F_v are minima.*

Assume that $F_v(e)$ is a minimum. Then

3. *$Kv = \{v' \in Gv : \|v\| = \|v'\|\}$ and*
4. *$G_v = (K_v)_{\mathbb{C}}$.*

We now draw some important consequences of this theorem.

If H is an algebraic subgroup of an algebraic group G, there is always a natural structure of G-variety on the homogeneous space G/H [**6**]. However, this variety may not be affine.

COROLLARY 6.13 (Matsushima's Theorem). *Let H be an algebraic subgroup of the reductive group G. Then the homogeneous space G/H is affine if and only if H is reductive.*

PROOF. If G/H is affine, then we may equivariantly embed G/H as a closed orbit $Gv \subseteq V$ for some G-module V. Relative to a K-invariant hermitian form on V, where $G = K_{\mathbb{C}}$, there exists a point $gv \in Gv$ where $\|gv\|$ is a minimum. So $G_{gv} = (K_{gv})_{\mathbb{C}}$ is reductive. Since H is conjugate to G_{gv}, it is also reductive. Conversely, if H is reductive, then $G/\!\!/H$ is an affine G-variety, and one can show that it has the defining properties of G/H. □

DEFINITION 6.14. Let V, etc. be as in Theorem 6.12. The set

$$M = \{v \in V : (dF_v)(e) = 0\}$$

is the *Kempf–Ness set* of V.

REMARKS 6.15.

1. M is the set of points $v \in V$ which are closest to 0 on the orbit Gv.
2. $M = \{v \in V : T_v(Gv) \perp v\}$.

3. M is a real K-stable algebraic cone in V.

If S is an affine G-variety, we call $M_S = S \cap M$ the *Kempf–Ness set of* S, where we assume S is embedded in V.

PROPOSITION 6.16. *The composition* $M_S \hookrightarrow S \to S/\!\!/G$ *is proper and induces a homeomorphism:* $M_S/K \cong S/\!\!/G$.

EXAMPLE 6.17. Let $K = \mathrm{U}_n(\mathbb{C})$, $G = \mathrm{GL}_n(\mathbb{C})$, and $V = \mathrm{M}_n(\mathbb{C})$ with the conjugation actions of K and G. Use the K-invariant Hermitian form: $(A, B) \mapsto \mathrm{Tr}(A\bar{B}^t)$, A, $B \in \mathrm{M}_n(\mathbb{C})$. Then for $A \in \mathrm{M}_n(\mathbb{C})$, the tangent space to $G \cdot A$ at A is $\{AB - BA \mid B \in \mathrm{M}_n(\mathbb{C})\}$. By 6.15 2 we have

$$\begin{aligned} M &= \{A : \mathrm{Tr}\big((AB - BA)\bar{A}^t\big) = 0,\ \forall B \in \mathrm{M}_n(\mathbb{C})\} \\ &= \{A : \mathrm{Tr}(AB\bar{A}^t) = \mathrm{Tr}(BA\bar{A}^t),\ \forall B \in \mathrm{M}_n(\mathbb{C})\} \\ &= \{A : \mathrm{Tr}(B\bar{A}^t A) = \mathrm{Tr}(BA\bar{A}^t),\ \forall B \in \mathrm{M}_n(\mathbb{C})\} \\ &= \{A : \bar{A}^t A = A\bar{A}^t\}. \end{aligned}$$

The last equality shows that M is the set of normal matrices. We have

$$\mathrm{M}_n(\mathbb{C})/\!\!/\,\mathrm{GL}_n(\mathbb{C}) \cong M/\,\mathrm{U}_n(\mathbb{C}).$$

6.18. Categorical quotients. Let S be an affine G-variety. Then $\pi_S \colon S \to S/\!\!/G$ is the categorical quotient in the category of varieties. That is, if $\phi : S \to Z$ is a morphism of varieties, constant on G-orbits, then $\phi = \overline{\phi} \circ \pi_S$, where $\overline{\phi} : S/\!\!/G \to Z$ is induced by ϕ. Luna [**11**] showed even more:

THEOREM 6.19.

1. π_S *is categorical in the category of Hausdorff topological spaces.*
2. π_S *is categorical in the category of complex analytic varieties.*

6.20. Neeman's retraction theorem [15].

THEOREM 6.21. *Let* V *be a* G*-module, where* $G = K_{\mathbb{C}}$. *Let* $\langle \cdot, \cdot \rangle$ *be a* K*-invariant Hermitian product on* V*, and let* M *denote the corresponding Kempf–Ness set. Then*

1. *There exists a* K*-equivariant deformation retraction* ϕ_t *of* V *to* M, $0 \le t \le 1$, *where* $\phi_0 = id$ *and* $\phi_1 : V \to M$.
2. ϕ_t *respects the fibers of* π_V: $\phi_t(v) \in Gv$, $0 \le t < 1$, *and* $\phi_1(v) \in \overline{Gv}$.

REMARK 6.22. If $S \subseteq V$ is G-stable and closed, then ϕ_t retracts S onto $M_S = M \cap S$.

COROLLARY 6.23. *The inclusion* $M_S \hookrightarrow S$ *is a* K*-equivariant homotopy equivalence.*

From the homeomorphism of $S/\!\!/G$ and M_S/K we obtain a homotopy equivalence of $S/\!\!/G$ and S/K. We can obtain results on the topology of quotients by G from results about the topology of quotients by K. For example, if S is acyclic, then by a result of Oliver [**16**], S/K is acyclic and therefore $S/\!\!/G$ is acyclic.

6.24. Finding closed orbits. Let H be a reductive subgroup of G. We denote the centralizer (resp. normalizer) of H in G by $C_G(H)$ (resp. $N_G(H)$). Since H is reductive, so are $C_G(H)$ and $N_G(H)$. Let V be a G-module. We try to find closed orbits of G by looking for closed orbits of $C_G(H)$ (resp. $N_G(H)$) in V^H.

THEOREM 6.25 (Luna [**10**]). *Let V be a G-module and H a reductive subgroup of G. Suppose that $v \in V^H$. Then*

1. *Gv is closed if and only if $C_G(H)v$ is closed, if and only if $N_G(H)v$ is closed.*
2. *The natural morphism $V^H /\!\!/ (N_G(H)/H) \to V /\!\!/ G$ is finite.*

PROOF. We show part of 1 where Kempf–Ness plays a role. Assume that $C_G(H)v$ is closed and set $F_v(G) = \|gv\|^2$, where $\|\cdot\|$ is K-invariant and $G = K_{\mathbb{C}}$. Let $\phi = (dF_v)_e$. We may assume that v is in the Kempf–Ness set of $C_G(H)$, so that $(dF_v)_e$ vanishes on $T_e\big(C_G(H)\big)$. Since $H \subseteq G_v$, ϕ is H-invariant. Hence ϕ vanishes on the complement N to $(T_eG)^H$ in $T_e(G)$. But $T_eG = (T_eG)^H \oplus N = T_e(C_GH) \oplus N$, so $\phi \equiv 0$. It follows that $v \in M$ and that Gv is closed. □

EXAMPLE 6.26. Let $n \geq 1$ and let $G = \mathrm{O}(n+1, \mathbb{C})$ act standardly on $V = \mathbb{C}^{n+1}$, where $\mathbb{C}^{n+1}$ has coordinates $e_1, e_2, \dots, e_{n+1}$. Let H be a copy of $\mathrm{O}(n, \mathbb{C})$ acting on the first n coordinates. Then $N_G(H)/H \simeq \mathbb{Z}/2\mathbb{Z}$, so that all G-orbits through $V^H = \mathbb{C} \cdot e_{n+1}$ are closed. In fact, all closed G-orbits pass through V^H.

7. Luna's slice theorem

References for this section are [**21**] and [**9**].

Suppose that G is a finite group and V a G-module with $v \in V$. In the classical topology, the slice at v would be a small ball S around v which is stable by G_v. In the Zariski topology, such small balls are not possible, and we have to give up the idea of proving a slice theorem as in the compact case. The slices are just too big. Things can be salvaged, however, by asking that $G *^{G_v} S \to GS \subset V$ only be an isomorphism fiberwise (over the quotients) and that the only non-isomorphism be of a mild kind on the quotients:

DEFINITION 7.1. Let $\phi : X \to Y$ be a morphism of complex varieties. We say that ϕ is *étale at $x \in X$* if ϕ induces a complex analytic isomorphism of neighborhoods of x and $\phi(x)$. We say that ϕ *is étale* if it is étale at every $x \in X$.

REMARKS 7.2.

1. This is not the standard definition.
2. If X, Y are smooth, then ϕ is étale at x if and only if $d\phi_x : T_xX \to T_{\phi(x)}Y$ is an isomorphism.
3. Étaleness is open in the Zariski topology.

DEFINITION 7.3. A G-equivariant morphism $\phi : X \to Y$ of affine G-varieties is called *excellent* if ϕ and the induced morphism $\phi /\!\!/ G : X/\!\!/G \to Y/\!\!/G$ are étale, and if $(\phi, \pi_x) :\to Y \times_{Y/\!\!/G} X/\!\!/G$ (fiber product) is an isomorphism. Excellent morphisms are the replacements for isomorphism in Luna's slice theorem.

Let $H \subset G$ be reductive and let Y be an affine H-variety. Then we can define the *twisted product* $G *^H Y$ as in 4.2. The canonical morphism $G *^H Y \to G/H$ is a fiber bundle with fiber Y in the sense of Definition 7.11 below.

DEFINITION 7.4. Let S be an affine G-variety. A subset Z of S is called *G-saturated* if $Z = \pi_S^{-1}(\pi_S(Z))$.

THEOREM 7.5 (Luna's Slice Theorem). *Let X be an affine G-variety and Gx a closed orbit. Set $H = G_x$. Then*

1. *There exists a locally closed affine H-stable saturated subvariety S containing x such that $U = GS$ is G-saturated affine and open in X, and the morphism*

$$\phi : G *^H S \to U \subseteq X,$$
$$[g, s] \to gs,$$

is excellent.

If X is smooth at x, we can write

$$T_x X = T_e(Gx) \oplus N_x$$

as H-modules (H is reductive by 6.13). Then

2. *S is smooth at x, and $T_x S \cong N_x$ as H-modules.*
3. *There exists an excellent surjective H-equivariant morphism $\psi : S \to N_f$, such that $x \mapsto 0$, inducing an excellent map*

$$\tau : G *^H S \to G *^H N_f,$$

where $f \in \mathcal{O}(N)^H$ and $f(0) \neq 0$. Here $N_f = \{y \in N \mid f(y) \neq 0\}$.

Up to excellent morphisms, $G *^H N_f$ is G-isomorphic to U. For X smooth and Gx closed, the representation (N_x, G_x) is called the *slice representation at x*.

7.6. Isotropy type and slice type. Let X be an affine G-variety and let H be a reductive subgroup of G. Set

$$(X/\!\!/G)^{(H)} = \{Gx : G_x \text{ is conjugate to } H \text{ in } G\}$$

and set

$$X^{(H)} = \pi_X^{-1}\big((X/\!\!/G)^{(H)}\big).$$

So, points in $(X/\!\!/G)^{(H)}$ are closed orbits of isotropy type (H).

If X is smooth, we can define the slice type strata $(X/\!\!/G)^{(W,H)}$ as before.

COROLLARY 7.7. *Let X be a G-variety, Gx a closed orbit and $H = G_x$. Then*

1. *There exists a neighborhood U of x such that for all $y \in U$, $(G_y) \leq (G_x)$.*
2. *If X is smooth and (W, H) is the slice representation at x, then $(X/\!\!/G)^{(W,H)}$ is open if and only if $W = W^H \oplus W'$, where $\mathcal{O}(W')^H = \mathbb{C}$.*

Note that the condition on (W, H) in 2 above is precisely that there are no closed orbits in W with isotropy group smaller than H.

PROPOSITION 7.8. *Let V be a G-module. Then*

1. *The stratification of V by isotropy type is finite.*
2. *The closure of $(V/\!\!/G)^{(H)}$ is*

$$\overline{(V/\!\!/G)^{(H)}} = \bigcup_{(L)\geq(H)} (V/\!\!/G)^{(L)} = \pi_V(V^H).$$

THEOREM 7.9. *Let X be an irreducible affine G-variety and (H) an isotropy class. Then*

1. *The isotropy type stratification of X is finite.*
2. *The following are equivalent:*
   (a) $(X/\!\!/G)^{(H)}$ *is open and dense in* $X/\!\!/G$.
   (b) (H) *is a minimal isotropy class.*
   (c) (H) *is the minimum isotropy class.*

PROOF. To obtain 1 it suffices to embed X in a G-module. We omit the proof of 2. □

The H in 7.9 2 is called a *principal isotropy group.*

REMARK 7.10. If H is a principal isotropy group of (V, G), then Theorem 6.25 2 takes the following stronger form: $V^H /\!\!/ (N_G(H)/H) \simeq V/\!\!/G$ (Luna–Richardson Theorem [**12**]). In Example 6.26, we have that $V^H /\!\!/ (N_G(H)/H) \simeq \mathbb{C}/\{\pm 1\} \simeq \mathbb{C}^{n+1} /\!\!/ \mathrm{O}_{n+1}(\mathbb{C})$.

Let V be a G-module. We call the fiber $\mathcal{N}(V) = \pi_V^{-1}(\pi_V(0))$ the *null cone* of V. For example, in 6.9, the null cone is $\{(x, y) \mid x = 0 \text{ or } y = 0\}$.

DEFINITION 7.11. Let E, B and F be affine varieties and $\pi \colon E \to B$ a morphism. We say that π *is a fiber bundle over* B *with fiber* F if there is an étale surjective morphism $B' \to B$ and an isomorphism φ such that the diagram

$$\begin{array}{ccccc} E & \longleftarrow & E \times_B B' & \xrightarrow[\sim]{\varphi} & F \times B' \\ \downarrow & & \downarrow & & \downarrow \mathrm{pr}_2 \\ B & \longleftarrow & B' & = & B' \end{array}$$

commutes.

THEOREM 7.12. *Let X be a smooth affine G-variety and assume that for $x \in X$, Gx is a closed orbit. Then for $H = G_x$ we have:*

1. $(X/\!\!/G)^{(H)}$ *is a locally closed smooth subvariety of* $X/\!\!/G$.
2. *Over the connected component of* $(X/\!\!/G)^{(H)}$ *containing* Gx, $X^{(H)} \to (X/\!\!/G)^{(H)}$ *is a fiber bundle with fiber* $G *^H \mathcal{N}(W)$, *where* (W, H) *is the slice representation at* x.

8. Applications to classification problems

We see how the slice theorem can help in problems of classifying "nice" representations of a reductive group G.

DEFINITION 8.1. Let V be a G-module.

1. V is called *coregular* if $\mathcal{O}(V)^G$ is a polynomial ring, i.e., if $V/\!\!/G$ is an affine space.
2. V is called *cofree* if $\mathcal{O}(V)$ is a free $\mathcal{O}(V)^G$-module.
3. V is called *equidimensional* if all fibers of $\pi_V : V \to V/\!\!/G$ have the same dimension.

REMARKS 8.2.

1. Homological algebra gives that cofree is equivalent to coregular plus equidimensional.
2. *Popov conjecture*: If G is connected and V is equidimensional, then V is cofree.

PROPOSITION 8.3. *Let V be a G-module.*

1. *Suppose that $V = V_1 \oplus V_2$ is coregular (resp. cofree, resp. equidimensional). Then both V_1 and V_2 are coregular (resp. cofree, resp. equidimensional).*
2. *Suppose that V is coregular (resp. cofree, resp. equidimensional). Then every slice representation of V is coregular (resp. cofree, resp. equidimensional).*

PROOF. We sketch a proof of 2 for coregularity. Let Gx be closed, $H = G_x$ with slice representation (N, H). We have two morphisms which are excellent:

$$G *^H S \to GS \subseteq V$$

and

$$G *^H S \to G *^H N_f$$

where S and f are as in 7.5. Now $S/\!\!/H \xrightarrow{\varphi} GS/\!\!/G$ is an étale surjection. Since $GS/\!\!/G \subset V/\!\!/G$ is smooth, étaleness of φ implies that $S/\!\!/H$ is also smooth. We have an étale surjective map from $S/\!\!/H$ to $N_f/\!\!/H$. From étaleness and smoothness of $S/\!\!/H$ if follows that $N_f/\!\!/H$ is smooth. Since $N/\!\!/H$ is a cone, it is smooth if and only if it is smooth at $\pi_N(0)$, and $\pi_N(0) \in N_f/\!\!/H$. Hence $N/\!\!/H$ is smooth, i.e., (N, H) is coregular. □

Here we give an example where one can use the slice theorem to show that a representation is not coregular:

EXAMPLE 8.4. Consider $\wedge^2\mathbb{C}^5 \oplus (\wedge^2\mathbb{C}^5)^* \oplus 2\mathbb{C}^5 \oplus 2(\mathbb{C}^5)^*$, a representation of $\mathrm{SL}_5(\mathbb{C})$. Choose one copy each of $\mathbb{C}^5$ and $(\mathbb{C}^5)^*$. Then the element $x = (e_5, e_5^*) \in \mathbb{C}^5 \oplus (\mathbb{C}^5)^*$ lies in a closed orbit with $\big(\mathrm{SL}_5(\mathbb{C})\big)_x = \mathrm{SL}_4(\mathbb{C})$. The corresponding slice representation is (see the proof of 4.8)

$$\big(\wedge^2\mathbb{C}^4 \oplus (\wedge^2\mathbb{C}^4)^* \oplus 2\mathbb{C}^4 \oplus 2(\mathbb{C}^4)^* \oplus \theta_4, \mathrm{SL}_4(\mathbb{C})\big).$$

Taking another slice using a similar point $x = (e_4, e_4^*) \in \mathbb{C}^4 \oplus (\mathbb{C}^4)^*$ yields the new slice representation $\big(3\mathbb{C}^3 \oplus 3(\mathbb{C}^3)^* \oplus \theta_8, \mathrm{SL}_3(\mathbb{C})\big)$. Ignoring the trivial eight dimensional factor, we have a minimal generating set consisting of at least 11 invariants: 9 contractions of the copies of $\mathbb{C}^3$ and $(\mathbb{C}^3)^*$ and 2 determinants for the three copies of $\mathbb{C}^3$ and $(\mathbb{C}^3)^*$. But the dimension of the quotient of $3\mathbb{C}^3 \oplus 3(\mathbb{C}^3)^*$ by $\mathrm{SL}_3(\mathbb{C})$ is $18 - 8 = 10$. Thus the slice representation is not coregular, hence neither was our original representation.

References

1. E. Bierstone, *Lifting isotopies from orbit spaces*, Topology **14** (1975), 245–252.
2. G. Bredon, *Introduction to compact transformation groups*, Academic Press, New York, 1972.
3. C. Chevalley, *Theory of Lie groups* I, Princeton University Press, Princeton, 1946.
4. J. Dadok and V. Kac, *Polar representations*, J. Algebra **92** (1985), 504–524.
5. G. Hochschild, *The structure of Lie groups*, Holden-Day, San Francisco, 1965.
6. J. E. Humphreys, *Linear algebraic groups*, Grad. Texts in Math., No. 21, Springer-Verlag, New York–Heidelberg–Berlin, 1975.
7. G. Kempf and L. Ness, *The length of a vector in a representation space*, Algebraic Geometry, Proc. Summer Meeting, Univ. Copenhagen (Copenhagen, 1978), Lecture Notes in Math., vol. 732, 1979, Springer, Berlin, pp. 233–244.
8. H. Kraft, *Geometrische Methoden in der Invariantentheorie*, Aspekte der Mathematik, vol. D1, Vieweg Verlag, Braunschweig/Wiesbaden, 1985.
9. D. Luna, *Slices étales*, Bull. Soc. Math. France, Paris, Mémoire, vol. 33, Soc. Math. France, Paris, 1973, pp. 81–105.

10. ———, *Adhérences d'orbite et invariants*, Invent. Math. **29** (1975), 231–238.
11. ———, *Fonctions différentiables invariantes sous l'opération d'un groupe réductif*, Ann. Inst. Fourier **26** (1976), 33–49.
12. D. Luna and R. W. Richardson, *A generalization of the Chevalley restriction theorem*, Duke Math. J. **46** (1979), 487–496.
13. L. N. Mann, *Finite orbit structure on locally compact manifolds*, Mich. Math. J. **9** (1962), 87–92.
14. G. D. Mostow, *Equivariant embedding in euclidean spaces*, Ann. of Math. (2) **65** (1957), 432–446.
15. A. Neemon, *The topology of quotient varieties*, Ann. of Math. (2) **122** (1985), 419–459.
16. R. Oliver, *A proof of the Conner conjecture*, Ann. of Math. (2) **103** (1976), 637–644.
17. R. S. Palais, *Imbedding of compact differentiable transformation groups in orthogonal representations*, J. Math. Mech. **6** (1957), 673–678.
18. C. Procesi and G. Schwarz, *Inequalities defining orbit spaces*, Invent. Math. **81** (1985), 539–554.
19. G. W. Schwarz, *The topology of algebraic quotients*, Topological Methods in Algebraic Transformation Groups (H. Kraft et al., eds.) (New Brunswick, 1988), Progr. Math., vol. 80, Birkhäuser Verlag, Basel-Boston, 1989, pp. 135–152.
20. ———, *Quotients of compact and complex algebraic groups*, Théorie des Invariants et Géométrie des Variétés Quotients, Travaux en Cours **6**1, Hermann, Paris, 2000, pp. 5–83.
21. P. J. Slodowy, *Der Scheibensatz für algebraische Transformationsgruppen*, Algebraic Transformation Groups and Invariant Theory, DMV Sem., vol. 13, Birkhäuser, Basel-Boston, 1989, pp. 89–113.
22. F. W. Warner, *Foundations of differentiable manifolds and Lie groups*, Grad. Texts in Math. **94** (1983), Springer-Verlag, New York–Heidelberg–Berlin.

Department of Mathematics, Brandeis Univ., P.O Box 549110, Waltham, MA 02454-9110, USA.

E-mail address: `Schwarz@Brandeis.edu`

Centre de Recherches Mathématiques
CRM Proceedings and Lecture Notes
Volume **35**, 2004

Notes on Invariant Rings of Divided Powers

J. Segal

ABSTRACT. These are notes on the talk on invariant rings of divided powers, given at the *Workshop on Invariant Theory*, Kingston, Ontario, April 8–19 2002.

We give some basic results on these rings, showing in particular that for permutation representations or representations in non-modular characteristic the graded vector space structure is the same as for polynomial rings; finding the symmetric invariants; some results on the $\mathrm{GL}(n, \mathbb{F})$ invariants; and showing that finite generation fails in a simple case. We find also that polynomial invariant rings of permutation representations admit divided power algebra structures.

1. Background

Let $\mathbb{F}$ be a field, V an n-dimensional $\mathbb{F}$-vector space, and V^* its dual space. There are several related funtors in which we are interested. The *tensor algebra* $T(V)$ is defined as

$$V \oplus (V \otimes V) \oplus (V \otimes V \otimes V) \oplus \cdots,$$

a graded ring with $V^{\otimes^d}$ in degree d. In each degree d, the symmetric group Σ_d operates via permutation of the tensor components. The quotient is called $S(V)$, the *polynomial algebra* (or *symmetric algebra*) on V. The subalgebra of $T(V)$ which is invariant under the group actions is called $D(V)$, the *algebra of divided powers*.

We give $D(V)$ the multiplicative structure inherited from the *shuffle product* on $T(V)$ (see below). This accounts for the fundamental difference between $S(V)$ and $D(V)$: in non-zero characteristic the latter is not multiplicatively finitely generated and has zero divisors. In characteristic zero the two are isomorphic, so we are interested in the case of characteristic p, with p a prime. See e.g. [**13**, Appendix 2] for more details.

Each of the above is a functor from vector spaces to algebras. Given a linear action of a group G on V (a representation), we thus obtain an action of G on each of the above algebras. Classical invariant theory is concerned with the elements of $S(V^*)$ invariant under such an action; see [**20**] for a good account of current knowledge. There is little on the invariants of $T(V)$; see [**3**]. In this paper we shall examine the G-invariants of $D(V)$.

2000 *Mathematics Subject Classification.* Primary: 13A50; Secondary: 16W30.
This is the final form of the paper.

There are a number of ways of looking at divided powers. Algebraically, they are the Hopf duals to polynomial rings; topologically, they are homology rings dual to polynomial cohomology rings. They appear as the algebra of partial differential operators on polynomial functions from a free R-module F to R. Each of these points of view gives us different tools in studying these rings, which are otherwise much less tractable the polynomials: divided power rings are not noetherian, or even finitely generated, depriving us of much of the standard machinery of commutative algebra used in invariant theory on polynomials.

In algebraic topology, an important tool is the *Steenrod algebra* $\mathcal{A}_p$, the algebra of all cohomology operations which commute with suspension. $\mathcal{A}_p$ has a natural action on polynomials, in some sense generalising the Frobenius. This action commutes with the G-action mentioned above, and it has played a vital role in recent invariant theory (see e.g. [**20**]). Amongst the vast body of work on $\mathcal{A}_p$ there is currently much interest in the so-called *hit problem*: when is a polynomial hit by a Steenrod algebra element acting on a polynomial of lower degree, or, equivalently, what is an $\mathcal{A}_p$-module generating set of polynomials? The advances made in this direction have used the action of $\mathcal{A}_p$ on the dual to $S(V^*)$, namely $D(V)$. Dualising the problem results in asking which elements of $D(V)$ are annihilated by $\mathcal{A}_p$—see e.g. [**4, 6, 12, 18**]. There are many examples of cross-fertilisation between invariant theory and algebraic topology via the Steenrod algebra—see e.g. [**19**].

It is natural in this connection to consider $S(V^*)$ as a *Hopf algebra*—that is, an algebra with a comultiplication and counit which satisfy various commutativity relations (see e.g. [**21**]). Such algebras have the advantage that their duals (algebra maps into the base field) are again Hopf algebras, with multiplication corresponding to the original *co*multiplication. The important paper [**16**] showed that $\mathcal{A}_p$ is also a Hopf algebra, with dual a polynomial ring on infinitely many generators. Recently, interesting work has been done relating this to the *Dickson algebra*, $S(V^*)^{\mathrm{GL}(n,\mathbb{F})}$ in characteristic 2—see [**15**]. Intriguingly, in [**23**] there is even a definition of the mod 2 Steenrod algebra as a mod 2 reduction of a divided powers algebra over $\mathbb{Z}$.

The Hopf dual to $S(V^*)$ is $D(V)$, and the group ring $\mathbb{F}G$ also has a Hopf dual (which is semi-simple independently of characteristic), allowing us to use the fundamental results of S. Montgomery, M. Cohen, and others (see e.g. [**5, 8–10**]) on Hopf algebra actions. An example of the connections between divided power and Hopf structures can be found e.g. in [**7**].

Divided power algebras also play an important role in the theory of *Schur functors* and thus in the representation theory of the general linear group and matrix algebras. An excellent exposition of these ideas can be found in [**1**], and explicitly for $\mathrm{GL}(n, \mathbb{F}_2)$ in [**11**]. See also [**2**], where connections with the Steenrod algebra are examined. Some work has also been done on *co*invariants of $D(V)$ in the case of $\mathrm{GL}(n, \mathbb{F}_2)$; that is $D(V)/(D(V)^{\mathrm{GL}(n,\mathbb{F}_2)})^+$; see [**22**].

2. Definition

Let V be an n-dimensional vector space over the field $\mathbb{F}$ with basis $x_1, \ldots, x_n$ and let V^* be the dual vector space with dual basis $y_1, \ldots, y_n$. For a graded object A, denote the homogeneous component of degree d by A_d.

Let $D(V)$ be the *graded* dual of $S(V^*)$. This is defined as $\oplus_d \mathrm{Hom}_{\mathbb{F}}(S(V^*)_d, \mathbb{F})$; we denote the element dual to a monomial $y_1^{a_1} \cdots y_n^{a_n} \in S(V^*)$ by $x_1^{(a_1)} \cdots x_n^{(a_n)}$, and call it a *divided power*. The multiplication induced on $D(V)$ by its inclusion in

$T(V)$ under the *shuffle product* gives it an algebra structure. The shuffle product is defined as follows:

$$T(V)_{(k)} \times T(V)_{(l)} \to T(V)_{(k+l)}$$

$$x_1 \otimes \cdots \otimes x_k \cdot x_{k+1} \otimes \cdots \otimes x_{k+l} \mapsto \sum x_{\sigma(1)} \otimes x_{\sigma(2)} \otimes \cdots \otimes x_{\sigma(k+l)}$$

where the sum is over all elements $\sigma \in \Sigma_{k+l}$ which satisfy $\sigma(1) < \sigma(2) < \cdots < \sigma(k)$, $\sigma(k+1) < \sigma(k+2) < \cdots < \sigma(k+l)$.

Considered as an element of $T(V)$, the notation $x_1^{(a_1)} \cdots x_n^{(a_n)}$ means the orbit sum of

$$x_1^{\otimes a_1} \otimes \cdots \otimes x_n^{\otimes a_n}$$

under the operation of the symmetric group $\Sigma_{a_1+\cdots+a_n}$: thus e.g. $x_1^{(3)} = x_1 \otimes x_1 \otimes x_1$, and $x_1^{(1)} x_2^{(1)} = x_1 \otimes x_2 + x_2 \otimes x_1$.

The rules for divided powers are obtained by interpreting "bracket powers" to mean "tensor this many times and then take the orbit sum" in $T(V)$.

This implies that each $x^{(a)}$ behaves "as if" it were $x^a/a!$ in a polynomial ring. Thus in particular

$$x^{(a)} x^{(b)} = \binom{a+b}{a} x^{(a+b)}.$$

With the further identities

$$(x+y)^{(a)} = \sum_{i=0}^{a} x^{(i)} y^{(a-i)}$$

and

$$(x^{(a)})^{(b)} = \frac{(ab)!}{(a!)^b b!} x^{(ab)}$$

the ring $D(V)$ is called an *algebra of divided powers*. It "looks like" a polynomial ring with the divided power relations above replacing the usual rules for powers. The "monomials" in divided powers form an additive basis of $D(V)$.

Note that $S = S(V^*)$ has the structure of a Hopf algebra, with comultiplication

$$\Delta : S \to S \otimes S$$

defined via $\Delta(y) = y \otimes 1 + 1 \otimes y$ if y is a basis element, and extended multiplicatively componentwise on products. Thus $\Delta(zy) = (z \otimes 1 + 1 \otimes z)(y \otimes 1 + 1 \otimes y) = zy \otimes 1 + z \otimes y + y \otimes z + 1 \otimes zy$. Using the Sweedler notation [**21**], a general f is mapped to $\sum f_{(1)} \otimes f_{(2)}$. The (graded) dual Hopf algebra S^* is a graded ring, with the vector space dual to $S(V^*)_{(r)}$ in degree r. Thus as a graded vector space it is the same as $D(V)$. It turns out that the multiplication on S^* induced from Δ coincides with that on $D(V)$, so $D(V) \cong S^*$ as (Hopf) algebras.

3. Graded vector space structure

Let G be a group acting on a ring A. The *ring of invariants* A^G is defined as $\{a \in A : ga = a \ \forall g \in G\}$.

If a finite group G acts on $S(V^*)$ via a linear representation, G also acts on $S(V^*)^* = D(V)$ naturally via

$$gf(a) := f(g^{-1}a)$$

for $g \in G$, $f \in D(V)$, $a \in S(V^*)$. Then $f \in D(V)^G$, the ring of invariants, if and only if $gf(a) = f(a)$, i.e. if and only if $f(g^{-1}a) = f(a)$ for all $g \in G$ and $a \in S(V^*)$.

This holds in general for a linear action of G on a ring A. Suppose that the order of G is invertible in A, i.e. that we are in the *non-modular* case. Then the map

$$\pi_G : A \to A^G, \quad a \longmapsto \frac{1}{|G|}\sum_{g\in G} ga$$

is a projection onto A^G. Suppose $\tilde{f} \in (A^G)^*$. Then we can extend $\tilde{f}$ to an element $f \in A^*$ via

$$f(a) := \tilde{f}\big(\pi_G(a)\big).$$

The restriction of f to A^G is $\tilde{f}$ as π_G acts as the identity on A^G. We claim that the set of such extended f is precisely $(A^*)^G$. For suppose $h \in (A^*)^G$. Then

$$h(a) = \Big(\frac{1}{|G|}\sum_{g\in G} g\Big)(h)(a) = h\Big(\frac{1}{|G|}\sum_{g\in G} ga\Big) = h\big(\pi_G(a)\big),$$

so h is the extension of $h|_{A^G}$ to A. In the other direction, if f is as above, then

$$gf(a) = f(g^{-1}a) = \tilde{f}\Big(\frac{1}{|G|}\sum_{g'\in G} g'g^{-1}(a)\Big) = \tilde{f}\Big(\frac{1}{|G|}\sum_{g'\in G} g'(a)\Big) = f(a)$$

for any $a \in A$ and $g \in G$, so $f \in (A^*)^G$ as claimed. Hence we have proved the following

LEMMA 1. *In the non-modular case, $(A^*)^G \cong (A^G)^*$ as graded vector spaces. In particular, $D(V)^G$ and $\big(S(V^*)^G\big)^* \cong S(V^*)^G$ have the same Poincaré series.*

In general, it might seem likely that the invariant divided powers have the graded vector space form of the invariant ring of the *transpose* representation: if $G < \mathrm{GL}(n,\mathbb{F})$, let G^T be the group consisting of all matrices in G transposed. Clearly $G \cong G^T$. It is easy to see that if G acts on V, it acts as transpose matrices on V^*. Indeed, by the map taking $y^a \mapsto x^{(a)}/a!$, where $x \in V$ is the dual element to $y \in V^*$, the two rings $S(V^*)$ and $D(V)$ can be seen to be *isomorphic* as rings and as G-modules *in degree* $< p$, the characteristic–so in particular the invariant rings $S(V^*)^{G^T}$ and $D(V)^G$ will be isomorphic up to this degree. However, this is not the case in higher degrees—see the section on the Dickson invariants for counterexamples, bearing in mind that the general linear group is closed under transposition.

PROPOSITION 2. *If G is a permutation representation, $D(V)^G \cong S(V^*)^G$ as graded vector spaces. In fact, $\sum \lambda_i x_1^{(a_{i_1})} \cdots x_n^{(a_{i_n})} \in D(V)^G$ if and only if $\sum \lambda_i y_1^{a_{i_1}} \cdots y_n^{a_{i_n}} \in S(V^*)^G$.*

PROOF. As G permutes the basis elements $x_1, \dots, x_n$ of V, it also permutes the natural monomial basis of $S(V^*)_{(d)}$. Let that basis be $b_1, \dots, b_r$, with the dual basis of $D(V)_{(d)}$ being $b_1^*, \dots, b_r^*$. Suppose $g \in G$ takes b_j to b_{gj}. Then by definition

$$\begin{aligned}(gb_i^*)(b_j) &= b_i^*(g^{-1}b_j)\\ &= b_i^*(b_{g^{-1}j})\\ &= \delta_{i,g^{-1}j}\\ &= \delta_{gi,j}\\ &= (b_{gi}^*)(b_j).\end{aligned}$$

Thus g also takes b_j^* to b_{gj}^*, and a sum of the b_j is invariant under G if and only if the same sum of b_j^* are invariant. □

COROLLARY 3. *Let G be a permutation representation. Then the ring of polynomial invariants $S(V^*)^G$ has* **two** *different algebra structures, given by the usual polynomial algebra structure and the structure obtained by multiplying polynomials "as if" they were divided powers.*

Recall that $D(V)$ consists of the symmetric elements of $T(V)$; and $S(V)$ (*not* $S(V^*)$) is the symmetric quotient of $T(V)$. This enables us to define a map $\eta : D(V) \to S(V)$ which is actually *G-equivariant*, for any group G acting linearly on V. Thus η induces a map which we shall also call

$$\eta : D(V)^G \to S(V)^G.$$

We define it as follows:

Let $\{x_1, \ldots, x_n\}$ be a basis for V. Consider a monomial $x_1^{(a_1)} x_2^{(a_2)} \ldots x_n^{(a_n)}$ in $D(V)$. This corresponds to an element of $T(V)$, namely the *orbit sum* s of

$$x_1 \otimes \cdots \otimes x_1 \otimes x_2 \otimes \cdots \otimes x_n$$

under the action of the symmetric group $\Sigma_{a_1+\cdots+a_n}$, where x_i appears a_i times in the tensor product. By the orbit-stabiliser theorem, this orbit sum has

$$\frac{|\Sigma_{a_1+\cdots+a_n}|}{|\Sigma_{a_1}||\Sigma_{a_2}|\cdots|\Sigma_{a_n}|} = \frac{(a_1+\cdots+a_n)!}{a_1!\cdots a_n!}$$

summands.

Now consider how we get $S(V)$ from $T(V)$—we take the quotient by the action of the symmetric group in each degree. So in degree $a_1 + \cdots + a_n$, each summand of s is taken to the same monomial in $S(V)$—call it $\bar{s}$. Thus the whole sum is taken to

$$\frac{(a_1+\cdots+a_n)!}{a_1!\cdots a_n!} \cdot \bar{s}.$$

Let η be the map taking $x_1^{(a_1)} x_2^{(a_2)} \cdots x_n^{(a_n)}$ in $D(V)$ to $(a_1+\cdots+a_n)!/(a_1!\cdots a_n!)\cdot\bar{s}$ in $S(V)$, and extend it additively to general elements of $D(V)$. Then it is clearly G-equivariant: the action of G on $D(V)$ and $S(V)$ is precisely that induced by the action on $T(V)$.

LEMMA 4. *Let V be a vector space with $\mathbb{F}$-basis $\{x_1, \ldots, x_n\}$, and let G act on V linearly. Then if $S(V)^G_{(n)} = 0$, and*

$$z = \sum_i x_1^{(a_{i1})} \cdots x_n^{(a_{in})} \in D(V)^G_{(n)},$$

then

$$\frac{(a_{i1}+\cdots+a_{in})!}{a_{i1}!\cdots a_{in}!} \equiv 0 \mod p \quad \textit{for all } i.$$

PROOF. As η is a degree zero graded map $D(V)^G \to S(V)^G$, if $\eta(z) = 0$ (as is forced by $S(V)^G_{(n)} = 0$), it must be that each summand of z is mapped to zero. The result follows from the definition of η. □

4. Σ_n invariants of divided powers

Looking for examples of invariant rings of divided powers, it is natural to start with the equivalent to the *symmetric polynomials*, and in fact we can find a nice generating set:

Let the symmetric group Σ_n act linearly on $D(V)$ by permuting the x_i.

PROPOSITION 5. *$D(V)^{\Sigma_n}$ is generated as an algebra by* $\{\sum_{i=1}^n x_i^{(a)} : a = 1, 2, \dots\}$.

PROOF. Every symmetric element is in the subalgebra A of $D(V)$ generated by $\{\sum_{i=1}^n x_i^{(a)} : a = 1, 2, \dots\}$ if and only if every Σ_n-orbit of a "monomial" is in A. We shall prove this second claim, by induction on the number d of variables involved in the monomial $x_{i_1}^{(a_1)} x_{i_2}^{(a_2)} \cdots x_{i_d}^{(a_d)}$. The claim is clear for $d = 1$, as the Σ_n-orbit of a monomial involving one variable $x_j^{(b)}$ is $\sum_{i=1}^n x_i^{(b)}$.

Suppose it holds for $d \leq m - 1$. Without loss of generality (by renumbering the variables if necessary), let $x_1^{(a_1)} x_2^{(a_2)} \cdots x_m^{(a_m)}$ be some monomial involving m variables.

Note that

$$\left(\sum_{i=1}^n x_i^{(a)}\right)\left(\sum_{j=1}^n x_j^{(b)}\right) = \binom{a+b}{a} \sum_{i=1}^n x_i^{(a+b)} + \sum_{i \neq j} x_i^{(a)} x_j^{(b)},$$

$$\left(\sum_{i=1}^n x_i^{(a)}\right)\left(\sum_{j=1}^n x_j^{(b)}\right)\left(\sum_{k=1}^n x_k^{(c)}\right)$$

$$= \left(\frac{(a+b+c)!}{a!b!c!}\right) \sum_{i=1}^n x_i^{(a+b+c)} + \binom{a+b}{a} \sum_{i \neq j} x_i^{(c)} x_j^{(a+b)}$$

$$+ \binom{a+c}{c} \sum_{i \neq j} x_i^{(b)} x_j^{(a+c)} + \binom{b+c}{b} \sum_{i \neq j} x_i^{(a)} x_j^{(b+c)} + \sum_{i \neq j \neq k} x_i^{(a)} x_j^{(b)} x_k^{(c)}.$$

In general for $m \in \mathbb{N}$

$$(*) \qquad \prod_{k=1}^m \left(\sum_{i=1}^n x_i^{(a_k)}\right) = \sum_{I \text{ partition of } m} \lambda_I \bar{x}^{(I)},$$

where $I = (I_1, I_2, \dots, I_r)$, with $I_j \subset \{1, \dots, m\}$, $\bigcap I_j = \emptyset$ and $\bigcup I_j = \{1, \dots, m\}$. Say $|I_j| = c_j$ and $I_j = \{a_{j,1}, \dots, a_{j,c_j}\}$. Then

$$\bar{x}^{(I)} := \sum_{i_1 \neq i_2 \neq \cdots \neq i_r} x_{i_1}^{(a_{1,1}+\cdots+a_{1,c_1})} x_{i_2}^{(a_{2,1}+\cdots+a_{2,c_2})} \cdots x_{i_r}^{(a_{r,1}+\cdots+a_{r,c_r})}$$

and, with the convention that $c_0 = 0$,

$$\lambda_I = \left(\frac{(a_{1,1}+a_{1,2}+\ldots+a_{1,c_1})!}{a_{1,1}!a_{1,2}!\cdots a_{1,c_1}!}\right)\left(\frac{(a_{2,1}+\ldots+a_{2,c_2})!}{a_{2,1}!\cdots a_{2,c_2}!}\right)\cdots\left(\frac{(a_{r,1}+\cdots+a_{r,c_r})!}{a_{r,1}!\cdots a_{r,c_r}!}\right)$$

$$= \left(\frac{\prod_{j=1}^r \left(\sum_{k=1}^{c_j} a_{j,k}\right)!}{\prod_{i=1}^m a_i!}\right).$$

Thus the term corresponding to $I = (\{1, \ldots, m\})$ is

$$\frac{(\sum_{k=1}^{m} a_k)!}{\prod_{k=1}^{m} a_k!} \sum_{i=1}^{n} x_i^{(\sum_{k=1}^{m} a_k)}$$

and if $m \leq n$, that corresponding to $I = (\{1\}, \{2\}, \ldots, \{m\})$ is

$$\sum_{i_1 \neq i_2 \neq \cdots \neq i_m} x_{i_1}^{(a_1)} x_{i_2}^{(a_2)} \cdots x_{i_m}^{(a_m)},$$

the Σ_n-orbit of $x_1^{(a_1)} x_2^{(a_2)} \cdots x_m^{(a_m)}$. The coefficient $\lambda_{(\{1\},\{2\},\ldots,\{m\})} = 1$, and every other term involves less than m variables, so from (*) it follows that

$$\sum_{i_1 \neq i_2 \neq \cdots \neq i_m} x_{i_1}^{(a_1)} x_{i_2}^{(a_2)} \cdots x_{i_m}^{(a_m)} = \prod_{k=1}^{m} \left(\sum_{i=1}^{n} x_i^{(a_k)} \right) - \sum_{I \neq (\{1\},\{2\},\ldots,\{m\})} \lambda_I \bar{x}^{(I)}$$

and by induction, everything on the right of the equation is in A. Thus the Σ_n-orbit of $x_1^{(a_1)} x_2^{(a_2)} \cdots x_m^{(a_m)} \in A$ as claimed. □

5. $D(V_1)^{C_{p-1}}$ is not D.P.-finitely generated

In the case of groups acting linearly on polynomials, it is well known that the invariant ring is finitely generated. An obvious question is whether this is also the case for divided powers. Here we clearly need a slightly different definition of "finitely generated", as even $D(V)$ is itself not multiplicatively finitely generated. Thus we must certainly allow not only products, but also divided powers: so we shall say a ring A is *generated with divided powers* or *d.p. generated* by some set F if any element of A can be written as a product of elements of F *or their divided powers*. Thus $D(V)$ is clearly d.p. generated by the set $\{x_1, \ldots, x_n\}$.

Unfortunately, even with this extended definition, rings of invariants are not in general finitely generated, as the following elementary example shows.

Let V_1 be a 1-dimensional vector space over $\mathbb{F}_p$, p a prime > 2, and G be C_{p-1}, the cyclic group of order $p-1$ with generator acting on V by multiplication by any non-zero element of the field. Then $D(V)$ is the divided power algebra on one variable, and it is clear (in particular from Lemma 1) that $D(V_1)^G$ is generated as a vector space by $\{x^{(a(p-1))} : a \in \mathbb{N}\}$.

PROPOSITION 6. *$D(V_1)^{C_{p-1}}$ is not finitely generated with divided powers.*

PROOF. We claim that for each $r \in \mathbb{N}$ the element $x^{(p^r(p-1))} \in D(V_1)^{C_{p-1}}$ cannot be hit by elements of lower degree.

There are two possibilities for such an element to be hit: either (a) $x^{(p^r(p-1))} = \lambda x^{(a)} x^{(b)}$ for some $x^{(a)}$, $x^{(b)} \in D(V_1)^{C_{p-1}}$ and $\lambda \in \mathbb{F}_p$, or (b) $x^{(p^r(p-1))} = \mu (x^{(a)})^{(b)}$ for some $x^{(a)} \in D(V_1)^{C_{p-1}}$ and $\mu \in \mathbb{F}_p$.

First (a):

$$x^{(m-n)} x^{(n)} = \binom{m}{n} x^{(m)}.$$

Thus for $x^{(p^r(p-1))}$ to be hit like this, there must be some n for which $\binom{p^r(p-1)}{n} \neq 0$ modulo p. For which n is this the case?

Let $\mathrm{Ord}_p(t)$ be the power of p that appears in the prime decomposition of t. Clearly, t is non-zero mod p precisely when $\mathrm{Ord}_p(t) = 0$. Let us write n in a p-adic decomposition as

$$n = \sum_{i=0}^{s} n_i p^i, \quad 0 \le n_i < p \quad \forall i,$$

which we write as $(n_0, n_1, \ldots, n_s)$, and similarly for m. Then

$$\text{(\#)} \qquad \mathrm{Ord}_p(n) = \frac{n - \sum n_i}{p-1}$$

(see e.g. [**14**], p. 46). Further,

$$\begin{aligned}\mathrm{Ord}_p \binom{m}{n} &= \mathrm{Ord}_p(m!) - \mathrm{Ord}_p(n!) - \mathrm{Ord}_p((n-m)!) \\ &= \frac{1}{p-1}\Big(\Big(m - \sum m_i\Big) - \Big(n - \sum n_i\Big) \\ &\qquad - \Big((m-n) - \sum (m-n)_i\Big)\Big).\end{aligned}$$

This is zero precisely when

$$\sum m_i - \sum n_i = \sum (m-n)_i$$

in other words when there is no *carrying* of p-adic terms in the subtraction $m - n$.

In the case of interest, we want to know $\mathrm{Ord}_p \binom{p^r(p-1)}{n}$. Here, the p-adic decomposition of $p^r(p-1)$ looks like $(0, 0, \ldots, 0, p-1)$ with r zeros before the $p-1$. Thus the only numbers n which can be subtracted from $p^r(p-1)$ *without carrying terms* are of the form $(0, 0, \ldots, 0, u)$ with r zeros—in other words, numbers of the form $p^r u$ with $u < p-1$. However, $p^r u$ is not divisible by $p-1$, so $x^{(p^r u)}$ is not in $D(V_1)^{C_{p-1}}$ for any $u < p-1$. Thus for any $x^{(n)} \in D(V)^G$,

$$x^{(n)} x^{(p^r(p-1)-n)} = 0,$$

and there can be no $x^{(a)}, x^{(b)} \in D(V_1)^{C_{p-1}}$ such that $x^{(p^r(p-1))} = \lambda x^{(a)} x^{(b)}$.

Now to (b):

$$\left(x^{(a)}\right)^{(b)} = (a, b) x^{(ab)},$$

where

$$(a, b) := \frac{(ab)!}{b!(a!)^b}.$$

We require $x^{(a)} \in D(V_1)^{C_{p-1}}$, so a must be of the form $d(p-1)$ for some d. Further, ab must equal $p^r(p-1)$, so d must be a power of p, and likewise b. Thus we want to know if $(x^{(p^s(p-1))})^{(p^t)} = \left(p^s(p-1), p^t\right) x^{(p^{s+t}(p-1))}$ is non-zero mod p for some s, t.

Using (#), we can calculate the power of p occuring in the prime decomposition of $\left(p^s(p-1), p^t\right)$, as $p^s(p-1) = (0, 0, \ldots, 0, p-1)$ and $p^t = (0, 0, \ldots, 0, 1)$ in the

p-adic decomposition:

$$\begin{aligned}
\operatorname{Ord}_p&\big(p^s(p-1),p^t\big)\\
&=\operatorname{Ord}_p\left(\frac{(p^{s+t}(p-1))!}{(p^t)!((p^s(p-1))!)^{p^t}}\right)\\
&=\operatorname{Ord}_p\big((p^{s+t}(p-1))!\big)-\operatorname{Ord}_p\big((p^t)!\big)-p^t\operatorname{Ord}_p\big((p^s(p-1))!\big)\\
&=\frac{1}{p-1}\big((p^{s+t}(p-1)-(p-1))-(p^t-1)-p^t\big(p^s(p-1)-(p-1)\big)\big)\\
&=\frac{1}{p-1}(p^{t+1}-2p^t-p+2).
\end{aligned}$$

This would be zero only when $p^{t+1}-2p^t-p+2=(p^t-1)(p-2)=0$, which cannot occur as $p>2$.

Thus $p|\big(p^s(p-1),p^t\big)$, and $(x^{p^s(p-1)})^{(p^t)}=0$ for all s and t, finishing the proof. □

6. The divided Dickson algebra

Over a finite field $\mathbb{F}_q$ with $q=p^r$ elements, p prime, the polynomial invariants of the general linear group $\mathrm{GL}(n,\mathbb{F}_q)$ are well known:

$$S(V^*)^{\mathrm{GL}(n,\mathbb{F}_q)}=\mathbb{F}_q[d_{n,n-1},\dots,d_{n,0}]$$

where the degree of the polynomial generator $d_{n,i}$ is q^n-q^i. It is called the *Dickson algebra* after its discoverer, and is a polynomial ring. We are interested in calculating the equivalent in the divided power algebra, $D(V)^{\mathrm{GL}(n,\mathbb{F}_q)}$.

LEMMA 7. *Elements of $D(V)^{\mathrm{GL}(n,\mathbb{F}_q)}$ are symmetric of the form*

$$\sum\lambda_i x_1^{(a_{i,1}(q-1))}\cdots x_n^{(a_{i,n}(q-1))}$$

where $\lambda_i\in\mathbb{F}_q$.

PROOF. The group $\mathrm{GL}(n,\mathbb{F}_q)$ is generated by the subgroup S of permutation matrices together with the matrices

$$A=\begin{pmatrix}\lambda & \mathbf{0}\\ \mathbf{0} & I_{n-1}\end{pmatrix}\quad\text{and}\quad B=\begin{pmatrix}1&1&\mathbf{0}\\0&1&\mathbf{0}\\ \mathbf{0}&\mathbf{0}&I_{n-2}\end{pmatrix}$$

where λ is a multiplicative generator of $\mathbb{F}_q^\times$, the multiplicative group of the field, I_m denotes the $m\times m$ identity matrix and $\mathbf{0}$ denotes a row or column of zeros. (This follows easily from e.g. the Bruhat decomposition of invertible matrices). Thus an element of $D(V)$ is in $D(V)^{\mathrm{GL}(n,\mathbb{F}_q)}$ if and only if it is symmetric and fixed under A and B.

We need only consider homogeneous elements. Suppose

$$\sum\lambda_i x_1^{(b_{i,1})}\cdots x_n^{(b_{i,n})}$$

is an invariant element of degree d. Then it is invariant under A, so

$$\sum\lambda_i x_1^{(b_{i,1})}\cdots x_n^{(b_{i,n})}=\sum\lambda^{b_{i,1}}\lambda_i x_1^{(b_{i,1})}\cdots x_n^{(b_{i,n})}.$$

Thus $\lambda^{b_{i,1}}=1$ for all $b_{i,1}$, so $b_{i,1}$ is divisible by $q-1$. By symmetry the same holds for the other $b_{i,m}$. □

Note that we do not claim that *all* these elements are invariant. The precise invariant ring is the intersection of the above with the invariant ring of the group generated by B, which is not yet calculated. We can however give some examples of elements which are also invariant under B:

LEMMA 8. *The set of monomials in* $D(V)^{\mathrm{GL}(n,\mathbb{F}_q)}$ *is* $\{\mu e_t : \mu \in \mathbb{F}_q, t = 1,2,3,\dots\}$, *where*

$$e_t := x_1^{(q^t-1)} x_2^{(q^t-1)} \cdots x_n^{(q^t-1)}.$$

PROOF. Fix t. Then e_t is clearly symmetric and invariant under A (as $(q-1) \mid (q^t-1)$). We thus need to show that e_t is invariant under B, i.e. under the map taking $x_1 \mapsto x_1 + x_2$ and fixing the other x_i. We find that

$$\begin{aligned}
B(e_t) &= (x_1+x_2)^{(q^t-1)} x_2^{(q^t-1)} \cdots x_n^{(q^t-1)} \\
&= \left(\sum_{j=0}^{q^t-1} x_1^{((q^t-1)-j)} x_2^{(j)} \right) x_2^{(q^t-1)} \cdots x_n^{(q^t-1)} \\
&= \sum_{j=0}^{q^t-1} x_1^{((q^t-1)-j)} \left(x_2^{(j)} x_2^{(q^t-1)} \right) \cdots x_n^{(q^t-1)} \\
&= x_1^{(q^t-1)} x_2^{(q^t-1)} \cdots x_n^{(q^t-1)} = e_t
\end{aligned}$$

as $x_2^{(j)} x_2^{(q^t-1)} = 0$ for all $0 < j \leq q^t - 1$: recall (see the proof to Proposition 6) that $x^{(a)}x^{(b)} = 0$ if and only if there is carrying in the p-adic expansion of the sum $a+b$. As $q = p^r$, $q^t - 1 = (p-1, p-1, \dots, p-1)$ p-adically, where $p-1$ appears rt times. Thus $x^{(a)}x^{(q^t-1)} = 0$ for any non-zero $a \leq q^t - 1$ as required.

This calculation also implies that the e_t are the only symmetric monomials with "bracket powers" divisible by $q-1$ which are invariant under B. For such a monomial $m = x_1^{(a(q-1))} \dots x_n^{(a(q-1))}$ must also satisfy the same kind of equation as above, implying that $a(q-1)$ is a number for which $x_2^{(j)} x_2^{(a(q-1))} = 0$ for all $0 < j \leq a(q-1)$.

This implies that $a(q-1) = p^u - 1$ for some $u \geq r$. So

$$a(p^r - 1) = p^u - 1 \implies ap^r - a = p^u - 1 \implies a \equiv 1 \mod p^r.$$

So $a = a_1 p^r + 1$, say. Thus

$$\begin{aligned}
(a_1p^r + 1)(p^r - 1) = p^u - 1 &\implies a_1p^{2r} - (a_1 - 1)p^r - 1 = p^u - 1 \\
&\implies a_1p^{2r} - (a_1-1)p^r = p^u \implies a_1p^r - (a_1 - 1) = p^{u-r} \\
&\implies a_1p^r - a_1 = p^{u-r} - 1 \implies a_1(p^r - 1) = p^{u-r} - 1.
\end{aligned}$$

Thus either $r = u$ and $a_1 = 0$ or if not, $a_1 \equiv 1 \mod p^r$, so $a_1 = a_2 p^r + 1$ and we can repeat the process. With each step the power of p on the right hand side of the equation decreases by r, until at some point it must reach zero. Thus r divides u, $a_t = 0$ for some t, and a is of the form $a = 1 + p^r + p^{2r} + \dots + p^{(t-1)r} = 1 + q + q^2 + \dots + q^{t-1}$. Hence

$$a(q-1) = (1 + q + q^2 + \dots + q^{t-1})(q-1) = q^t - 1$$

as claimed. □

Note that some of the e_t are examples of elements of $D(V)^{\mathrm{GL}(n,\mathbb{F}_q)}$ which appear in degrees in which there are *no* elements in $S(V^*)^{\mathrm{GL}(n,\mathbb{F}_q)}$. For example, suppose that $n < q^{n-1}$ (this holds for all pairs (n,q) except $(1,q)$ and $(2,2)$). Then e_1 has degree $n(q-1)$ which is strictly less than $q^{n-1}(q-1)$, the degree of the smallest Dickson invariant in $S(V^*)^{\mathrm{GL}(n,\mathbb{F}_q)}$.

References

1. K. Akin, D. A. Buchsbaum, and J. Weyman, *Schur functors and Schur complexes*, Adv. Math. **44** (1982), 207–278.
2. M. A. Alghamdi, M. C. Crabb, and J. R. Hubbuck, *Representations of the homology of* BV *and the Steenrod algebra.* I, Adams Memorial Symposium on Algebraic Topology **2** (1990), 217–234; London Math. Soc. Lecture Note Ser. **176**, Cambridge Univ. Press, 1992.
3. D. J. Anick, *On the homogeneous invariants of a tensor algebra*, Algebraic Topology, (Evanston, 1988), Contemp. Math., vol. 96, Amer. Math. Soc., Providence, RI, 1989, pp. 15–17.
4. D. J. Anick and F. P. Peterson, *A_2-annihilated elements in $H_*(\Omega\Sigma RP^2)$*, Proc. Amer. Math. Soc. **117** (1993), no. 1, 243–250.
5. R. J. Blattner, M. Cohen, and S. Montgomery, *Crossed products and inner actions of Hopf algebras*, Trans. Amer. Math. Soc. **298** (1986), no. 2, 671–711.
6. D. P. Carlisle and R. M. Wood, *The boundedness conjecture for the action of the Steenrod algebra on polynomials*, Adams Memorial Symposium on Algebraic Topology **2** (Manchester, 1990); London Math. Soc. Lecture Note Ser. **176**, Cambridge Univ. Press, 1992, pp. 203–216.
7. W. Chin and D. Quinn, *Actions of divided power Hopf algebras*, J. Algebra **144** (1991), no. 2, 371–389.
8. M. Cohen and D. Fishman, *Hopf algebra actions*, J. Algebra **102** (1986), no. 2, 363–379.
9. M. Cohen, D. Fishman, and S. Montgomery, *Hopf Galois extensions, smash products, and Morita equivalence*, J. Algebra **133** (1990), no. 2, 351–372.
10. M. Cohen and S. Montgomery, *Group-graded rings, smash products and group actions*, Trans. Amer. Math. Soc. **282** (1984), no. 1, 237–258.
11. M. C. Crabb and J. R. Hubbuck, *Representations of the homology of* BV *and the Steenrod algebra.* II, Algebraic Topology: New Trends in Localisation and Periodicity, Progr. Math., vol. 136, Birkhäuser, Basel, Verlag, 1996, pp. 145–154.
12. M. D. Crossley, *$A(p)$-annihilated elements in $H_*(CP^\infty \times CP^\infty)$*, Math. Proc. Cambridge Philos. Soc. **120** (1996), no. 3, 441–453.
13. D. Eisenbud, *Commutative algebra with a view toward algebraic geometry*, Graduate Texts in Mathematics, vol. 150, Springer Verlag, New York, 1995.
14. G. H. Hardy and E. M. Wright, *An introduction to the theory of numbers*, Oxford, Clarendon Press XVI, 1938.
15. I. Llerena and N. H. V. Hu'ng, *The complete Steenrod algebra and the generalised Dickson algebra* (C. Broto, C. Casacuberta, and G. Mislin, eds.), 1994 BCAT, Progr. Math., vol. 136, Birkhäuser, Basel, 1996, pp. 271–284.
16. J. Milnor, *The Steenrod algebra and its dual*, Ann. of Math. **67** (1958), 150–171.
17. M. Nicollerat, *Homology of symmetric algebras*, Comm. Algebra **7** (1997), 657–676.
18. J. Repka and P. Selick, *On the subalgebra of $H_*\big((\mathbb{R}P^\infty)^n;\mathbb{F}_2\big)$ annihilated by Steenrod operations*, J. Pure Appl. Algebra **127** (1998), no. 3, 273–288.
19. J. Segal, *Pointwise conjugate representations of groups and modules over the Steenrod algebra*, Ph.D. Thesis, Göttingen, 1999; (`http://webdoc.sub.gwdg/diss/2000/segal`).
20. L. Smith, *Polynomial invariants of finite groups*, Research Notes in Mathematics, vol. 6, A. K. Peters Ltd., Wellesley, MA, 1995.
21. M. E. Sweedler, *Hopf algebras*, W. A. Benjamin Inc., New York, 1969.
22. P. R. Turner, *Dickson coinvariants and the homology of QS^0*, Math. Z. **224** (1997), no. 2, 209–228.
23. R. M. Wood, *Differential operators and the Steenrod algebra*, Preprint, University of Manchester, 1995.

University of Kent at Canterbury, IMS, Canterbury, Kent CT2 7NZ, UK.
E-mail address: `joel@berlin.com`

Centre de Recherches Mathématiques
CRM Proceedings and Lecture Notes
Volume **35**, 2004

Classical Covariants and Modular Invariants

R. J. Shank

There is a relationship between the covariants of binary forms, a central topic in classical invariant theory, and the invariants of modular representations of cyclic groups of prime order. This relationship was identified by Gert Almkvist [**1**] and used implicitly in both [**15**] and [**17**]. In this note we investigate the relationship and provide a progress report on an application. Our primary motivation is a desire to construct nice generating sets for the rings of invariants of modular representations of cyclic groups of prime order.

1. Modular invariants

Let p denote a prime number, let $\mathbb{Z}/p$ denote the cyclic group of order p, and let $\mathbb{F}$ denote a field of characteristic p. A representation of a cyclic group is determined by the Jordan canonical form of (the image of) the generator. If $n \leq p$ then the $n \times n$ matrix over $\mathbb{F}$ consisting of a single Jordan block with eigenvalue 1, has order p and determines an indecomposable representation of $\mathbb{Z}/p$ which we denote by V_n. (For $n > p$, the order of the matrix is greater than p.). Note that there are no non-trivial p^{th} roots of unity in $\mathbb{F}$. Thus 1 is the only eigenvalue for the image of a generator of $\mathbb{Z}/p$ under a representation over $\mathbb{F}$. Therefore, up to isomorphism, the only indecomposable $\mathbb{F}\mathbb{Z}/p$-modules are $V_1, V_2, \ldots, V_p$.

Let V be any finite dimensional vector space over an arbitrary field $\mathbf{k}$. We choose a basis, $\{x_1, \ldots, x_n\}$, for the dual, V^*, of V and consider a subgroup G of $\mathrm{GL}(V)$. The action of G on V induces an action on V^* which extends to an action by algebra automorphisms on the symmetric algebra of V^*, $\mathbf{k}[V] := \mathbf{k}[x_1, \ldots, x_n]$. The ring of invariants of G is the subring of $\mathbf{k}[V]$ given by

$$\mathbf{k}[V]^G := \{f \in \mathbf{k}[V] \mid g \cdot f = f \text{ for all } g \in G\}.$$

If G is a finite group and $|G|$ is not invertible in $\mathbf{k}$ then we say the representation of G on V is *modular*. If $|G|$ is invertible in $\mathbf{k}$ then V is called a *non-modular* representation. We will be primarily interested in the case $G = \mathbb{Z}/p$. Note that V_n^* and V_n are isomorphic $\mathbb{Z}/p$-modules and we will usually choose the basis for V_n^* so that the generator of $\mathbb{Z}/p$ is in Jordan canonical form.

2000 *Mathematics Subject Classification.* 13A50.

I thank Eddy Campbell and David Wehlau for organising the *Workshop on Invariant Theory*. I thank Medicis (CNRS/École Polytechnique) for access to computer resources including Magma version 2.8.

This is the final version of the paper.

For a finite G, the *transfer* is defined by:

$$Tr^G: \mathbf{k}[V] \longrightarrow \mathbf{k}[V]^G$$
$$f \longmapsto \sum_{g\in G} g\cdot f$$

and is a homomorphism of $\mathbf{k}[V]^G$-modules. For non-modular representations, Tr^G is surjective. For modular representations, the image of the transfer, I^G, is a proper non-zero ideal of $\mathbf{k}[V]^G$. For proofs of this fact and other general properties of the modular transfer see [**16**]. For $f \in \mathbf{k}[V]$, we define the *norm* of f to be the product over the orbit of f, $N(f) := \prod_{h\in G\cdot f} h$.

The central problem of invariant theory is to find (nice) generators for the algebra $\mathbf{k}[V]^G$. In practice, this problem is much harder in the modular setting. Even for representations of $\mathbb{Z}/p$, finding manageable generating sets for the ring of invariants can be difficult. Hughes and Kemper [**10**] have given an upper bound on the degrees of the generators for any representation of $\mathbb{Z}/p$. Therefore by taking all homogeneous invariants with degree less than or equal to the upper bound we do get a finite generating set. However such generating sets are far from manageable. Minimal generating sets for $\mathbb{F}[V_2]^{\mathbb{Z}/p}$ and $\mathbb{F}[V_3]^{\mathbb{Z}/p}$ can be found in Dickson's Madison Colloquium [**5**]. Finite SAGBI bases[1] (see Section 4 for definitions) for $\mathbb{F}[V_4]^{\mathbb{Z}/p}$ and $\mathbb{F}[V_5]^{\mathbb{Z}/p}$ can be found in [**15**]. The problem of finding a nice generating set for $\mathbb{F}[V_n]^{\mathbb{Z}/p}$ for $n > 5$ remains open. The results of Section 6 represent preliminary work for $n = 6$. Even when the invariants of the indecomposable summands are understood, it can be difficult to construct generating sets for decomposable representations. Campbell and Hughes, in [**3**], describe a generating set for $\mathbb{F}[mV_2]^{\mathbb{Z}/p}$ which is refined to a minimal generating set in [**18**]. A SAGBI basis is given for $\mathbb{F}[V_2 \oplus V_3]^{\mathbb{Z}/p}$ in [**17**]. I understand that Brandon Fodden, a student at Queen's University (Canada), has constructed a SAGBI basis for $\mathbb{F}[2V_3]^{\mathbb{Z}/p}$. Example 5.4 is related to this problem and Example 5.5 is related to $\mathbb{F}[V_3 \oplus V_4]^{\mathbb{Z}/p}$.

2. Covariants of binary forms

Consider the contragradiant action of $\mathrm{SL}_2(\mathbb{C})$ on the span of x and y, and extend this to an action by algebra automorphisms on $\mathbb{C}[x, y]$. There is a linear action of $\mathrm{SL}_2(\mathbb{C})$ on the span of $\{a_0, a_1, \ldots, a_m\}$ with respect to which the binary form

$$f := \sum_{i=0}^{m} \binom{m}{i} a_i x^{m-i} y^i$$

is invariant. Extend the given actions of $\mathrm{SL}_2(\mathbb{C})$ to actions by algebra automorphisms on $\mathbb{C}[a_0, \ldots, a_m]$ and $\mathbb{C}[x, y, a_0, \ldots, a_m]$. An element of $\mathbb{C}[a_0, \ldots, a_m]^{\mathrm{SL}_2(\mathbb{C})}$ is called an *invariant* of the binary form while an element of $\mathbb{C}[x, y, a_0, \ldots, a_m]^{\mathrm{SL}_2(\mathbb{C})}$ is called a *covariant* of the binary from. Following classical terminology, for $h \in \mathbb{C}[x, y, a_0, \ldots, a_m]$, the *degree* of h is its degree as a polynomial in the a_i's while the *order* of h is its degree as a polynomial in x and y. Thus, for example, f has degree 1 and order m. The action of $\mathrm{SL}_2(\mathbb{C})$ on $\mathbb{C}[x, y, a_0, \ldots, a_m]$ preserves both degree and order. Since an arbitrary covariant is the sum of its homogeneous components, we need only consider elements of $\mathbb{C}[x, y, a_0, \ldots, a_m]^{\mathrm{SL}_2(\mathbb{C})}$ which are homogeneous with respect to both degree and order. The *weight* of a monomial

[1]A SAGBI basis is a particularly nice generating set.

$a_0^{e_0}a_1^{e_1}\cdots a_m^{e_m} \in \mathbb{C}[a_0,\ldots,a_m]$ is defined to be $e_1 + 2e_2 + \cdots + me_m$. The *source* of a homogeneous covariant of order k is the coefficient of x^k and is a homogeneous element of $\mathbb{C}[a_0,\ldots,a_m]$. The covariant can be recovered from its source [**9**, pp. 41–43]. Furthermore, any homogeneous isobaric element of $\mathbb{C}[a_0,\ldots,a_m]$ which is invariant under the upper-triangular unipotent subgroup (a seminvariant), is the source of a homogeneous covariant ([**13**, Theorem 9.45]). If the source has degree d, and weight ω, then the resulting covariant has order $k = m\cdot d - 2\omega$ ([**9**, pp. 41–43], [**8**, §31]). Define

$$\sigma := \begin{pmatrix} 1 & 1 \\ 0 & 1 \end{pmatrix}.$$

The action of σ on $\mathbb{C}[a_0,\ldots,a_m]$ is given by $\sigma(a_i) = \sum_{j=0}^{i}\binom{i}{j}a_j$ and the seminvariants are precisely the elements of $\mathbb{C}[a_0,\ldots,a_m]^\sigma$. Take $m = n-1$ and choose a basis, $\{x_1,\ldots, x_n\}$, for the span of $\{a_0,\ldots,a_{n-1}\}$ so that σ is in Jordan canonical form, i.e., $\sigma(x_1) = x_1$ and $\sigma(x_i) = x_i + x_{i-1}$ for $1 < i \leq n$. For $n \leq 6$, a suitable change of basis is given by $a_0 = x_1$, $a_1 = x_2$, $a_2 = 2x_3 + x_2$, $a_3 = 6x_4 + 6x_3 + x_2$, $a_4 = 24x_5 + 36x_4 + 14x_3 + x_2$, $a_5 = 120x_6 + 240x_5 + 150x_4 + 30x_3 + x_2$.

3. The connection: integral invariants

Define an algebra automorphism, σ, on $\mathbb{Z}[x_1,\ldots, x_n]$ by $\sigma(x_1) = x_1$ and $\sigma(x_i) = x_i + x_{i-1}$ for $1 < i \leq n$. Reducing coefficients modulo p gives a surjection from $\mathbb{Z}[x_1,\ldots, x_n]$ to $\mathbb{F}_p[x_1,\ldots, x_n]$ which induces a ring homomorphism from $\mathbb{Z}[x_1,\ldots, x_n]^\sigma = \mathbb{Z}[x_1,\ldots, x_n]^{\mathbb{Z}}$ to $\mathbb{F}_p[x_1,\ldots, x_n]^\sigma = \mathbb{F}_p[x_1,\ldots, x_n]^{\mathbb{Z}/p}$. We will refer to elements in the image of this homomorphism as *rational* invariants.

The inclusion of $\mathbb{Z}$ into $\mathbb{C}$ induces a ring monomorphism from $\mathbb{Z}[x_1,\ldots, x_n]^\sigma$ to $\mathbb{C}[x_1,\ldots, x_n]^\sigma = \mathbb{C}[x_1,\ldots, x_n]^{\mathbb{Z}}$, the ring of seminvariants.

THEOREM 3.1. *The image of $\mathbb{Z}[x_1,\ldots, x_n]^\sigma$ in $\mathbb{C}[x_1,\ldots, x_n]^{\mathbb{Z}}$ is a generating set and the image of $\mathbb{Z}[x_1,\ldots, x_n]^\sigma$ in $\mathbb{F}_p[x_1,\ldots, x_n]^{\mathbb{Z}/p}$ is surjective in low degrees.*

PROOF. This is essentially [**1**, Theorem 2.5]. The proof relies on comparing Hilbert series. □

We are primarily interested in using this result to construct elements of $\mathbb{F}_p[V_n]^{\mathbb{Z}/p}$ from covariants of the binary $(n-1)$-form. If we start with a homogeneous covariant, we may easily identify the source. The source is an element of $\mathbb{C}[x_1,\ldots, x_n]^\sigma$. If the source lies in $\mathbb{Q}[x_1,\ldots, x_n]^\sigma$, we may clear the denominators to get an element of $\mathbb{Z}[x_1,\ldots, x_n]^\sigma$ which then projects to an element of $\mathbb{F}_p[V_n]^{\mathbb{Z}/p}$. It is a consequence of the above theorem that we may choose the generators of $\mathbb{C}[x_1,\ldots, x_n]^\sigma$ to lie in $\mathbb{Z}[x_1,\ldots, x_n]^\sigma$. The results of Section 5 below will give us a process for explicitly constructing generating sets for $\mathbb{C}[x_1,\ldots, x_n]^\sigma$ which lie in $\mathbb{Z}[x_1,\ldots, x_n]^\sigma$. However, such a generating set will not necessarily project to a generating set for the subring of rational invariants. There are a number of serious questions about the coefficients that have yet to be addressed. Despite this, the process has been used successfully in a number of cases.

In [**15**, §6], I conjectured that, for an indecomposable representation V, the ring $\mathbb{F}_p[V]^{\mathbb{Z}}$ is generated by rational invariants, the image of the transfer and the norm of the generator of the $\mathbb{Z}/p$-module V^*. I believe that this should hold for decomposable representations as well, as long as the norms of a generating set for the $\mathbb{Z}/p$-module V^* are included. Note that for representations of $\mathbb{Z}/p^i$, elements in the image of the relative transfer would need to be added. Also, in this case, the concept of rational invariant, which is essentially a large prime approximation, is less useful.

Since the degrees of a generating set for the rational invariants are independent of the prime, the conjecture implies that for all but a finite (possibly empty) set of primes, an upper bound on the degrees of a minimal generating set is given by an upper bound on the degrees of the generators for the image of the transfer. The same conclusion is reached in [**10**, §2.4] using the 'periodicity' of $\mathbb{F}_p[V]$.

4. SAGBI bases

A SAGBI basis for a subalgebra of $\mathbf{k}[x_1, \ldots, x_n]$ is a **S**ubalgebra **A**nalog to a **G**öbner **B**asis for **I**deals and as such is a particularly nice generating set. SAGBI bases were introduced independently by Robbiano and Sweedler [**14**] and Kapur and Madlener [**11**]. Unfortunately, even a finitely generated subalgebra does not necessarily have a finite SAGBI basis. In fact, as demonstrated by the ring of invariants of the canonical representation of the alternating group on three letters, even the ring of invariants of a finite group may fail to have a SAGBI basis (see [**6**, Lemma 2.1], [**7**] or [**20**, Example 11.2]).

We use the convention that a monomial is a product of variables and that a term is a monomial with a non-zero coefficient. See [**4**, Chapter 2] for a detailed discussion of monomial orders. For $f \in \mathbf{k}[x_1, \ldots, x_n]$, we use $\mathrm{LT}(f)$ to denote the lead term of f and $\mathrm{LM}(f)$ to denote the lead monomial of f. Suppose that R is a subalgebra of $\mathbf{k}[x_1, \ldots, x_n]$. Let $\mathrm{LT}(R)$ denote the vector space spanned by the lead terms of elements of R. Then $\mathrm{LT}(R)$ is a subalgebra of $\mathbf{k}[x_1, \ldots, x_n]$. If C is a subset of R then let $\mathrm{LM}(C)$ denote the set of lead monomials of elements of C. If C is a subset of R such that $\mathrm{LM}(C)$ generates the algebra $\mathrm{LT}(R)$ then C generates R and C is called a SAGBI *basis* for R. Note that $\mathrm{LT}(R)$ is a graded algebra. If the subalgebra generated by $\mathrm{LM}(C)$ coincides with $\mathrm{LT}(R)$ in degrees less than or equal to d, we say that C is a SAGBI basis through degree d. For a detailed discussion of SAGBI bases see [**14**], [**11**] or [**20**, Chapter 11].

Taking $C = R$ gives a SAGBI basis for R. Thus every subalgebra has a SAGBI basis. However, if $\mathrm{LT}(R)$ is not finitely generated then R does not have a finite SAGBI basis (at least using the given monomial order). Although the characterization of subalgebras which admit a finite SAGBI basis remains an important open problem, there are some circumstances which guarantee the existence of a finite SAGBI basis.

THEOREM 4.1. [**18**, Lemma 3.1] *Suppose $\{h_1, \ldots, h_n\}$ is a homogeneous system of parameters for $\mathbf{k}[x_1, \ldots, x_n]$ with $\mathrm{LM}(h_i) = x_i^{d_i}$. If $A \subseteq \mathbf{k}[x_1, \ldots, x_n]$ is a subalgebra with $\{h_1, \ldots, h_n\} \subseteq A$, then A has a finite* SAGBI *basis.*

Suppose that V is a modular representation of $\mathbb{Z}/p$. Choose our basis for V^* so that the generator, say σ, is in Jordan canonical form, i.e., σ is represented by an upper-triangular unipotent matrix. Choose a monomial order with $x_1 < x_2 < \cdots < x_n$. Then $\mathrm{LM}(N(x_i)) = x_i^{[G:G_{x_i}]}$, where G_{x_i} is the isotropy subgroup of x_i, and $\{N(x_1), N(x_2), \ldots, N(x_n)\}$ is a homogeneous system of parameters for $\mathbb{F}[V]$. Therefore, using this basis and order, $\mathbb{F}[V]^{\mathbb{Z}/p}$ has a finite SAGBI basis (see [**18**, Theorem 3.3, Corollary 3.4]).

Suppose that $C \subseteq \mathbf{k}[V]$. A *tête-à-tête* (over C) is the analogue of an S-polynomial and consists of two factorisations of a monomial over $\mathrm{LM}(C)$. We will refer to a tête-à-tête as *trivial* if the two factorisations have a common factor greater than 1. A tête-à-tête is given by two subsets $\Lambda_1, \Lambda_2 \subseteq C$ and positive integers e_s for $s \in \Lambda_1$ and d_h for $h \in \Lambda_2$ such that

$$\prod_{s\in\Lambda_1} \mathrm{LM}(s^{e_s}) = \prod_{h\in\Lambda_2} \mathrm{LM}(h^{d_h}).$$

It is then possible to choose $c_1, c_2 \in \mathbf{k}$ so that

$$c_1 \prod_{s\in\Lambda_1} \mathrm{LT}(s^{e_s}) = c_2 \prod_{h\in\Lambda_2} \mathrm{LT}(h^{d_h}).$$

The difference

$$c_1 \prod_{s\in\Lambda_1} s^{e_s} - c_2 \prod_{h\in\Lambda_2} h^{d_h}$$

is either zero or has a smaller lead monomial. Despite the ambiguity, we will sometimes refer to this difference as the tête-à-tête.

Subduction is the analogue of reduction. For a homogeneous $f \in \mathbf{k}[V]$ of positive degree, if $\mathrm{LM}(f)$ has a factorisation over $\mathrm{LM}(C)$ then, there exits a finite subset $\Lambda \subseteq C$, a coefficient $c \in \mathbf{k}$ and positive integers e_h for $h \in \Lambda$ such that

$$\mathrm{LT}(f) = c \prod_{h\in\Lambda} \mathrm{LT}(h)^{e_h}.$$

The difference,

$$f - c\prod_{h\in\Lambda} h^{e_h},$$

is called a primary subduction of f and is either zero or has a lead monomial less than $\mathrm{LM}(f)$. A full subduction of f consists of iterating this process as long as the lead monomial has a factorisation over $\mathrm{LM}(C)$. If C is a SAGBI basis for the subalgebra, then subduction provides a test for subalgebra membership: f is an element of the subalgebra if and only if f subducts to zero [**14**, 16.6]. Subduction also provides a SAGBI basis test: C is a SAGBI basis for the subalgebra generated by C if and only if every non-trivial tête-à-tête subducts to zero [**14**, 2.6]. (Note that we work exclusively with homogeneous polynomials.).

For a finite subset $C \subseteq \mathbf{k}[V]$, define $d := |C|$ and $A_C := \mathbf{k}[t_1, \ldots, t_d]$. Further define $\phi : A_C \to \mathbf{k}[V]$ by $\phi(t_h) = \mathrm{LT}(h)$ and $\Phi : A_C \to \mathbf{k}[V]$ by $\Phi(t_h) = h$. The kernel of ϕ, I_C, is a toric ideal whose binomial[2] generators correspond to tête-à-têtes over C. Let $\mathcal{T}_C$ be a finite generating set for I_C consisting of binomials.

THEOREM 4.2. *C is a* SAGBI *basis for the subalgebra generated by C if and only every element of $\Phi(\mathcal{T}_C)$ subducts to zero.*

PROOF. This is essentially [**20**, Corollary 11.5]. The elements of $\mathcal{T}_C$ correspond to a set of generating tête-à-têtes. □

ALGORITHM 4.3. *Given a finite set C of homogeneous polynomials of positive degree in $\mathbf{k}[V]$, and a positive integer d, return a* SAGBI *basis through degree d for the subalgebra generated by C.*

[2]Our convention is that a binomial is a linear combination of two distinct monomials.

1. *Define* $m := \min\{\text{degree}(h) \mid h \in C\}$. *Set* G *to be the set of elements of* C *of degree* m *and set* $i := m+1$.
2. *Compute a Gröbner basis of binomials through degree* i *for the kernel of* $\phi : A_G \to \mathbf{k}[V]$. *Call this basis* B.
3. *Subduct each element of* $\Phi(B)$ *against the set* C *and adjoin non-zero full subductions to* C.
4. *If* $i = d$, *return* C. *Otherwise set* $i := i+1$.
5. *Adjoin elements of degree* i *from* C *to* G *and go to Step* 2.

REMARK 4.4. The Magma [**2**] command "Groebner(S,d)" was introduced with version 2.7 and allows the computation of a partial Gröbner basis through degree d. David Wehlau and I have written a Magma script implementing subduction. Thus the above algorithm has essentially been implemented in Magma. We note that the algorithm, as described here, does not produce a minimal SAGBI basis.

5. Transvectants

The transvection process can be used to produce new covariants from old. Define algebra homomorphisms $\rho : \mathbb{C}[x, y, a_0, \ldots, a_m] \to \mathbb{C}[X, Y, x, y, a_0, \ldots, a_m]$ and $\pi : \mathbb{C}[X, Y, x, y, a_0, \ldots, a_m] \to \mathbb{C}[x, y, a_0, \ldots, a_m]$ by $\rho(a_i) = a_i$, $\rho(x) = X$, $\rho(y) = Y$, $\pi(a_i) = a_i$, $\pi(X) = x$, $\pi(Y) = y$, $\pi(x) = x$ and $\pi(y) = y$. The transvectant of two covariants is defined using Cayley's Ω-operator,

$$\Omega := \frac{\partial^2}{\partial x \partial Y} - \frac{\partial^2}{\partial y \partial X}.$$

The rth transvectant of h and g is $(h, g)^r := \pi\big(\Omega^r\big(h \cdot \rho(g)\big)\big)$ (see [**21**], [**8**, §48], [**13**, Ch. 5] and [**19**, §4.3]). Note that if h is homogeneous of degree d_1 and order k_1, and g is homogeneous of degree d_2 and order k_2, then $(h, g)^r$ is homogeneous of degree $d_1 + d_2$ and order $k_1 + k_2 - 2r$. Also note that to combat coefficient bloat, Grace and Young scale the transvectant by a coefficient which depends on the degree and order of h and g.

The transvectant of two covariants is again a covariant and a generating set for the ring of covariants can be constructed by starting with the binary form and iteratively constructing transvectants. This is a key element in Gordan's proof that the ring of covariants is finitely generated [**8**, Chapter VI].

EXAMPLE 5.1. **The Cubic**. A fundamental set of covariants is given by the form f, the Hessian $H := (f, f)^2$, $t := (f, H)^1$ and $\Delta := (f, t)^3$ (see [**8**, §88], [**9**, p. 68], [**19**, Proposition 3.7.7], [**13**, p. 39] and [**12**, §6.4]). The source of f gives x_1. The source of H gives an invariant with lead monomial x_2^2. The source of t gives an invariant with lead monomial x_2^3. The source of Δ gives an invariant with lead monomial $x_2^2 x_3^2$. Comparing with [**15**, §4], we see that we have a SAGBI basis for the ring of rational invariants.

EXAMPLE 5.2. **The Quartic**. A fundamental set of covariants is given by f, $H = (f, f)^2$, $i := (f, f)^4$, $t = (f, H)^1$, and $j := (f, H)^4$ [**8**, §89]. The lead monomials of the corresponding rational invariants are x_1, x_2^2, x_3^2, x_2^3 and x_3^3. Subducting the tête-à-tête formed from the third and fifth invariants gives an invariant with lead monomial $x_4^2 x_3^2 x_2^2$. Comparing with [**15**, §5], we see that we have a SAGBI basis for the ring of rational invariants.

The next three examples involve simultaneous covariants of a system of binary forms. Although Sections 2 and 3 dealt only with covariants of a single form, there are natural generalisations to a system of forms. Furthermore, a generating set for the ring of simultaneous covariants for a finite set of binary forms can be constructed by starting with the binary forms and iteratively constructing transvectants [**8**, Chapter VIII].

EXAMPLE 5.3. A linear form ℓ and a quadratic form f. A fundamental set of covariants is given by ℓ, f, the discriminant $\Delta := (f,f)^2$, $(f,\ell)^1$, and $(f,\ell^2)^2$ [**8**, §138 A] [3]. If we use $\{x_1, y_1, x_2, y_2, z_2\}$ as our basis for $V_2^* \oplus V_3^*$, the lead monomials of the corresponding rational invariants are x_1, x_2, y_2^2, x_2y_1 and $x_2y_1^2$. Comparing with [**17**, §5], we see that we have a SAGBI basis for the ring of rational invariants.

EXAMPLE 5.4. Two quadratics: f_1 and f_2. A fundamental set of covariants is given by f_1, f_2, $(f_1,f_1)^2$, $(f_2,f_2)^2$, $(f_1,f_2)^1$ and $(f_1,f_2)^2$ [**8**, §139]. If we use $\{x_1, y_1, z_1, x_2, y_2, z_2\}$ as our basis for $V_3^* \oplus V_3^*$, the lead monomials of the corresponding rational invariants are x_1, x_2, y_1^2, y_2^2, x_2y_1 and x_2z_1. The only non-trivial tête-à-tête is formed from the second, third and fifth generators. This subducts to 0. Therefore we have a SAGBI basis for the rational invariants. This is consistent with unpublished work of Brandon Fodden on $\mathbb{F}_p[2V_3]^{\mathbb{Z}/p}$.

EXAMPLE 5.5. A quadratic ϕ and a cubic f. A fundamental set of covariants is given in [**8**, §140]. We use $\{x_1, y_1, z_1, x_2, y_2, z_2, w_2\}$ as our basis for $V_3^* \oplus V_4^*$. Below we describe the fundamental covariants along with the lead monomials of the corresponding rational invariants. The covariant $c_{i,j}$ has degree i and order j. The fundamental covariants:

$$\begin{array}{ll} \phi,\ x_1; & f,\ x_2; \\ D := c_{2,0} := (\phi,\phi)^2,\ y_1^2; & H := c_{2,2} := (f,f)^2,\ y_2^2; \\ T := c_{3,3} := (f,H)^1,\ y_2^3; & \Delta := c_{4,0} := (H,H)^2,\ z_2^2y_2^2; \\ c_{2,3} := (\phi,f)^1,\ x_2y_1; & c_{2,1} := (\phi,f)^2,\ x_2z_1; \\ c_{3,2} := (\phi,H)^1,\ y_2^2y_1; & c_{3,1} := (\phi^2,f)^3,\ x_2z_1y_1; \\ c_{3,0} := (\phi,H)^2,\ y_2^2z_1; & c_{4,1} := (\phi,T)^2,\ y_2^3z_1; \\ c_{5,1} := (\phi^2,T)^3,\ y_2^3z_1y_1; & c_{5,0} := (\phi^3,f^2)^6,\ x_2^2z_1^3; \\ c_{7,0} := (\phi^3,fT)^6,\ y_2^3x_2z_1^3. & \end{array}$$

For a covariant c, let $\mathrm{inv}(c)$ denote a corresponding rational invariant. The tête-à-tête given by $\left(\mathrm{inv}(c_{2,1})\,\mathrm{inv}(H), \mathrm{inv}(f)\,\mathrm{inv}(c_{3,0})\right)$ subducts to an invariant with lead monomial $y_2^3y_1$. A Magma [**2**] calculation based on Theorem 4.2 verifies that we now have a SAGBI basis for the $\mathbb{Q}[x_1, y_1, z_1, x_2, y_2, z_2, w_2]^{\mathbb{Z}}$.

6. The binary quintic

A fundamental set of covariants for the binary quintic can be found either in [**8**, Ch. VII] or [**21**]. The descriptions are slightly different but yield the same lead monomials for the corresponding rational invariants. Grace and Young describe each covariant as a transvectant in terms of f, i, H and t while Sylvester uses various lower degree covariants in his description. There is a certain elegance to

[3]Beware the typographical error in the last line of [**8**, §138] : $(f,\ell)^2 = 0$.

the Grace and Young description but Sylvester's description is more computationally efficient—a Magma [**2**] construction using the Grace and Young description takes four times longer than one based on Sylvester's construction. Below we give Sylvester's description of the fundamental covariants along with the lead monomial of the corresponding rational invariant. The covariant $c_{i,j}$ has degree i and order j. Sylvester's covariants:

$$f,\ x_1;\quad H = (f,f)^2,\ x_2^2;\quad i := (f,f)^4,\ x_3^2;\quad t = (f,H)^1,\ x_2^3;$$

$$\begin{array}{ll}
c_{3,5} := (i,f)^1,\ x_3^2x_2; & c_{3,3} := (i,f)^2,\ x_3^3; \\
c_{4,6} := (c_{3,3},f)^1,\ x_3^3x_2; & c_{4,4} := (c_{3,3},f)^2,\ x_3^4; \\
c_{4,0} := (i,i)^2,\ x_4^2x_3^2; & c_{5,7} := (c_{4,4},f)^1,\ x_3^4x_2; \\
c_{5,3} := (c_{3,3},i)^1,\ x_4^2x_3^2x_2; & c_{5,1} := (c_{3,3},i)^2,\ x_4^2x_3^3; \\
c_{6,4} := (c_{4,4},i)^1,\ x_4^2x_3^3x_2; & c_{6,2} := (c_{3,3},c_{3,3})^2,\ x_4^2x_3^4; \\
c_{7,5} := (c_{4,4},c_{3,3})^1,\ x_4^2x_3^4x_2; & c_{7,1} := (c_{4,4},c_{3,5})^4,\ x_4^3x_3^4; \\
c_{8,2} := (c_{4,4},c_{4,6})^4,\ x_4^3x_3^5; & c_{8,0} := (c_{4,4},c_{4,4})^4,\ x_4^4x_3^4; \\
c_{9,3} := (c_{6,2},c_{3,3})^1,\ x_4^3x_3^6; & c_{11,1} := (c_{5,1},c_{6,2})^1,\ x_4^5x_3^6; \\
c_{12,0} := (c_{6,2},c_{6,2})^2,\ x_4^6x_3^6; & c_{13,1} := (c_{7,1},c_{6,2})^1,\ x_4^6x_3^7; \\
c_{18,0} := (c_{13,1},c_{5,1})^1,\ x_5^2x_4^5x_3^{11}. &
\end{array}$$

For comparison, the Grace and Young degree 18 covariant is $(i^7, ft)^{14}$.

A SAGBI basis for $\mathbb{Q}[x_1,x_2,x_3,x_4,x_5,x_6]^{\mathbb{Z}}$ through degree 25 was computed using a variation of Algorithm 4.3 in Magma [**2**] on Medicis[4]. The lead monomials for the 60 generators:

$$x_1,\ x_2^2,\ x_3^2,\ x_2^3,\ x_3^2x_2,\ x_3^3,\ x_4^2x_2^2,\ x_3^3x_2,\ x_4^2x_3^2,\ x_4^2x_2^3,\ x_4^2x_3^2x_2,\ x_4^2x_3^3,\ x_4^3x_2^3x_4^2x_3^3x_2,$$
$$x_4^3x_3^2x_2,\ x_4^3x_2^4,\ x_4^4x_2^3,\ x_4^3x_3^2x_2^2,\ x_4^3x_3^3x_2,\ x_4^3x_3^4,\ x_4^5x_2^3,\ x_4^3x_3^3x_2^2, x_5^2x_4^2x_3^2x_2^2,\ x_4^3x_3^5,$$
$$x_4^5x_2^4,\ x_4^5x_3^2x_2^2,\ x_4^6x_2^4,\ x_4^5x_3^3x_2^2,\ x_4^6x_3^2x_2^2,\ x_4^7x_2^4, x_4^6x_3^3x_2^2,\ x_5^2x_4^2x_3^6x_2,\ x_4^8x_2^4, x_5^2x_4^2x_3^7x_2,$$
$$x_5^2x_4^2x_3^8,\ x_4^8x_2^5,\ x_5^2x_4^2x_3^9,\ x_4^9x_2^5,\ x_4^8x_3^3x_2^3,\ x_4^{10}x_2^5, x_4^9x_3^3x_2^3,\ x_4^{11}x_2^5,\ x_4^9x_3^5x_2^2,\ x_4^{11}x_2^6,$$
$$x_4^{10}x_3^4x_2^3,\ x_5^2x_4^5x_3^9x_2,\ x_4^{12}x_2^6,\ x_4^{11}x_3^3x_2^4,\ x_5^2x_4^5x_3^{11},\ x_4^{13}x_2^6, x_4^{14}x_2^6,\ x_4^{14}x_2^7,\ x_4^{13}x_3^4x_2^4,$$
$$x_4^{15}x_2^7,\ x_4^{14}x_3^3x_2^5,\ x_4^{14}x_3^4x_2^4,\ x_4^{16}x_2^7,\ x_4^{14}x_3^5x_2^4,\ x_4^{17}x_2^7, x_4^{17}x_2^8.$$

CONJECTURE 6.1. *Using the graded reverse lexicographic order with* $x_1 < x_2 < x_3 < x_4 < x_5 < x_6$,

(i) *For a given* $k > 1$, *the smallest* j *such that* $x_4^k x_2^j$ *is the lead monomial of a rational invariant is given by* $j = \lfloor k/3 \rfloor + 2$;
(ii) *The rational invariants do not have a finite* SAGBI *basis.*

Reviewing the SAGBI basis through degree 25 verifies (i) for $k \leq 17$. The second part of the conjecture is easily seen to be a consequence of the first.

The preceding analysis of the binary quintic originated in an attempt to construct of a finite SAGBI basis for $\mathbb{F}_p[V_6]^{\mathbb{Z}/p}$. We know that $\mathbb{F}_p[V_6]^{\mathbb{Z}/p}$ does have a finite SAGBI basis. However, the rôle that the subring of rational invariants will play in the description of this SAGBI basis is unclear.

---

[4]`http://www.medicis.polytechnique.fr/medicis/cri-eng.html`.

References

1. G. Almkvist, *Invariants, mostly old ones*, Pacific J. Math. **86** (1980), no. 1, 1–13.
2. W. Bosma, J. J. Cannon, and C. Playoust, *The Magma algebra system. I. The user language*, J. Symbolic Comput. **24** (1997), 235–265.
3. H. E. A. Campbell and I. P. Hughes, *Vector invariants of* $\mathrm{U}_2(\mathbb{F}_p)$: *a proof of a conjecture of Richman*, Adv. Math. **126** (1997), 1–20.
4. D. Cox, J. Little, and D. O'Shea, *Ideals, varieties, and algorithms*, An introduction to computational algebraic geometry and commutative algebra, Undergrad. Texts Math., Springer-Verlag, New York, 1992.
5. L. E. Dickson, *On invariants and the theory of numbers*, Dover Publications, Inc., New York, 1966.
6. M. Göbel, *Computing bases for rings of permutation-invariant polynomials*, J. Symbolic Comput. **19** (1995), 285–291.
7. ______, *A constructive description of* SAGBI *bases for polynomial invariants of permutation groups*, J. Symbolic Comput. **26** (1998), 261–272.
8. J. H. Grace and A. Young, *The algebra of invariants*, Cambridge Univ. Press, Cambridge, 1903.
9. D. Hilbert, *Theory of algebraic invariants*, (B. Sturmfels, ed.), Cambridge Univ. Press, Cambridge, 1993.
10. I. Hughes and G. Kemper, *Symmetric powers of modular representations, Hilbert series and degree bounds*, Comm. Algebra **28** (2000), 2059–2088.
11. D. Kapur and K. Madlener, *A completion procedure for computing a canonical basis of a k-subalgebra*, Computers and Mathematics (Cambridge, 1989) (E. Kaltofen and S. Watt, eds.), vol. 89, Springer, New York, 1989, pp. 1–11.
12. J. P. S. Kung and G.-C. Rota, *The invariant theory of binary forms*, Bull. Amer. Math. Soc. **10** (1984), no. 1, 27–92.
13. P. J. Olver, *Classical invariant theory*, London Math. Soc. Stud. Texts, vol. 44, Cambridge University Press, Cambridge, 1999.
14. L. Robbiano and M. Sweedler, *Subalgebra bases*, Commutative Algebra (Salvador, 1988), Lecture Notes in Math., vol. 1430, Springer, Berlin, 1990, pp. 61–87.
15. R. J. Shank, SAGBI *bases for rings of formal modular seminvariants*, Commentarii Mathematici Helvetici **73** (1998), no. 4, 548–565.
16. R. J. Shank and D. L. Wehlau, *The transfer in modular invariant theory*, J. Pure Appl. Algebra **142** (1999), no. 1, 63–77.
17. ______, *Noether numbers for subrepresentations of cyclic groups of prime order*, Bull. London Math. Soc. **34** (2002) 438–450.
18. ______, *Computing modular invariants of p-groups*, J. Symbolic Comput. **34** (2002), no. 5, 307–327.
19. B. Sturmfels, *Algorithms in invariant theory*, Texts Monogr. Symbol. Comput., Springer-Verlag, Vienna, 1993
20. ______, *Gröbner bases and convex polytopes*, Univ. Lecture Ser., vol. 8, Amer. Math. Soc., Providence, RI, 1996.
21. J. J. Sylvester, *A synoptical table of the irreducible invariants and covariants to a binary quintic, with a scholium on a theorem in conditional hyperdeterminants*, Amer. J. Math. **1** (1878), 370–378.

Institute of Mathematics and Statistics, University of Kent at Canterbury, Canterbury, Kent CT2 7NF, UK.

E-mail address: `R.J.Shank@ukc.ac.uk`

Centre de Recherches Mathématiques
CRM Proceedings and Lecture Notes
Volume **35**, 2004

Classification of Nearly Closed Orbits for the Action of Semisimple Complex Linear Groups on the Projective Spaces

A. V. Smirnov

NOTATION. We denote, as it is usually done, the tangent algebra of a Lie group which is denoted by a capital Latin letter by the corresponding small Gothic letter:

$k = \mathbb{C}$: the field of complex numbers;
G: a connected semisimple Lie group over $\mathbb{C}$;
T: a fixed maximal algebraic torus in G;
$\mathfrak{t}(\mathbb{R})$: the real form of $\mathfrak{t}$ which consists of elements having real-valued eigenvectors;
$\mathbb{E} = \mathfrak{t}(\mathbb{R})^*$:
B: a fixed maximal Borel subgroup of G which contains T;
P_u: the unipotent radical of a group P;
U: the maximal unipotent subgroup of G lying in B;
$\Phi = \Phi(V) \subset \mathbb{E}$: the system of weights of G in V;
$V_\lambda \subset V$: the weight space corresponding to the weight $\lambda \in \Phi_V$;
$M = \operatorname{conv}\{\psi | V_\psi \neq \emptyset\} \subset \mathbb{E}$: the weight polytope of G in V;
$\Delta = \Delta_G$: the system of roots of G;
$\Delta^+ = \Delta_G^+$: the set of positive roots ($\mathfrak{b} = \langle \mathfrak{t}, e_\alpha | \alpha \in \Delta^+ \rangle$);
$\mathbb{P}(V)$: the projectivization of a vector space V;
$\langle v \rangle$: the point of a projective space which corresponds to the line spanned on v;
$n = \operatorname{rk}(G, G)$: the semisimple rank of G;
$\{\alpha_1, \dots, \alpha_n\}$: the system of simple roots of $\mathfrak{g}$;
$\{\pi_1, \dots, \pi_n\}$: the system of fundamental weights of $\mathfrak{g}$ (see [**15**]).

In this paper I would like to present my recent results, a complete version of which has just been published [**18**]. The difference between the two papers is the absence of proofs in the text that follows. The reader will find here general explanations and some examples.

Consider a finite-dimensional linear representation $G : V$ of the group G and its induced representation in the projective space $\mathbb{P}(V)$. It is well-known that closed orbits of this action are exactly the orbits of points $\langle u \rangle$, where $u \in V$ is the highest

2000 *Mathematics Subject Classification.* Primary: 14E25, 20G05; Secondary: 14L30, 14M17.
This is the final form of the paper.

weight vector of G. In particular, if G acts on V irreducibly, the closed orbit is unique.

It is clear that the closure of any G-orbit in $\mathbb{P}(V)$ contains a closed orbit. Let us call an orbit *nearly closed* if its closure consists exactly of two orbits (one of which is closed).

For the first time nearly closed orbits were considered by E. B. Vinberg and D. N. Akhiezer in 1977. They obtained some results which have never been published, however, due to their incompleteness.

On the other hand, abstract two-orbit varieties were studied in many papers. By definition, a two-orbit variety is a connected projective algebraic G-variety which consists of two orbits (one of which is open and the other is closed).

It is usually supposed that two-orbit varieties are also normal (one of the reasons is the so-called Luna's conjecture: "all normal two-orbit varieties are spherical"). Normal two-orbit varieties were earlier classified under some assumptions. D. N. Akhiezer [**1**] and independently A. Hackleberry and D. Snow [**10**] classified two-orbit varieties with the closed orbit having codimension one. M. Brion [**5, 6**] reproduced those results by means of the Luna–Vust theory of homogeneous embeddings [**5, 6**]. Then D. Feldmuller described all two-orbit varieties with the closed orbit having codimension two. Finally, in 2000, S. Cupit-Foutou obtained a classification of all two-orbit normal varieties in her Ph.D. thesis and published in [**7**].

My paper [**18**] contains a complete classification of nearly closed orbits (and thereby two-orbit varieties). No results of the above listed papers are used in it.

Here I am going only to present my main results and general ideas.

For simplicity, we consider a slightly more general situation, where G is a connected reductive algebraic linear group. However, it is easy to show that the connected center of G acts trivially on all its nearly closed orbits, so the classification of nearly closed orbits of reductive groups is reduced to the classification of nearly closed orbits of semisimple groups.

Let

$$V = \bigoplus_{\lambda \in \Phi} V_\lambda$$

be the weight decomposition of V. The convex hull of Φ in $\mathbb{E}$ will be called a *weight polytope* of G and denoted by M.

For each vector $v = \sum_{\lambda \in \Phi} v_\lambda$ let us define the set of its weights $\Phi_v = \{\lambda \in \Phi | v_\lambda \neq 0\}$. The convex hull of this set will be called the *support* of the vector v and will be denoted as $\operatorname{supp} v$.

In the first part of my work [**18**] I develop and use the method of supports (see, for instance, [**17**]) to obtain essential restrictions for nearly closed orbits. Here I shall illustrate it with the help of some examples:

1. $G = A_1$. As is well-known, all representations of A_1 can be interpreted as spaces of homogeneous forms of two variables x and y. Moreover, the closed orbit in $\mathbb{P}$("forms of degree n") is the orbit of $\langle y^n \rangle$ (or $\langle x^n \rangle$, due to the action of the Weil group). It is easy enough to show that nearly closed orbits are the projectivizations of the orbits of all other monomials, i.e. $x^i y^{n-i}$ for all natural i smaller than n. Clearly, the orbits of $\langle x^i y^{n-i} \rangle$ and $\langle y^i x^{n-i} \rangle$ coincide, but one can show that neither two of the other orbits do. Note that the supports of these monomials are points.

2. $G = A_2$. Let the highest weight be equal to $n\pi_1$. Then the representation can be interpreted as the space of homogeneous forms of degree n of three variables x, y and z. As in the previous example, the closed orbit in $\mathbb{P}$("forms of degree n") is the orbit of $\langle y^n \rangle$ (or $\langle x^n \rangle$, $\langle z^n \rangle$). As my work shows, all nearly closed orbits here are again the orbits of $\langle x^i y^{n-i} \rangle$ (or $\langle x^i z^{n-i} \rangle$, $\langle y^i z^{n-i} \rangle$). The weight polytope is a triangle and the weights of the points $x^i y^{n-i}$ are on its edges. This means that the orbit is induced (what will be explained below) from the orbit of the previous example. Intuitively the construction of parabolic induction can be interpreted as the "pushing from the face" of the weight polytope.

3. We can generalize the previous example in the following way. Let us take such a representation that the polytope M has an edge l containing at least three integral weights ("almost all" representations satisfy this condition). Then the restriction $V|_l = \bigoplus_{\lambda \in \Phi \cap l} V_\lambda$ of V to l is a representation of A_1 which contains a nearly closed orbit. Consequently, we can induce an orbit from this one, so "almost all" representations have nearly closed orbits constructed in this way. So, to simplify the classification, we have to introduce the notion of a simple nearly closed orbit.

Let us consider a G-orbit in the space V. Let Γ be the minimal face of M containing the support of any vector v lying in this orbit. Let $\widetilde{\mathfrak{g}}$ denote the subalgebra of $\mathfrak{g}$ generated by $\mathfrak{t}$ and root vectors, which correspond to roots parallel to Γ, and $\widetilde{G}$ denote the corresponding connected subgroup of G. The subspace

$$\widetilde{V} = \bigoplus_{\lambda \in \Gamma \cap \Phi} V_\lambda$$

is obviously invariant under $\widetilde{G}$. We prove that $Gv \cap \widetilde{V} = \widetilde{G}v$ and that the orbit $G\langle v \rangle \subset \mathbb{P}(V)$ is nearly closed if and only if the orbit $\widetilde{G}\langle v \rangle \subset \mathbb{P}(\widetilde{V})$ is nearly closed. In the last case we will say that the nearly closed orbit $G\langle v \rangle$ is *induced* from the face Γ. As we show in [**18**], the face Γ satisfying the conditions formulated above is defined up to the action of the Weil group.

A nearly closed orbit is called *simple* if it is impossible to induce it from a proper face.

In accordance with what was formulated above, the study of arbitrary nearly closed orbits is reduced to the study of simple nearly closed orbits. Simple nearly closed orbits are "much more rare". For example, if the group is simple, its rank is more than two and the highest weight is strictly dominant, then such a representation has no simple nearly closed orbits (this is an *a posteriori* result).

Furthermore, let $G\langle v \rangle$ be a nearly closed orbit and $u \in V$ such a highest weight vector that $G\langle u \rangle \subset \overline{G\langle v \rangle}$. The irreducible G-invariant space $V^{\mathrm{pr}} \subset V$ generated by u will be called the *main irreducible component* of V (with respect to this nearly closed orbit). Let v^{pr} be the projection of v onto this component. We prove that:

1. The weight polytopes of G in spaces V and V^{pr} coincide;
2. The orbit $G\langle v^{\mathrm{pr}} \rangle$ is nearly closed;
3. $\operatorname{supp} v = \operatorname{supp} v^{\mathrm{pr}}$;
4. The stabilizers of the points $\langle v \rangle$ and $\langle v^{\mathrm{pr}} \rangle$ in $\mathfrak{g}$ coincide.

In view of this, the classification of nearly closed orbits is substantially reduced to the case where G acts on V irreducibly.

In my work [**18**], using the method of supports, I managed to prove that any nearly closed orbit contains such a point $\langle v \rangle$ that $\operatorname{supp} v$ is either a segment whose end-points are vertices of M and which is parallel to none of the roots of G, or a point (this result, in a weakened form, was first obtained by E. B. Vinberg). Correspondingly, we shall call the orbit $G\langle v\rangle$ an orbit of the first type, or of the second one. A deeper analysis enables us to obtain a relatively short list of semisimple complex linear Lie groups which can have nearly closed orbits, and to describe as well the supports and stabilizers of suitably chosen points of these orbits.

All the examples listed above belong to the first type. Let us consider an example of the second type:

4. $G = A_2$, $V = \mathrm{Ad}$, $M = \mathrm{conv}\{\alpha_1, \alpha_2, \alpha_1 + \alpha_2, -\alpha_1, -\alpha_2, -\alpha_1 - \alpha_2\}$—a regular hexagon, $v = e_{\alpha_1} + e_{\alpha_2}$, $\mathrm{supp}(v) = [\alpha_1, \alpha_2]$—a diagonal (not the longest one) of the hexagon.

The goal of the second part of my work [**18**] is to prove that all the described possibilities really exist. Here the fact that all the orbits of the obtained list are spherical plays a crucial role, because it enables us to apply the theory of spherical embeddings [**2, 4, 5, 14**]. For each of the obtained homogeneous spaces we find all complete two-orbit (normal) spherical embeddings (for all cases, but one, there is a unique embedding of the sort). Then for each of the obtained spherical embeddings X we examine the section spaces of all possible linear G-bundles on it. Thus we find all linear representations $G \to \mathrm{GL}(V)$, for which there is such an orbit in $\mathbb{P}(V)$ that the normalization of its closure is G-equivariantly isomorphic to X. After this we have two lists: the list of possible orbits obtained via the method of supports and the list of orbits we can construct obtained via the method of spherical varieties. Surprisingly, these lists coincide, so we obtain the classification and also come to an *a posteriori* proof of the Luna conjecture.

The following table summarizes the obtained classification of nearly closed orbits. In the second column of it we specify the type of $\mathfrak{g}$; in the third one—the type of the Lie subalgebra $\mathfrak{k}$ of the stabilizer of a point of the nearly closed orbit; in the sixth column—the vertices of the convex hull of a suitably chosen point of this orbit. In the fourth column we specify the highest weight vector Λ of the main irreducible component of the G-representation, where the nearly closed orbit is contained. The parameters a and b are arbitrary natural numbers (they appear as the coefficients at the primitive divisors, the fact that they are positive means that the corresponding bundle is ample).

As additional components, one can take any irreducible G-representations with highest weight vectors contained in the set $\big(\mathbf{P} \cap \big(\Lambda + \mathbf{X}(G/H)\big)\big)\backslash\Lambda$ described in the fifth column, where $\mathbf{X}(G/H)$ is the rank lattice of G/H, and $\mathbf{P}$ is the convex polytope, whose vertices are specified in the fourth column of the table. Nearly closed orbits corresponding to the data in a row form a collection with a number of parameters equal to the number of additional irreducible components. Specifically, an irreducible representation can contain no more than a single nearly closed orbit of this type.

In the last column we specify the codimension d of the closed orbit in the closure of the open one.

Numbers specified in the first column correspond to the numbers which are assigned to the nearly closed orbits in the paper [**18**]. Orbits having numbers from 1 to 9 belong to the first type, and the rest to the second type. The orbit number 0 is the only case where we have $\mathrm{rk}\,\mathfrak{g} = 1$.

TABLE 1

N	$\mathfrak{g}$	$\mathfrak{k}$	Λ	vertices of $\mathbf{P}$	vertices of $\operatorname{supp} v$	d
0	A_1	$\mathfrak{t}$	$(a+b)\pi_1$	$(a-b)\pi_1,\ (a+b)\pi_1$	$(a-b)\pi_1$	1
1	$A_1 \oplus A_1$	$\operatorname{diag}(A_1)$	$a(\pi_1+\pi_2)$	$0, a(\pi_1+\pi_2)$	$a(-\pi_1+\pi_2),\ a(\pi_1-\pi_2)$	1
2	A_2	$\mathbb{C}$	$a(\pi_1+\pi_2)$	$a(\pi_1+\pi_2),\ a\pi_1,\ a\pi_2$	$2a\pi_1 - a\pi_2,\ -a\pi_1 + 2a\pi_2$	2
3	B_2	$\mathbb{C}$	$a(\pi_1+\pi_2)$	$a(\pi_1+\pi_2),\ (5a/4)\pi_1,\ (5a/3)\pi_2$	$2a\pi_1 - a\pi_2,\ -a\pi_1 + 3a\pi_2$	2
4	G_2	$\mathbb{C}$	$a(\pi_1+\pi_2)$	$a(\pi_1+\pi_2),\ (7a/3)\pi_1,\ (7a/5)\pi_2$	$4a\pi_1 - a\pi_2, -a\pi_1 + 2a\pi_2$	2
5	B_3	G_2	$a\pi_3$	$0, a\pi_3$	$a(\pi_1-\pi_3),\ a(\pi_3-\pi_1)$	1
6	C_3	$A_1 \oplus \mathbb{C}$	$a\pi_2$	$a\pi_2,\ (a/2)\pi_2,\ (a/3)(\pi_1+\pi_3)$	$a(\pi_3-\pi_1),\ a(\pi_1+\pi_2-\pi_3)$	4
7	D_n	B_{n-1}	$a\pi_1$	$0, a\pi_1$	$a(\pi_{n-1}-\pi_n),\ a(\pi_n-\pi_{n-1})$	1
8	F_4	$G_2 \oplus \mathbb{C}$	$a\pi_1$	$a\pi_1,\ (a/2)\pi_1,\ (a/3)\pi_2$	$a(\pi_2-\pi_4),\ a(\pi_1+\pi_4-\pi_2)$	8
9	A_n	$A_{n-1} \oplus \mathbb{C}$	$a\pi_1 + b\pi_n$	a) $(a-b)\pi_1,\ a\pi_1 + b\pi_n$	$(a-b)\pi_1$	1
				b) $(b-a)\pi_n,\ a\pi_1 + b\pi_n$	$(b-a)\pi_n$	1
10	B_n	D_n	$a\pi_1$	$0, a\pi_1$	0	1
11	B_n	$A_{n-1} \oplus \mathbb{C}$	$a\pi_1 + b\pi_n$	$b\pi_n, a\pi_1 + b\pi_n$	$b\pi_n$	1
12	B_n	$A_{n-2} \oplus \mathbb{C} \oplus \mathbb{C}$	$a\pi_1 + b\pi_n$	a) $a\pi_1 + b\pi_n,\ (a-b)\pi_1 + b\pi_n,$ $(a-b/2)\pi_1 + (b/2)\pi_{n-1}$	$(a-b)\pi_1 + b\pi_n$	n
				b) $a\pi_1 + b\pi_n,\ a\pi_{n-1} + (-2a+b)\pi_n,\ b\pi_n$	$a\pi_{n-1} + (b-2a)\pi_n$	n
				c) $a\pi_1 + b\pi_n,\ (b-a)\pi_1 + (2a-b)\pi_n,\ b\pi_n,$ $(a-b/2)\pi_1 + (b/2)\pi_{n-1}$	$(b-a)\pi_1 + (2a-b)\pi_n$	n

TABLE 1 (continued)

N	$\mathfrak{g}$	$\mathfrak{k}$	Λ	vertices of $\mathbf{P}$	vertices of $\operatorname{supp} v$	d
13	C_n	$C_{n-1} \oplus C_1$	$a\pi_2$	$0,\ a\pi_2$	0	1
14	C_n	$C_{n-1} \oplus \mathbb{C}$	$b\pi_1 + a\pi_2$	$b\pi_1,\ b\pi_1 + a\pi_2$	$b\pi_1$	1
15	C_n	$A_{n-2} \oplus \mathbb{C} \oplus \mathbb{C}$	$a\pi_1 + b\pi_n$	a) $a\pi_1 + b\pi_n,\ (a-b)\pi_1 + b\pi_n,\ (a-b/2)\pi_1 + (b/2)\pi_{n-1}$	$(a-2b)\pi_1 + b\pi_n$	n
				b) $a\pi_1 + b\pi_n,\ a\pi_{n-1} + (-a+b)\pi_n,\ b\pi_n$	$a\pi_{n-1} + (-a+b)\pi_n$	n
				c) $a\pi_1 + b\pi_n,\ (2b-a)\pi_1 + (a-b)\pi_n,\ b\pi_n,$ $(a-b)\pi_1 + b\pi_{n-1}$	$(2b-a)\pi_1 + (a-b)\pi_n$	n
16	G_2	$\mathfrak{t}$	$a\pi_1 + b\pi_2$	a) $a\pi_1 + b\pi_2,\ (a+b)\pi_1,\ (a-b)\pi_1 + b\pi_2$	$(a-b)\pi_1 + b\pi_2$	2
				b) $a\pi_1 + b\pi_2,\ (a+b)\pi_1,\ b\pi_2,\ (b-a)\pi_1 + b\pi_2$	$(b-a)\pi_1 + b\pi_2$	2
17	G_2	$\mathfrak{t}$	$a\pi_1 + b\pi_2$	a) $a\pi_1 + b\pi_2,\ (b+a/3)\pi_2,\ 2a\pi_1 + (b-a)\pi_2$	$2a\pi_1 + (b-a)\pi_2$	2
				b) $a\pi_1 + b\pi_2,\ (a+b)\pi_1,\ b\pi_2,\ (a-3b)\pi_1 + 2b\pi_2$	$(a-3b)\pi_1 + 2b\pi_2$	2
				c) $a\pi_1 + b\pi_2,\ (a+b)\pi_1,\ ((a/3)+b)\pi_2,$ $(3b-a)\pi_1 + (a-b)\pi_2$	$(3b-a)\pi_1 + (a-b)\pi_2$	2
18	G_2	$A_1 \oplus \mathbb{C}$	a) $a\pi_2 + b\pi_1$	$a\pi_2 + b\pi_1,\ (a+b)\pi_1$	$(a+b)\pi_1$	1
			b) $a\pi_2$	$a\pi_2,\ a\pi_1$	$a\pi_1$	2
19	G_2	A_2	$a\pi_1$	$0,\ a\pi_1$	0	1
20	F_4	B_4	$a\pi_1$	$0,\ a\pi_1$	0	1
21	F_4	$A_1 \oplus A_1$	$b\pi_1 + a\pi_4$	a) $b\pi_1 + a\pi_4,\ (b/2)\pi_3 + (a-(b/2))\pi_4,\ b\pi_2 + (a-b)\pi_4$	$b\pi_2 + (a-b)\pi_4$	6
				b) $b\pi_1 + a\pi_4,\ (b-a)\pi_1 + a\pi_2,\ (b-2a)\pi_1 + a\pi_3$	$(b-2a)\pi_1 + a\pi_3$	6
				c) $b\pi_1 + a\pi_4,\ (b/2)\pi_3 + (a-b/2)\pi_4,\ (b-a)\pi_1 + a\pi_2,$ $(2a-b)\pi_2 + (b-a)\pi_3$	$(2a-b)\pi_2 + (b-a)\pi_3$	6

Acknowledgements

I would like to express my gratitude to the organizers of the Workshop on Classical Invariant Theory, Dr. Eddy Campbell and Dr. David Wehlau, from Queen's University, Kingston, as well as to Prof. Alexander Kirillov from the University of Pennsylvania, Philadelphia, whose joint financial support made my participation in the Workshop a reality.

References

1. D. Akhiezer, *Equivariant completion of homogeneous algebraic varieties be homogeneous divisors*, Ann. Global Anal. Geom. **1** (1983), no. 1, 49–78.
2. ———, *Spherical varieties*, Forschungsschwerpunkt Komplexe Mannigfaltigkeiten, Schriftenreihe. Ht. **199** (1993), Bochum.
3. N. Bourbaki, *Groupes et algèbres de Lie.* IV–VI, Paris, Hermann, 1972.
4. M. Brion, *Variétés sphériques*, Notes de la session de la S.M.F. Opérations Hamiltoniennes et Opérations de Groupes Algébriques, Institut Fournier, Grenoble, 1997.
5. ———, *On spherical varieties of rank one*, Group Actions and Invariant Theory (Montreal, 1988), CMS Conf. Proc., vol. 10, Amer. Math. Soc. Providence, RI, 1989, pp. 31–41.
6. ———, *A note on two-orbit varieties*, Topology Hawaii (Honolulu, 1990), World Sci. Publishing, River Edge, NJ, 1992, pp. 35–40.
7. St. Cupit-Foutou, *Classification of two-orbit varieties*, Comment. Math. Helv. **78** (2003), no. 2, 227–244.
8. ———, *La classification des variété-à-deux-orbites*, Ph.D., 2000.
9. D. Feldmüller, *Two-orbit varieties with smaller orbit of codimension two*, Arch. Math. (Basel) **54** (1990), no. 6, 582–593.
10. A. Huckleberry and D. Snow, *Almost-homogeneous Kähler manifolds with hypersurface orbits*, Osaka J. Math. **19** (1982), no. 4, 763–786.
11. J. E. Humphreys, *Reflection groups and Coxeter groups*, Cambridge Studies in Advanced Mathematics, vol. 29, Cambridge University Press, Cambridge, 1990.
12. ———, *Linear algebraic groups*, 2nd ed., Graduate Texts in Mathematics, vol. 21, New York, Heidelberg, Berlin, Springer-Verlag, 1981.
13. M. Krämer, *Sphärishe Untergruppen in kompakten zusammenhängenden Lieghruppen*, Compositio Math. **38** (1979), no. 2, 129–153.
14. D. Luna and T. Vust, *Plongements d'espaces homogènes*, Comment. Math. Helv. **58** (1983), no. 2, 186–245.
15. A. L. Onishchik and E. B. Vinberg, *Lie groups and algebraic groups*, Springer Series in Soviet Mathematics, Springer-Verlag, Berlin, 1990.
16. D. I. Panyushev, *Complexity and nilpotent orbits*, Manuscripta Math. **83** (1994), no. 3–4, 223–237.
17. V. L. Popov and E. B. Vinberg, *Invariant theory*, Algebraic Geometry IV, Encyclopedia Math. Sci. **55** (1994), Springer, pp. 123–284.
18. A. V. Smirnov, *Nearly closed orbits in projective representations of semisimple complex Lie groups*, Trudy Moskovskogo Matematicheskogo Obschestva **64** (2003), 213–270.

CHAIR OF HIGH ALGEBRA, DEPARTMENT OF MATHEMATICS, MOSCOW STATE UNIVERSITY, VOROBJOVY GORY, MOSCOW, 119899, RUSSIA.

E-mail address: asmirnov@rdm.ru

Centre de Recherches Mathématiques
CRM Proceedings and Lecture Notes
Volume **35**, 2004

Convex Cones and SAGBI Bases of Permutation Invariants

N. M. Thiéry and S. Thomassé

Abstract. Let G be a permutation group acting on $\{1,\dots,n\}$, and $<$ be any admissible term order on the polynomial ring $\mathbb{K}[x_1,\dots,x_n]$. We prove that the invariant ring $\mathbb{K}[x_1,\dots,x_n]^G$ of G has a finite SAGBI basis if, and only if, G is generated by reflections.

1. Introduction

Let $\mathbb{K}$ be a field (or a ring). Let G be a permutation group on $\{1,\dots,n\}$. Let $V := \mathbb{K}^n$, and $x := (x_1,\dots,x_n)$ be the canonical basis of the dual of V. The invariant ring $\mathbb{K}[x]^G$ of G is the graded subalgebra of the polynomials of $\mathbb{K}[x] := \mathbb{K}[x_1,\dots,x_n]$ which are invariant for the natural action of G on the variables. Given a vector $u \in \mathbb{N}^n$, we denote by x^u the monomial $x_1^{u_1}\cdots x_n^{u_n}$. The orbit sum $\mathrm{o}(x^u)$ of a monomial x^u is the sum of all the monomials in the G-orbit of x^u. Given a subset S of $\{1,\dots,n\}$, we denote by $\mathfrak{S}(S)$ the symmetric group of all permutations of S and by $e_d(S)$ the dth *elementary symmetric polynomial* in the corresponding variables $\sum_{(i_1<\dots<i_d)\in S^d} x_{i_1}\cdots x_{i_d}$.

We recall the definition of a SAGBI basis (Subalgebra Analog of a Gröbner Basis for Ideals) [**6, 9**], which provides a useful device in the computational study of invariant rings of permutation groups [**12**]. Given an admissible term order $<$ on $\mathbb{K}[x]$, we denote by $\mathrm{in}_<(\mathbb{K}[x]^G)$ the monoid of all initial monomials of invariants in $\mathbb{K}[x]^G$. For a permutation group, this can also be defined as the monoid of all monomials x^u such that $x^u \geq g.x^u$ for all $g \in G$. By abuse of notations, we also denote by $\mathrm{in}_<(\mathbb{K}[x]^G)$ the initial algebra of $\mathbb{K}[x]^G$, which is the linear span of those monomials. A SAGBI basis of $\mathbb{K}[x]^G$ is a subset S of invariants in $\mathbb{K}[x]^G$ whose initial monomials generates $\mathrm{in}_<(\mathbb{K}[x]^G)$ as an algebra. It follows in particular that S generates $\mathbb{K}[x]^G$ as an algebra. A monomial $x^u \in \mathrm{in}_<(\mathbb{K}[x]^G)$ is *irreducible* if it cannot be decomposed as a product of two non-trivial monomials of $\mathrm{in}_<(\mathbb{K}[x]^G)$ (this definition generalizes for any monoid). If x^u is irreducible, then any SAGBI basis of $\mathrm{in}_<(\mathbb{K}[x]^G)$ must contain a polynomial having x^u as initial monomial. The set of all orbit sums of irreducible monomials is actually the minimal reduced SAGBI

2000 *Mathematics Subject Classification.* Primary: 13A50; Seconday: 52B20.

This research was partially funded by Queen's University and the Royal Military College of Kingston (Ontario, Canada).

This is the final form of the paper.

basis of $\text{in}_<(\mathbb{K}[x]^G)$. We refer for example to [**1, 10**] for further background on invariant rings, term orders, and SAGBI bases.

As opposed to Gröbner bases, SAGBI bases need not be finite, and it is an important problem to classify the algebras which have a finite SAGBI basis for some term order [**10**]. In this article we complete this classification for invariant rings of permutation groups in their permutation basis.

Let $S_1, \dots, S_r$ be the orbits of G on the variables, and E be the familly of the n elementary symmetric polynomials

$$(e_1(S_1), \dots, e_{|S_1|}(S_1), \dots, e_1(S_r), \dots, e_{|S_r|}(S_r)). \tag{1.1}$$

Assume that G is generated by reflections (i.e. by transpositions); otherwise said, assume that G is the direct product $\mathfrak{S}(S_1) \times \cdots \times \mathfrak{S}(S_r)$ of the symmetric groups on its transitive components. Then, it is well known that not only $\mathbb{K}[x]^G$ is the polynomial ring in E, but E is a *comprehensive* SAGBI basis of $\mathbb{K}[x]^G$, that is a SAGBI basis of E for any admissible term order.

In [**2**], M. Göbel proved that these are the only permutation groups with a finite SAGBI basis with respect to the lexicographic term order.

THEOREM 1.1. *Let G be a permutation group acting on $\{1, \dots, n\}$, and $<_{\text{lex}}$ be the lexicographic term order on $\mathbb{K}[x]$. Then, the invariant ring $\mathbb{K}[x]^G$ has a finite* SAGBI *basis with respect to $<_{\text{lex}}$ if, and only if, G is generated by reflections.*

In [**3–5**], M. Göbel further conjectures that this result extends to any admissible term order, and proves it for the alternating groups A_n, and a few other groups.

In this paper, we confirm this conjecture by proving the following theorem.

THEOREM 1.2. *Let G be a permutation group acting on $\{1, \dots, n\}$, and $<$ be any admissible term order on $\mathbb{K}[x]$. Then, the invariant ring $\mathbb{K}[x]^G$ has a finite* SAGBI *basis with respect to $<$ if, and only if, G is generated by reflections.*

This result was also obtained independently by Z. Reichstein [**8**], as a corollary of a more general theorem on multiplicative invariants, as well as by Shigeru Kuroda [**7**], with some further generalizations. The techniques used in each cases are essentially the same, our proof being by far the shortest.

Note that Theorem 1.2 does not preclude the existence of a finite SAGBI basis for $\mathbb{K}[x]^G = \mathbb{K}[V]^G$ when the dual basis of V is not one of its permutation basis. Consider, for example, the cyclic group $C_3 = A_3$ acting by permutation on the variables x_1, x_2, x_3 in characteristic 3. In the basis $y_1 := x_1$, $y_2 := x_2 - x_1$, $y_3 := x_3 - 2x_2 + x_1$, the invariant ring $\mathbb{F}_3[x]^{C_3} = \mathbb{F}_3[V]^{C_3} = \mathbb{F}_3[y_1, y_2, y_3]^{C_3}$ has a finite SAGBI basis w.r.t. the degree reverse lexicographic order with $y_1 < y_2 < y_3$ [**11**]. Thus, classifying the invariants rings of permutation groups which have a finite SAGBI basis for some term order in some basis remains an open problem.

The paper is organized as follows. In Section 2, we show that, if a convex cone C of $\mathbb{R}^{+n}$ is not closed for the usual Euclidean topology, then its monoid of *integer vectors* (vectors with nonnegative integer coordinates) contains infinitely many irreducible elements. In Section 3, we describe the monoid $\text{in}_<(\mathbb{K}[x]^G)$ as the monoid of integer vectors of a suitable convex cone $C_<(G)$ of $\mathbb{R}^{+n}$, and prove Theorem 1.2 by showing that this convex cone is not closed when G is not generated by reflections. This last step relies on an explicit and fairly simple construction of a line segment in C with an extremity not in C.

2. Convex cones

Let $\mathbb{R}^+$ be the set of non-negative real numbers. We call *convex cone* a subset C of ${\mathbb{R}^+}^n$ such that $\lambda u + \mu v \in C$ for all u, v in C and λ, μ in $\mathbb{R}^+$. In particular, C is a monoid for $+$. Its *integral cone* $M(C) := C \cap \mathbb{N}^n$ is the submonoid of the integer vectors of C. An element of $M(C)$ is *irreducible* if it cannot be decomposed as the sum of two non-trivial elements of $M(C)$. Obviously, the set of irreducible elements of $M(C)$ is a minimal generating set for $M(C)$.

LEMMA 2.1. *Let C be a convex cone which spans $\mathbb{R}^n$ as a vector space, and is not closed for the usual topology of $\mathbb{R}^n$. Then, its integral cone $M(C)$ has infinitely many irreducible elements.*

PROOF. Let $v_1, \ldots, v_n$ be a subset of $M(C)$ which spans $\mathbb{R}^n$ as a vector space. Assume $M(C)$ is generated by a finite set S of integer vectors, and let $\langle S \rangle$ be the convex subcone of C spanned by S. This subcone is a closed subset of $\mathbb{R}^n$. Hence, $C \backslash \langle S \rangle$ is non-empty. Let v_0 be a vector of $C \backslash \langle S \rangle$, and consider the convex hull $O := \{\lambda_0 v_0 + \lambda_1 v_1 + \cdots + \lambda_n v_n, 0 < \lambda_i < 1\}$, which is both open and contained in C. Since v_0 is in the closure of O, and $\langle S \rangle$ is closed, $O \backslash \langle S \rangle$ is non-empty, open and included in C. Hence, $C \backslash \langle S \rangle$ has a non-empty interior.

Now, we can take an open ball $B(u, \epsilon)$ of center u and radius $\epsilon > 0$ in $C \backslash \langle S \rangle$. The open ball $B(\sqrt{n}/\epsilon u, \sqrt{n})$ still lies in $C \backslash \langle S \rangle$, but also contains an integer vector p. This vector p cannot be generated by S, which is the desired contradiction. $\square$

As a first application, we verify one of the very first results on SAGBI basis, namely that the invariant ring $\mathbb{K}[x]^{A_3}$ of the alternating group A_3 on 3 elements has no finite SAGBI basis with respect to the lexicographic term order with $x_1 > x_2 > x_3$. Let $x^u := x_1^{u_1} x_2^{u_2} x_3^{u_3}$ be a monomial. Then, x^u is initial if, and only if,

$$(u_1, u_2, u_3) \geq_{\text{lex}} (u_2, u_3, u_1) \quad \text{and} \quad (u_1, u_2, u_3) \geq_{\text{lex}} (u_3, u_1, u_2), \tag{2.1}$$

which we can rewrite as

$$\begin{aligned} &(u_1 - u_2, u_2 - u_3, u_3 - u_1) \geq_{\text{lex}} (0,0,0) \quad \text{and} \\ &(u_1 - u_3, u_2 - u_1, u_3 - u_2) \geq_{\text{lex}} (0,0,0). \end{aligned} \tag{2.2}$$

We define the *initial cone* C to be the convex cone of all vectors of ${\mathbb{R}^+}^3$ satisfying those inequations, so that $\text{in}_{\text{lex}}(\mathbb{K}[x]^{A_3})$ is the integral cone of C. Note that C spans $\mathbb{R}^n$, since it contains the independent vectors $(1,0,0)$, $(1,1,0)$, and $(1,1,1)$.

Now, proving that $\mathbb{K}[x]^{A_3}$ has no finite SAGBI basis for lex is immediate: $(1,0,1)$ is not in C, whereas $(1,0,1) + \epsilon(1,0,0)$ is in C for any $\epsilon > 0$; hence, C is not closed, and by Lemma 2.1 $\text{in}_{\text{lex}}(\mathbb{K}[x]^{A_3})$ has infinitely many irreducible elements.

The general proof will follow exactly the same line.

3. Infiniteness of SAGBI basis of permutation invariants

Let $<$ be an admissible term order on $\mathbb{K}[x]$; without loss of generality, we may assume that $x_1 > x_2 > \cdots > x_n$. By the classical characterization of admissible term orders (see [**1**]), there exists n linearly independent linear forms $l_1, \ldots, l_n$ in $\mathbb{R}^n \mapsto \mathbb{R}$ such that $x^u > x^v$ if, and only if,

$$(l_1(u), \ldots, l_n(u)) >_{\text{lex}} (l_1(v), \ldots, l_n(v)). \tag{3.1}$$

Then, a monomial x^u is in the initial algebra $\mathrm{in}_<(\mathbb{K}[x]^G)$ if, and only if, $x^u \geq g.x^u$, for all $g \in G$, that is

$$(3.2)\qquad \big(l_1(u), \dots, l_n(u)\big) \geq_{\mathrm{lex}} \big(l_1(g.u), \dots, l_n(g.u)\big), \quad \text{for all } g \in G.$$

We define the *initial convex cone* $C := C_<(\mathbb{K}[x]^G)$ to be the convex cone of $\mathbb{R}^{+n}$ defined by those inequations:

$$(3.3)\qquad \big(l_1(u - g.u), \dots, l_n(u - g.u)\big) \geq_{\mathrm{lex}} (0, \dots, 0), \quad \text{for all } g \in G,$$

so that $\mathrm{in}_{\mathrm{lex}}(\mathbb{K}[x]^G)$ is the integral cone of C. It is well known that any *non-increasing* monomial (monomial x^u such that $u_1 \geq u_2 \geq \cdots \geq u_n$) is in C; it is the initial monomial of the corresponding symmetric function. Hence, $(1, 0, \dots, 0)$, $(1, 1, 0, \dots, 0)$, $(1, \dots, 1)$ are in C, and C spans $\mathbb{R}^n$ as a vector space.

We now turn to the proof of Theorem 1.2.

PROOF. Let G be a permutation group. Assume that C is closed, while G is not generated by reflections. Then, there exists $a < b$ such that the transposition (a, b) is not in G, while a is in the G-orbit of b. Choose such a pair $a < b$ with b minimal. We claim that there is no transposition (a', b) in G with $a' < b$. Otherwise, a and a' are in the same G-orbit, and by minimality of b, $(a, a') \in G$; thus, $(a, b) = (a, a')(a', b)(a, a') \in G$. Pick $g \in G$ such that $g.b = a$, and define

$$(3.4)\qquad u_t := \big((n-1)t, (n-2)t, \dots, (n-b+1)t, n-b, (n-b-1)t, \dots, t, 0\big).$$

Note that $u_0 < g.u_0$, so $u_0 \notin C$, whereas $u_1 \in C$. Furthermore, the entries of u_t are all distinct, except when $t = n - b/n - a'$ for some $a' < b$, in which case the a'th and bth entries are equal. Since $(a', b) \notin G$, for any t, $0 < t \leq 1$, the orbit of u_t is of size $|G|$, and there exists a unique permutation $\sigma_t \in G$ such that $\sigma_t.u_t$ is in C.

Let $t_0 = \inf\{t \geq 0, u_t \in C\}$. If $u_{t_0} \notin C$, then u_{t_0} is in the closure of C, but not in C, a contradiction. Otherwise, $u_{t_0} \in C$, and $t_0 > 0$ because $u_0 \notin C$. For any permutation σ, $\{\sigma.u_t, t \geq 0\}$ is a half-line; so, C being convex and closed, $I_\sigma := \{t, \sigma.u_t \in C\}$ is a closed interval $[x_\sigma, y_\sigma]$. For example, $I_{\mathrm{id}} = [t_0, 1] \subsetneq [0, 1]$. Since the interval $[0, 1]$ is the union of all the I_σ, there exists $\sigma \neq \mathrm{id}$ such that $t_0 \in I_\sigma$. This contradicts the uniqueness of σ_{t_0}. □

4. Conclusion

Let a be a real number. The convex cone in $\mathbb{R}^{+2}$ of all (x, y) such that $y > ax$ is a very simple geometrical object; yet, its integer monoid has a rich structure. The very same phenomenon appears for SAGBI basis of permutation groups, and explains the tediousness of the ad hoc constructions of infinite sequences of irreducible elements in [**3**–**5**]. The proof of Theorem 1.2 suggests that looking at the geometry of the initial convex cone is the proper tool to obtain further information on SAGBI basis of permutation groups.

References

1. D. Cox, J. Little, and D. O'Shea, *Ideals, varieties, and algorithms*, An Introduction to Computational Algebraic Geometry and Commutative Algebra, 2nd ed., Undergrad. Texts Math., Springer-Verlag, New York, 1997.
2. M. Göbel, *A constructive description of* SAGBI *bases for polynomial invariants of permutation groups*, J. Symbolic Comput. **26** (1998), no. 3, 261–272.
3. ———, *The optimal lower bound for generators of invariant rings without finite* SAGBI *bases with respect to any admissible order*, Discrete Math. Theor. Comput. Sci. **3** (1999), no. 2, 65–70 (electronic).

4. ———, *The "smallest" ring of polynomial invariants of a permutation group which has no finite* SABBI *bases* w.r.t. *any admissible order*, Theoret. Comput. Sci. **225** (1999), no. 1-2, 177–184.
5. ———, *Rings of polynomial invariants of the alternating group have no finite* SAGBI *bases with respect to any admissible order*, Inform. Process. Lett. **74** (2000), no. 1-2, 15–18.
6. D. Kapur and K. Madlener, *A completion procedure for computing a canonical basis for a k-subalgebra*, Computers and Mathematics (Cambridge, 1989), Springer, New York, 1989, pp. 1–11.
7. S. Kuroda, *The infiniteness of the* SAGBI *basis for certain invariant rings*, Osaka J. Math. **39** (2002), 665–680.
8. Z. Reichstein, SAGBI *basis in rings of multiplicative invariants*, Comment. Math. Helv. **78** (2003), no. 1, 185–202.
9. L. Robbiano and M. Sweedler, *Subalgebra bases*, Commutative Algebra (Salvador, 1988), Springer, Berlin, 1990, pp. 61-87.
10. B. Sturmfels, *Gröbner bases and convex polytopes*, Univ. Lecture Ser., vol. 8, Amer. Math. Soc., Providence, RI, 1996.
11. R. James Shank and D. Wehlau, *Computing modular invariants of p-groups*, J. Symbolic Comput. **34** (2002), no. 5, 307–327.
12. N. M. Thiéry, *Computing minimal generating sets of invariant rings of permutation groups with* SAGBI-*Gröbner basis*, Discrete Models: Combinatorics, Computation, and Geometry (Paris, 2001); Discrete Math. Theor. Comput. Sci. Proc., AA, Maison Inform. Math. Discrèt. (MIMD), Paris, 2001, pp. 315–328 (electronic).

LaPCS, Université Claude Bernard Lyon I, Bâtiment recherche [B] 50, avenue Tony-Garnier, Domaine de Gerland, 69366 Lyon Cedex 07, France.

E-mail address, Nicolas M. Thiéry: `nthiery@users.sf.net`

E-mail address, Stéphan Thomassé: `Stephan.Thomasse@univ-lyon1.fr`

Centre de Recherches Mathématiques
CRM Proceedings and Lecture Notes
Volume **35**, 2004

Some Problems in Invariant Theory

D. L. Wehlau

ABSTRACT. We present summaries of a number of research problems in invariant theory. These problems were posed during the organized problem sessions held during the April 2002, Kingston, Ontario, Invariant Theory Workshop and Conference.

1. Introduction

During the Invariant Theory Workshop held in Kingston, Ontario in April 2002 there were four hour long organized problem sessions. During these sessions a number of the workshop participants posed problems for future research. V. Popov and Z. Reichstein have written articles which appear elsewhere in this proceedings where they present detailed descriptions of the problems they posed during the workshop. Here we briefly summarize the other problems that were posed during the workshop. Two of these problems have been solved since they were posed. Notably, Question 2.1 was solved during the workshop. The other problem which has been settled since the workshop is Conjecture 13.2. At the time of this writing all the other problems remain open.

A few of the participants wrote the sections bearing their name which describe the problem(s) they posed. For the other sections, I wrote the description based upon my notes from the problem session. Many of these descriptions have been benefited from the comments of the original problem posers. I, however, bear the responsibility for any errors that may have have crept through. I thank everyone who participated in this project, and particularly acknowledge the anonymous referee for a great many useful comments and corrections.

Before proceeding to the summaries of the problems, we provide a few definitions and some notation which will prove useful in a number of the sections.

Let $\mathbb{F}$ be a field and consider a finitely generated Noetherian graded $\mathbb{F}$-algebra $A = \oplus_{i=0}^{\infty} A_i$. We denote by $\mathcal{H}(A,t)$ the Hilbert series of A, $\mathcal{H}(A,t) := \sum_{i=0}^{\infty} (\dim A_i) t^i$. We define the *Noether Number* of A, denoted $\beta(A)$, to be the maximum degree of an element of a homogeneous minimal generating set for A. It is not hard to show that $\beta(A)$ is independent of the choice of homogeneous minimal generating set. Indeed, if we consider the artinian algebra $B := A/(A_+)^2$ where $A_+ := \oplus_{i=1}^{\infty} A_i$ is

2000 *Mathematics Subject Classification.* Primary: 13A50; Secondary: 20G15.
This is the final version of the paper.

the *augmentation ideal* of A, then $\beta(A)$ is the largest degree in which the graded algebra B is non-zero.

For an ideal I of A we let $\beta(I)$ denote the largest degree of an element in a homogeneous minimal set of ideal generators for I. Again considering I/I^2 we may show that $\beta(I)$ is well-defined.

We call a representation V of G a *permutation representation* if there exists a basis $B = \{v_1, \ldots, v_n\}$ of V which is stabilized by every element g of G. Thus V is a permutation representation if and only if G is (conjugate to) a subgroup of the symmetric group $S_n \subset \mathrm{GL}(V)$. If V is a permutation representation then G also permutes $B^* = \{x_1, \ldots, x_n\}$, the basis of V^* which is dual to B. For $f \in \mathbb{F}[V]$ we define the orbit of f, $G \cdot f := \{g \cdot f \mid g \in G\}$. Further for $f \in \mathbb{F}[V]$ we define the *orbit sum* of f as $\sum_{h \in G \cdot f} h$. It is well known that for a permutation representation the set of orbit sums of the monomials (with respect to $\{x_1, x_2, \ldots, x_n\}$) in $\mathbb{F}[V]$ forms a vector space basis for $\mathbb{F}[V]^G$.

Let V be a G-module. The *Hilbert ideal*, J, is the ideal of $\mathbb{F}[V]$ generated by the augmentation ideal, $(\mathbb{F}[V]^G)_+$, of the algebra of G-invariants.

2. Karin Baur

Let $V = \mathbb{C}^{n+1}$ be the standard representation of $\mathrm{SL}_{n+1}(\mathbb{C})$. Denote by $\varepsilon_1, \varepsilon_2, \ldots, \varepsilon_{n+1}$ the weights of $\mathrm{SL}_{n+1}(\mathbb{C})$ acting on V with corresponding weight vectors $e_1, e_2, \ldots, e_{n+1}$. Write $\omega_i := \varepsilon_1 + \epsilon_2 + \cdots + \varepsilon_i$ and let $v_i = v_{\omega_i} := e_1 \wedge e_2 \wedge \cdots \wedge e_i$ denote a highest weight vector in $\wedge^i V$. Let $i < j$ and consider $\wedge^i V \otimes \wedge^j V$. This decomposes as a direct sum of irreducible $\mathrm{SL}_{n+1}(\mathbb{C})$ representations:

$$\wedge^i V \otimes \wedge^j V = V_{\omega_i + \omega_j} \oplus V_\lambda \oplus \cdots \oplus V_\mu.$$

We are interested in knowing whether every decomposable tensor in the Cartan component, $V_{\omega_i+\omega_j}$, is contained in the orbit closure $\overline{\mathrm{SL}_{n+1}(\mathbb{C}) \cdot (v_i \otimes v_j)}$. The answer to this question is "yes" if $(i,j) = (1,n), (1,2)$ or $(n-1,n)$. The answer is "no" if $j - i \geq 2$ and $(i,j) \neq (1,n)$.

QUESTION 2.1. Does every decomposable tensor in $V_{\omega_i+\omega_{i+1}}$ belong to

$$\overline{\mathrm{SL}_{n+1}(\mathbb{C}) \cdot (v_i \otimes v_{i+1})}$$

when $i \notin \{1, n-1\}$?

During the Kingston workshop Christian Ohn found the solution (for SL_n) to Question 2.1 which had been posed by Karin Baur. Later, Baur was able to generalize Ohn's result and classify all tensor products $V_\lambda \otimes V_\mu$ of irreducible representations of a reductive group G where the set of decomposable tensors in the Cartan component $V_{\lambda+\mu}$ is equal to the orbit closure $\overline{Gv_\lambda \otimes v_\mu}$ (orbit of the tensor of two highest weight vectors).

The general result [**3**] is the following:

THEOREM 2.2. *Let G be a connected reductive complex algebraic group, let λ and μ be dominant integral weights (with respect to a fixed Borel subgroup).*

The set of decomposable tensors in the Cartan component $V_{\lambda+\mu}$ of $V_\lambda \otimes V_\mu$ is equal to the orbit closure $\overline{Gv_\lambda \otimes v_\mu}$ if and only if the following two conditions hold:

(i) $\overline{L_\lambda v_\mu} = \langle L_\lambda v_\mu \rangle$;

(ii) $\overline{L_\mu v_\lambda} = \langle L_\mu v_\lambda \rangle$ (*where* $L_\mu \subset G$ *is the Levi subgroup generated by the maximal torus and the root subgroups of the roots perpendicular to* μ, *and* $\langle Gv \rangle$ *is the linear span of the orbit* Gv, *i.e., the representation it generates*).

3. Harm Derksen

Let V be an $\mathbb{F}$-vector space and let G be a linearly reductive subgroup of $\mathrm{GL}(V)$ such that G is a Zariski closed subset of $\mathrm{End}(V)$. Let W be a G-module.

DEFINITION 3.1. The *complexity* of W is denoted $c(W)$ and defined by

$$c(W) := \min\{d \mid \text{every irreducible subrepresentation of} W \text{ appears in } \oplus_{i=0}^{d} (\otimes_{j=1}^{i} V)\}.$$

EXAMPLE 3.2. If the characteristic of $\mathbb{F}$ is zero then $G = \mathrm{SL}(n, \mathbb{F})$ is linearly reductive. Let $\lambda = (\lambda_1, \ldots, \lambda_{n-1})$ be a highest weight and let $V(\lambda)$ be the corresponding irreducible $\mathrm{SL}(n, \mathbb{F})$-module. Then we have $c\big(V(\lambda)\big) = |\lambda| = \sum_{i=1}^{n-1} i\lambda_i$.

DEFINITION 3.3. We also define $\mathcal{C}_d(W) :=$ smallest set of covariants, C, such that

1. If Z is an irreducible G-module with $c(Z) \le d$ and $\phi : W \to Z$ is a linear covariant, then $\phi \in C$;
2. If $\phi_1, \phi_2 : W \to Z$ with $\phi_1, \phi_2 \in C$ then $\lambda_1\phi_1 + \lambda_2\phi_2 \in C$ for all scalars λ_1, λ_2; and
3. If $\phi_1 : W \to Z_1$ and $\phi_2 : W \to Z_2$ are two covariants with $\phi_1, \phi_2 \in C$ and $\psi : Z_1 \otimes Z_2 \to Z_3$ with $\mathcal{C}(Z_3) \le d$ then $\psi \circ (\phi_1 \otimes \phi_2) : W \to Z_3$ lies in C.

For a covariant $\phi : W \to Z$ we define the complexity $c(\phi)$ of ϕ as the smallest d such that $\phi \in \mathcal{C}_d(W)$. The constant $\mu(\mathbb{F}[W]^G) = \min\{d \mid c(f) \le d \;\forall f \in \mathbb{F}[W]^G\} < \infty$ gives a measure for the complexity of the invariant ring $\mathbb{F}[W]^G$.

PROBLEM 3.4. *Give bounds for* $\mu(\mathbb{F}[W]^G)$.

EXAMPLE 3.5. View the algebraic group $\mathrm{GL}(V)$ as a Zariski closed subset of $\mathrm{End}(V \oplus V^*)$.

Let $W = \mathrm{End}V \oplus \cdots \oplus \mathrm{End}V$. The algebra $\mathbb{F}[W]^{\mathrm{GL}(V)}$ is generated by traces of products, i.e., by elements of the form $\mathrm{Tr}(A_{i_1} \cdot \cdots \cdot A_{i_r})$. First observe that $c(\mathrm{End}(V)) = 2$ because $\mathrm{End}(V) \subset (V \oplus V^*) \otimes (V \oplus V^*)$. We can view A_{i_j} as a covariant $W \to \mathrm{End}(V)$ of complexity 2. It is now easy to see using the composition $\mathrm{End}(V) \times \mathrm{End}(V) \to \mathrm{End}(V)$ (which is also a covariant) that every product $A_{i_1} \cdots A_{i_r}$ defines a covariant of complexity 2. Finally the invariant $\mathrm{Tr}(A_{i_1} \cdots A_{i_r})$ is an invariant of complexity at most 2. This shows that $\mu\big(\mathbb{F}[W]^{\mathrm{GL}(V)}\big) \le 2$.

A positive answer to the following question would imply that the isomorphism of two graphs can be tested in polynomial time (which is a famous still open problem).

QUESTION 3.6. The symmetric group S_n can be viewed as the set of permutation matrices in $\mathrm{End}(\mathbb{F}^n)$. Does there exist a constant m such that for all n, $\mu(\mathbb{F}[\mathrm{S}^2(\mathbb{F}^n)]^{S_n}) \le m$?

QUESTION 3.7 (Popov). With G, V as before, find all G-modules W such that $\mu(\mathbb{F}[W]^G) = 1$.

4. Gene Freudenberg

Let $G_a = (\mathbb{C}, +)$. There exist non-linear actions of G_a on $\mathbb{C}^n$ for $n \geq 5$ with non-finitely generated rings of invariants. For linear actions of G_a the ring of invariants is always finitely generated by Weitzenböck's theorem.

QUESTION 4.1. What happens in positive characteristic?

There exists a linear G_a^3 action on $\mathbb{C}^{18}$ with a non-finitely generated ring of invariants (Mukai).

QUESTION 4.2. What about linear actions of G_a^2? Do these always have a finitely generated ring of invariants?

QUESTION 4.3 (G. Kemper). For which groups G, is $\mathbb{F}[V]^G$ finitely generated for all G-modules V? Perhaps this class of groups consists only of those groups whose unipotent radical is trivial together with $(\mathbb{F}, +)$?

PROBLEM 4.4 (Nagata's second problem). *Let K be a field with $\mathbb{C} \subset K \subset \mathbb{C}(x_1, x_2, \ldots, x_n)$ such that $K \cap \mathbb{C}[x_1, x_2, \ldots, x_n]$ is not finitely generated. Can K have transcendence degree* 3?

5. Julia Hartmann

Let G be a finite subgroup of $\mathrm{GL}(V)$ and let J denote the Hilbert ideal. The ring of coinvariants is $\mathbb{F}[V]_G := \mathbb{F}[V]/J$. Assume that $\mathbb{F}[V]^G$ is a polynomial ring. If V is a non-modular G-representation then the action of G on $\mathbb{F}[V]_G$ is equivalent to the regular representation. Conversely if the representation is modular, then $\mathbb{F}[V]_G$ is never the regular representation. What representation is it?

EXAMPLE 5.1 (Jim Shank). Let V_2 be the 2 dimensional non-trivial representation of the cyclic group, C_p, of order p over the field $\mathbb{F}_p$ of order p. Then $\mathbb{F}_p[V_2]^{C_p} = \mathbb{F}_p[x, Ny]$ is a polynomial ring with generators x of degree 1 and Ny of degree p. Also $\mathbb{F}_p[V_2]_{C_p}$ is the trivial representation of dimension p.

6. Loek Helminck

Let G be a connected reductive group over $\mathbb{C}$ and let $\theta, \sigma : G \to G$ be two commuting involutions of G. Denote the fixed point groups of σ and θ by $H := G^\sigma$ and $K := G^\sigma$. Consider the double coset space $H\backslash G/K$. Let $\beta\colon G \to G$, $g \mapsto g\theta(g)^{-1}$ and let Q denote Image(β). Then β induces an isomorphism $G/K \xrightarrow{\sim} Q$. If $h \in H$ and $x \in G$ we let $h{*}x$ denote $hx\theta(h)^{-1}$. Then, via β, the natural left action of H on G/K becomes its $*$-action on Q. We want to compute the invariant theoretic quotient $Q/\!\!/H$ of Q by the $*$-action of H. In [**8**] we established a version of the CHEVALLEY–LUNA–RICHARDSON Theorem [**15**] showing that $Q/\!\!/H \simeq A/W_H^*(A)$ where A denotes a maximal θ and σ split torus in Q and $W_H^*(A) = N_H^*(A)/Z_H^*(A)$ is the natural finite group acting on A, which comes from the twisted action $*$ on A. Here $N_H^*(A) = \{h \in H \mid h * A = A\}$ and $Z_H^*(A) = \{h \in H \mid h * a = a$ for every $a \in A\}$. We also established the existence of slices, showed that the orbit type stratifications of $A/W_H^*(A)$ and $Q/\!\!/H$ agree on connected components and gave a characterization of the group $W_H^*(A)$, which is essentially a combination of a reflection group together with a set of translations.

Assume now that G, H and K are defined over $\mathbb{R}$ and denote their sets of $\mathbb{R}$-rational points by $G_\mathbb{R}$, $H_\mathbb{R}$ and $K_\mathbb{R}$. Can we find similar results in the real case?

QUESTION 6.1. Does there exist a CHEVALLEY-LUNA-RICHARDSON type characterization for $H_{\mathbb{R}}\backslash G_{\mathbb{R}}/K_{\mathbb{R}}$, similar to the one above for all triples $(G_{\mathbb{R}}, H_{\mathbb{R}}, K_{\mathbb{R}})$? If $H_{\mathbb{R}}\backslash G_{\mathbb{R}}/K_{\mathbb{R}} \simeq \cup_{i\in I} E(A^i)_{\mathbb{R}}/W^*(A^i)$ with the $\{A^i \mid i \in I\}$ a set of maximal σ and θ-split $\mathbb{R}$-tori and $E(A^i) \subset A^i$, then can we find a characterization of $\{A^i \mid i \in I\}$ and of the finite groups $W^*(A^i)$? Do there exist slices and do we have an orbit type stratification? The answer to these questions is yes in the two extreme cases: If $G_{\mathbb{R}}$ is compact, then $H_{\mathbb{R}}\backslash G_{\mathbb{R}}/K_{\mathbb{R}} \simeq A_{\mathbb{R}}/W^*_{H_{\mathbb{R}}}(A)$ and if θ is a Cartan involution of $G_{\mathbb{R}}$, then $H_{\mathbb{R}}\backslash G_{\mathbb{R}}/K_{\mathbb{R}} \simeq A^2_{\mathbb{R}}/W_{(H\cap K)_{\mathbb{R}}}(A)$. Here $A^2_{\mathbb{R}} := \beta(A_{\mathbb{R}}) = \{a^2 \mid a \in A_{\mathbb{R}}\} \simeq \mathbb{R}^+)^l$ for some l.

7. Wilberd van der Kallen

We would like to call this problem the Cohomological Mumford Problem.

Let G be a reductive linear algebraic group defined over a field $\mathbb{F}$ of finite characteristic. Let A be a commutative $\mathbb{F}$-algebra on which G acts rationally by $\mathbb{F}$-algebra automorphisms. Let I be a G invariant ideal of A. We ask if $H^{\text{even}}(G, A/I)$ is integral over the image of $H^{\text{even}}(G, A)$. Here we restrict to the even part of the rational cohomology [**10**, I, Ch. 4] for reasons of terminology: We do not want to explain the notion of integral extensions for noncommutative rings.

The case where G equals SL_2 follows, after reduction to the case where A is finitely generated, from [**11**, Thm 4.12] applied to the symmetric algebra $S^*_A(A/I)$.

Recall that the Mumford Conjecture for G is equivalent to the statement that $H^0(G, A/I)$ is integral over the image of $H^0(G, A)$ for all such A and I. In fact the Mumford Conjecture, proved by Haboush [**10**, II, Prop. 10.7], concerns the following special case. Let V be a finite dimensional G-module with a nonzero invariant vector v. Let L be the line spanned by v. Take for A the ring $\mathbb{F}[V]$ of polynomial functions on V and take for I the ideal of the functions that vanish on L. Note that $\mathbb{F}[L]$ is just a polynomial ring in one variable. Therefore, to say that $\mathbb{F}[L] = H^0(G, A/I)$ is integral over the image of $H^0(G, A)$ is equivalent to saying that the image in $\mathbb{F}[L]$ is more than $\mathbb{F}$. Now Nagata has shown [**17**, Lemma 5.1B] that the case of general A, I follows from the special case.

8. Dikran Karaguezian

The G-module structure of $\mathrm{S}^*(V)$ should yield significant information about the invariant ring $\mathrm{S}^*(V)^G$. Thus, one might hope to obtain results in invariant theory by proving results for the module structure of $\mathrm{S}^*(V)$.

This general program has so far worked in practice in the following way. Given a G-module V of some special type, one proves a module-structure theorem for $\mathrm{S}^*(V)$ with specific bounds on the degrees of a homogeneous system of parameters and bounds on the degrees in which various modules appear. Then consequences can be deduced for $\mathrm{S}^*(V)^G$.

Examples of results whose proof follows this program can be found in [**13**, Theorem 16.4], and in [**12**, Section 8]. These results suggest two questions. First, can module-structure theory be used to prove further results in invariant theory? Second, are there general theorems describing the relationship between the module-structure of $\mathrm{S}^*(V)$ and the properties of the invariant ring $\mathrm{S}^*(V)^G$?

9. Gregor Kemper

Recall that Knop [14] showed that in characteristic zero, $\deg \mathcal{H}(\mathbb{F}[V]^G, t) \leq -\dim \mathbb{F}[V]^G$. Is this true in characteristic $p > 0$? Is it true for G a finite group in characteristic $p > 0$? It is shown in [9] that this bound is valid if p^2 does not divide $|G|$.

Let G be a finite group whose order is divisible by the characteristic of $\mathbb{F}$. Let R denote the ring of invariants $R := \mathbb{F}[V]^G$. Take $f_1, f_2, \ldots, f_n$ a homogeneous system of parameters for R and put $A = \mathbb{F}[f_1, f_2, \ldots, f_n]$. Then, R is a finitely generated A-module. Let $g_1, g_2, \ldots, g_m$ be homogeneous invariants which generate R as an A-module. Such a set of invariants is called a set of homogeneous *secondary invariants*. We order these secondary invariants such that $\deg g_1 \leq \deg g_2 \leq \cdots \leq \deg g_m$ and we assume that m is the minimum possible, i.e., $g_1, g_2, \ldots, g_m$ minimally generate R as an A-module.

QUESTION 9.1. Is it always true that g_m does not occur in any of the generating A-linear relations among the g_i?

Note that this question is motivated by observation of a number of computations. Also, note that an affirmative answer implies that $A \cdot g_m$ is a free summand of R. Furthermore, an affirmative answer would also imply that $\deg g_m \leq (\sum_{i=1}^n \deg f_i) - n$ by an argument of Bram Broer.

10. Hanspeter Kraft

The group GL_n (or SL_n) acts on $M_{n\times n}$ the set of $n \times n$ matrices by conjugation. The invariants here are generated by the n functions $A \mapsto \mathrm{Tr}(A^k)$ for $k = 1, 2, \ldots, n$ where $\mathrm{Tr}(X)$ denotes the trace of the matrix X. The null cone for this representation is $\mathcal{N}(M_{n\times n})$ = the set of nilpotent matrices.

Next consider $2M_{n\times n} := M_{n\times n} \oplus M_{n\times n}$. The invariants here are generated by the functions $(A, B) \mapsto \mathrm{Tr}(A^{i_1} B^{j_1} \cdots A^{i_k} B^{j_k})$. Also, the null cone $\mathcal{N}(2M_{n\times n}) = \{(A, B) \mid \exists g \in \mathrm{GL}_n$ such that gAg^{-1} and gBg^{-1} are both strictly upper triangular matrices$\}$. Consider $\mathcal{M}_n := \{(A, B) \in 2M_{n\times n} \mid \lambda A + \mu B$ is nilpotent for all scalars $\lambda, \mu\}$.

For $n = 2$, $\mathcal{M}_2 = \mathcal{N}(2M_{2\times 2})$.

For $n = 3$, $\mathcal{M}_3 \supsetneq \mathcal{N}(2M_{3\times 3})$. To see this consider $A = \begin{pmatrix} 0&1&0\\0&0&1\\0&0&0 \end{pmatrix}$ and $B = \begin{pmatrix} 0&0&0\\1&0&0\\0&-1&0 \end{pmatrix}$. Here $\lambda A + \mu B = \begin{pmatrix} 0&\lambda&0\\ \mu&0&\lambda\\ 0&-\mu&0 \end{pmatrix}$ is nilpotent for all λ and μ. But $(A, B) \notin \mathcal{N}(2M_{3\times 3})$ since if $(X, Y) \in \mathcal{N}(2M_{3\times 3})$ then XY is nilpotent. But $AB = \begin{pmatrix} 1&0&0\\0&-1&0\\0&0&0 \end{pmatrix}$.

DEFINITION 10.1. If we have a function $p(v)$ and we write $p(\lambda v_1 + \mu v_2) = \sum_i \lambda^i \mu^{d-i} p_i(v_1, v_2)$ then the functions p_i are *polarizations* of p.

Note that $\mathcal{M}_n$ = the common zero set of all the polarizations of all the functions $A \mapsto \mathrm{Tr}(A^k)$ for $k = 1, 2, \ldots, n$.

Consider again $n = 3$. The polarizations of $\mathrm{Tr}\, A$ are the two functions $\mathrm{Tr}\, A$ and $\mathrm{Tr}\, B$. The polarizations of $\mathrm{Tr}\, A^2$ are the three functions $\mathrm{Tr}\, A^2$, $\mathrm{Tr}\, AB$ and $\mathrm{Tr}\, B^2$ (since $\mathrm{Tr}\, AB = \mathrm{Tr}\, BA$). The polarizations of $\mathrm{Tr}\, A^3$ are the four functions $\mathrm{Tr}\, A^3$, $\mathrm{Tr}\, A^2B$, $\mathrm{Tr}\, AB^2$ and $\mathrm{Tr}\, B^3$. But there is another generating invariant, $\mathrm{Tr}(ABAB) = \mathrm{Tr}(BABA)$ which is not a polarization.

The representation $2M_{3\times 3}$ is 18 dimensional. The common zero set of all the polarizations, $\mathcal{M}_3$, has dimension 9. Thus these 9 functions arising as polarizations form a regular sequence in the ring of polynomial functions on $2M_{3\times 3}$. Also $\dim \mathcal{N}(2M_{3\times 3}) = 9$. Therefore, $\mathcal{N}(2M_{3\times 3})$ is a component of $\mathcal{M}_3$.

QUESTION 10.2. There are $2+3+\cdots+(n+1) = \binom{n+2}{2} - 1$ polarizations of the n functions $\operatorname{Tr} A^k$ for $k = 1, 2, \ldots, n$. Also $\operatorname{codim} \mathcal{N}(2M_{n\times n}) = \binom{n+2}{2} - 1$. Question: Does this set of polarizations always form a regular sequence in the ring of polynomial functions on $2M_{n\times n}$?

Note that this has been proved for $n = 4$. Also, note that $\mathcal{N}(2M_{n\times n})$ is always irreducible.

10.3. Bialynicky–Birula theory.

THEOREM 10.4 (Bialynicky–Birula). *Let X be a smooth projective variety on which $\mathbb{C}^*$ acts algebraically, i.e., via a morphism $\lambda : \mathbb{C}^* \to \operatorname{Aut}(X)$. Then the fixed points $F := X^{\mathbb{C}^*}$ form a smooth variety with components $F = \cup_i F_i$. For the set $X_i := \{x \in X \mid \lim_{t\to 0} \lambda(t)\cdot x \in F_i\}$ we have the morphism $p_i : X_i \to F_i$ (by taking limits). Every fibre of p_i is a vector space. Also for all affine open subsets $U \subset F_i$, we have $p_i^{-1}(U) \cong U \times \mathbb{C}^d$.*

QUESTION 10.5 (George Lusztig). Is $p_i : X_i \to F_i$ an algebraic vector bundle? It is if $d = 1$.

11. Joel Segal

Let $G = S_d \times \mathrm{GL}(V)$ where V is an $\mathbb{F}$-vector space of dimension n. Consider the natural G-representation afforded by $\underbrace{V \otimes V \otimes \cdots \otimes V}_{d \text{ copies}} = V^{\otimes d}$. What is $\mathbb{F}[V \otimes \cdots \otimes V]^G$? Of particular interest is the case where $q = p^r$ and $\mathbb{F}$ is the field of order q.

A (hopefully) easier question is to determine the fixed subspace $(V \otimes \cdots \otimes V)^G$. For example, as a partial answer, it can be shown that this fixed subspace is nonzero only if $(q-1)$ divides d, using the fact that

$$D(V) := \bigoplus_{d=1}^{\infty} (V^{\otimes d})^{S_d}$$

can be given a Hopf algebra structure.

The question can be generalized: what is $D(V)^H$, where $H \leq \mathrm{GL}(V)$? It is not in general finitely generated as an algebra—can we find other finiteness properties that will make a 'finite description' of the invariants possible?

12. Nicolas Thiery

PROBLEM 12.1. *Consider the $\binom{n}{2}$ dimensional vector space V_n whose basis is indexed by the pairs $\{i,j\}$ of $\{1,\ldots,n\}$ together with the natural permutation representation on V_n of the symmetric group S_n: $\sigma\cdot\{i,j\} = \{\sigma(i),\sigma(j)\}$. Find $\beta(\mathbb{F}[V_n]^{S_n})$.*

Here are some known values: $\beta(\mathbb{F}[V_3]^{S_3}) = 3$, $\beta(\mathbb{F}[V_4]^{S_4}) = 5$ and $\beta(\mathbb{F}[V_5]^{S_5}) = 9$. A partial computation shows that $\beta(\mathbb{F}[V_6]^{S_6}) \leq 11$, and equality is likely to hold.

Suppose V is a permutation representation of G. Since orbit sums always form a vector space basis for $\mathbb{F}[V]^G$, the Hilbert Series of $\mathbb{F}[V]^G$, $\mathcal{H}(\mathbb{F}[V]^G, t)$ is independent of the characteristic p of $\mathbb{F}$. Furthermore, algebraic properties like the degree bound or the size of a (homogeneous) minimal generating set only depends on the characteristic, not on the field $\mathbb{F}$ itself. The degree bound $\beta(\mathbb{F}[V]^G)$ may depend on the characteristic; take for example the cyclic group C_5: the degree bound is 5 in characteristic 0, and 7 in characteristic 5. Is this only a modular problem? More precisely, assume that the characteristic p does not divide the size of G (the non-modular case), and let respectively S_0 be a minimal generating set of $\mathbb{Q}[V]^G$ composed of orbit sums, and S_p be a minimal generating set of $\mathbb{F}[V]^G$. Here is a series of three conjectures concerning the non-modular case by order of likelihood:

CONJECTURE 12.2 (Folklore). $\beta(\mathbb{F}[V]^G) = \beta(\mathbb{Q}[V]^G)$.

CONJECTURE 12.3 (Folklore). *The number of generators of each degree in S_p and S_0 coincide.*

CONJECTURE 12.4 (Folklore). *The set of orbit sums S_0 is a minimal generating set of $\mathbb{F}[V]^G$.*

Positive answers would be of great help for practical computations!

For the next question, let $\mathbb{F}$ be a field of characteristic 0 and suppose V is a permutation representation of a finite group G. Let $\tau(d)$ denote the number of elements of degree d in a homogeneous minimal generating set for $\mathbb{F}[V]^G$.

QUESTION 12.5. Characterize those permutation G-modules V such that τ is unimodular, i.e., such that there exits a positive integer m with $\tau(1) \leq \tau(2) \leq \cdots \leq \tau(m) \geq \tau(m+1) \geq \tau(m+2) \geq \cdots$.

This property has been tested on all transitive permutation groups for dimension $n \leq 6$. It seems to hold for most permutation groups. The alternating group, A_n is always a counter example for $n \geq 4$. Otherwise, the smallest counter example for a transitive group is for $n = 6$; in the computer algebra program GAP, this group is given by "TransitiveGroup(6,8)".

13. David Wehlau

Let $\mathbb{F}$ be a field of characteristic p where $p \neq 0$. Let G be a finite group whose order is divisible by p.

CONJECTURE 13.1 (Wehlau). *If U is a G-submodule of the G-module V then $\beta(\mathbb{F}[U]^G) \leq \beta(\mathbb{F}[V]^G)$.*

This conjecture is easy to verify if U is a summand of V, i.e., if there exists a G-stable complement to U in V. As a consequence, the conjecture holds if V is a non-modular representation of G, (even if G is not finite). The main result in [**23**] is a proof that the conjecture holds when G is the cyclic group of order p.

One may also ask whether $\beta(\mathbb{F}[V/U]^G) \leq \beta(\mathbb{F}[V]^G)$. It would be very useful to develop relationships between $\beta(\mathbb{F}[V \otimes W]^G)$ and $\beta(\mathbb{F}[V]^G)$ and $\beta(\mathbb{F}[W]^G)$. Similarly we may ask about $\beta(\mathbb{F}[V \otimes W]^{G\times H})$ versus $\beta(\mathbb{F}[V]^G)$ and $\beta(\mathbb{F}[W]^H)$ where V is a G-module and W is an H-module.

We define an $\mathbb{F}[V]^G$-module homomorphism, called the *transfer* or *trace* map, $\mathrm{Tr}^G : \mathbb{F}[V] \to \mathbb{F}[V]^G$ by $\mathrm{Tr}^G(f) = \sum_{g \in G} g \cdot f$. We wish to consider the image of the transfer. We denote this image by $\operatorname{Im}\operatorname{Tr}^G$. Since Tr^G is an $\mathbb{F}[V]^G$-module homomorphism, $\operatorname{Im}\operatorname{Tr}^G$ is an ideal of $\mathbb{F}[V]^G$.

Similarly, if H is a subgroup of G we may define the relative transfer homomorphism $\operatorname{Tr}^G_H : \mathbb{F}[V]^H \to \mathbb{F}[V]^G$. This homomorphism is defined using a sum over a collection of left coset representatives for H in G: $\operatorname{Tr}^G_H(f) = \sum_{g\in G/H} g \cdot f$. Thus $\operatorname{Tr}^G = \operatorname{Tr}^G_{\{e\}}$.

False Conjecture 13.2. $\beta(\mathbb{F}[V]^G/\operatorname{Im}\operatorname{Tr}^G) \leq |G|$.

This conjecture is now known to be false due to a counter example due to Fleischmann, Kemper and Shank (see [**7**, Remark 3.5]). A related conjecture which is still open is the following.

Conjecture 13.3. $\beta(\mathbb{F}[V]^G/ - \operatorname{Im}\operatorname{Tr}_{<G}) \leq |G|$ *where* $\operatorname{Im}\operatorname{Tr}_{<G}$ *is the ideal generated by all* $\operatorname{Im}\operatorname{Tr}^G_H$ *as* H *varies over all proper subgroups of* G.

Here are two more open conjectures concerning Noether numbers.

Conjecture 13.4 (Kemper). *If* $\mathbb{F}[V]^G$ *is Cohen Macaulay then* $\beta(\mathbb{F}[V]^G) \leq |G|$.

Conjecture 13.5. (Derksen and Kemper [**6**, 3.8.6]). $\beta(J) \leq |G|$ *where* J *is the Hilbert ideal in* $\mathbb{F}[V]^G$.

Conjecture 13.5 has been shown to be true if p does not divide $|G|$.

Finally, here is a conjecture which relates the structure of the ring of invariants to the transfer homomorphism.

Conjecture 13.6 (Shank and Wehlau). *Let* G *be a* p*-group. Then* $\mathbb{F}[V]^G$ *is a polynomial ring if and only if* $\operatorname{Im}\operatorname{Tr}^G$ *is a principal ideal.*

Let $\mathbb{F}_p$ denote the field of prime order p. Shank and Wehlau [**22**] proved that if $\mathbb{F}_p[V]^G$ is a polynomial ring then $\operatorname{Im}\operatorname{Tr}^G$ is principal. Broer [**4**] later extended this result by showing that for any field $\mathbb{F}$ of characteristic p, if $\mathbb{F}[V]^G$ is a polynomial ring then $\operatorname{Im}\operatorname{Tr}^G$ is principal.

References

1. K. Akin and D. A. Buchsbaum, *Characteristic-free representation theory of the general linear group* II. *Homological considerations*, Adv. in Math. **72** (1988), 171–210.
2. D. J. Anick, *On the homogeneous invariants of a tensor algebra*, Algebraic Topology (Evanston, 1988), Contemp. Math. vol. 96, Amer. Math. Soc., Providence, RI, 1989, pp 15–17.
3. K. Baur, *Cartan components and decomposable tensors*, Preprint.
4. A. Broer, *Remarks on invariant theory of finite groups*, Preprint, Université de Montréal, 1997.
5. D. Daigle and G. Freudenburg, *A counterexample to Hilbert's fourteenth problem in dimension five*, J. Algebra **221** (1999), 353–371.
6. H. Derksen and G. Kemper, *Computational invariant theory*, Encyclopaedia Math. Sci., vol. 130, Springer-Verlag, Berlin, Heidelberg, New York 2002.
7. Peter Fleischmann, Gregor Kemper, and R. James Shank, *On the depth of cohomology modules*, Report UKC/IMS/03/17, IMS, University of Kent, UK, 2003.
8. A. G. Helminck and G. W. Schwarz, *Orbits and invariants associated with a pair of commuting involutions*, Duke Math. J. **106** (2001), no. 2, 237–279.
9. Ian Hughes and Gregor Kemper, *Symmetric powers of modular representations for groups with a Sylow subgroup of prime order*, J. of Algebra **241** (2001), 759–788.
10. J.-C. Jantzen, *Representations of algebraic groups, pure and applied mathematics*, Pure Appl. Math., vol. 131, Academic Press, Inc., Boston, MA, 1987.

11. W. van der Kallen, *Cohomology with Grosshans graded coefficients*, these proceedings.
12. D. B. Karagueuzian and P. Symonds, *The module structure of a group action on a polynomial ring: examples, generalizations, and applications*, these proceedings.
13. ———, *The module structure of a group action on a polynomial ring: A finiteness theorem*, Preprint, 2002.
14. F. Knop, *Der kanonische Modul eines Invariantenringes*, J. Algebra **127** (1989), 40–54.
15. D. Luna and R. W. Richardson, *A generalization of the Chevalley restriction theorem*, Duke Math. J. **46** (1979), 487–496.
16. M. Nagata, *On the 14th problem of Hilbert*, Amer. J. Math. **81** (1959), 766–772.
17. ———, *Invariants of a group in an affine ring*, J. Math. Kyoto Univ. **3** (1963/1964), 369–377.
18. M. Nicollerat, *Homology of symmetric algebras*, Comm. Algebra **7** (1997), 657–676.
19. N. M. Thiéry, *Algebraic invariants of graphs: A study based on computer exploration*, SIGSAM Bulletin **34** (2000), no. 3, 9–20.
20. ———, *Invariants algébriques de graphes et reconstruction; une étude expérimentale*, Ph.D. thesis, Université Lyon I, 1999.
21. ———, *Computing minimal generating sets of invariant rings of permutation groups with* SAGBI-*Gröbner basis*, Discrete Models: Combinatorics, Computation, and Geometry (R. Cori, J. Mazoyer, M. Morvan and R. Mosseri, eds.) (Paris, 2001), Discrete Math. Theor. Comput. Sci. Proc. AA, vol. 3, MIMD, Paris, 2001, pp. 315–329 (electronic).
22. R. J. Shank and D. L. Wehlau, *The transfer in modular invariant theory*, J. Pure Appl. Algebra **142** (1999), no. 1, 63–77.
23. ———, *Noether numbers for subrepresentations of cyclic groups of prime order*, Bull. London Math. Soc. **34** (2002), no. 4, 438–450.

Department of Mathematics and Computer Science, Royal Military College, P.O. Box 17000, STN Forces, Kingston, ON K7K 7B4, Canada.

E-mail address: `wehlau@rmc.ca`

Centre de Recherches Mathématiques
CRM Proceedings and Lecture Notes
Volume **35**, 2004

The Peterson Conjecture for Algebras of Invariants

R. M. W. Wood

In this article we shall confine our attention to algebras of invariants $\mathbf{P}(n)^G$ over the field of two elements, where $\mathbf{P}(n) = \mathbb{F}_2[x_1, \ldots, x_n]$ is the polynomial algebra in n variables and $G \subset \mathrm{GL}(n, \mathbb{F}_2)$ is a subgroup of the general linear group acting on the right of $\mathbf{P}(n)$ by matrix substitution [**8**]. The Steenrod algebra $\mathcal{A}$ acts on the left of $\mathbf{P}(n)$ and commutes with the action of G. In particular $\mathbf{P}(n)^G$ is a graded left $\mathcal{A}$-module. The grading on $\mathbf{P}(n)$ is by degree of homogeneous polynomials, where the variables x_i are in degree 1. The symmetric group Σ_n is identified as the subgroup of $\mathrm{GL}(n, \mathbb{F}_2)$ consisting of matrices with a single 1 in each row and column and zeros elsewhere. Any subgroup of Σ_n is called a permutation group.

We are concerned with the problem of finding a minimal generating set for $\mathbf{P}(n)^G$ as an $\mathcal{A}$-module. This problem is generally regarded as difficult for arbitrary n, even when G is the trivial group. However, there is a statement that can be made about the case $\mathbf{P}(n)$, which we refer to as the Peterson conjecture in honour of Frank Peterson who first formulated the problem and solved it in the case $n = 2$ [**5, 6**]. We wish to extend the Peterson conjecture to $\mathbf{P}(n)^G$ when G is a permutation group. First of all we recall the definition of $\mu(d)$, a numerical function defined on a non-negative integer d as the smallest number k for which d can be written the form $d = \sum_{i=1}^{k}(2^{\lambda_i} - 1)$ for numbers $\lambda_i \geq 0$ and $k > 0$. The original Peterson conjecture stated that all homogeneous elements of degree d in the $\mathcal{A}$-module $\mathbf{P}(n)$ are decomposable if and only if $\mu(d) > n$. The conjecture was proved in a stronger form for the first time in [**11**]. The following invariant version of the theorem comes out of recent joint work in progress with Helen Weaver.

THEOREM 1. *Let G be a permutation group. In any minimal generating set for $\mathbf{P}(n)^G$ as a module over the Steenrod algebra, there are no generators in degree d if and only if $\mu(d) > n$.*

One case of this result has recently appeared in [**3**], namely $G = \Sigma_n$. In this situation the algebra of invariants $\mathbf{P}(n)^{\Sigma_n}$ can be interpreted as the cohomology $\mathbf{H}^*(\mathrm{BO}(n), \mathbb{F}_2)$ of the classifying space of the orthogonal group $O(n)$. The original problem for the trivial group G was settled in [**10–12**]. In the next sections we shall explain how certain ingredients in [**3**] can be modified to apply to the proof our main Theorem 1. In the last section we shall make a few comments about the

---

2000 *Mathematics Subject Classification.* 55S10.
This is the final form of the paper.

nature of the main result and why it does not extend to arbitrary subgroups of $\mathrm{GL}(n, \mathbb{F}_2)$.

1. The Steenrod algebra

We recall that $\mathcal{A}$ is a Hopf algebra generated as an algebra by the Steenrod squares Sq^i, for $i \geq 0$, subject to the condition $\mathrm{Sq}^0 = 1$ and the Adem relations [**9**], with coproduct defined on generators by

$$\psi(\mathrm{Sq}^n) = \sum_{0 \leq i \leq n} \mathrm{Sq}^i \otimes \mathrm{Sq}^{n-i}.$$

Modules $\mathbf{M}$ over the Steenrod algebra which arise from the cohomology of topological spaces or from invariant theory are themselves graded algebras and have a special property called *unstability*, which links the product in $\mathbf{M}$ with the coproduct in $\mathcal{A}$. The following statement summarises the basic properties of these modules and we shall refer to them as *Hopf modules.*

PROPOSITION 2. *For a Hopf module* $\mathbf{M}$ *over the Steenrod algebra and homogeneous elements* $f, g \in \mathbf{M}$ *we have*

1. $\mathrm{Sq}^k(f) = 0$ *if* $\deg(f) < k$.
2. $\mathrm{Sq}^k(f) = f^2$ *if* $\deg(f) = k$.
3. *The Cartan formula*

$$\mathrm{Sq}^k(fg) = \sum_{0 \leq r \leq k} \mathrm{Sq}^r(f)\mathrm{Sq}^{k-r}(g).$$

In principle the above rules permit the evaluation of a Steenrod operation on any polynomial in $\mathbf{P}(n)$ by a reductive process using the Cartan formula.

As a Hopf algebra the Steenrod algebra admits a *conjugation* operator χ. This is an anti-automorphism of order 2 and we denote the conjugate of an element $\Theta \in \mathcal{A}$ by $\chi(\Theta) = \widehat{\Theta}$.

The following result goes back to Kraines [**4**].

PROPOSITION 3. *Let* g *be a homogeneous polynomial in* $\mathbf{P}(n)$ *and suppose* $\deg(g) < \mu(k)$ *for some* k. *Then* $\widehat{\mathrm{Sq}^k}(g) = 0$.

2. Hit equations

Let $\mathbf{M}$ be a graded $\mathcal{A}$-module. An element $f \in \mathbf{M}$ is reducible or *hit* if it satisfies a hit equation

$$f = \sum_{i>0} \mathrm{Sq}^i(h_i),$$

where $h_i \in \mathbf{M}$ and the summation is taken over a finite collection of squaring operations in positive grading.

The following formula provides a canonical hit equation which is crucial in the development of the hit problem [**1, 7, 10–12**].

PROPOSITION 4. *Let* $\mathbf{M}$ *be an Hopf-module over* $\mathcal{A}$. *Then for homogeneous elements* $f, g \in \mathbf{M}$ *we have*

$$f\mathrm{Sq}^k(g) - \widehat{\mathrm{Sq}^k}(f)g = \sum_{i>0} \mathrm{Sq}^i\big(\widehat{\mathrm{Sq}^{k-i}}(f)g\big).$$

A proof is supplied after Proposition 3.3 in [**3**]. An immediate consequence, known as the *χ-trick*, is that $f\mathrm{Sq}^k(g)$ is hit in $\mathbf{M}$ if and only if $\widehat{\mathrm{Sq}^k}(f)g$ is hit. This applies in one significant case.

PROPOSITION 5. *Let f, g be homogeneous polynomials in $\mathbf{P}(n)$ and suppose $\deg(g) < \mu(k)$ for some k. Then $f\mathrm{Sq}^k(g)$ is canonically hit by the equation*

$$f\mathrm{Sq}^k(g) = \sum_{j>0} \mathrm{Sq}^j\big(\widehat{\mathrm{Sq}^{k-j}}(f)g\big).$$

The proof follows immediately from Proposition 4 and Proposition 3.

3. *G*-orbits

Given a monomial f and permutation group G we can form the "smallest" G-invariant polynomial $\sigma_G(f)$ containing f as follows. An element π in G acts on a monomial by permuting the variables. Then the G-orbit of f is defined by $\sigma_G(f) = \sum_{i=1}^{t} f\pi_i$, where $\pi_1, \ldots, \pi_t$ run through a set of left coset representatives for the stabiliser of f in G. Such G -orbits generate $\mathbf{P}(n)^G$ additively [**8**]. A monomial $x_1^{d_1} \cdots x_n^{d_n}$ is called a *spike* if every exponent d_i is of the form $2^{\lambda_i} - 1$. The spikes must always appear in any set of generators of the module $\mathbf{P}(n)$ and their G-orbits in any generating set for $\mathbf{P}(n)^G$. The reason is that a spike can never appear as a term in the image of a squaring operator [**12**]. If $\mu(d) > n$ then there are no spikes in degree d. The statement of Theorem 1 can be paraphrased by saying that all elements of $\mathbf{P}(n)^G$ are hit in a degree for which no spikes exist. Work on the original Peterson conjecture for $\mathbf{P}(n)$ and related topics can be found in [**5, 6, 10–12**]. An element of $\mathbf{P}(n)^G$ may be hit in $\mathbf{P}(n)$ but in general there is no obvious way of converting a hit equation in $\mathbf{P}(n)$ into a hit equation in $\mathbf{P}(n)^G$ for this element. The next result gives one circumstance in which this is possible.

THEOREM 6. *Let G be a permutation group and f be a monomial in $\mathbf{P}(n)$. We write $f = pq^2$ where $p = x_1 \cdots x_e$ is the square free part of f and $e = \deg(p)$. If $\deg(p) < \mu\big(\deg(q)\big)$ then $\sigma_G(f)$ is hit in $\mathbf{P}(n)^G$.*

PROOF. Since G is a permutation group, we see that any element $g \in G$ which fixes f must fix p and q individually and therefore fixes $\Theta(p)q$ for any element Θ in the Steenrod algebra because the action of g commutes with the action of Θ. Consequently, for any set π_i, $1 \le i \le t$, of coset representatives for the stabiliser of f in G the element $\sum_{i=1}^{t}\big(\Theta(p)q\big)\pi_i$ is G-invariant. We note from Proposition 2 that $f = p\mathrm{Sq}^k(q)$, where $k = \deg(q)$. Since $\deg(p) < \mu\big(\deg(q)\big)$ we have from Proposition 5 the canonical hit equation

$$f = p\mathrm{Sq}^k(q) = \sum_{j>0} \mathrm{Sq}^j\big(\widehat{\mathrm{Sq}^{k-j}}(p)q\big).$$

Hence the equation

$$\sigma_G(f) = \sum_{i=1}^{t} f\pi_i = \sum_{j>0}\sum_{i=1}^{t} \mathrm{Sq}^j\big(\widehat{\mathrm{Sq}^{k-j}}(p)q\pi_i\big)$$

is a G-invariant hit equation as required. □

4. Proof of the main result

The proof of Theorem 1 is in two directions. Suppose first of all that we are in a degree d for which $\mu(d) \leq n$. Then there exists a spike f in $\mathbf{P}^d(n)$. For any permutation group $G \subset \Sigma_n$, the G-orbit $\sigma_G(f) = \sum_i f\gamma_i$ is formed from coset representatives γ_i of the stabiliser of f in G. In particular $\sigma_G(f)$ is a sum of distinct spikes. It is a non-zero element in $\mathbf{P}^d(n)^G$ which cannot be hit. This shows that a minimal generating set for $\mathbf{P}(n)^G$ as an $\mathcal{A}$-module must contain elements in degree d whenever $\mu(d) \leq n$. Conversely, suppose $\mu(d) > n$ and consider a monomial f in $\mathbf{P}(n)^G$ of degree d. As in Theorem 6 we write $f = pq^2$, where $p = x_1 \cdots x_e$ is the square-free part of f and $e = \deg(p)$.

LEMMA 7. *The condition* $\deg(p) < \mu(\deg(f))$ *implies* $\deg(p) < \mu(\deg(q))$.

PROOF. To prove the lemma, suppose to the contrary that $\deg(p) \geq \mu(\deg(q))$. We have $\deg(q) = \sum_{i=1}^{r}(2^{\lambda_i} - 1)$, where $\deg(p) \geq r$, by definition of the μ-function. Then

$$\begin{aligned}\deg(f) &= \deg(p) + 2\deg(q)\\ &= \deg(p) + \sum_{i=1}^{r}(2^{\lambda_i+1} - 2)\\ &= \sum_{i=1}^{r}(2^{\lambda_i+1} - 1) + \sum_{i=r+1}^{\deg(p)}(2 - 1).\end{aligned}$$

Hence $\mu(\deg(f)) \leq \deg(p)$, contrary to assumption. This proves the lemma.

The condition $\mu(d) > n$ implies that

$$\deg(p) \leq n < \mu(d) = \mu(\deg(f)).$$

Hence by Lemma 7 we have

$$\deg(p) < \mu(\deg(q)).$$

Then an application of Theorem 6 completes the proof that f is hit in $\mathbf{P}^d(n)G$. This shows that no elements in a minimal generating set for the $\mathcal{A}$-module $\mathbf{P}^d(n)G$ lie in degrees d for which $\mu(d) > n$, as required in the main Theorem 1. □

5. Remarks

The Peterson conjecture is certainly not true for all subgroups of $\mathrm{GL}(n, \mathbb{F}_2)$ if $n > 2$. For example, taking $n = 3$ and $G = \mathrm{GL}(3, \mathbb{F}_2)$, the algebra of invariants $\mathbf{P}(3)^G$ is the Dickson algebra $\mathbb{F}_2[Q_2, Q_1, Q_0]$, which is the polynomial algebra on the basic Dickson invariants Q_i in degree $2^3 - 2^i$. In particular there are no non-zero elements of $\mathbf{P}(3)^G$ in degree 1, which is always a spike degree. Furthermore, it is shown in [**2**] that Q_2^3, which lies in degree 12, is not hit, although $\mu(12) = 4 > 3$. An elementary fact also worth noting is that the Peterson conjecture is true in one direction for $\mathbf{P}(n)^G$ whenever G has odd order.

PROPOSITION 8. *Let* $G \subset \mathrm{GL}(n, \mathbb{F}_2)$ *be a group of odd order. Then in any minimal generating set for* $\mathbf{P}(n)^G$ *as a module over the Steenrod algebra, there are no generators in degree* d *if* $\mu(d) > n$.

PROOF. The proof is just a matter of observing that if $\mu(d) > n$ and f is G-invariant, then by the main Theorem 1 there is a hit equation

$$f = \sum_{i>0} \mathrm{Sq}^i(h_i),$$

where the h_i are some polynomials in $\mathbf{P}(n)$. We can now apply the classical transfer $\tau = \sum_{g\in G} g$ to obtain the new equation

$$f = f\tau = \sum_{i>0} \mathrm{Sq}^i(h_i\tau),$$

which is a hit equation in $\mathbf{P}(n)^G$, because the elements $h_i\tau$ are G-invariant and f occurs an odd number of times in $f\tau$. It follows immediately that all elements of $\mathbf{P}(n)^G$ in degree d are hit in $\mathbf{P}(n)^G$ whenever $\mu(d) > n$, which proves the proposition. □

It can happen in the case of a group of odd order that a minimal generating set for $\mathbf{P}(n)^G$ has no elements in a degree where there are spikes. Consider for example the general linear group $\mathrm{GL}(2, \mathbb{F}_2)$, which is isomorphic to the symmetric group Σ_3. Up to conjugacy, G has only two proper subgroups, the symmetric group Σ_2 and the Sylow 3-subgroup, which is cyclic of order 3. In the first case the Peterson conjecture holds by the main Theorem 1. It also holds for the second case in one direction by the above Proposition 8. It is easy to check, however, that in degree 4, where $\mu(4) = 2$, the only non-zero invariant is $x_1^4 + x_2^4 + x_1^2x_2^2$, which is hit by $\mathrm{Sq}^2(x_1^2 + x_2^2 + x_1x_2)$. As a matter of interest, for $G = \mathrm{GL}(2, \mathbb{F}_2)$ itself, the basic Dickson invariants are $Q_1 = x_1^2 + x_2^2 + x_1x_2$ and $Q_0 = x_1^2x_2 + x_1x_2^2$. It is shown in [**2**] that the powers $Q_1^{2^\lambda-1}$ form a minimal generating set as a module over the Steenrod algebra. These lie in degrees $2(2^\lambda - 1)$ whose μ values are 2. It follows that Proposition 8 is true for all subgroups $\mathrm{GL}(2, \mathbb{F}_2)$ but for rather coincidental reasons. It would be interesting to find the correct analogue of the Peterson conjecture for arbitrary subgroups of $G \subset \mathrm{GL}(n, \mathbb{F}_2)$. The answer is presumably G-dependent when G is not a permutation group.

References

1. M. D. Crossley, *H^*V is a bounded type over $\mathcal{A}_p$*, Group representations: Cohomology, Group Actions and Topology (Adem, Alejandro, et al., eds.) (Seattle, 1996), Summer research institute on cohomology, representations, and actions of finite groups, Amer. Math. Soc. Proc. Symp. Pure Math., vol 63, 1998, pp. 183–190.
2. N. H. V. Hung and F. P. Peterson, *$\mathcal{A}$-generators for the Dickson algebra*, Trans. Amer. Math. Soc. **347** (1995), 4687–4728.
3. A. S. Janfada and R. M. W. Wood, *The hit problem for symmetric polynomials over the Steenrod algebra*, Math. Proc. Cambridge Philos. Soc. **133** (2002), 295–303.
4. D. Kraines, *On excess in the Milnor basis*, Bull. London Math. Soc. **3** (1971), 363–365.
5. F. P. Peterson, *Generators of $\mathbf{H}^*(RP^\infty \wedge RP^\infty)$ as a module over the Steenrod algebra*, Abstracts Amer. Math. Soc. **833** (1987), 55–89.
6. ———, *$\mathcal{A}$-generators for certain polynomial algebras*, Math. Proc. Cambridge Philos. Soc. **105** (1989), 311–312.
7. W. Singer, *On the action of Steenrod squares on polynomial algebras*, Proc. Amer. Math. Soc. **111** (1991), 577–583.
8. L. Smith, *Polynomial invariants of finite groups*, Res. Notes Math., vol. 6, A. K. Peters Ltd, Wellesley, MA, 1995.

9. N. E. Steenrod and D. B. A. Epstein, *Cohomology operations* (Phillip A. Griffiths, John N. Mather, and Elias M. Stein, eds.), Ann. of Math. Stud., vol. 50, Princeton University Press, Princeton, NJ, 1962.
10. R. M. W. Wood, *Steenrod squares of polynomials, advances in homotopy theory*, Advances in Homotopy Theory (S. Salamon, B. Steer. and W. Sutherland, eds.), London Math. Soc. Lecture Note Ser., vol. 139, 1989, Cambridge University Press, Cambridge, pp. 173–177.
11. ———, *Steenrod squares of polynomials and the Peterson conjecture*, Math. Proc. Camb. Phil. Soc. **105** (1989), 307–309.
12. ———, *Problems in the Steenrod algebra*, Bull. London Math. Soc. **30** (1998), 449–517.

DEPARTMENT OF MATHEMATICS, UNIVERSITY OF MANCHESTER, OXFORD ROAD, MANCHESTER M13 9PL, UK.

E-mail address: `reg@ma.man.ac.uk`

Centre de Recherches Mathématiques
CRM Proceedings and Lecture Notes
Volume **35**, 2004

Weakly Symmetric and Weakly Commutative Spaces

O. Yakimova

ABSTRACT. This article is devoted to some special kinds of homogeneous spaces. We describe their most interesting properties and state some open problems.

1. Weakly symmetric spaces

Weakly symmetric spaces were introduced by Selberg in 1956 in his famous article on the trace formula. It seems that they were forgotten for almost 40 years. But recently they have become popular again. Their geometric properties were studied in several papers of Berndt, Prüfer, and Vanhecke. Akhiezer and Vinberg found a connection between weakly symmetric and spherical spaces. As a result of this, the classification of some particular kind of weakly symmetric spaces was obtained.

Let G be a real Lie group and $K \subset G$ a compact subgroup of G. Assume that the homogeneous space $M = G/K$ is connected. Denote by $\mathrm{Aut}(G, K)$ the set of the automorphisms of G preserving K. For any $\sigma \in \mathrm{Aut}(G, K)$ there is a natural action of σ on M given by $\sigma(gK) = \sigma(g)K$.

DEFINITION 1. The homogeneous space M is called *weakly symmetric* with respect to σ if the following equivalent conditions hold:

1. $\forall x,\, y \in M\ \exists g \in G$ such that $gx = \sigma(y)$, $gy = \sigma(x)$;
2. $\forall \xi \in T_{eK}M\ \exists k \in K$ such that $(\mathrm{Ad}\, k)\xi = -d\sigma(\xi)$.

The equivalence of these conditions was proved in 1995 [**3**]. The idea of the proof is quite clear from the geometric point of view. Because K is compact one can introduce a G-invariant Riemannian metric on M, which automatically appears to be σ-invariant. Both conditions are equivalent to the following one: if γ is a geodesic line in M and $o \in \gamma$, then there is an isometry s such that $s(o) = o$ and ds acts on $T_o\gamma$ as -1.

2000 *Mathematics Subject Classification.* Primary: 14L30; Secondary: 22F30, 32U10.
This is the final version of the paper.

The homogeneous space M is called weakly symmetric if it is weakly symmetric with respect to some σ. Clearly any symmetric space is weakly symmetric. But the converse is not true. The example was given by Selberg himself:

EXAMPLE 1 (see [**10**]). Let $G = \mathrm{SL}_2(\mathbb{R}) \times \mathrm{SO}_2$ and $K = \mathrm{SO}_2$ be a subgroup of G embedded diagonally. Then $M = G/K$ is weakly symmetric with respect to the σ defined by

$$\sigma\big((g,h)\big) := \left(\begin{pmatrix} 1 & 0 \\ 0 & -1 \end{pmatrix} g \begin{pmatrix} 1 & 0 \\ 0 & -1 \end{pmatrix}, h^{-1}\right).$$

In the reductive case the situation becomes more interesting if we consider complex groups instead of real ones. Let $G(\mathbb{C})$ be a complex reductive Lie group, $H \subset G(\mathbb{C})$ a reductive Lie subgroup. Then the quotient $X = G(\mathbb{C})/H$ is an affine algebraic variety. Assume that $\sigma \in \mathrm{Aut}\big(G(\mathbb{C}), H\big)$ is an automorphism of $G(\mathbb{C})$ preserving H. It acts on X in a natural way. Consider the following actions of $G(\mathbb{C})$ and σ on $X \times X$:

$$g(x,y) = (gx, gy), \quad \sigma(x,y) = \big(\sigma(y), \sigma(x)\big).$$

DEFINITION 2. The homogeneous space X is called *weakly symmetric* with respect to σ, if σ acts trivially on the algebra of invariants $\mathbb{C}[X \times X]^{G(\mathbb{C})}$. X is called weakly symmetric if it is weakly symmetric with respect to some σ.

The action of $G(\mathbb{C})$ on $X \times X$ is stable by a theorem of Luna, see [**6**]. Hence σ preserves generic $G(\mathbb{C})$-orbits. In other words, condition 1 of the first definition is satisfied for generic points x and y of X.

It turns out that real weakly symmetric spaces are just the real forms of complex ones. Let $G(\mathbb{C})$ be the complexification of G and $H = K(\mathbb{C})$ the corresponding subgroup of $G(\mathbb{C})$.

THEOREM 1 (Akhiezer–Vinberg [**1**]). *M is weakly symmetric if and only if X is weakly symmetric. Moreover, if $\sigma \in \mathrm{Aut}\, G(\mathbb{C})$, $\sigma(G) = G$ and $\sigma(H) = H$, then M is weakly symmetric with respect to σ if and only if X is weakly symmetric with respect to σ.*

This article of Akhiezer and Vinberg contains one more striking result.

THEOREM 2. *X is weakly symmetric if and only if X is spherical. If X is spherical then it is weakly symmetric with respect to some Weyl involution of $G(\mathbb{C})$ preserving the real form G.*

The automorphisms which can appear in the definition of weakly symmetric space can be described in algebraic terms. Let $X = G(\mathbb{C})/H$ be an affine spherical algebraic variety and

$$\mathbb{C}[X] = \bigoplus_{\lambda \in \Lambda(X)} V_\lambda,$$

the decomposition of $\mathbb{C}[X]$ into irreducibles.

THEOREM 3. *Suppose that $\sigma \in \mathrm{Aut}\big(G(\mathbb{C}), H\big)$. Then X is weakly symmetric with respect to σ if and only if $\sigma(V_\lambda) = V_\lambda^*$ for all $\lambda \in \Lambda(X)$.*

Here we will prove it in a particular case that illustrates the general situation but requires an easier technique. The complete proof of this theorem can be found in [**13**].

PROOF.
($\Longrightarrow$) Consider the action of $G(\mathbb{C})$ on the algebra

$$\mathbb{C}[X \times X] = \mathbb{C}[X] \otimes \mathbb{C}[X] = \bigoplus_{\lambda,\mu \in \Lambda(X)} V_\lambda \otimes V_\mu.$$

It is known that each summand of the form $V_\lambda \otimes V_\lambda^*$ contains a non-trivial $G(\mathbb{C})$-invariant vector. Thus, we have

$$\forall \lambda \in \Lambda(X) \quad V_\lambda \otimes V_\lambda^* = \sigma(V_\lambda \otimes V_\lambda^*) = \sigma(V_\lambda^*) \otimes \sigma(V_\lambda).$$

In particular, $\sigma(V_\lambda) = V_\lambda^*$.

($\Longleftarrow$) This part we prove in particular case when for all $\lambda \in \Lambda(X)$ we have $V_\lambda = V_\lambda^*$ and σ is an element of finite order. One can assume that the action $G(\mathbb{C}) : X$ is effective, i.e., $G(\mathbb{C}) \subset \operatorname{Aut} X$. The automorphism σ can be considered as an element of the group $\operatorname{Aut} X$. The subgroup $\widehat{G} = \langle G(\mathbb{C}), \sigma \rangle \subset \operatorname{Aut} X$ generated by $G(\mathbb{C})$ and σ is reductive. The variety X is a spherical homogeneous space of $\widehat{G}$. More precisely, $X = \widehat{G}/\widehat{H}$, where $\widehat{H} = \langle H, \sigma \rangle$.

We prove that the representation of $\widehat{G}$ in V_λ is self-dual and orthogonal for every $\lambda \in \Lambda(X)$. If the representation of $\widehat{G}$ in some V_λ were not self-dual, then the linear space V_λ^* would occur in $\mathbb{C}[X]$ as well. But the representations of $G(\mathbb{C})$ in V_λ and V_λ^* are isomorphic. Since $G(\mathbb{C})/H$ is spherical, we would have a contradiction. Suppose $\widehat{G} : U$ is a symplectic representation of $\widehat{G}$. There exists then a non-degenerate skew-symmetric form on $U^{\widehat{H}}$: it is the restriction of similar form on U. This gives $\dim U^{\widehat{H}}$ is even. Hence such a representation cannot occur in $\mathbb{C}[X]$.

Now we can describe the algebra of $\widehat{G}$ invariants

$$\mathbb{C}[X \times X]^{\widehat{G}} = \bigoplus_{\lambda \in \Lambda(X)} S^2 V(R)^{\widehat{G}}$$

and $\dim S^2 V(R)^{\widehat{G}} = 1$. Since

$$\mathbb{C}[X \times X]^{G(\mathbb{C})} = \bigoplus_{\lambda \in \Lambda(X)} (V_\lambda \otimes V_\lambda^*)^{G(\mathbb{C})}$$

we have $\mathbb{C}[X \times X]^{G(\mathbb{C})} = \mathbb{C}[X \times X]^{\widehat{G}}$.

This algebraic approach leads to a complete classification of real weakly symmetric homogeneous spaces of reductive groups. One has to take a list of all spherical spaces from [**4**] or [**7**] and then find their real forms. Of course, each of them has a compact real form. It was shown in [**14**] that all non-compact weakly symmetric spaces can be obtained in the following way: $G = (G_{\mathbb{C}}^{\tau\varphi})^0$ and $K = H^\tau$. Here H is a reductive spherical subgroup of $G(\mathbb{C})$ and φ is an involution of $G_{\mathbb{C}}$ such that its restriction to H is the identity mapping, τ is a compact real structure commuting with φ and preserving H. That article contains also a complete list of real weakly symmetric homogeneous spaces of semisimple Lie groups.

2. (Weakly) commutative spaces

Now our ground field is $\mathbb{R}$ again. The group G is not necessarily reductive. We denote the Lie algebras of Lie groups by corresponding small gothic letters, for example, $\mathfrak{g} = \operatorname{Lie} G$ and $\mathfrak{k} = \operatorname{Lie} K$. Let $\mathcal{D}(M)$ stand for the algebra of differential operators on M. There is a G-invariant filtration on $\mathcal{D}(M)$. The ith component of $\mathcal{D}(M) = \cup \mathcal{D}^i(M)$ consists of all operators of order less than or equal to i. It can be shown that $\operatorname{gr}(\mathcal{D}(M)^G) = (\operatorname{gr}\mathcal{D}(M))^G = S(\mathfrak{g}/\mathfrak{k})^K$, see for example [**12**]. It is a Poisson algebra whose bracket is induced by the commutator in $\mathcal{D}(M)$; or, in the last interpretation, it has a Poisson structure as the factor of $S(\mathfrak{g})^K$ by the Poisson ideal $(S(\mathfrak{g})\mathfrak{k})^K$.

Definition 3. The homogeneous space M is said to be *commutative* if the algebra of invariant differential operators $\mathcal{D}(M)^G$ is commutative.

It was proved in [**11**] that the notion of a commutative Riemannian homogeneous space is equivalent to a well known notion of a Gelfand pair.

In Selberg's consideration, commutativity is the most important property of weakly symmetric spaces. He proved that they do have it and pointed out that his "trace formula" is true not only for weakly symmetric spaces but for all commutative ones as well. He did not know if the second class of spaces is strictly larger, but as was shown by Lauret in [**5**] this is really the case. Lauret's example is a homogeneous space of a non-reductive group. This is not an accident. A commutative space of a reductive group is always weakly symmetric, see [**1**].

Definition 4. The homogeneous space M is said to be *weakly commutative* if $\operatorname{gr}\mathcal{D}(M)^G$ (or equivalently $S(\mathfrak{g}/\mathfrak{k})^K$) is a commutative Poisson algebra.

Obviously, each commutative space is weakly commutative. The converse statement was an open problem until this year. First, it was proved in several particular cases: by Mikityuk for homogeneous spaces of reductive groups [**7**], and by Vinberg for weakly commutative spaces of Heisenberg type (the definition will be given below) [**12**]. Finally, Leonid Rybnikov proved it in the general case, see [**9**].

The inclusion of various types of homogeneous spaces considered above is illustrated by the following diagram.

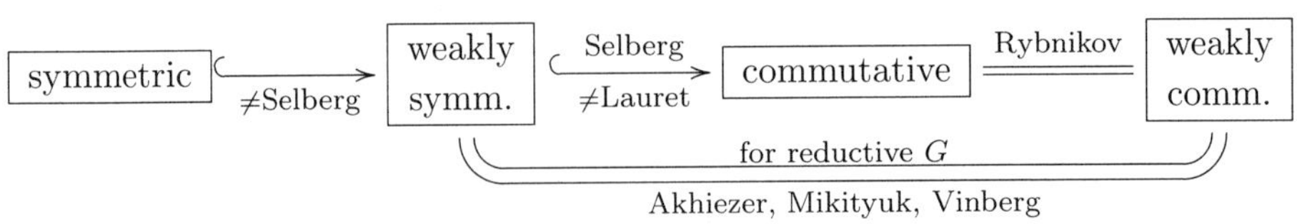

It is a challenging problem to classify all weakly symmetric and commutative spaces. There is a theorem describing structure of (weakly) commutative spaces. Therefore, one can hope to obtain a list of the commutative homogeneous spaces and then find weakly symmetric ones among them.

First of all, if M is a commutative homogeneous space of G, then G has a Levi decomposition $G = N \rtimes L$ such that K is the subgroup of the reductive group L.

Set $\mathfrak{l} = \operatorname{Lie} L$, $\mathfrak{n} = \operatorname{Lie} N$ and $\mathfrak{m} := \mathfrak{l}/\mathfrak{k}$. If a reductive Lie group F acts on a linear space V then we denote by $F_*(V)$ a generic stabilizer for this action.

THEOREM 4. *The homogeneous space M is (weakly) commutative if and only if the following conditions hold:*

(a) $\big[[\mathfrak{n},\mathfrak{n}],\mathfrak{n}\big] = 0$;
(b) $\mathbb{R}[\mathfrak{n}]^L = \mathbb{R}[\mathfrak{n}]^K$;
(c) *the homogeneous space $L_*(\mathfrak{n})/K_*(\mathfrak{n})$ is commutative*;
(d) *the homogeneous space $\big(N \leftthreetimes K_*(\mathfrak{m})\big)/K_*(\mathfrak{m})$ is commutative.*

Vinberg proved in [**12**] that conditions (a), (b) are satisfied for the weakly commutative spaces. The proof for (c), (d) and of the 'sufficient' part was given in [**15**].

Let's discuss conditions (a)–(d) in more details. The first one is rather simple. For example, it is always true for a commutative N.

Because the orbits of compact group K in $\mathfrak{n}$ are separated by polynomial invariants, L and K have the same invariants in $\mathbb{R}[\mathfrak{n}]$ if and only if they have the same orbits. In other words, condition (b) means that there is a decomposition $L = L_*(\mathfrak{n})K$. All nontrivial decompositions of reductive groups into product of two reductive subgroups are classified by Onishchik [**8**]. So one can check (b) for all spaces.

The group $L_*(\mathfrak{n})$ is reductive, so in (c) we can replace the word 'commutative' by 'weakly symmetric' and look up the given homogeneous space in the list.

But (d) is rather tricky. Weakly commutative homogeneous space $(N \leftthreetimes L)/K$ is called a space of a *Heisenberg type* if $L = K$. For this type of spaces $\mathfrak{g}/\mathfrak{k} = \mathfrak{n}$, so Definition 3 means that the Poisson algebra $S(\mathfrak{n})^K$ is commutative. Clearly, if N is commutative, then d) holds. Another series of examples are obtained in the following way. Let $K \subset \mathrm{U}_n$ and $N = H_n$ be a Heisenberg group with Lie algebra $\mathfrak{h}_n = \mathbb{C}^n \oplus \mathbb{R}$. Here K acts on $\mathbb{C}^n$ as a subgroup of U_n and on $\mathbb{R}$ trivially. The complexification of $\mathfrak{h}_n$ is of the form $W \oplus W^* \oplus \mathbb{C}$ as a $K(\mathbb{C})$-module. The corresponding homogeneous space of the Heisenberg type $(H_n \leftthreetimes K)/K$ is commutative if and only if the representation of $K(\mathbb{C})$ in $\mathbb{C}[W]$ is multiplicity free, see [**2**]. Article [**12**] contains a classification of commutative spaces of Heisenberg type with additional restriction that the representation of K in $\mathfrak{n}/[\mathfrak{n},\mathfrak{n}]$ is irreducible. Several important facts concerning the structure of those spaces are also proved in that article. But still the complete classification is not known yet.

Suppose one has a Riemannian homogeneous space $(N \leftthreetimes L)/K$ and wants to find out if it is commutative. This problem can be reduced to a simpler one or even solved in particular cases by Theorem 4. The following example shows how this theorem can be applied.

EXAMPLE 2. Take $L = \mathrm{U}_{2n}$, $K = \mathrm{Sp}_n$, $N = H_{2n}$. Then $(N \leftthreetimes L)/K$ is a (weakly) commutative space. Let's verify conditions (a)–(d).

The first one is satisfied by the very definition of a Heisenberg group. We can consider only $L : \mathbb{C}^{2n}$ in order to verify (b). It is well known, that $\mathbb{R}[\mathbb{C}^{2n}]^{\mathrm{U}_{2n}} =$

$\mathbb{R} = \mathbb{R}[\mathbb{C}^{2n}]^{\mathrm{Sp}_n}$. Another way to see this is the following: $L_*(\mathbb{C}^{2n}) = \mathrm{U}_{2n-1}$ and from [**8**] we get $\mathrm{U}_{2n} = \mathrm{U}_{2n-1}\,\mathrm{Sp}_n$.

The generic stabilizer for $K : \mathfrak{n}$ is equal to $\mathrm{Sp}_n \cap \mathrm{U}_{2n-1} = \mathrm{Sp}_{n-1}$. The homogeneous space $\mathrm{U}_{2n-1}/\mathrm{Sp}_{n-1}$ is a compact real form of the complex spherical space $\mathrm{GL}_{2n-1}(\mathbb{C})/\mathrm{Sp}_{2n-2}(\mathbb{C})$, and, hence, is weakly symmetric. So, only d) is left. Here we have $\mathfrak{m} = \mathfrak{u}_{2n}/\mathfrak{sp}_n = \bigwedge \mathbb{C}^{2n}$. It is a classical result that $K_*(\bigwedge \mathbb{C}^{2n}) = \mathrm{SU}_2 \times \cdots \times \mathrm{SU}_2$ $-a$ product of n copies of SU_2. As a $K_*(\mathfrak{m})$-module $\mathfrak{n} = \mathfrak{v}_1 \oplus \cdots \oplus \mathfrak{v}_n \oplus \mathbb{R}$, where $\mathfrak{v}_i = \mathbb{C}^2$ for every i. Each $\mathfrak{v}_i$ is acted upon by its own SU_2. Note that $[\mathfrak{v}_i, \mathfrak{v}_j] = 0$ for $i \neq j$. For $K_* := K_*(\mathfrak{m})$-invariants we have $S(\mathfrak{n})^{K_*} = \mathbb{R}[t_1, \dots, t_n] \otimes \mathbb{R}[\xi]$, where each t_i is a SU_2 invariant in $S^2(\mathfrak{v}_i)$. Evidently, t_i and t_j commute as elements of Poisson algebra $S(\mathfrak{n})^K$, and ξ lies in the center of $\mathfrak{n}$.

Let us discuss some difficulties concerning classification problem. First, assume that N is commutative. Then one has to find all triples (L, K, V), where K and L are compact groups, $K \subset L$, V is a space of a linear representation of L and $L = L_*(V)K$. It is a rather technical procedure. The main obstacle here is that the description of decompositions in [**8**] is rather complicated. But once we have the list for this type of spaces, all that will be left is just to find the structures of Lie algebra on V, compatible with the action of L, satisfying (a) and (d).

The situation with weakly symmetric spaces is even more unclear. Here the analogue of Theorem 4 is not true. It is shown by the following example. Note that it is also a new example of commutative but not weakly symmetric space.

EXAMPLE 3. Set $L = \mathrm{SO}_8$, $K = \mathrm{Spin}_7$ and let N be a simply connected commutative Lie group with $\mathfrak{n} = \mathbb{R}^2 \otimes \mathbb{R}^8$. The notation here means that $\mathbb{R}^8$ is an irreducible Spin_7-module. Denote by M the homogeneous space $(N \leftthreetimes L)/K$. We have $L_*(\mathfrak{n}) = \mathrm{SO}_6$ and $K_*(\mathfrak{n}) = \mathrm{SU}_3$. The real homogeneous space $\mathrm{SO}_6/\mathrm{SU}_3$ is weakly symmetric, because it is a real form of the complex spherical space $\mathrm{SO}_6(\mathbb{C})/\mathrm{SL}_3(\mathbb{C})$. The condition (c) also holds, see [**8**]. The homogeneous space $(N \leftthreetimes K)/K$ with commutative N is always weakly symmetric with respect to identity mapping. Hence, this space is commutative, but it is not weakly symmetric. Let's prove it now.

Assume that M is weakly symmetric with respect to some automorphism $\sigma \in \mathrm{Aut}\, G$, preserving K. Then $-\sigma$ preserve the K-orbits in $\mathfrak{g}/\mathfrak{k}$. The normalizer of K in $\mathrm{Aut}\, L$ is the subgroup of the inner automorphisms corresponding to the elements of K. Hence, we have $\sigma = \sigma' \,\mathrm{Ad}(k)$, where $\mathrm{Ad}(k)$ stands for an inner automorphism of G corresponding to $k \in K$, the automorphism σ' acts on L trivially and on $\mathfrak{n}$ trivially or as -1. More over, because the image of K in $\mathrm{GL}(\mathfrak{n})$ contains -1, we can assume that σ' acts on $\mathfrak{n}$ as -1. Evidently, M is weakly symmetric with respect to σ'. Let $v := \xi + \eta_1 + \eta_2$ be a vector of $\mathfrak{g}/\hat{\mathfrak{k}}$, here $\xi \in \mathfrak{l}/\mathfrak{k} \cong \mathbb{R}^7$, and η_1, η_2 are linear independent vectors of $\mathbb{R}^8$. If $\mathrm{Ad}(k)\sigma'(v) = -v$, then $\mathrm{Ad}(k)$ preserves η_1 and η_2, i.e., $k \in (\mathrm{Spin}_7 \cap \mathrm{SO}_6) = \mathrm{SU}_3$. Denote by ζ a SU_3-invariant non-trivial vector in $\mathbb{R}^7$. Take $\xi = \zeta$ then, as we have shown, $-v$ and $\sigma(v)$ are not in the same K-orbit. This completes the proof.

□

Acknowledgements

I am very grateful to the organizers of the Kingston conference Eddy Campbell and David Wehlau for the invitation and the warm hospitality.

References

1. D. N. Akhiezer and E. B. Vinberg, *Weakly symmetric spaces and spherical varieties*, Transform. Groups **4** (1999), 3–24.
2. C. Benson, J. Jenkins, and G. Ratcliff, *On Gelfand pairs associated with solvable Lie groups*, Trans. Amer. Math. Soc. **321** (1990), 85–116.
3. J. Berndt, F. Prüfer, and L. Vanhecke, *Symmetric-like Riemannian manifolds and geodesic symmetries*, Proc. Royal Soc. Edinburgh, Sect. A **125** (1995), 265–282.
4. M. Brion, *Classification des espaces homogènes sphériques*, Compositio Math. **63** (1987), no. 2, 189–208.
5. J. Lauret, *Commutative spaces which are not weakly symmetric*, Bull. London Math. Society **30** (1998), no. 1, 29–37.
6. D. Luna, *Sur les orbites fermées des groupes algèbriques réductifs*, Invent. Math. **16** (1972), 1–5.
7. I. V. Mikityuk (Mykytyuk), *On the integrability of invariant Hamiltonian systems with homogeneous configuration space*, Math. USSR Sbornik **57** (1987), 527–546 (in Russian).
8. A. L. Onishchik, *Inclusion relations between transitive transformation groups*, Tr. Mosk. Mat. Obshch. **11** (1962), 199–242 (in Russian).
9. L. Rybnikov, *On commutativity of weakly commutative Riemannian homogeneous spaces*, Funkt. Analiz i ego prilozh. **37** (2003), no. 2, 41–51 .
10. A. Selberg, *Harmonic analysis and discontinuous groups in weakly symmetric Riemannian spaces with applications to Dirichlet series*, J. Indian Math. Soc. **20** (1956), 47–87.
11. E. G. F. Thomas, *An infinitesimal characterization of Gel'fand pairs*, Conference in Modern Analysis and Probability (New Haven, 1982), Contemp. Math., vol. 26, Amer. Math. Soc., Providence, RI, 1984, pp. 379–385.
12. E. B. Vinberg, *Commutative homogeneous spaces and coisotropic actions*, Uspekhi Matem. Nauk **56** (2000), no. 1, 3–62 (in Russian).
13. O. S. Yakimova, *Righteous isometries of weakly symmetric spaces*, Math. Sbornik **193** (2002), 143–156 (in Russian).
14. ______, *Weakly symmetric spaces of semisimple Lie groups*, Vestnik Moscow Univ. (2002), no. 2, 57–60 (in Russian).
15. ______, *On weakly commutative homogeneous spaces*, Uspihi matem. nauk **57** (2002), no. 3, 171–173 (in Russian).

M. V. Lomonosov Moscow State University, Department of Mathematics and Mechanics Chair of High Algebra, Vorobjovy Gory, Moscow, 119899, Russia.

E-mail address: yakimova@mccme.ru

Titles in This Series

35 **H. E. A. Eddy Campbell and David L. Wehlau, Editors,** Invariant theory in all characteristics, 2004

34 **P. Winternitz, J. Harnad, C. S. Lam, and J. Patera, Editors,** Symmetry in physics, 2004

33 **André D. Bandrauk, Michel C. Delfour, and Claude Le Bris, Editors,** Quantum control: Mathematical and numerical challenges, 2003

32 **Vadim B. Kuznetsov, Editor,** The Kowalevski property, 2002

31 **John Harnad and Alexander Its, Editors,** Isomonodromic deformations and applications in physics, 2002

30 **John McKay and Abdellah Sebbar, Editors,** Proceedings on moonshine and related topics, 2001

29 **Alan Coley, Decio Levi, Robert Milson, Colin Rogers, and Pavel Winternitz, Editors,** Bäcklund and Darboux transformations. The geometry of solitons, 2001

28 **J. C. Taylor, Editor,** Topics in probability and Lie groups: Boundary theory, 2001

27 **I. M. Sigal and C. Sulem, Editors,** Nonlinear dynamics and renormalization group, 2001

26 **J. Harnad, G. Sabidussi, and P. Winternitz, Editors,** Integrable systems: From classical to quantum, 2000

25 **Decio Levi and Orlando Ragnisco, Editors,** SIDE III—Symmetries and integrability of difference equations, 2000

24 **B. Brent Gordon, James D. Lewis, Stefan Müller-Stach, Shuji Saito, and Noriko Yui, Editors,** The arithmetic and geometry of algebraic cycles, 2000

23 **Pierre Hansen and Odile Marcotte, Editors,** Graph colouring and applications, 1999

22 **Jan Felipe van Diejen and Luc Vinet, Editors,** Algebraic methods and q-special functions, 1999

21 **Michel Fortin, Editor,** Plates and shells, 1999

20 **Katie Coughlin, Editor,** Semi-analytic methods for the Navier-Stokes equations, 1999

19 **Rajiv Gupta and Kenneth S. Williams, Editors,** Number theory, 1999

18 **Serge Dubuc and Gilles Deslauriers, Editors,** Spline functions and the theory of wavelets, 1999

17 **Olga Kharlampovich, Editor,** Summer school in group theory in Banff, 1996, 1998

16 **Alain Vincent, Editor,** Numerical methods in fluid mechanics, 1998

15 **François Lalonde, Editor,** Geometry, topology, and dynamics, 1998

14 **John Harnad and Alex Kasman, Editors,** The bispectral problem, 1998

13 **Michel C. Delfour, Editor,** Boundaries, interfaces, and transitions, 1998

12 **Peter C. Greiner, Victor Ivrii, Luis A. Seco, and Catherine Sulem, Editors,** Partial differential equations and their applications, 1997

11 **Luc Vinet, Editor,** Advances in mathematical sciences: CRM's 25 years, 1997

10 **Donald E. Knuth,** Stable marriage and its relation to other combinatorial problems: An introduction to the mathematical analysis of algorithms, 1997

9 **D. Levi, L. Vinet, and P. Winternitz, Editors,** Symmetries and integrability of difference equations, 1996

8 **J. Feldman, R. Froese, and L. M. Rosen, Editors,** Mathematical quantum theory II: Schrödinger operators, 1995

7 **J. Feldman, R. Froese, and L. M. Rosen, Editors,** Mathematical quantum theory I: Field theory and many-body theory, 1994

6 **Guido Mislin, Editor,** The Hilton Symposium 1993: Topics in topology and group theory, 1994

TITLES IN THIS SERIES

5 **D. A. Dawson, Editor,** Measure-valued processes, stochastic partial differential equations, and interacting systems, 1994

4 **Hershy Kisilevsky and M. Ram Murty, Editors,** Elliptic curves and related topics, 1994

3 **Rémi Vaillancourt and Andrei L. Smirnov, Editors,** Asymptotic methods in mechanics, 1993

2 **Philip D. Loewen,** Optimal control via nonsmooth analysis, 1993

1 **M. Ram Murty, Editor,** Theta functions: From the classical to the modern, 1993